AF885011

REGIONS OF RECENT STAR FORMATION

ASTROPHYSICS AND SPACE SCIENCE LIBRARY

A SERIES OF BOOKS ON THE RECENT DEVELOPMENTS
OF SPACE SCIENCE AND OF GENERAL GEOPHYSICS AND ASTROPHYSICS
PUBLISHED IN CONNECTION WITH THE JOURNAL
SPACE SCIENCE REVIEWS

Editorial Board

J. E. BLAMONT, *Laboratoire d'Aeronomie, Verrières, France*

R. L. F. BOYD, *University College, London, England*

L. GOLDBERG, *Kitt Peak National Observatory, Tucson, Ariz., U.S.A.*

C. DE JAGER, *University of Utrecht, The Netherlands*

Z. KOPAL, *University of Manchester, England*

G. H. LUDWIG, *NOAA, National Environmental Satellite Service, Suitland, Md., U.S.A.*

R. LÜST, *President Max-Planck-Gesellschaft zur Förderung der Wissenschaften, München, F.R.G.*

B. M. McCORMAC, *Lockheed Palo Alto Research Laboratory, Palo Alto, Calif., U.S.A.*

H. E. NEWELL, *Alexandria, Va., U.S.A.*

L. I. SEDOV, *Academy of Sciences of the U.S.S.R., Moscow, U.S.S.R.*

Z. ŠVESTKA, *University of Utrecht, The Netherlands*

VOLUME 93
PROCEEDINGS

REGIONS OF RECENT STAR FORMATION

PROCEEDINGS OF THE SYMPOSIUM ON
"NEUTRAL CLOUDS NEAR HII REGIONS –
DYNAMICS AND PHOTOCHEMISTRY",
HELD IN PENTICTON, BRITISH COLUMBIA, JUNE 24–26, 1981

Edited by

R. S. ROGER and P. E. DEWDNEY
Dominion Radio Astrophysical Observatory,
Herzberg Institute of Astrophysics,
National Research Council of Canada, Penticton, B.C.

D. REIDEL PUBLISHING COMPANY
DORDRECHT : HOLLAND / BOSTON : U.S.A.
LONDON : ENGLAND

Library of Congress Cataloging in Publication Data
Main entry under title:

Regions of recent star formation.

(Astrophysics and Space Science Library. Proceedings ; v. 93)
"Sponsored by the Herzberg Institute of Astrophysics, National Research Council of Canada"–Verso t.p.
Includes indexes.
1. Stars–Formation–Congresses. 2. HII regions (Astrophysics)–Congresses. 3. Cosmochemistry–Congresses. I. Roger, R. S. II. Dewdney, P. E. (Peter E.) III. Herzberg Institute of Astrophysics. IV. Series.
QB806.R45 523.8 81-22684

ISBN-13: 978-94-009-7780-8 e-ISBN-13: 978-94-009-7778-5
DOI: 10.1007/978-94-009-7778-5

Published by D. Reidel Publishing Company,
P.O. Box 17, 3300 AA Dordrecht, Holland.

Sold and distributed in the U.S.A. and Canada
by Kluwer Boston Inc.,
190 Old Derby Street, Hingham, MA 02043, U.S.A.

In all other countries, sold and distributed
by Kluwer Academic Publishers Group,
P.O. Box 322, 3300 AH Dordrecht, Holland.

D. Reidel Publishing Company is a member of the Kluwer Group.

All Rights Reserved
Copyright © 1982 by D. Reidel Publishing Company, Dordrecht, Holland
Softcover reprint of the hardcover 1st edition 1982
No part of the material protected by this copyright notice may be reproduced or utilized in any form or by any means, electronic or mechanical including photocopying, recording or by any informational storage and retrieval system, without written permission from the copyright owner

TABLE OF CONTENTS

Preface xi

List of Participants xiii

THE DYNAMICAL EVOLUTION OF HII REGIONS IN NON-UNIFORM ENVIRONMENTS (REVIEW)
G. Tenorio-Tagle 1

HII Regions in Collapsing Massive Molecular Clouds 15
H.W. Yorke, P. Bodenheimer , G. Tenorio-Tagle

The Dynamical Evolution of an HII Region 25
R.H. Harten, M. Felli

Radio Continuum Observations of W1 31
H.E. Matthews

Interaction of the HII Region S236 with the Surrounding Medium 39
A. Falchi, G. Tofani, R.H. Harten

Optical and Millimeter Wavelength Study of the Complex Sh2-147/Sh2-153 43
M. Heydari-Malayeri, C. Kahane, R. Lucas, G. Testor

The HII Region W40 and its Molecular Cloud 53
Richard M. Crutcher, You-Hua Chu

HII Bubbles and Shocks in Molecular Clouds 61
T.J. Mazurek

A Multi-purpose Scanning Fabry-Perot Interferometer System 67
J.R. Roy, R. Arsenault, G. Joncas

Fine Structure Lines in HII Regions Interacting with Molecular Clouds 73
D.A. Naylor, R. Emery, B. Fitton, I. Furniss, R.E. Jennings, K.J. King

Excited OH in Absorption Towards W3(OH) 81
C.M. Walmsley, A. Winnberg, T.L. Wilson, A. Baudry

Ultra-compact HII Regions Embedded in Infrared Sources 83
Sun Kwok

DYNAMICAL EFFECTS OF STELLAR WINDS AND SHOCKED GAS ON INTERSTELLAR CLOUDS (REVIEW) 91
J. Michael Shull

A Calculation of Molecule Abundances Behind Slow Shocks 107
George F. Mitchell, Terry J. Deveau

Gravitational Instabilities in Shock Compressed Gas Layers 117
Gary L. Welter

Magnetic Fields and the Evolution of Shocked Gas Clouds 123
Johann Nittmann

Radiation-Hydrodynamics of HII Regions and Molecular Clouds 129
Maxwell T. Sandford II, Rodney W. Whitaker, Richard I. Klein

High Velocity Gas in Molecular Clouds 133
Ronald L. Snell, Suzan Edwards

High Velocity CO Emission Around T Tauri Stars 141
Suzan Edwards, Ronald L. Snell

Observations and Interpretation of the Line Profiles of Excited H_2 in Orion 147
T.R. Geballe, D. Nadeau

High Velocity H_2 Line Emission in the NGC 2071 Region 155
S.E. Persson, T.R. Geballe, Theodore Simon, Carol J. Lonsdale, F. Baas

Infrared Atomic Hydrogen Line Formation in Luminous Stars 161
Julian H. Krolik, Howard A. Smith

Atomic Hydrogen Zones Associated with HII Regions 167
R.S. Roger

The NGC 7538 Region: The Distribution and Dynamics of Molecules Compared with those of HI and H^+ 175
Hélène R. Dickel, John R. Dickel, William J. Wilson

Atomic and Ionized Hydrogen in Cepheus OB3 181
P.E. Dewdney

eutral Hydrogen Observations of the Puppis Window 185
J. Gregory Stacy, Peter D. Jackson,

Neutral Hydrogen Towards Tycho's Supernova Remnant 193
J.S. Albinson, S.F. Gull

The Velocities of the Neutral and Ionized Components of HII Regions 201
M. Fich, R.R. Treffers, L. Blitz

A CO Survey of 372 Optical HII Regions 209
Leo Blitz, Antony A. Stark

HI and CO Observations of Distant HII Regions in the Galactic Anticenter 213
Edwin J. Grayzeck, Peter D. Jackson, J.R. Sewall

Dynamics of CO Clouds Around HII Regions in the Outer Galaxy 221
P.D. Jackson, J.R. Sewall

CO J=2-1 Observations of Southern Galactic Plane HII Regions 231
Glenn J. White, J.P. Phillips

Observations of CO J=3→2 Emission From Molecular Clouds 237
Glenn J. White, J.P. Phillips, Graeme D. Watt

Properties of Giant Molecular Clouds in the Galactic Molecular Ring 245
W.L.H. Shuter, A. Szabo

Changes of the Star Formation Rate and the Initial Mass Function with Galactic Radius 249
J.L. Puget, R. Gispert, G. Serra

INFRARED AND MASER SOURCES IN REGIONS OF STAR FORMATION 251
Reinhard Genzel, Dennis Downes (REVIEW)

High Velocity CO Line Wings and the Dynamics of Star Forming Molecular Cloud Cores 287
John Bally

Kinematics of Molecular Gas in Orion from Observations of the ^{13}CO J=2→1 Line 295
P.F. Goldsmith, R. Arquilla, F.P. Schloerb, N.Z. Scoville

Molecular Hydrogen Emission From Broad Wing Cloud Cores 301
Adair P. Lane, John Bally

Asymmetric Broad HCO^+ Line Wings in Cores of Molecular Clouds 307
Aa. Sandqvist, A. Wootten, R.B. Loren, P. Friberg and Å. Hjalmarson

Some of the Problems Raised by CO and HCO^+ Observations in the Rho Ophiuchi Cloud 315
M. Pérault, E. Falgarone

CO J=3→2 and Far Infrared Continuum Observations of L1551, Orion KL and IRC +10216 323
J.P. Phillips, Glenn J. White, P.A.R. Ade, C.T. Cunningham, E.I. Robson, G.D. Watt

CO in the Horsehead Nebula 329
Antony A. Stark, John Bally

Small Scale Clumping in the Orion Molecular Cloud 335
P. Bastien, J. Bieging, C. Henkel, R.N. Martin, T. Pauls, C.M. Walmsley, T.L. Wilson, L.M. Ziurys

The ON-1 CO Cloud Complex -- Onset of Star Formation 337
F.P. Israel, H.A. Wootten

CO Observations of the Molecular Cloud Encompassing Sharpless 222 343
R.A. Christie, W.H. McCutcheon, C.P. Chan

Formation of a B0.5 Star Due to the Interaction of a Shock Wave with a Molecular Cloud in IC1805 349
V.A. Hughes

CHEMISTRY RELEVANT TO MOLECULAR CLOUDS NEAR HII REGIONS 357
W.D. Watson, C.M. Walmsley (REVIEW)

Selective Photodestruction of CO Isotopic Species 379
William D. Langer, John Bally

X-Ray Ionization and the Chemistry of the Orion Molecular Cloud 385
Julian H. Krolik, Timothy R. Kallman

Methyl Cyanide as a Probe of the Temperature and Density in $SgrB_2$; Quasi-Equilibrium in Molecular Rotational Levels 391
Richard A. Linke, Sally E. Cummins, Sheldon Green, Patrick Thaddeus

High Spatial Resolution Observations of HCN in S 235B 399
G. Sandell, B. Höglund, A.G. Kislyakov

A Model for the Formaldehyde Maser Near NGC 7538 - IRS 1 407
Wilfried Boland, Teije de Jong

H_2CO Near Compact HII Regions - New WSRT Results 409
J.R. Forster, W.M. Goss, H.R. Dickel

Is a BN-Type Object the Energy Input to the NH_3 Cloud in NGC 2071? 419
G. Calamai, M. Felli

VLA Observations of OH Masers and Associated Ultracompact Continuum Sources 425
B.E. Turner

Carbon Monoxide in the Magellanic Clouds 433
F.P. Israel, Th. de Graauw, S. Lidholm, H. van de Stadt, C. de Vries

Molecular Line Mapping of OMC-1 439
F.P. Schloerb, P.F. Goldsmith, N.Z. Scoville

An Upper Limit to the Atomic Carbon Abundance in the Orion Plateau 445
C.A. Beichman, T.G. Phillips, H.A. Wootten, M. Frerking

Observations of Neutral Carbon in the NGC 1977 Bright Rim 453
Alwyn Wootten, T.G. Phillips, C.A. Beichman, M. Frerking

Observations of CO in TMC 1 463
R. Braun, W.H. McCutcheon, W.L.H. Shuter

CO Emission Associated with Cold Neutral Hydrogen 469
W.L. Peters III, F.N. Bash

On the Correlation of CH Abundance and Extinction in Dark Nebulae 479
G. Sandell, L.E.B. Johansson

Author Index 485

Index of Astronomical Objects 487

Subject Index 491

PREFACE

The symposium on "Neutral Clouds near HII Regions" was prompted by an obvious need to bring together workers specifically interested in the dynamical and photochemical effects in regions showing clear evidence of on-going star formation. This is currently an area of considerable research activity with much new observational material over the wavelength range from X-ray to radio. Furthermore, the field is beginning to mature. No longer is molecular spectroscopy concerned only with the search for new lines and with preliminary surveys. No longer are evolving HII regions modelled with the naive assumption of constant density. Similarly, ideas of successive star formation, "champagne" and "blister" models of HII regions, and refinements to abundance calculations are examples which show that theoretical initiative is keeping pace.

We were both surprised and gratified by the number of contributed papers and the extent to which they addressed the subject matter. In the proceedings we have grouped these papers near the most appropriate of the four invited review papers. The subjects of these reviews are in the general areas of "Evolution of HII Regions", "Dynamical Interactions", "Chemistry in Active Regions" and "Infrared and Maser Sources". The symposium comprised 42 orally presented papers and 23 poster papers. All but two are reproduced in this volume.

Although discussion was invited both for orally presented papers and, in written form, for poster papers, only the former was forthcoming. The contents of the discussion have been severely distilled to remove, for example, questions and comments on points which are made clear in the written text, questions and comments on subjects which were in the oral presentation but mysteriously absent in the text, dialogue in which the participants were obviously speaking at cross-purposes, and finally, comments and questions which

R. S. Roger and P. E. Dewdney (eds.), Regions of Recent Star Formation, xi–xii.
Copyright © 1982 by D. Reidel Publishing Company.

appear to add little of value to the contribution. Undoubtedly those who can remember exactly the words that were uttered will occasionally wince at our editorial audacity.

We thank the several people who contributed to the organization of the symposium, especially Erika Rohner, Walter Gully and Serge Pineault. Tom Landecker and Roy Hamilton arranged the recording and documenting of the discussion. Dorothy Stewart, Elizabeth Jones and Erika Rohner contributed substantially to the preparation of the manuscripts. Perhaps it is stating the obvious to say that we are most indebted to the many authors for their prompt preparation of the manuscripts.

Finally it is a pleasure to thank Dr. J.L. Locke, Director of the Herzberg Institute of Astrophysics and Drs. John Galt and Lloyd Higgs, the retiring and current directors of the Dominion Radio Astrophysical Observatory for their encouragement in this venture.

R.S. Roger
P.E. Dewdney
Dominion Radio Astrophysical Observatory
Herzberg Institute of Astrophysics
National Research Council of Canada
Penticton, B.C., Canada

LIST OF PARTICIPANTS

John Bally	Bell Laboratories, Holmdel
Alan H. Barrett	Massachusetts Inst. of Technology, Cambridge
Pierre Bastien	Astronomische Inst. der Univ. Bonn
Charles Beichman	California Inst. of Technology, Pasadena
Leo Blitz	University of California, Berkeley
Wilfried Boland	University of Amsterdam, Amsterdam
R. Braun	University of British Columbia, Vancouver
Christopher Chan	University of British Columbia, Vancouver
Richard A. Christie	University of British Columbia, Vancouver
Carman H. Costain	Dominion Radio Astrophysical Observatory, HIA, Penticton
Richard M. Crutcher	University of California, Berkeley
Sue Derebey	University of California, Berkeley
Terry J. Deveau	St. Mary's University, Halifax
Peter E. Dewdney	Dominion Radio Astrophysical Observatory, HIA, Penticton
Hélène R. Dickel	University of Illinois Observatory, Urbana
John R. Dickel	University of Illinois Observatory, Urbana
Eli Dwek	University of Maryland, College Park
Suzan Edwards	Smith College, Northampton
Ambretta Falchi	Observatorio di Arcetri, Florence
Edith Falgarone	Observatoire de Paris
Michel Fich	University of California, Berkeley
James R. Forster	The Netherlands Foundation For Radio Astronomy, Dwingeloo

John A. Galt	Dominion Radio Astrophysical Observatory, HIA, Penticton
T. Geballe	Mt. Wilson Observatories, Pasedena
Reinhard Genzel	University of California, Berkeley
Paul F. Goldsmith	University of Massachusetts, Amherst
Mark A. Gordon	National Radio Astronomy Observatory, Tucson
E.J. Grayzeck	University of Nevada, Las Vegas
Steve F. Gull	Cambridge University
Ronald H. Harten	The Netherlands Foundation for Radio Astronomy, Dwingeloo
Carl Heiles	University of California, Berkeley
M. Heydari-Malayeri	Observatoire de Meudon
Lloyd A. Higgs	Dominion Radio Astrophysical Observatory, HIA, Penticton
V.A. Hughes	Queen's University, Kingston
F.P. Israel	European Space Agency - ESTEC, Noordwijk
P.D. Jackson	University of Maryland, College Park
Gilles Joncas	Université Laval, Quebec
I. Kazès	Observatoire de Paris
D.W. Keenan	York University, Toronto
Julian H. Krolik	Massachussets Inst. of Technology, Cambridge
Sun Kwok	Herzberg Institute of Astrophysics, Ottawa
T.L. Landecker	Dominion Radio Astrophysical Observatory, HIA, Penticton
Adair P. Lane	University of Massachusetts, Amherst
William D. Langer	Princeton Universit,
Richard B. Larson	Yale University, New Haven

Richard A. Linke	Bell Laboratories, Holmdel
Carol Lonsdale	Institute for Astronomy, Honolulu
Robert B. Loren	McDonald Observatory, Fort Davis
Henry E. Matthews	Herzberg Institute of Astrophysics, Ottawa
T.J. Mazurek	State University of New York, Stony Brook
George Mitchell	St. Mary's University, Halifax
D.A. Naylor	European Space Agency - ESTEC, Noordwijk
Johann Nittmann	The University of Leeds
Sven Eric Persson	Mt. Wilson & Las Campanas Observatories, Pasadena
William L. Peters III	University of Texas, Austin
J.P. Phillips	Queen Mary College, London
Serge Pineault	Dominion Radio Astrophysical Observatory, HIA, Penticton
J.L. Puget	Centre National de la Recherche Scientifique, Paris
R.S. Roger	Dominion Radio Astrophysical Observatory, HIA, Penticton
G. Sandell	Helsinki University
M.T. Sandford II	Los Alamos National Laboratory, Los Alamos
Aage Sandqvist	Stockholm Observatory
F.P. Schloerb	University of Massachusetts, Amherst
E.G. Schmidt	University of Nebraska, Lincoln
J. Michael Shull	University of Colorado, Boulder
William L.H. Shuter	University of British Columbia, Vancouver
Ronald Snell	University of Massachusetts, Amherst
Lewis E. Snyder	University of Illinois, Urbana

Antony A. Stark	Bell Laboratories, Holmdel
Mary Stevens	University of California, Berkeley
G. Tenorio-Tagle	Max Planck Institut für Astrophysik, Munich
Barry E. Turner	National Radio Astronomy Observatory, Charlottesville
Anne B. Underhill	Goddard Space Flight Center, Greenbelt
J.F. Vaneldik	University of Alberta, Edmonton
Melvyn Roger Viner	Queen's University, Kingston
C.M. Walmsley	University of Illinois, Urbana
William D. Watson,	University of Illinois, Urbana
Gary L. Welter	S.F.B. Radioastronomie, Bonn
Glenn J. White	Queen Mary College, London
H. Alwyn Wootten	California Inst. of Technology, Pasadena
Harold W. Yorke	Universitats-Sternwarte, Göttingen

THE DYNAMICAL EVOLUTION OF HII REGIONS IN NON-UNIFORM ENVIRONMENTS

G. Tenorio-Tagle
Max-Planck-Institut für Astrophysik,
Garching bei München

I. INTRODUCTION

The perfect balance established between the ionizing photon output from an early type star and the recombinations in a volume of gas surrounding it, was first envisaged by Strömgren (1939). This principle allowed him to deduce, under the assumption of constant density, values for the size of the ionized region - termed the Strömgren sphere - and to infer the existence of a sharp discontinuity between ionized and neutral matter - termed the ionization front (I-front). Strömgren's idea gave rise to the "classical" theory for the evolution of the HII regions, that which predicts the expansion of the hot ionized gas into the surrounding cold neutral matter, due to the pressure difference generated by ionization. The "classical" theory was developed by many authors. Kahn (1954), Axford (1961) and Goldsworthy (1961) classified the various possible types of I fronts expected to occur during the formation and evolution of a Strömgren sphere. They also predicted the formation of a shock front, moving into the neutral matter, ahead of D type I fronts. Analytical treatments of the problem (Savedoff and Greene 1955; Newman and Axford 1968) demonstrated that a small separation exists between the two fronts, and that a fair amount (1%) of the stellar radiative energy could be converted, through the shock, into kinetic energy of the compressed neutral gas (Spitzer, 1968).

Time-dependent solutions of the problem appeared during the sixties. Mathews (1965) investigated the formation and early expansion of the Strömgren sphere, while Lasker (1966) studied the late evolution. Similar calculations with new difference schemes were performed more recently (Manfroid 1975, Tenorio-Tagle 1976) to obtain a more detailed and consistent picture. As a consequence of this work the theoretical understanding of the evolution of HII regions could be summarized according to the following sequence of events:

a) Formation - As the early type star switches itself on by moving onto the main sequence (MS), its ionizing radiation generates a spherical supersonic (weak R) I front that rushes through the gas leaving it hot and ionized but almost undisturbed. However, recombinations (in the ion-

R. S. Roger and P. E. Dewdney (eds.), Regions of Recent Star Formation, 1–14.
Copyright © 1982 by D. Reidel Publishing Company.

ized volume) and geometrical dilution of the ionizing radiation constantly reduce the speed of the I front as it moves away from the parent star. Both effects eventually lead to a much slower weak R front, which develops within its own structure an increasingly larger pressure gradient. This pressure gradient due essentially to the temperature gradient in a gas of varying degrees of ionization, leads to mass motions. Thus dynamically speaking the weak R I front behaves like a compression wave. This phase ends when the I front reaches the Strömgren radius (R_O).

b) Expansion - As the formation phase comes to an end, and the I front slows down even further, the pressure gradient within the I front structure evolves into a shock front. The shock front moves ahead of the I front, while the I front evolves into a (subsonic) D type front. Expansion is the longest evolutionary phase and is characterized by an outward moving shell of compressed neutral matter between the two fronts, pushed by the expansion of the ionized hot gas. The expansion gives rise to rarefaction waves in the ionized volume and this lowers the number of recombinations and allows more photons at the I front. However, because of the shock and the compressed region ahead of the I front, the front only slowly advances (mass wise) through the neutral gas. During this phase the I front behaves like a rarefaction wave which slowly transfers material from the dense shell into the low density HII region. This phase ends when the pressure of the ionized region becomes comparable to the outside pressure (equilibrium Strömgren sphere) or when the exciting star moves off the MS, resulting in a recombination phase.

Although, the development of this general picture was followed by other detailed studies of the problem (Manfroid 1976, a recombining H II region) and the inclusion of other relevant physical processes (Mathews 1969, - dusty H II regions; Tenorio Tagle 1977, the ionization of globules, etc.) it became evident that a huge discrepancy had grown between the theory's expectations and reality!

The controversy originated from a large number of observations which demonstrated that:

1) Ionized nebulae have a complicated non-uniform density distribution.

2) Many H II regions are density bounded, not ionization bounded.

3) There is no clear evidence of expansion away from a central position. For example (in Orion) OIII is blue shifted with respect to O II, and O II with respect to S II (see Osterbrock 1974).

4) In several positions the emission lines are split by more than 20 km s^{-1}, implying supersonic velocity differences along a line of sight, and

5) Neighboring lines of sight show no systematic trend in the radial velocity peaks of the lines, as would be expected from the expansion theory, and do not show the expected variation of the surface brightness across the nebula.

Careful analysis of complete sets of observations led eventually to accurate descriptions of the present morphology and structure of various ionized nebulae. The best known of these is that of Zuckerman (1973) who established the facts that the Orion nebula was neither spherical nor expanding away from the "central" trapezium stars. He also noted that the ionized gas is blue-shifted with respect to the molecular cloud material, and proposed to study them as interacting systems rather than separate, neighboring, coexisting entities. Zuckerman mentions other authors with similar ideas: Münch and Wilson (1962), Wurm (1964), and Terzian and Balick (1974). Later Balick et al (1974, for Orion) and Gras-dalen (1974, for NGC 2024) postulated a cavity in the neutral cloud re-filled with ionized gas, while Dopita et al (1974, for Orion) proposed I fronts advancing in various directions into the neutral cloud.

Studies of the spatial association of H II regions (blisters) and molecular clouds (Wilson et al 1974, Dickinson et al 1974, Blair et al 1975) and their kinematic data, Israel (1976, 1978), Shaver (1977) have shown that Zuckerman's idea and the above mentioned models could describe a large number of optically observed nebulae. All of these empirical models could roughly describe the gaseous configurations presently established in various nebulae. However, none of them attempted to explain how this configuration could arise.

II THE CHAMPAGNE MODEL

Recently, numerical calculations for which the assumption of constant density was relaxed, were performed in one (Tenorio-Tagle, 1979) and two (Bodenheimer et al, 1979) dimensions. The models assumed a high density (n_c) cloud medium, birth place of the exciting star, to be initially in pressure equilibrium with a less dense (n_{ic}) and hotter intercloud (ic) gas. The time evolution showed that under such assumptions the "classical" evolutionary sequence is abruptly interrupted by the appearance of a new evolutionary phase. This phase begins whenever the I front overtakes the cloud's edge and ionizes the low density ic gas. Therefore, the time at which that occurs is a function of the cloud density and the depth (R_*) at which star formation took place. If the star is close enough to the cloud's edge to allow the I front to overtake it during the formation of the initial Strömgren sphere ($R_* < R_o$), a supersonic weak R I front will overrun the density jump. If the star is born deeper into the cloud ($R_* > R_o$) this will occur during the expansion phase. Thus a D type I front and its leading shock will overtake the discontinuity. Both cases (R and D), were investigated and shown to lead to a large pressure gradient ($A = P_{cloud}/P_{ic} = n_c/n_{ic}$) between the ionized cloud and newly ionized ic gas. In case D, the I front requires a longer time to reach this situation as it must first ionize the layer of shocked gas. This is only possible when the leading shock decays and stops compressing the material ahead of it, which happens as the shock enters the ic medium and moves away from its piston (the I front). In either case (R or D), once the density transition is ionized the I front becomes a fast weak R front rushing through the ic gas, and leaving behind a partially density-bounded nebula.

a) The Champagne phase. - During this phase the flow tries to readjust itself to the new gradient in pressure (A), by developing two gas-dynamical disturbances moving in opposite directions. First, the ionized cloud material reacts to A by streaming towards the ionized ic gas (the champagne shower). For large values of A (> 10,) the streaming reaches supersonic velocities which generate an isothermal shock (IS) in the ionized ic gas, supported by a continuous shower of material away from the cloud. The outward flow is regulated by the second disturbance, a rarefaction wave (RW) that moves into the cloud.

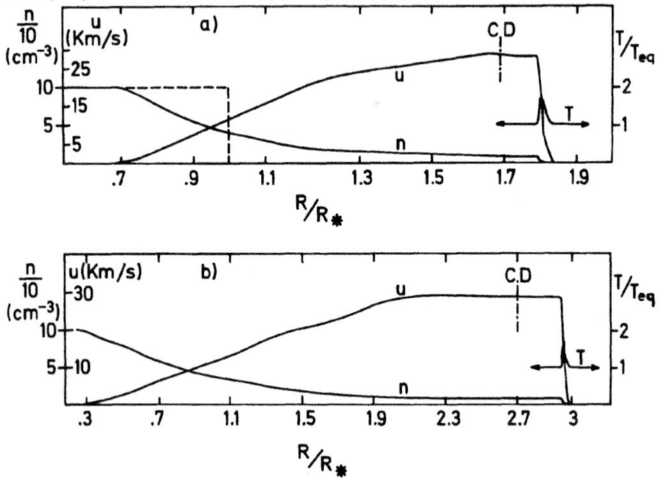

Figure 1. Case R. The Champagne shower. Run of density (n), velocity (u) and temperature (T), at ta = 2.2×10^5yr, tb = 5×10^5yr, $R_o = 2.84 \times 10^{19}$cm, CD marks the position of a contact discontinuity which separates shocked ionized ic gas from ionized material spewed from the cloud. Dashed line = initial density distribution. (From Tenorio-Tagle, 1979).

Both disturbances remain in the flow for as long as a pressure gradient exists. The RW on the one hand would lower the pressure in the cloud by accelerating material outwards, leading to an extended H II region with an exponential density distribution. The density profile matches at both ends of the RW, head and tail, the cloud and the shocked intercloud density, respectively. At the same time the IS would raise the pressure of the ionized ic gas by compressing it by a factor of M^2, where M is the shock mach number (in the frame of reference of the undisturbed ionized gas ahead of it) given by $M^2 \exp((M^2-1)/M) = A$ (see Bedijn and Tenorio-Tagle, 1981). This implies for large values of A (say > 100, typical for an n_c/n_{ic} ratio) values of M larger than 3, and a compression factor ~ 10. Figure 1 shows a well-developed champagne flow from a plane-parallel calculation of case R in a cloud of density $n_c = 10^2 cm^{-3}$, $T_c = 10K$, $n_{ic} = 1\ cm^{-3}$, $T_{ic} = 10^3K$. and $R_* = R_o$. The initial density distribution of the neutral gas (dashed line in 1a) is shown, and the exciting star is to the left and out of the pictures at a dimensionless radius $R/R_* = 0$. The displacement in time of both disturbances (IS and RW) can be clearly seen by comparing Fig. 1a with 1b.

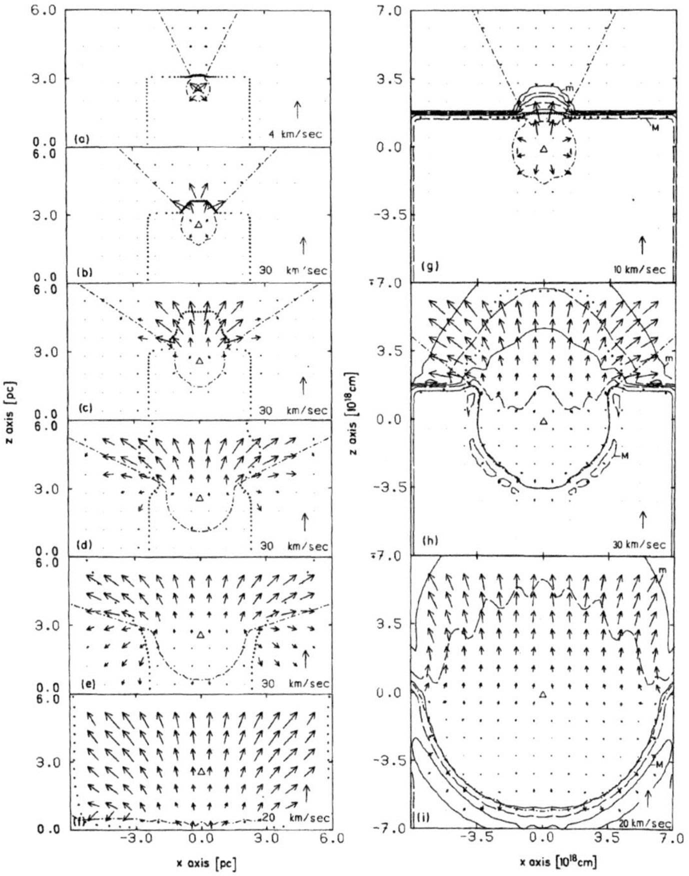

FIG. 2

Figure 2. Case R. Evolutionary sequence. Dashed-dot line = boundary of ionized region. Δ = position of star. crosses = boundary of material originally inside molecular cloud. Solid alternating with dashed lines (figs. g, h, and i) equidensity contours with m and M corresponding to the minimum and maximum density shown. Solid arrows = velocity vectors with length proportional to speed-scale in lower right hand corner. (From Bodenheimer et al, 1979).

The IS shows up by compressing and accelerating the ionized ic gas and by heating the newly swept up material to a temperature $T/T_{eq} > 1$ (T_{eq} = 8100K). The shocked gas however, soon radiates away this extra energy and achieves the equilibrium temperature (T_{eq}) determined by photo-ionization and mainly

[O II] cooling. The RW extends from the region (behind the IS) of constan density, constant velocity to the undisturbed ionized cloud. Throughout the RW (and the whole flow) conservation of mass is upheld, implying a larger velocity for a decrease in density. At time t_b, the IS is already $3R_o$ away from the edge of the original cloud ($R/R_* = 1$) and the head of the RW is approaching the star. Note that no disturbance precedes the RW on its way into the cloud. Therefore the other side of the nebula, the ionization bounded side (IBS), does not know about the champagne phase and continues to expand into the neutral cloud according to the "classical formulation. It is only when the RW crosses the stellar position that material will flow from the IBS into the density bounded side (DBS) of the nebula and the further evolution of the IBS will follow a different track. The rarefaction of the material in the IBS of the nebula, allows more photons at the I front that moves into the cloud. The I front and its leading shock can then reach velocities $\sim$ 15-20 kms^{-1}. Detailed numerical calculations in one dimension of this latter part of the champagne phase, and analytical expressions for the propagation of the I front into the cloud are given for a variety of cases (that account for different initial values of R_*, and the possible motion of the ionizing source into the cloud) by Bedijn and Tenorio-Tagle (1981).

b) The two dimensional (2D) approach. -

A less accurate but more versatile numerical approach is due to Bodenheimer et al. (1979), Tenorio-Tagle et al (1979) and Yorke et al (1981). The evolution of the three physical regions of the problem (cloud, ic and H II region) was followed with a 2D axisymmetric hydrodynamical code. A detailed radiative transfer was included to determine in all directions as a function of time the dimensions of the H II region. All gas contained within this zone (either cloud or ic gas) was assumed to be in ionization equilibrium and to have a temperature equal to T_{eq}. Otherwise the dense cloud matter was assigned a temperature T_c that kept it in pressure equilibrium with a thinner but hotter (T_{ic}) gas. The pressure gradient between ionized and neutral (either molecular or atomic) gas led to mass motions and this to a larger H II region. Although, in this approach, the resolution is poorer and the physics are approximated (e.g. a rough ionization and energy balance) detailed tests for the constant density case led to the same results as obtained for the "classical" model with one dimensional codes (see Bodenheimer et al, Fig. 1). The successful tests allowed us to calculate cases R and D and to obtain a more complete picture of the new evolutionary phase.

Figure 2 shows the results for case R, from the moment the I front crosses the cloud's edge (Fig. 2a); through the champagne phase up to the time at which the RW catches up with the inward moving front (Fig. 2f). Here one can indeed appreciate how the ionization and the density bounded sides of the nebula evolve with time. The two main features that characterize the champagne phase (the IS and the RW), although dependent on the geometry of the problem, are present in the flow. Figure 2 displays the supersonic ($v > 30$ kms^{-1}) expansion of the rarefied cloud gas, while it is dispersed over large distances inside a continuously wider cone of ionized ic gas. One can also track the motion (in all directions) of the RW as it advances into the ionized cloud, simply by noticing where in the flow gas begins to move outwards. (e.g. At t - 2.85 $\times 10^5$ yr, Figs 2e and i, the RW has clearly gone past the star).The calculation was stopped when the champagne phase led to the straight evaporation of the cloud, i.e. once the RW catches up with the inward moving I front, all newly ionized gas streams then outwards past the stellar position, while approaching its maximum speed.

Case D (shown to lead to similar results, see Fig. 3 of Bodenheimer et al 1979), and many more cases have been calculated in an attempt to show the general features that can be obtained during the champagne phase under a wide variety of circumstances. The parameters were: The stellar position (inside or outside the cloud), the mass and shape of the parent cloud, a static or a moving ionizing source, the possible presence of neighboring clouds and lately (see the contribution of Yorke at this meeting) the cloud's self-gravity and rotation.

III. APPELLATION CONTROLÉE

The hydrodynamical studies described in section II led us to propose the champagne phase as the physical process through which a compact (radio) H II region evolves into an extended (also optically observable) nebula. The physical association between molecular clouds and H II regions became then a by-product of such an evolutionary phase showing also how, as expected, the measured dynamical age (size/c_{HII}) is only a lower limit to the age of the HII region (Shaver, 1977). Here, I would like to list other properties (and predictions) that naturally emanate from our models. Whenever possible, a reference to relevant observational work will be given.

a) Predictions.

First of all, the existence of the champagne phase implies the formation of stars inside dense clouds, at such a distance from the cloud's edge (R_*) as to allow the I front to overrun the density discontinuity before the star moves off the MS (i.e. before 5×10^6 yr). From the studies of Smith et al (1978) we know that the mean lifetime of a compact radio H II region is 5×10^5 yr, which implies an average distance (Israel 1976), $R_* \sim 3$pc(if $n_c = 10^3 cm^{-3}$ and the stellar photon output is $\sim 10^{49} s^{-1}$). i.e. star formation occurs mainly on the borders of dense clouds and the champagne phase is unavoidable.

The ionization of the edge of the cloud turns the ionization bounded nebula into a partially density bounded H II region, limited by the flow of cloud material and the swept up ionized ic gas. Through this side of the nebula a large fraction of the stellar photon output escapes and produces a supersonic I front in the ic gas. The interaction of this I front with a second neighbouring cloud has been shown to lead to structures which could be identified as bright rims. Tenorio-Tagle and Bedijn (1981) have shown how behind the I front - which after the interaction with the second cloud evolves into a D type front with a leading shock - a density peak forms and is constantly supported by a secondary shock that follows the I front. Our calculations show values of the peak density (irrespectiv of the density in the second cloud) of the order of those found in the H II region that undergoes the champagne phase. This implies a stronger association between the bright rims and the exciting star than what we first thought (Pottasch, 1956), i.e. bright-rims are not only facing the star, but also, their densities are equal to the density of the ionized primary cloud that undergoes the champagne phase, times a geometrical dilution factor. This accounts for the density-distance relation found by Pottasch (1958), without the need of having denser globules closer to the star. Other calculations with a constant photon flux (i.e. without a champagne phase) lead to the rapid decay of the density peak. Calculations of the problem in other symmetries are now under way.

During the champagne phase, material from the parent cloud is continuously spewed into a conical volume of ionized ic gas. In our models two disturbances control this motion. The edge of the density bounded side is limited by the IS. This, although strong (mach > 3), might only lead to a tenuous compressed shell of ionized ic gas, undetectable when compared to the emitting dense ionized cloud gas. On the other hand, the champagne RW should be easy to trace as its motion leads to an exponential density gradient of the rarefied cloud gas while giving it a continuously larger velocity of expansion (reaching a maximum speed $\sim$ 3 times the speed of sound $\sim$40 km s^{-1}). Champagne flows with these genera characteristics have been recently recognized by several authors eg. Mufson et al 1981, Roger and Pedlar 1981, Donati-Falchi et al 1980, Harten and Felli 1981, Heydari-Malayeri et al 1980, Heydari-Malayeri and Testor 1981, and Deharveng (1980). Similar flows are expected to be taking place in giant H II regions where their ionized core-halo configuration make evident the existence of the pressure gradient required to generate a champagne flow (see Balick et al 1980).

The production of the extended blister H II region results then from the propagation of the RW into the ionized cloud. This allows more photons at the I front moving into the cloud, and therefore to the further disruption of the parent cloud. From our models, we obtained an average disruption rate of 2 - 3 × $10^{-3} M_{\odot} yr^{-1}$, which limits the lifetime of molecular clouds to some 10^7yr (Yorke et al, 1981) especially if clouds, as it seems to be the case, are prevented from gravitational collapse (Blitz and Shu, 1980).

It is worth noticing that through the champagne phase the star deposits some 10^{50} ergs into the ISM, in the form of kinetic energy of the fast moving ionized gas (see Bodenheimer et al, 1979). This so far has been neglected when considering the events that lead to the violent ISM (McCray and Snow 1979).

b) Other brands

Several attempts to explain the molecular cloud - H II region association exist now in the literature. Steady-state models are due to Kandel and Sibille (1978). A semi-analytical study of case R was performed by Whitworth (1979) who derived a similar disruption rate ($3 \times 10^{-3} M_\odot yr^{-1}$). Another approach is due to Icke (1979) and Icke et al (1980), who also calculated steady-state configurations on assumed exponential density distributions. Such a gas distribution is a by-product in our models. However, even if stars are born in a density gradient, the H II region will still develop a champagne-type flow (Welter, 1980) - This would only persist if, as in our models, the assumption of spherical symmetry is relaxed. Icke's et al radio continuum results seem to resemble some observations. However we believe it is rather far-fetched to attempt to explain phenomena like 30 Doradus with such a crude configuration. Also their explanation for the observed line splitting seems to be based on a misunderstanding of the ionization process (see Bedijn and Tenorio-Tagle 1981).

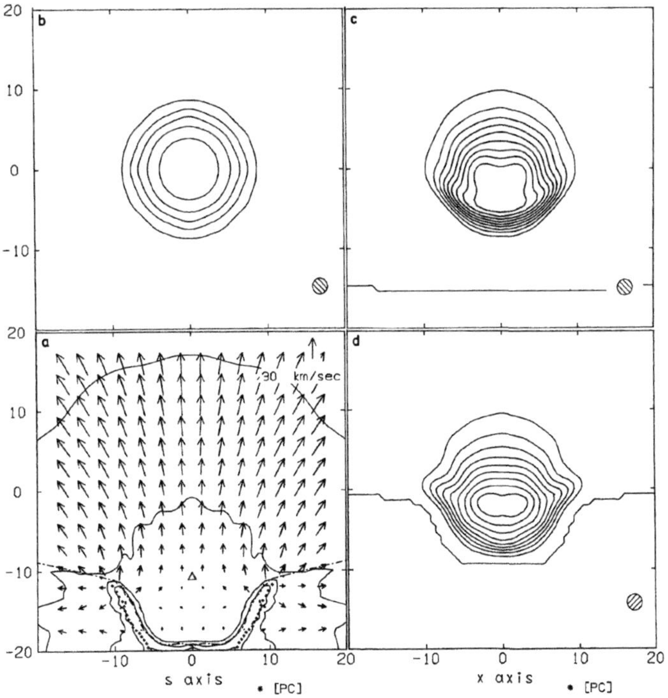

Figure 3. Radio continuum maps. 11 cm emission from model R (at $t = 4 \times 10^6$yrs, figure a) at b = 0°, c = 60° and d = 90° in a linear scale. In the maps the star is always located at position (0,0). Beam size is also indicated.

IV. THE CELLAR

Only a taste. - A comparison of our models with the observations has only been possible through the recognition of the flow parameters (the run of density and velocity). However, radio continuum maps and line profiles as a function of time and line of sight will soon be available, at least for the general cases. Figure 3, for example, shows the recently calculated radio continuum (11 cm) emission along three different lines of sight (0°, 60° and 90°) obtained from model R (figure 3a). The star, always located at position (0,0), is off the main centre of emission when observed at an angle $\neq 0^o$. Note also that the linear scale allows us to detect only the brightest part of the nebula. We believe that a careful study of these configurations (as a function of angle and time) will help towards a better understanding of the structure, orientation and morphology of observed nebulae.

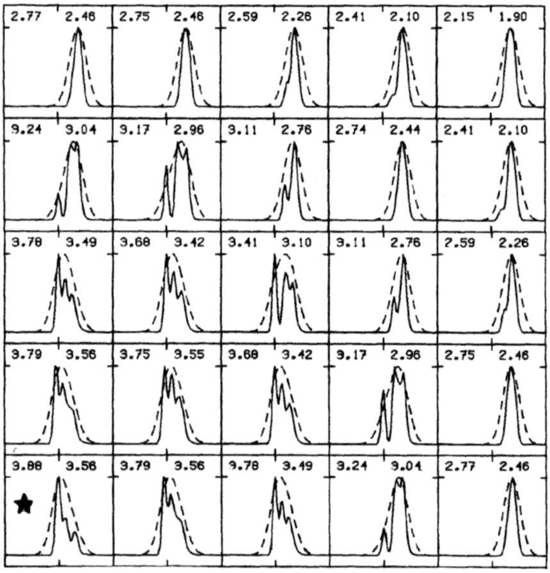

Figure 4. Normalized Hydrogen (dashed line) and Oxygen ([OII] + [OIII], solid line) line profiles from model R (see Figure 3a) seen from 0°. Velocity range = $\pm$ 60 km/s with zero indicated in every figure. The log of the absolute intensity is written in the upper left and right hand corners for H and O, respectively.

On the other hand, our dynamical models offer the possibility of studying the velocity field, and to compare it with the variety of line profiles observed in H II regions. Figure 4 shows the oxygen and Hα line profiles calculated for model R (at $t = 4 \times 10^6$yr) at an angle of 0°.

The full size of our observing grid is 20 pc^3 centred at the star, with selected line of sight positions every 2.5 pc, a beam size of 1 pc and a velocity range $\pm$ 60 kms^{-1}. The stellar position (*) is indicated in the figure, and only one quadrant is shown as our models are axis-symmetric. The figure clearly shows the spatial behaviour of the lines. Note the velocity shift, the broadening and even the splitting in oxygen. A careful study of the line emitting regions (along a line of sight) has shown the splitting to be produced by long paths of matter at different velocities rather than by small dense features. Line profiles from other models (such as model 1 from Tenorio Tagle et al, 1979) present up to 40 km s^{-1} splitting in oxygen, and very broad H_α lines.

Recently, Deharveng (1980) has made an attempt to classify supersonic motions in H II regions, finding two main types of split lines: One of them occurs in a scale of the order of 0.05 pc and is of the order of 20 km s^{-1}. When seen in [O III] it shows components of equal intensity, while in [OII] or [NII] they are uneven. The second kind of splitting could be as large as 60 kms^{-1}, and occurs over larger distances (1-20pc) with comparable intensities in the various components. This splitting is commonly found in the neighbourhood of neutral intrusions or in the vicinity of what observers call ionization fronts. From our results (see Figure 4) we can now postulate that the latter splitting is due to a well developed champagne flow.

Cheers!

ACKNOWLEDGEMENTS

I am grateful to Drs. H.W. Yorke and P. Bodenheimer for their comments and suggestions, and for allowing me to present some of our results prior to publication.

REFERENCES

Axford, W.I.: 1961, Phil. Trans. Roy. Soc. London, A, 253, 301.
Balick, B., Gammon, R.H., Doherty, L.H.: 1974, Astrophys. J. 188, 45.
Balick, B., Boeshaar, G.O., Gull, T.R.: 1980, Astrophys. J. 242, 584.
Bedijn P.J., Tenorio-Tagle, G.: 1981, (Paper IV) Astron. Astrophys. 98, 85.
Blair, G.N., Peters, W.L., Vanden Bout, P.A.: 1975, Astrophys.J. 200, L161.
Blitz, L., Shu, F.H.: 1980, Astrophys. J. 238, 148.
Bodenheimer, P., Tenorio-Tagle, G., Yorke, H.W.: 1979, (Paper II) Astrophys. J. 233, 85.
Deharveng, L.: 1980, Ph.D. Dissertation, University of Marseille.
Dickinson, D.F., Frogel, J.A., Persson, S.E.: 1974, Astrophys. J. 192, 347.
Donati-Falchi, A., Felli, M., Tofani, G.: 1980, Astron. Astrophys. 89, 363.
Dopita, M.A., Dyson, J.E., Meaburn, J.: 1974, Astrophys. Space Sci. 28, 61.
Goldsworthy, F.A.: 1961, Phil. Trans. Roy. Soc. London, A, 253, 277.
Grasdalen, G.L.: 1974, Astrophys. J. 193, 373.
Harten, R., Felli, M.: 1981, Astron. Astrophys. in press.
Heydari-Malayeri, M., Testor, G.: 1981, Astron. Astrophys., in press.
Heydari-Malayeri, M., Testor, G., Lortet, M.C.: 1980, Astron. Astrophys. 84, 154.
Icke, V.: 1979, Astron. Astrophys. 78, 352.
Icke, V., Gatley, I., Israel, F.P.: 1980, Astrophys. J. 236, 808.
Israel, F.P.: 1976, Unpublished Ph.D. Thesis, University of Leiden.
Israel, F.P.: 1978, Astron. Astrophys. 70, 769.
Kahn, F.D.: 1954, Bull. Astron. Inst. Neth. 12, 187.
Kandel, R.S., Sibille, F.: 1978, Astron. Astrophys. 68, 217.
Lasker, B.: 1966, Astrophys. J. 143, 700.
Manfroid, J.: 1975, Ph.D. Thesis, University of Liege.
Manfroid, J.: 1976, Astron. Astrophys. 46, 31.
Mathews, W.G.: 1965, Astrophys. J. 142, 1120.
Mathews, W.G.: 1969, Astrophys. J. 157, 583.
McCray, R., Snow, T.P.: 1979, Ann. Rev. Astron. Astrophys. 17, 213.
Mufson, S.L., Fountain, W.F., Gary, G.A., Howard, W.E., O'Dell, C.R., Wolff, M.T.: 1981, Astrophys. J., in press.
Münch, G., Wilson, O.C.: 1962, Zs. f. Ap. 56, 149.
Newman, R.C., Axford, W.I.: 1968, Astrophys. J. 153, 595.

Osterbrock, D.E.: 1974, "Astrophysics of Gaseous Nebulae" (Chapter VI), Freeman and Co., San Fransisco.
Pottasch, S.: 1956, Bull. Astron. Inst. Neth. 13, 77.
Pottasch, S.: 1958, Bull. Astron. Inst. Neth. 14, 29.
Roger, R. S., Pedlar, A.: 1981, Astron. Astrophys. 94, 238.
Savedoff, M.P., Greene, J.: 1955, Astrophys. J. 122, 477.
Shaver, P.A.: 1977, in "Topics in Interstellar Matter" ed. H. van Woerden, D. Reidel, Dordrecht.
Smith, L.F., Biermann, P., Mezger, P.G.: 1978, Astron. Astrophys. 66, 65.
Spitzer, L.: 1968, "Diffuse Matter in Space," Interscience, New York.
Strömgren, B.: 1939, Astrophys. J. 89, 526.
Tenorio-Tagle, G.: 1976, Astron. Astrophys. 53, 411.
Tenorio-Tagle, G.: 1977, Astron. Astrophys. 61, 189.
Tenorio-Tagle, G.: 1979, (Paper I), Astron. Astrophys. 71, 59.
Tenorio-Tagle, G., Yorke, H.W., Bodenheimer, P.: 1979, (Paper III) Astron. Astrophys. 80, 110.
Tenorio-Tagle, G., Bedijn, P.J.: 1981, (Paper V) Astron. Astrophys. (in press).
Terzian, Y., Balick, B.: 1974, "Fundamentals of Cosmic Phsyics" 1, 301.
Welter, G.L.: 1980, Astrophys. J. 240, 514.
Whitworth, A.: 1979, M.N.R.A.A. 186, 59.
Wilson, W.J., Schwartz, P.R., Epstein, E.E., Johnson, W.A. Etcheverry, R.D., Mori, T.T., Berry, G.G., Dyson, H.B.: 1974, Astrophys. J. 191, 357.
Wurm, K.: 1964, AFCRL AF61(052)-259 Tech. Report.
Yorke, H.W., Bodenheimer, P., Tenorio-Tagle, G.: 1981, (paper VI), Astron. Astrophys. (in press).
Zuckerman, B.: 1973, Astrophys. J. 183, 863.

DISCUSSION FOLLOWING REVIEW BY TENORIO-TAGLE

LINKE: While your model appears to apply nicely to star formation at the edge of a cloud, this does not support your contention that star formation occurs <u>only</u> at the edge.

TENORIO-TAGLE: The observations have shown us that star formation occurs <u>mainly</u> near the edges of dense clouds. All we claim is that champagne flows also imply a small initial distance from the stellar position to the cloud's edge.

GOLDSMITH: Could you elaborate briefly on the effect of the flow of molecular gas on the ionization-bounded side of the HII region - will there be significant velocity perturbations on it before it is eaten away by the advancing region?

TENORIO-TAGLE: From one-dimensional calculations (Bedijn and Tenorio-Tagle, 1981) we show that when the rarefaction wave crosses the stellar position, first there are more photons in the ionization front causing it to move faster into the shocked material. Because of this there are more absorbing collisions, and the ionization front then slows down. However, the rarefaction wave is still advancing and throwing material out. So the ionization front and its preceding shock wave display an oscillation in velocity. Velocity variations of 10 to 20 km s^{-1} appear in our calculations.

ISRAEL: One of the problems of 'blister' HII regions has always been the small velocity difference (of order 5 km s^{-1}) between the ionized and neutral material. In your diagrams the velocity seems to increase away from the interface and be very low near it. This might explain the observed situation. Is this velocity change a direct consequence of the model?

TENORIO-TAGLE: Yes, the rarefaction wave moving into the cloud leaves behind an exponential density distribution while accelerating the gas outwards. The product of density and velocity is constant, thus the denser the gas (i.e. near the cloud) the lower the velocity.

KROLIK: In your example illustrated in Figure 1 you used an initial intercloud temperature of 1000 K. Whether you have an isothermal shock or not depends on the strength of the cooling behind the shock which in turn depends on the post-shock temperature determined by the initial intercloud temperature. It seems to me that if you chose a more realistic temperature, the shock might not cool strongly enough to be isothermal, and you might have substantially different dynamics.

TENORIO-TAGLE: We have performed the calculations with a range of values (e.g. initial intercloud temperatures of 10000 K) and always get the same result. It is hard to suppress the discontinuity once the ionization breaks out. However, in all our calculations the assumption of an isothermal shock is there.

HII REGIONS IN COLLAPSING MASSIVE MOLECULAR CLOUDS

H.W. Yorke
Universitaets-Sternwarte Goettingen
P. Bodenheimer
Lick Observatory, UC Santa Cruz
G. Tenorio-Tagle
Max-Planck-Institut fuer Astrophysik, Garching b. Muenchen

ABSTRACT

Results of two-dimensional numerical calculations of the evolution of HII regions associated with self-gravitating, massive molecular clouds are presented. Depending on the location of the exciting star, a champagne flow can occur concurrently with the central collapse of a nonrotating cloud. Partial evaporation of the cloud at a rate of about 0.005 $M_\odot$/yr results. When 100 O-stars are placed at the center of a freely falling cloud of 3 10^5 $M_\odot$ no evaporation takes place. Rotating clouds collapse to disks and the champagne flow can evaporate the cloud at a higher rate (0.01 $M_\odot$/yr). We conclude that massive clouds containing OB-stars have lifetimes of no more than 10^7 yr.

1. INTRODUCTION

The effects of non-uniform environments on the evolution of HII regions has been considered in numerical models only rather recently (e.g. review by G. Tenorio-Tagle in this volume and references therein). First of all, the process of OB-star formation will modify the spatial distribution and chemistry of molecular cloud material in the immediate vicinity (10^{18} cm) of the OB-stars, which in turn will influence the way compact HII regions form (Yorke 1980a,b). Later during the evolution of HII regions the ionization front (I-front) could cross the boundary of the molecular cloud within which is formed. When the boundary region becomes ionized and heated to 10^4 K its pressure increases by more than 3 orders of magnitude, resulting in a rapid outward acceleration of the ionized material. Expansion velocities in excess of 30 km/s are possible in this "champagne flow" (Tenorio-Tagle 1979; Bodenheimer, Tenorio-Tagle, Yorke 1979). The interaction of an I-front moving into a molecular cloud from the outside can also lead to supersonic champagne flows and at least partial disruption of the cloud (Tenorio-Tagle, Yorke, Bodenheimer 1979; Bedijn and Tenorio-Tagle 1981; Tenorio-Tagle and Bedijn 1981). It is hard to imagine a mixture of molecular clouds and O-stars for which the "champagne phenomenon" will not play a major evolutionary role.

R. S. Roger and P. E. Dewdney (eds.), Regions of Recent Star Formation, 15–23.
Copyright © 1982 by D. Reidel Publishing Company.

The champagne phenomenon can be an important mechanism for the input of kinetic energy into the interstellar medium and limit the lifetimes of molecular clouds. The latter point is especially interesting in view of the controversy regarding the ages of molecular clouds. Solomon and Sanders (1979) and Scoville and Hersh (1979) have argued that molecular clouds and cloud complexes must be rather long-lived ($>10^8$ yr), based on an inferred predominance of molecular to atomic gas in the region 4 - 8 kpc from the galactic center. Blitz and Shu (1980) have questioned this result, however, arguing that the "observed" abundance ratio of H_2 to ^{13}CO in the solar neighborhood (Dickman 1978) is actually about a factor of 5 lower than that adopted by Solomon and Sanders. Taking into account metallicity gradients which are of the order of those reported by Mezger et al.(1979) for oxygen, the value of $N(H_2)/N(^{13}CO)$ appropriate for the 4 - 8 kpc "molecular" region should be even 30% lower. Blitz and Shu conclude that the observed surface mass density of CO emitting clouds (see e.g. Scoville and Solomon 1975; Gordon and Burton 1976; Cohen and Thaddeus 1977) is consistant with ages of giant molecular complexes $<3\ 10^7$ yr.

Stenholm (1980a,b,c) has demonstrated that masses of molecular clouds derived from CO radio maps can be uncertain by a factor of 10 to 100. Here and in other papers of the series (Morfill and Stenholm 1979; Stenholm, Hartquist, Morfill 1981) a distinction is made between different types of molecular clouds (see also Rowan-Robinson 1979). Type I clouds do not display the symptoms of active OB-star formation: OH masers, strong IR sources and compact HII regions; type II clouds do. With this distinction in mind the question of the lifetime of molecular clouds should be reposed: What are the lifetimes of the two types? Do the different types (and subtypes) simply represent different evolutionary stages of development? Little theoretical work has been done regarding the second question. The first question has been posed and discussed in a recent theoretical investigation (Yorke, Bodenheimer, Tenorio-Tagle 1981, hereafter referred to as YBT). We wish to supplement the results of YBT with further evidence.

YBT have argued that the champagne flow is an effective mechanism (perhaps the most effective) for destroying type II molecular clouds (producing OB-stars) and nearby type I clouds. Contrary to earlier studies of the champagne phenomonon, YBT have included the binding effects of gravity by considering rather massive clouds with surface escape velocities of the order of or larger than the isothermal sound speed of ionized 10^4 K gas. The evolutionary time scales were often much larger than the free-fall time scale of the molecular cloud. Magnetic fields were not considered.

2. THE NUMERICAL MODEL

The numerical procedure used for the two-dimensional, axially symmetric hydrodynamic calculations is described by YBT and the references cited therein. The initial conditions have been chosen to cover a wide variety of physical conditions. Common to each choice of parameters or "case" are the following assumptions: The ratio of temperature to mean molecular weight for the three phases of gas (molecular, intercloud, and ionized gas) were fixed at $T/\mu = 5$, 3000 and 16200, respectively. The choice of initial density, 600 cm^{-3} for molecular gas and 1 cm^{-3} for the intercloud gas, assured that these two phases were initially in pressure equilibrium. The dusty gas was assumed to have an opacity 200 cm^2/g for hydrogen ionizing photons.

The molecular clouds were assumed to be initially cylindrical with height equal to diameter. The basic parameters which were varied include the mass of the molecular cloud, its rotational velocity, the distance of the ionizing source from the cloud center, its velocity along the symmetry axis and its ionizing photon luminosity. Fourteen different cases are discussed in YBT. Here we give details of three particular examples.

3. RESULTS

Using the same notation as YBT we consider the cases C100, S5 and R3, illustrated in figures 1-3, respectively. For cases C100 and S5 a non-rotating cloud of diameter 30 pc, mass 2.9 10^5 $M_\odot$, and surface escape velocity 13 km/s was assumed. In case R3 a smaller cloud (diameter 20 pc, mass 8.5 10^4 $M_\odot$, surface escape velocity 8 km/s) was adopted, uniformly rotating at 1.5 10^{-14} rad/s. For cases S5 and R3 a hydrogen-ionizing photon luminosity of 7.6 10^{48} s^{-1} was adopted, typical for a single 05 star. The initial radius of the Strömgren sphere in the dusty molecular cloud was 0.72 pc. In Case C100 the photon luminosity was larger by a factor of 100 and the initial Strömgren sphere had a radius of 2.5 pc.

3.1 Model C100 (Fig. 1)

The evolution was followed for 2.64 million years, assuming a powerful UV source at the center (with a flux comparable to the output of about 100 05 stars). The HII region first expands until the infalling material is able to stop and reverse the outward flow. At an evolutionary time t = 1.496 10^6 yr the HII region has its largest geometrical extent; a shock front surrounding the HII region is evident. During the subsequent evolution this region of maximum density appears to break up into fragments as the boundary of the HII region recedes due to compression of the ionized material. The clumps coalesce, forming two collapsing condensations about 3 pc above and below the equator. Note that even for this extreme case of a highly luminous UV source the HII region was not able to halt the central collapse or to cross the cloud-intercloud boundary and initiate a champagne flow.

3.2 Model S5 (Fig. 2)

The ionizing source was placed 5 pc from the top of the cylindrical

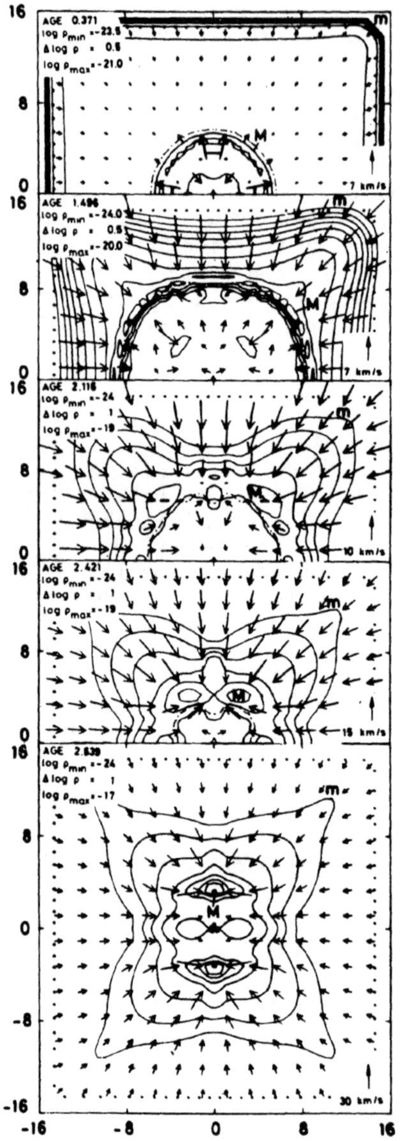

Figure 1. Meridional planes of a collapsing molecular cloud (case C100) at five evolutionary times, "AGE", as given in the top left corner of each segment in units of 10^6 yr. The coordinate axes give cylindrical (R,Z) distances in pc. The equator (Z = 0) is a plane of symmetry; the Z-axis (R = 0) is the symmetry axis. Iso-density contours and velocity vectors (length proportional to speed) are plotted. For the first four times only the top half-plane is shown. The location of the ionizing source is indicated by a triangle, the location of the I-front by a dash-dotted line. Small crosses denote the original location of the cloud-intercloud boundary. The minimum contour level in each segment is indicated by an "m", the maximum by an "M". The corresponding numerical values for "$\log \rho_{min}$" and "$\log \rho_{max}$" are given in the upper left corner of each segment along with the logarithmic spacing of the contour intervals, "$\Delta \log \rho$" (cgs units).

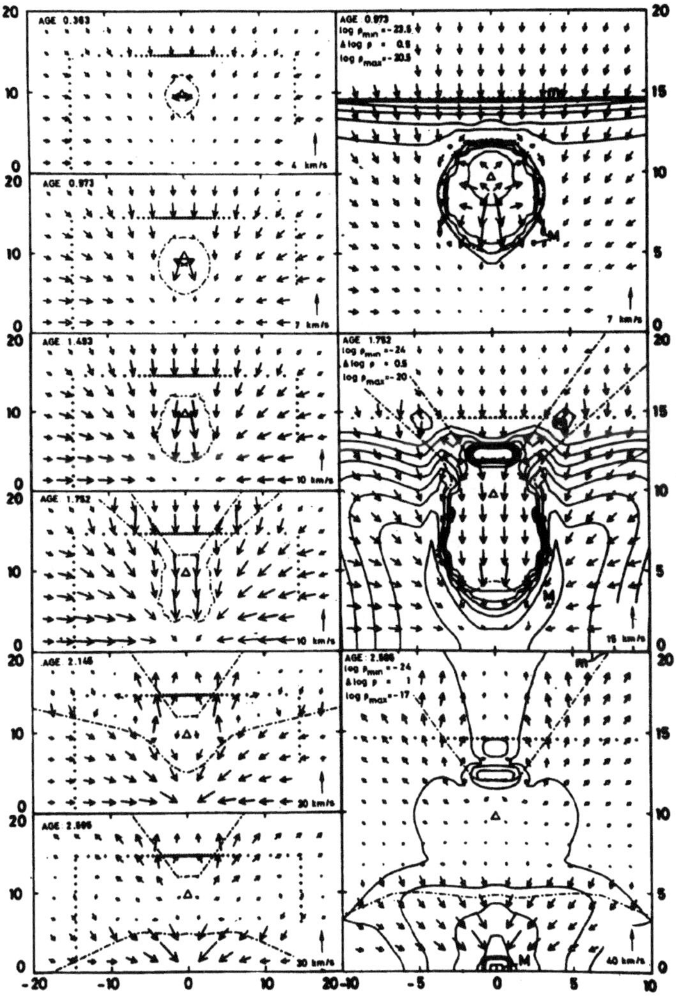

Figure 2. The evolution of model S5. Symbols and lines as in Fig. 1.

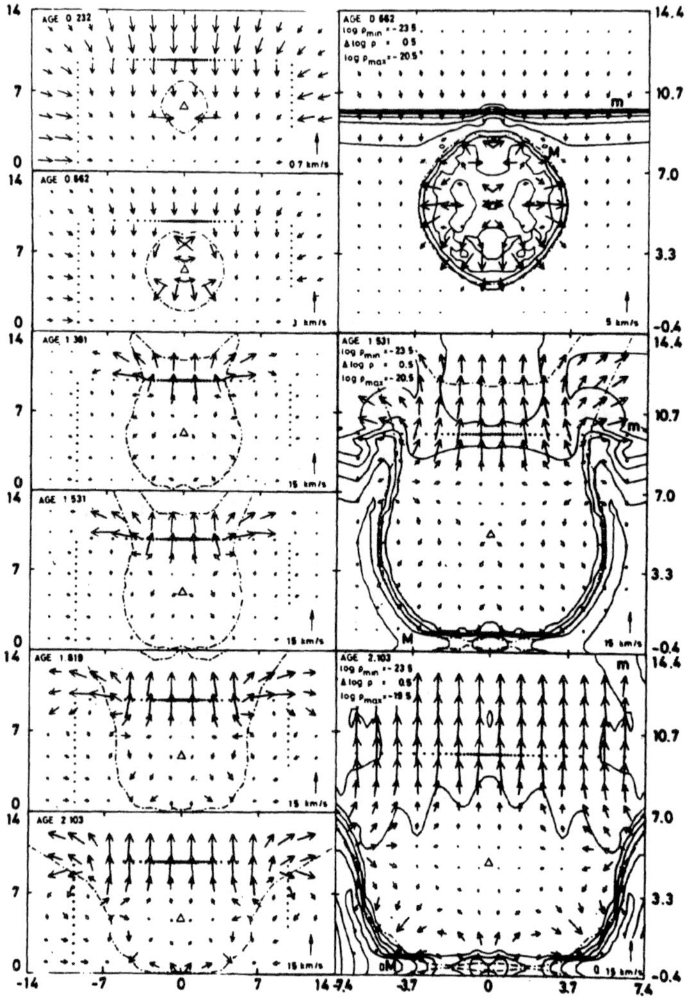

Figure 3. The evolution of model R3. Symbols and lines as in Fig. 1.

cloud, 10 pc from the cloud center. During the gravitational collapse of the cloud a condensation formed above the stationary star, where infalling gas collided with the expanding HII region. The I-front was able to break out to the sides of the condensation at t = 1.6 million years and initiate a champagne flow. The final configuration shown in Fig. 2, t = 2.595 million years, displays a flattened, dense 50 $M_\odot$ globule about 2.5 pc above the exciting star. Most of the cloud material (2.7 10^5 $M_\odot$) has collected in the central regions at this stage, but about 7% of the original mass has been expelled for a mass loss rate of 0.008 $M_\odot$/yr.

3.3 Model R3 (Fig. 3)

The single O star was assumed to move at a constant velocity of 0.5 km/s towards the cloud center, starting 4 pc from the cloud edge (6 pc from the center). Initially, the rotational energy of the cloud was about 10% of the gravitational potential energy. Because of rotation the cloud collapses preferentially parallel to the rotation axis. The intercloud material, however, is not assumed to rotate and falls inward radially. At 0.9 million years the ionization front crosses the cloud-intercloud boundary to the sides of a slight density condensation which forms above the star. This globule is dispersed as the champagne flow becomes more developed.

At the center of the cloud, a disk-like condensation forms which is supported partially by rotation and does not collapse further. Its density (10^{-21} g cm^{-3}) is much lower than the central condensation in case S5 (10^{-17} g cm^{-3}). It appears as if the cloud is being dispersed by the champagne flow at an average mass loss rate of 0.01 $M_\odot$/yr.

4. CONCLUSIONS

To summarize the main results of the calculations we note the following points:

1) Champagne flows can occur in massive (10^5 $M_\odot$) strongly gravitationally bound molecular clouds and can result in evaporation of the cloud at an average rate of 0.005 to 0.01 $M_\odot$/yr per ionizing source.

2) The total amount of evaporation that can result from a single source depends strongly on the initial position of the source in the cloud, but not so much on the cloud's mass. The champagne phase appears to be more effective when the exciting star is located near the edge of the cloud. If the source is located at the center of a massive non-rotating cloud, no evaporation is possible.

3) The conclusions of (2) are modified if rotation of the cloud is considered. The rotation stops the collapse of the central regions to high density and allows a higher rate of evaporation.

4) These results suggest that total disruption of a massive molecular cloud by O-stars can occur on a time scale of 10^7 years if a few of these sources are initially located near the edge of the cloud so that champagne flows occur and if the central regions of the cloud are supported against gravitational collapse.

ACKNOWLEDGEMENTS

We thank the Max-Planck-Institut für Astrophysik, Garching for generously providing support, travel funds and use of its facilities. Travel funds to this conference for H.W.Y. have been provided by the Deutsche Forschungsgemeinschaft.

REFERENCES

Bedijn, P.J., Tenorio-Tagle, G.: 1981 Astron. Astrophys. 98, 85.
Blitz, L., Shu, F.H.: 1980, Astrophys. J. 238, 148.
Bodenheimer, P., Tenorio-Tagle, G., Yorke, H.W.: 1979 Astrophys. J. 233, 85.
Cohen, R.S., Thaddeus, P.: 1977, Astrophys. J. (Letters) 217, L155.
Dickman, R.L.: 1978, Astrophys. J. Suppl. 37, 407.
Gordon, M.A., Burton, W.B.: 1976, Astrophys. J. 208, 346.
Mezger, P.G., Pankonin, V., Schmid-Burgk, J., Thum, C., Wink, J.: 1979 Astron. Astrophys. 80, L3.
Morfill, G.E., Stenholm, L.G.: 1980, Astron. Astrophys. 90, 134.
Rowan-Robinson, M.: 1979, Astrophys. J. 234, 111.
Scoville, N.Z., Hersh, K.: 1979, Astrophys. J. 229, 578.
Scoville, N.Z., Solomon, P.M.: 1975, Astrophys. J. (Letters) 199, L105.
Solomon, P.M., Sanders, D.B.: 1979, in"Giant Molecular Clouds in the Galaxy",ed. P.M. Solomon and M.G. Edmunds, Oxford: Pergamon Press.
Stenholm, L.G.: 1980a, Astron. Astrophys. Suppl. 42, 23.
Stenholm, L.G.: 1980b, Astron. Astrophys. 89, 264.
Stenholm, L.G.: 1980c, Astron. Astrophys. 92, 142.
Stenholm, L.G., Hartquist, T.W., Morfill, G.E.: 1981, Astrophys. J., submitted.
Tenorio-Tagle, G.: 1979, Astron. Astrophys. 71, 59.
Tenorio-Tagle, G., Yorke, H.W., Bodenheimer, P.: 1979, Astron. Astrophys. 80, 110.
Tenorio-Tagle, G., Bedijn, P.J.: 1981, Astron. Astrophys., in press.
Yorke, H.W.: 1980a, Astron. Astrophys. 85, 215.
Yorke, H.W.: 1980b, Astron. Astrophys. 86, 286.
Yorke, H.W., Bodenheimer, P., Tenorio-Tagle, G. (YBT): 1981, preprint (MPI-PAE/Astro 269), Astron. Astrophys., submitted.

DISCUSSION FOLLOWING PAPER BY YORKE ET AL.

HARTEN: The effect of gravity in a cloud appears to be to impede the formation of a large density build-up between the star and the cloud center. The formation of a globule between the star and the edge of the cloud is opposite to that expected in the Elmegreen-Lada model of sequential star formation.

YORKE: Yes, this is indicated by case S5 and in other cases calculated which display globule formation. However, the numerical models should not be over-interpreted as refuting the sequential star formation model. The fact that globule formation occurs at all simply indicates that the cloud is unstable to globule formation.

BALLY: If all the molecular clouds in the Galaxy were in free-fall collapse such as you have described, star formation would be considerably more vigorous than it actually is. Thus, clouds must be supported by turbulence, rotation, magnetic fields or other mechanisms. How would this affect your conclusions?

YORKE: By not including gravity in our earlier models, we were implying some sort of support. What we show now is that in an extreme case we can evaporate material away from the cloud even with gravity.

BALLY: Computations with support are not quite the same as computations with gravity ignored altogether, since the gravitational field still affects the dynamics of the HII region.

TENORIO-TAGLE: In the cases where there is rotation inhibiting the collapse, the evaporation rates through champagne flow are even larger because the densities are lower and the material doesn't shield itself.

PHILLIPS: Recently we have found another example of a conical dark cloud in an HII region with star formation at the apex. There seems to be a massive condensation at the interface of the HII region and the molecular cloud, leading to a "shadow" zone in which the cloud remains neutral. Is this consistent with your model?

YORKE: Definitely.

THE DYNAMICAL EVOLUTION OF AN HII REGION

R.H. Harten, Radiosterrenwacht, Dwingeloo, The Netherlands
M. Felli, Arcetri Observatory, Florence, Italy

SUMMARY

Radio aperture synthesis observations combined with single dish measurements of a large sample (77) of HII regions allow one to test the applicability of models of star formation and HII region evolution. Strong evidence is found in support of the 'champagne flow' model of HII region evolution. There is no strong statistical evidence in support of the theory of successive star formation proposed by Elmegreen and Lada although there are several individual cases which do support it. This discrepancy is also discussed.

INTRODUCTION

During the past few years models for both the detailed evolution of HII regions as well as for the triggering mechanism for star formation have been proposed. Unfortunately only a limited sample of well studied HII regions is available with which these models can be compared. Using this sample it is possible to show the feasibility of the models in a limited number of cases but it is impossible to state how common the mechanism is on the whole. What is needed is a large sample of HII regions with which to compare the various models.

To help resolve this problem, the authors undertook a survey of 77 Sharpless HII regions using the Westerbork synthesis radio telescope at a frequency of 5 GHz (Felli et al. 1978; Felli and Harten 1981a, 1981b). The purpose of the survey was to study the relationship between the emission from the diffuse and compact (or low and high density) components in an HII region. This was done by comparing the small scale flux densities, sizes and structures mapped with the synthesis instrument with the total flux densities and sizes measured with the single dish (lower resolution) instruments. The relationship between these two components, as well as any correlations with other indicators such as IR and molecular line emission provides important information about the process of star formation and the evolution of an HII region.

R. S. Roger and P. E. Dewdney (eds.), Regions of Recent Star Formation, 25–30.
Copyright © 1982 by D. Reidel Publishing Company.

The main assumption made in the following analysis is that on the large scale star formation is occurring at a uniform rate in time. This means that the relative numbers of objects in the different phases of their evolution can be interpreted in terms of the period of time a given region will spend in a given phase or stage of its evolution. This then allows us to draw some firm conclusions about the evolutionary time scales of an HII region.

RESULTS

One of the more interesting results of the statistical analysis of the survey was a plot of the fraction of the total radio flux density contained in the small scale components vs the mean electron density of the entire HII region. A plot of this relationship for all sources in the survey plus a few well studied objects is given in Figure 1. Figure 1 shows that essentially all of the total detected radio flux density of an HII region is contained in small scale components ($\theta < 1'$) until the mean density drops below $n_e \approx 200$ cm^{-3}. At lower densities, the fraction of the total flux density contained in the small scale components becomes progressively smaller as the mean density decreases. The large range in distances and sizes in the survey insures that there is no serious distance or resolution effect in the diagram. The shape of the curve implies that as the nebula expands, the flux density is well contained until the mean density reaches about 200 cm^{-3}. After this point in the evolution, the radio emission is coming from an increasingly more diffuse structure. The abrupt change in the slope at this density would seem to imply a rapid change in the rate of change of the mean n_e.

This is in excellent agreement with the predictions of the champagne flow model. Examination of case 2 of Tenorio-Tagle et al. (1979) and case 1 of Bodenheimer et al. (1979) show that this break occurs when the shell of higher density material formed by the cloud material behind the shock moving into the inter-cloud medium, is dispersed by the rapid expansion of the shock (30 km s^{-1}) and the refraction wave which begins to propagate back from the shock towards the star. This phase occurs between the ages 1.5 to 3 x 10^5 years (depending on the particular case). Also the maximum gas velocities also occur during this phase. Both the lack of extensive small scale structures and the mean density are in agreement with the observations.

Another interesting relationship is the plot of $N(n_e)$ vs n_e shown in Figure 2. To minimize any distance or luminosity effects only sources which had a derived total flux of ionizing photons in the range $10^{47.5}$ to $10^{48.4}$ photons sec^{-1} (spectral types O9 to B0) were plotted. This diagram can be interpreted as the frequency of occurrence of the mean electron density of an HII region during its evolution and clearly shows that the change of n_e is much higher during the high density phase than during the lower density phase. For the classical model of HII region evolution the rate of change of n_e for a uniform density HII region can

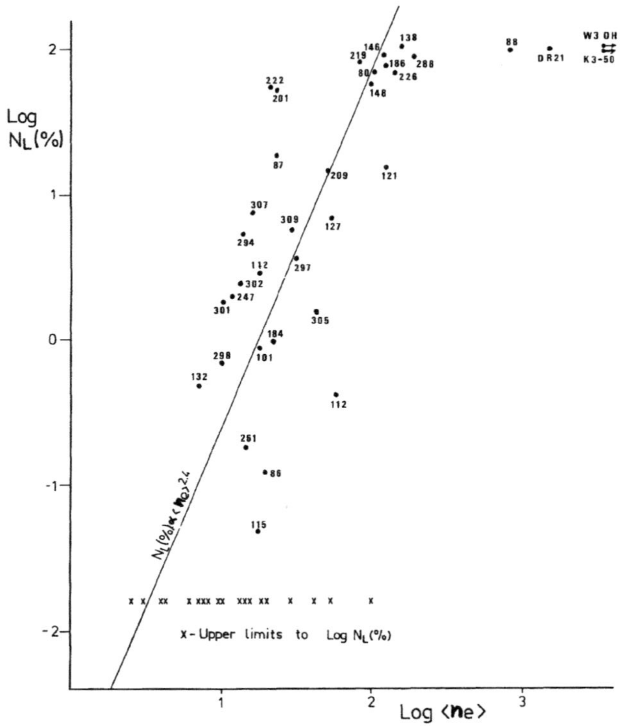

Figure 1. Plot of the fraction of the total radio flux density contained in the small scale components, $N_L(\%)$, versus the mean electron density of the HII region.

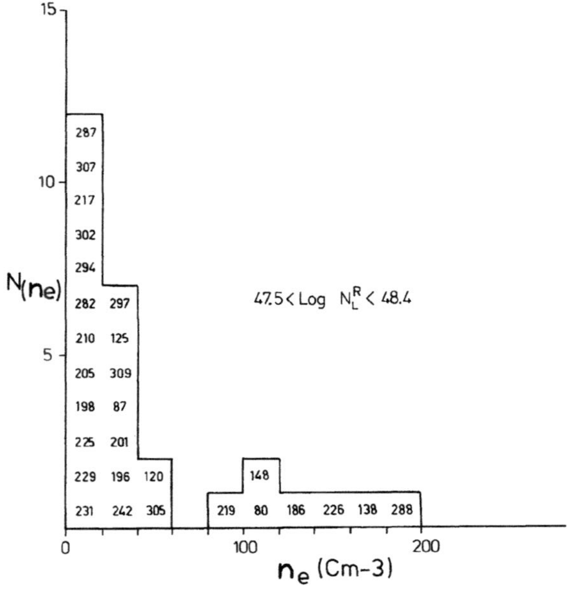

Figure 2. Histogram of the number of regions having a mean n_e within a given range versus the mean n_e for HII regions requiring exciting stars of spectral between B0 and O9 to account for the observed total radio flux density.

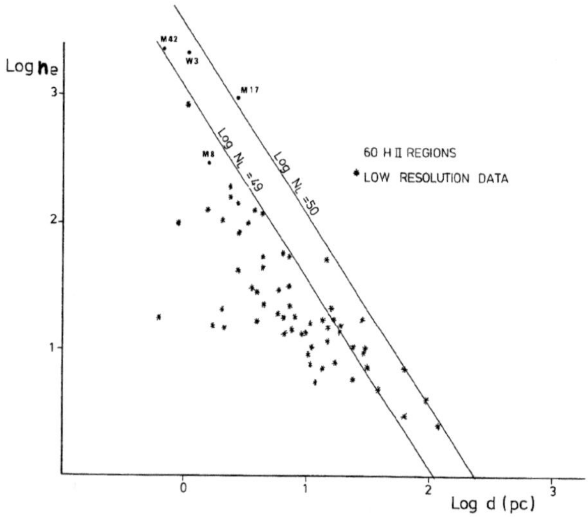

Figure 3. Log $\langle n_e \rangle$ - log diameter plot for the HII regions in the sample. Lines of constant ionizing photon flux, log N_L, are plotted for reference.

be expressed as

$$n_e(t) = n_e(o)(1 + \frac{7}{4}\,\frac{C_{II}t}{r_o})^{-6/7}$$

where r_o is the initial Stromgren radius and C_{II} is the sound speed in the HII region (Spitzer 1968). Based on this simple model, the ratio of the time needed to reach a density of 10 cm^{-3} to that required to reach a density of 100 cm^{-3} is about 14. If we compare the number of HII regions in our sample in the range $10<n_e<100$ to those in the range $n_e>100$ we obtain a ratio of 4.5. This difference is roughly the factor of 3 faster evolution predicted by the 'champagne effect' (Bodenheimer et al. 1979).

A plot of the log of the diameter versus the log of the mean electron density of an HII region is shown in Figure 3. Most of the points lie in the lower part of the diagram and there is a noticeable lack of objects at medium densities (n_e = 50 - 500) to the right of the line, log N_L = 49. A similar plot by Habing and Israel (1979) for a different sample shows a similar effect. In principle an HII region should evolve along a line of constant N_L, since the supply of ionizing photons should remain constant over the main sequence life time of the star. The lack of points to the right of the line log N_L = 49 would imply that the HII regions of earlier spectral type stars evolve much faster through these intermediate density phases. This is in agreement with the theory since the larger ionizing photon flux of these stars will speed up the initial expansion and onset of the champagne phase. The later phases (lower density) of the expansion are much slower and the main effect of the increased ionizing photon flux would be the ability to ionize a larger region, a trend which can be seen in the diagram.

The survey provided more than statistical evidence for comparison with the models. Several individual HII regions in the sample seem to fit the different model cases quite well. S201 is an excellent example of a bipolar nebula which has evolved from a thin molecular cloud (Martin and Barrett 1978). It's structure is quite similar to the thin slab model of Tenorio-Tagle et al. (1979) and Bodenheimer et al. (1979). Recent observations by the authors show that the agreement is quite good even in the finer scale structure. S140 is a good example of a nebula formed at the edge of a cloud. The depth of HII 'hole' in the cloud also seems to suggest that the star may be moving towards the cloud. S115 is a good example of a region in which the ionization front has passed out of the high density part of the cloud and ionized the remaining lower density (10 cm^{-3}) cloud material before the higher density checked material has reached it (Harten and Felli 1980).

The theory of succesive star formation of Elmegreen and Lada (1977) predicts the formation of new stars near the edges of existing HII regions. Using a typical cloud density of 10^3 cm^{-3}, the model predicts that the next generation of star formation will occur about 10-15 pc away and about 2.5 million years later. If this was a quite common phenomenon, one would expect to find compact features near the edges of evolved HII regions. In our sample which contained 39 complex or diffuse HII regions 13 of which had diameters larger than 19 pc only one object (S184) had a possible candidate for successive star formation. Thus we found no statistical evidence for this being a common phenomena in the interstellar medium.

There is evidence, however, that this process is at work in several regions. Recent studies of the S155-CEP OB3 region (Sargent, 1977; Felli et at., 1978; Harten et al., 1981) and W1 (Harten et al., 1981) have found structures which are in good agreement with the model's predictions. Yet in the regions around W3, W75 and W1 it is clear that much larger scale features, possible a spiral density wave, seem to be working.

There are several possible explanations for lack of statistical support for the model.

1) The process of successive star formation may be effective only under limited circumstances such as cloud size or shape or the type of ionizing stars formed.
2) The time scale for the evolution of a compact HII region is so short that the chance of finding diffuse and compact ones in the same region is quite small.

3) The time scale for the next generation, 2-3 x 10^6 years, is comparable with the expansion time of an HII region so that by the time the new generation of stars has formed, the original HII region is indistinguishable from the general background.
4) It is possible that successive generations may be of later type stars which do not produce easily detectable HII regions. Infrared studies might be needed to detect these objects.

CONCLUSION

Observations of a large sample of HII regions provide good statistical support for the champagne flow model of HII region evolution. Detailed studies of several regions show good agreement with this model. We find little statistical evidence to suggest that the mechanism for successive star formation is widespread. However, there is good evidence that the mechanism is operating in several regions.

ACKNOWLEDGEMENTS

The Westerbork Synthesis Radio Telescope is operated by the Netherlands Foundation for Radio Astronomy with the financial support of the Netherlands Organization for the Advancement of Pure Research (ZWO).

REFERENCES

Bodenheimer, P., Tenorio-Tagle, G., Yorke, H.W.: 1979, Astrophys. J. 233, 85
Elmegreen, B.G., Lada, C.J.: 1977, Astrophys. J. 214, 725
Felli, M., Tofani, G., Harten, R.H., Panagia, N.: 1978, Astron. Astrophys. 69, 199
Felli, M., Harten, R.H., Habing, H.J., Israel, F.P.: 1978, Astron. Astrophys. Suppl. 32, 423
Felli, M., Harten, R.H.: 1981a, Astron. Astrophys. in press
Felli, M., Harten, R.H.: 1981b, Astron. Astrophys. in press
Habing, H.J., Israel, F.P.: 1979, Ann. Rev. Astron. Astrophys. 17, 345
Harten, R.H., Felli, M.: 1980, Astron. Astrophys. 89, 140
Harten, R.H., Thum, C., Felli, M.: 1981a, Astron. Astrophys. 94, 231
Harten, R.H., Goss, W.M., Matthews, H.E., Israel, F.P.: 1981, submitted Astron. Astrophys.
Martin, R.N., Barrett, A.H.: 1978, Astrophys. J. Suppl. 36, 1
Sargent, A.: 1977, Astrophys. J. 218, 736
Spitzer, L.: 1968 "Diffuse Matter in Space," J. Wiley and Sons, New York
Tenorio-Tagle, G., Yorke, H.W., Bodenheimer, P.: 1979, Astron. Astrophys. 80, 110

RADIO CONTINUUM OBSERVATIONS OF W1

H.E. Matthews
Herzberg Institute of Astrophysics, National Research Council of Canada, Ottawa, Canada K1A 0R6

ABSTRACT

Observations are presented of the continuum emission from the thermal radio source W1 (S171, NGC7822). The complete region has been accurately mapped at λ11cm using the 100-m telescope in both total power and polarized radiation. The central part of the object also has been observed with the Westerbork Synthesis Radio Telescope at λ49, λ21, and λ6cm. In the light of these observations, the question of the existence of a non-thermal component of W1 is discussed, and indications for secondary star formation processes are examined.

1. INTRODUCTION

The well-known radio source W1 is identified with the bright HII region S171, situated at a distance of 845 pc in Cepheus. S171 lies within the Cep IV OB association which itself consists of some 40 OB stars distributed in a compact cluster (Berkeley #59) and a more dispersed population spread over about 5 degrees. MacConnell (1968) has studied the region in detail and derives an age of about 2×10^6 years for the association. A dark cloud complex extends across much of the southern and western parts of the region and contains a number of T Tauri stars and reflection nebulae.

S171 itself contains two main components, G118.6+4.8 to the east, and G118.1+5.0 in the west, which are most clearly seen in radio continuum maps. G118.1+5.0 is heavily obscured over much of its surface area. Associated with S171 is a broad distribution of diffuse emission extending over about 3 degrees of sky. Most notable is the almost complete 'Cepheus loop' of nebulosity to the north of S171. This contains the bright rim structure NGC7822, interpreted by Elmegreen et al. (1978) as possibly heated by magnetic field compression. A further small HII region (S170) lies about 2 degrees to the south of S171, but appears to be unrelated.

R. S. Roger and P. E. Dewdney (eds.), Regions of Recent Star Formation, 31–38.
Copyright © 1982 by D. Reidel Publishing Company.

For some years, since the radio observations of Churchwell and Felli (1970) showed a non-thermal spectral index for the region, there has been considerable discussion as to whether or not W1 is partly supernova remnant. The 'Cepheus loop' has been cited as additional evidence for this, and Bonsignori-Facondi and Tomasi(1979) have been the most recent authors to argue in this direction. More recently, however, Rossano et al. (1980) have mapped the complete region, including the loop emission at λ9cm and λ6cm and find no compelling evidence for a non-thermal component to W1 of any significance. The difficulty in reaching a conclusion in this respect has stemmed from the inability to map weak extended radio structures with confidence, and the comparison of flux densities derived from maps covering differing areas of sky.

Here we present continuum observations made of the W1 region with the Effelsberg 100-m telescope in both total power and polarization at λ11cm, and with the Westerbork Synthesis Radio Telescope (WSRT) with high resolution at three different frequencies.

2. OBSERVATIONS AND REDUCTION

The Effelsberg observations employed the dual-channel λ11-cm correlation receiver operating at the Gregorian focus of the 100-m telescope. At the centre frequency of 2695 MHz the halfpower beamwidth is 4.4 arc min and 1 Jy corresponds to 2.6K of main-beam brightness temperature. Two sets of observations were made: in the first the receiver was operated in its correlation radiometer mode (in which the power difference between the main horn and a sky horn is measured) and an area 4 × 4 deg. around W1 was mapped. The second set of measurements was made with the receiver in its polarimeter mode, giving two correlated polarization channels (Q and U) and two uncorrelated total power channels. In the latter mode an area (3 × 3 deg.) covering essentially all the continuum emission of W1 was mapped, together with an additional one square degree around S170. Instrumental polarization was determined to be less than 0.5%. In all cases orthogonal scans in a 'local' RA, Dec system (where the point of origin is the centre of the map) were made, with scans spaced by 2 arc min in both dimensions. Observations of this type are discussed in greater detail in Sieber et al. (1979).

The observations were reduced using the NOD2 program system (Haslam, 1974). The maps were set on a consistent baselevel using a program which evaluates all the crossing points of a given scan with equivalent points in the orthogonal data set. In this manner it is possible to approach the theoretically-expected noise level and map weak, diffuse structures with considerable confidence.

Four sets of aperture synthesis observations of W1 were made with the WSRT. The observing frequency (ν), field centre (RA, Dec), synthesized beam, and primary beamwidth θ_p to half-power are given below:

Table 1	ν(GHz)	RA(1950)	Dec(1950)	$\theta_{RA} \times \theta_{Dec}$	θ_p
	0.610	$00^h00^m00^s$	67°16'00"	50" × 55"	83'
	1.415	00 02 30	66 49 44	22 × 24	38
	1.415	23 58 42	67 04 30	22 × 24	38
	4.995	23 58 42	67 04 30	6.6 × 7.2	10.6

The typical limiting flux density in all maps is ∿4 mJy. Further details will be given elsewhere (Harten et al. 1981). The data were analyzed using standard software, and each map was 'cleaned' using the procedure described by Högbom (1974).

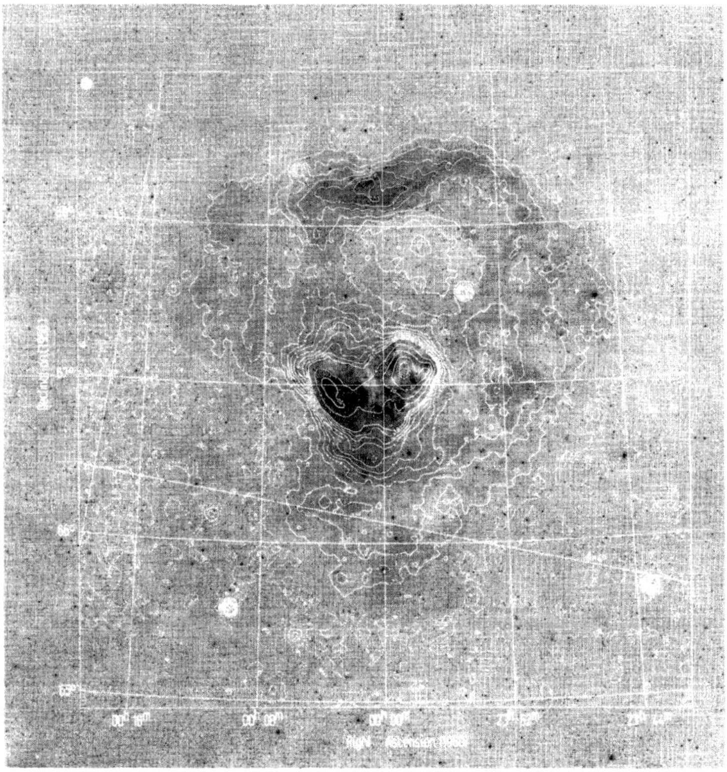

Fig. 1. W1 mapped at λ11cm with the 100-m telescope and the results overlaid on the red print of the region taken from the POSS. The telescope beam is shown by the filled circle in the upper left. The contour interval is 200, between 50 (dashed) and 2050, and thereafter 1000 (all in mK T_b)

3. RESULTS

The λ11-cm total power map of W1 made with the 100-m telescope is shown in Fig. 1 superimposed on the red print of the region taken from the Palomar Sky Survey. The r.m.s. noise level in the map is 40 mK equivalent main beam brightness temperature (T_b). The brightest point in the map is 6.2K T_b occurring in the region of G118.1+5.0. The radio contours follow the optical emission in the W1 region very closely in general, even to quite faint levels (the r.m.s. noise level in terms of emission measure for free-free radiation originating in gas at an electron temperature of 10000K is ∿90 $cm^{-6}pc$). At the same time, the map shows a wealth of detail in the diffuse emission. The lack of such diffuse emission to the east of G118.6+4.8 however strongly suggests that the nebula is ionization-bounded at least in this direction. A summation over the whole area of the map yields a total flux density at 2695 MHz of 290 ± 10 Jy. Background (presumably extragalactic) sources in the field contribute about 4 Jy to this total.

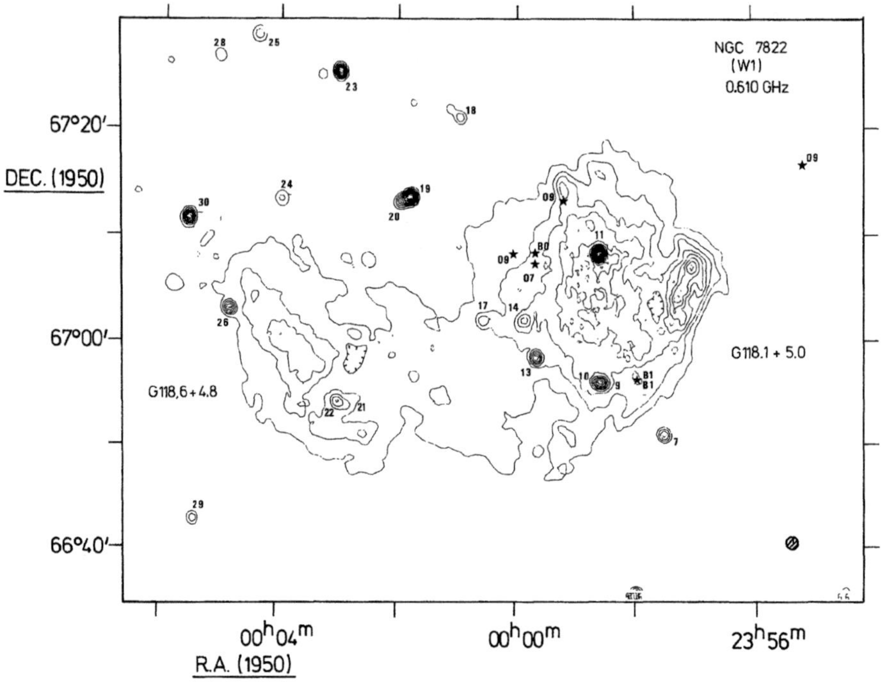

Fig. 2. The central portion of W1 mapped with the WSRT at λ49cm. Contours (uncorrected for primary beam attenuation) begin at 15, with an interval of 5 (mJy per beam). The beam appears in the lower right corner. Known exciting stars and (presumably) background sources (Harten et al. 1981) are indicated.

The λ11-cm polarized intensity map of W1 showed no emission above 3σ (45mK Tb in $\{Q^2 + U^2\}^{\frac{1}{2}}$) at any point in the map. Areas of ≥ 10% polarization should be apparent for all regions (i.e. most) of the nebula for which the continuum brightness temperature exceeds 300 mK.

The total power observations of the separate HII region S170 give a total flux density at λ11 cm of 3.32 ± 0.10Jy and a deconvolved size of 12.5 arc min. No polarization was detected in this region. Kallas et al. (1980) give a flux density for S170 of 3.2 ± 0.1 at λ21 cm, so that the object is almost certainly thermal.

The WSRT observations at 610 MHz of the central part of W1 are shown in Fig. 2, in which the two main concentrations of radio emission appear in detail. G118.6+4.8 has a complex ridge-like form and G118.1+5.0 shows a plateau of emission with a small ridge somewhat further to the west. In the 1415-MHz maps of these two regions still finer detail is evident. G118.6+4.8 (Fig. 3) shows the ridge to be a

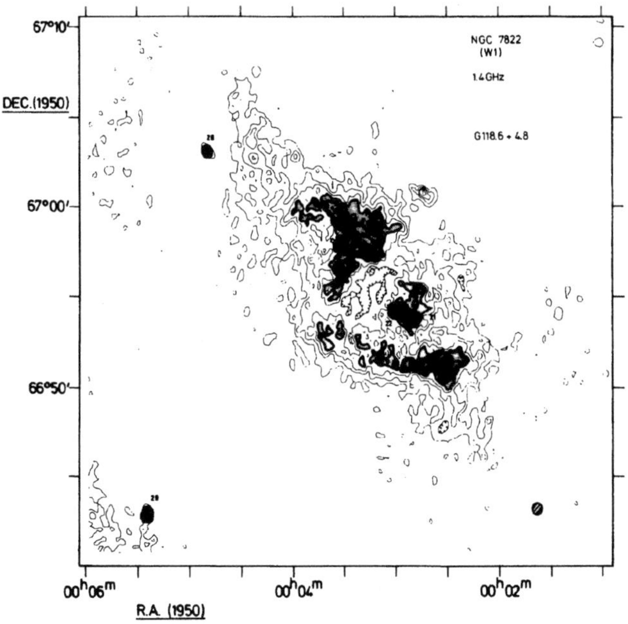

Fig. 3. G118.6+4.8 (W1-east) mapped at λ21cm with WSRT. Contours (corrected for the primary beam) begin at 1.5, with an interval of 1.5 (mJy per beam). The shaded portion delineates regions discussed in the text. Other details as for Fig. 2.

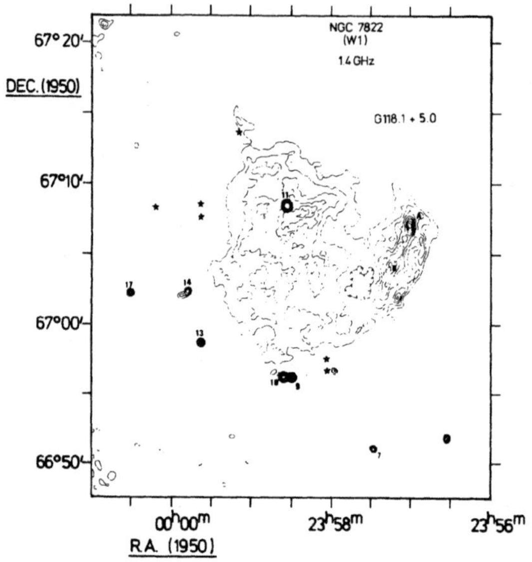

Fig. 4. G118.1+5.0 (W1-west) as seen at λ21cm with the WSRT. The contour levels are 2.5 to 25 in steps of 2.5, and in steps of 25 (mJy per beam, uncorrected for the primary beam) thereafter. Other details as for Fig. 3.

broken arc of emission with a complex condensation at its centre. In G118.1+5.0 (Fig. 4) the western ridge shows two condensations (A and B) and indications of another partial ridge still further west. The plateau shows surprisingly little structure, except for source #11 (see Fig. 4), which possesses a non-thermal spectral index. This source is essentially the only feature in the 4995-MHz observations of the region, and appears most likely to be an unrelated background source.

4. DISCUSSION

4.1. A non-thermal component to W1?

The 100-m observations at λ11cm reported here provide the best-determined total flux density measurement of W1 to date and taken together with the comparable observations at λ9cm and λ6cm of Rossano et al. (1980) strengthen their conclusion that there is no significant non-thermal component to the flux density of W1. As the latter authors point out, a map of the spectral index distribution in the diffuse material is subject to some uncertainty, but there also appears to be no grounds to consider any major part of W1 to be non-thermal in origin. Our failure to detect polarized emission from W1 strengthens this view, as does the detection by Pedlar (1980) of H166α recombination lines from most of the body of the nebula. The Cepheus loop emission to the north is thus more readily explained in terms of a 'champagne flow' (Tenorio-Tagle, this volume), rather than being the result of a supernova explosion.

4.2. Ionization balance.

Using a total flux density of 250 Jy at λ6 cm for W1 and assuming an electron temperature of 10^4K means that about 1.6×10^{49} Lyman continuum photons sec^{-1} are required to produce the observed region. The known stars in the Cep IV OB association total between 1 and 2×10^{49} photons sec^{-1}, but a significant fraction of these are expected to escape. Thus in addition to the known stars indicated on Figs. 2, 3 and 4 there may be others embedded in the denser obscured regions.

4.3. Star formation in W1.

The youthful age of the Cep IV OB association, their dispersion and their apparent existence near the edges of the obscuring cloud suggest a large-scale process of star formation has operated in this vicinity, perhaps a triggered by a spiral density wave. In addition, there is evidence for secondary processes of star formation in the present data. For instance, the eastern region G118.6+4.8 appears to be mostly excited by the Berkeley #59 cluster, except for the region in the neighbourhood of the break in the main ridge (in Fig. 3 the upper contours are shaded to accentuate the detailed form of the ridge). Here the radio continuum contours and the observation of wider H110α recombination lines (Rossano et al, 1980) and strong [OIII] emission (Parker et al, 1979) are suggestive of the presence of at least one further OB star in the region of sources #21 and #22 (see Fig. 3). This may be an example of induced star formation as envisaged by Elmegreen and Lada (1977).

As a second example of possible secondary star formation, solid angle arguments indicate that Berkeley #59 is insufficient to ionize all of the western complex G118.1+5.0, and that at least one further 09 star is required. Once again, wide recombination lines have been observed (Rossano et al.) in this region. The well-defined ridge to the west of G118.1+5.0 may be further evidence for the presence of obscured sources of ionization within this area.

ACKNOWLEDGEMENTS

The Effelsberg observations and reduction were carried out in conjunction with C.G.T. Haslam, C.J. Salter and D.L. Hills. The Westerbork measurements reported in this paper result from a collaboration with R.H. Harten, W.M. Goss and F.P. Israel. The Westerbork Synthesis Radio Telescope is operated by the Netherlands Foundation for Radio Astronomy with the financial support of Z.W.O.

REFERENCES

Bonsignori-Facondi, S.R., Tomasi, P.: 1979, Astron. Astrophys. 77, 93

Churchwell, E., Felli, M.: 1970, Astron. Astrophys. 4, 309

Elmegreen, B.G., Lada, C.J.: 1977, Astrophys. J. 214, 725

Elmegreen, B.G., Dickinson, D.F., Lada, C.J.: 1978 Astrophys. J. 220, 853

Harten, R.H., Goss, W.M., Matthews, H.E., Israel, F.P.: 1981, Astron. Astrophys., in press

Haslam, C.G.T.: 1974, Astron, Astrophys. Suppl. 15, 333

Högbom, J.A.: 1974, Astron. Astrophys. Suppl. 15, 417

Kallas, E., Reich, W.: 1980, Astron. Astrophys. Suppl. 42, 227

MacConnell, D.J.: 1968, Astrophys. J. Suppl. 16, 275

Parker, R.A.R., Gull, T.R., Kirschner, R.P.: 1979, *An Emission-Line Survey of the Milky Way*, NASA SP-434

Pedlar, A.: 1980, Mon. Not. Roy. Astron. Soc. 192, 179

Rossano, G.S., Angerhofer, P.E., Grayzeck, E.J.: 1980, Astron. J. 85, 716

Sieber, W., Haslam, C.G.T., Salter, C.J.: 1979, Astron. Astrophys. 74, 361

DISCUSSION FOLLOWING PAPER BY MATTHEWS

GRAYZECK: Recent photometric and spectroscopic observation of the Berkeley 59 cluster of stars indicates an age of 5×10^5 years. This is an excellent example of a region undergoing successive stages of star formation.

INTERACTION OF THE HII REGION S236 WITH THE SURROUNDING MEDIUM

A. Falchi, G. Tofani
Arcetri Astrophysical Observatory, Florence
R.H. Harten
Radiosterrenwacht, Dwingeloo

ABSTRACT The HII region S236 has been observed in the continuum at 50 cm wavelength and in the recombination line H166α at 10 selected positions. A mean electron temperature T_e = 7600 K is derived for the nebula and the measured velocity pattern is explained qualitatively in terms of a gas flow from the molecular cloud.

The structure of the evolved HII region S236 and its interaction with the surrounding medium is analyzed by means of radio continuum and line observations.

S236 is a faint extended nebula located in the Galactic anticenter region (ℓ = 173.6, b = -1.7) at a photometrically estimated distance of 3.4 kpc. A cluster of OB stars (NGC 1893) is coincident with the position of S236. The cluster is rich in O-type stars; seven have been found near the brighter inner part of the nebulosity.

The nebula has been measured at low resolution in the radio continuum at several wavelengths. The radio spectrum, although not accurate due to the presence of a complex background, is typical of a low density region. From the total flux (39.4 Jy at ν = .75 GHz), distance and size (25') of the region the mean electron density is about 15 cm^{-3}. With the same data the total mass of the ionized hydrogen M_{HII} is about 8×10^3 $M_\odot$. The excitation parameter, deduced from this measured radio flux, is U = 96.5 pc cm^{-2}. The presence in the cluster of an O4V star (α = 5h 19m 12.04s, δ = 33° 28' 0.42"), which has an excitation parameter U = 126 pc cm^{-2} (Panagia 1973), indicates that S236 is a density bounded HII region.

The area of S236 has been investigated in order to find molecular emission. The only evidence of a cold cloud associated with S236 is from measures of L. Blitz (1980), who detected, in a survey made with the Columbia University Telescope, $^{12}C^{16}O$ line concentrated in a small blob of approximately the same size as the beam (~8') in the position

R. S. Roger and P. E. Dewdney (eds.), Regions of Recent Star Formation, 39–42.
Copyright © 1982 by D. Reidel Publishing Company.

α = 5h 19m 06s δ = 33° 18' 00". This small cloud coincides with a patch of obscuration visible in the red plate of the PSS in the center of the area of S236.

In order to investigate the structure and dynamics of the HII gas, a map was made with the Westerbork Telescope at λ50cm. This map is still under reduction, but a preliminary "uncleaned" version suggests a general agreement between the optical and radio pictures of the nebula. At the same time the H166α recombination line was observed in many points of the nebula with the Arecibo radiotelescope. The instrument was used at 1.4 GHz, in the frequency-switching mode, with the 1008 channel autocorrelator backend in double-Nyquist mode. The spectral resolution is 9.76 kHz and the r.m.s. noise on the spectra ~0.01K. Ten positions were selected on the basis of the preliminary Westerbork map at λ50cm.

The electron temperatures were derived from the line parameters on the basis of the usual assumptions (Donati Falchi et al., 1980). The results do not show any systematic trend of the temperature in the nebula. A mean value of 7600±1800K has been derived. The gradient of electron temperature with the Galactic radius computed by Churchwell (1980) would predict for S236 a value of about 9000K. The lower result of the Arecibo data is in better agreement with the lower values of T_e which have been found by Pedlar (1980) in extended low brightness HII regions.

The following radial velocities V_{LSR} are related to S236:

1) a mean value 0.0 km s^{-1} from Hα (Georgelin, 1975)

2) a V_{LSR} = -7.2±0.4 km s^{-1} at the peak of the CO emission (Blitz, 1980)

3) the known radial velocities of the exciting stars are in the range from -8 to -27 km s^{-1}.

The mean value of V_{LSR} from our measures is about -7 km s^{-1}, in good agreement with the CO radial velocity. From a look at the distribution of the radial velocity in the nebula (see Figure 1), it appears that the ionized gas has a range of velocities between -3.5 km s^{-1} and -13.5 km s^{-1}. These variations are greater than the experimental errors. If this distribution of radial velocities is referred to the foreground ^{12}CO cloud, we can divide the ionized gas into two regions: one, North-East, which is red-shifted with respect to the CO by about 3 km s^{-1} and the other, South-West, which is blue-shifted by about 1 km s^{-1}. The most plausible explanation of this pattern is an ordered streaming of the ionized gas. In the context of the evolution of an HII region undergoing the "champagne phase", Tenorio-Tagle et al. (1979) treated the flow of ionized gas

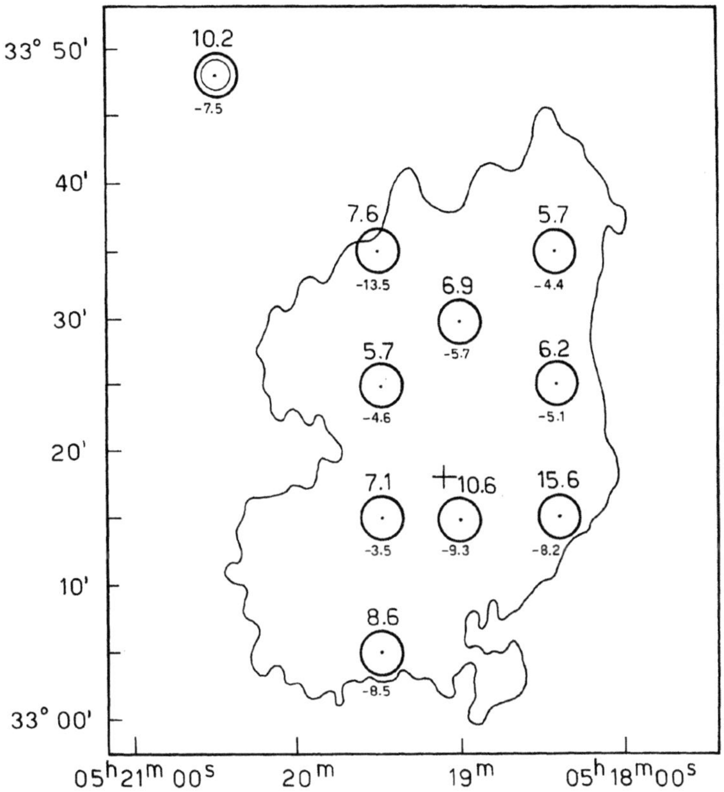

Figure 1. Sketch of the S236 region. Circles represent the HPBW of the Arecibo dish at the observed positions. The cross is the CO peak. The thin line indicates the WSRT radio emission. Large numbers are electron temperatures (10^3K). Small numbers are LSR velocities of the centroid in km s^{-1}.

from a slab of molecular material. They considered the case of a star initially located off center in a flat cylindrical cloud. After $5x10^5$ to $1x10^6$ yr, depending on the initial parameters of the cloud, the gas streamed in two opposite directions with different densities and velocities.

In S236 the situation seems to be more complex due to the presence of several O-type stars. The ionized gas may have had many ways to flow out with different velocities depending on the density and on the time at which the streaming began. We can then suppose that this region experienced at least two "champagne phases": a first one in the N-E direction and a second one in the S-W direction. The angle between the line-of-sight and the velocity of expanding gas should be greater than 60° - 70° to reproduce the observed pattern.

Three observational facts support the idea that the "champagne phase" first began in the N-E direction:

i) O-type stars are grouped on the N-E side of the CO remnant (probably they were born nearer to the E boundary of the parent molecular cloud).

ii) The ionized gas in this part of the nebula is more diffuse (ie. has had more time to expand).

iii) The gas is receding from the CO with a velocity greater than that of material in the S-W part.

ACKNOWLEDGEMENTS

We are grateful for the assistance of the staff of the Arecibo Radio Telescope during the observations and data reduction. This research has been supported by CNR Contract 78.01835.63 in the framework of the Arectri-Cornell Cooperation Agreement. The Arecibo Observatory is part of the National Astronomy and Ionoshpere Center, which is operated by Cornell University under contract with the National Science Foundation.

REFERENCES

Blitz, L.: 1980, private communication
Churchwell, E.: 1980, in "Radio Recombination Lines", Ed. P.A. Shaver, D. Reidel, Dordrecht p. 225
Donati Falchi, A., Felli, M., Tofani, G.: 1980, Astron. Astrophys. 89, p. 363
Georgelin, Y.M.: 1975, Ph. D. Thesis, Université de Provence, Marseille, France
Panagia, N.: 1973, Astron. J. 78, p. 929
Pedlar, A.: 1980, Monthly Notices Roy. Astron. Soc. 192, p. 179
Tenorio-Tagle, G., Yorke, H.W., Bodenheimer, P.: 1979, Astron. Astrophys. 80, p.110

OPTICAL AND MILLIMETER WAVELENGTH STUDY OF THE COMPLEX Sh2-147/Sh2-153

M. Heydari-Malayeri[+], C. Kahane[x], R. Lucas[x], G. Testor[+]
[+]Observatoire de Meudon, D.A.F., F-92190 France
[x]C.E.R.M.O., B.P. 53 X, F-38041, Grenoble Cedex, France

INTRODUCTION

Sh2-147/Sh2-153 is a vast HII region-molecular cloud complex of dimension 1°.5 located in the Perseus arm at $\ell \approx 109°$. This cloud, mapped for the first time by Israel (1980) in ^{12}CO emission, embodies the HII regions Sh2-147, 148, 149, 152 and 153. In this direction were detected several H_2O and OH masers (Lo et al., 1975; Baudry et al., 1977; Le Squeren, 1980), a number of infrared sources (Bergeat et al., 1975; Price and Walker, 1976) and a supernova remnant (Hughes et al., 1981; Gregory and Fahlman, 1980).

The study of this complex is interesting from several viewpoints, mainly:

1) This complex is considered as a region of recent or ongoing star formation;

2) Given the relative "smallness" and "simplicity" of the HII regions formed in the complex, it is suitable for studying the interaction of ionized gas with neutral matter;

3) In this connection, it is a good opportunity to confront the "champagne models" (Tenorio-Tagle, 1979; Bodenheimer, et al., 1979) with observations and determine the stage of evolution of the HII regions.

4) The fact that the HII regions of the complex are close together, and have brightness differences on the red PSS print, suggests a detailed study of the problem of sequential formation of the HII regions (Elmegreen and Lada, 1977).

5) The interesting, newly-discovered SNR, G109.1-1.0, calls for investigation of eventual changes due to the explosion in the molecular cloud.

R. S. Roger and P. E. Dewdney (eds.), Regions of Recent Star Formation, 43–51.
Copyright © 1982 by D. Reidel Publishing Company.

We have carried out millimeter observations in the molecular lines ^{13}CO, HCO^+, HCN and $H^{13}CO^+$, of the complex as well as detailed monochromatic photography and spectrography of some of the HII regions.

We present here the ^{13}CO map and also optical results on two of the HII regions of the complex: Sh2-152 and 148. Section II is a brief description of the observations. Section III presents large-scale features of the complex and Section IV deals with the main characteristics of each of the objects in the complex.

OBSERVATIONS

a) The optical observations were carried out at different periods from September, 1976 to February, 1980 at the Haute-Provence Observatory using the 193cm telescope coupled with the Lallemand electronographic camera and the Pellet nebular spectrograph. For details on the optical observations and the data processing see Heydari-Malayeri and Testor (1981).

b) The millimeter wavelength observations were made in December, 1980, and January and February, 1981 with the 2.5m antenna at the Bordeaux Observatory (Baudry et al., 1980). The ^{13}CO ($J = 1 \rightarrow 0$) line (110.20137 GHz) was observed. The Half-Power Beam Width (HPBW) was 5 arcmin. The single-sideband noise temperature was about 1000K. The receiver back-end is composed of 256 channels, 100 kHz wide (0.27 km s^{-1} velocity resolution). Frequency switching at 5 MHz was used.

MAIN FEATURES OF THE COMPLEX

The ^{13}CO emission has been measured at positions on a grid with a one-beamwidth spacing. Figure 1 shows a map of peak ^{13}CO equivalent brightness temperature (T_A^*). It shows good similarity with the CO map of Israel (1980) obtained at a lower resolution. The molecular cloud presents an elongated structure running approximately parallel to the Galactic plane and at least 1° long, corresponding to a linear dimension of 70pc, if a distance of 3.5 kpc (Israel, 1977) is assumed. A low-brightness extension (δ) is detected north of the compact X-ray source. The two main maxima (α and β) are clearly associated with the two groups of HII regions Sh2-152/153 and Sh2-148/149.

The line velocities range from −54 km s^{-1} at the SW end of the cloud to −49 km s^{-1} at the NE end (−47 km s^{-1} for the small cloud ε lying 40' north of Sh2-152). As Figure 2 shows, the line widths are moderately large (2.5 to 4.5 km s^{-1}).

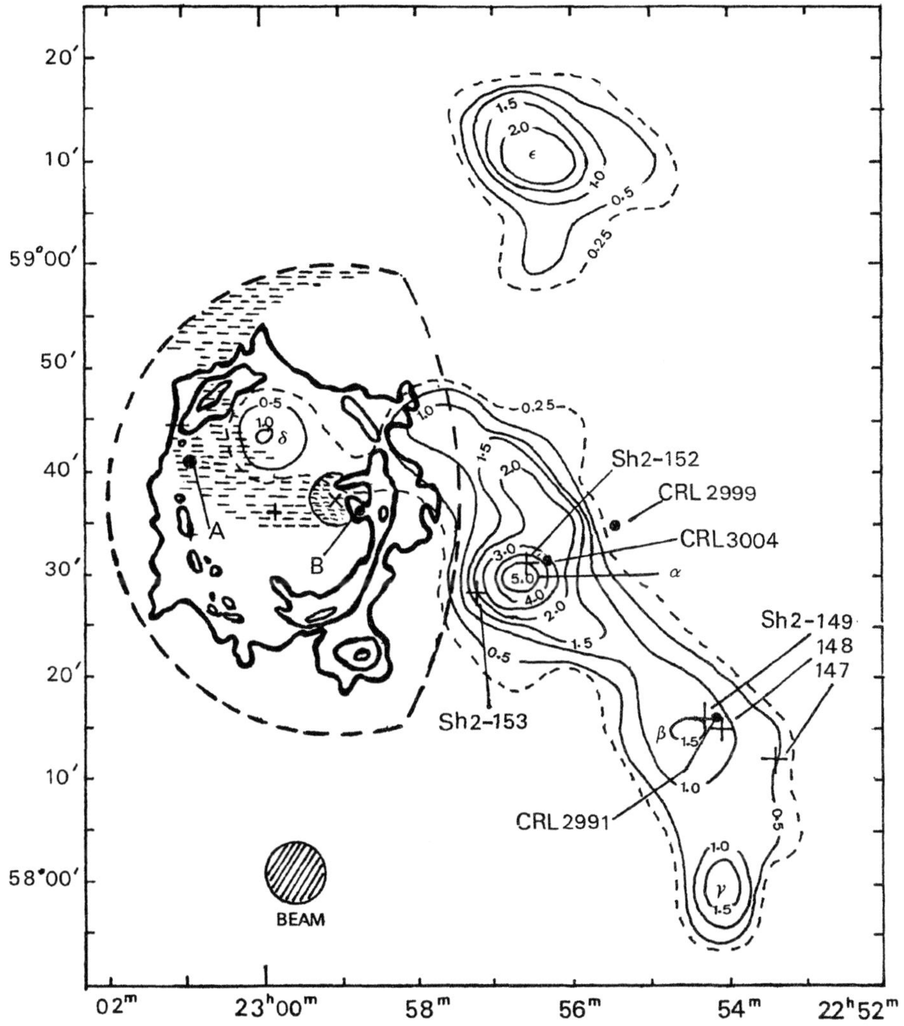

Fig. 1 Map of the complex Sh2-147/Sh2-153 and its associated SNR G109.1-1.0. i) Thin contours: Distribution of the ^{13}CO (J=1→0) emission observed at the Bordeaux Observatory. The levels represent the temperatures (T_A^*). Dashed contours are the detection limit. The corresponding HPBW (5') is shown. Crosses indicate the HII regions and dots other objects. Greek letters indicate temperature peaks. ii) Heavy contours: 0.610 GHz radio map of the SNR observed by Hughes et al. (1981). The center is indicated by +. A and B are two compact radio sources. iii) Heavy dashed lines: extent of the X-ray emission detected by Gregory and Fahlman (1980). iv) Hatched zone: X-ray jet. The compact central X-ray source is indicated by X.

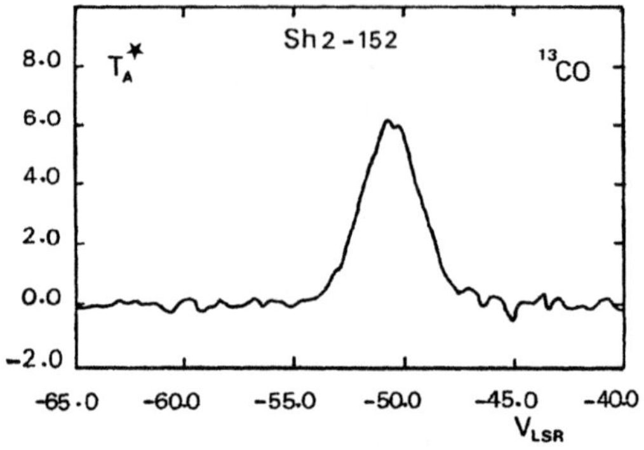

Fig. 2 Profile of the ^{13}CO line observed toward Sh2-152. The integration time is 10 min.

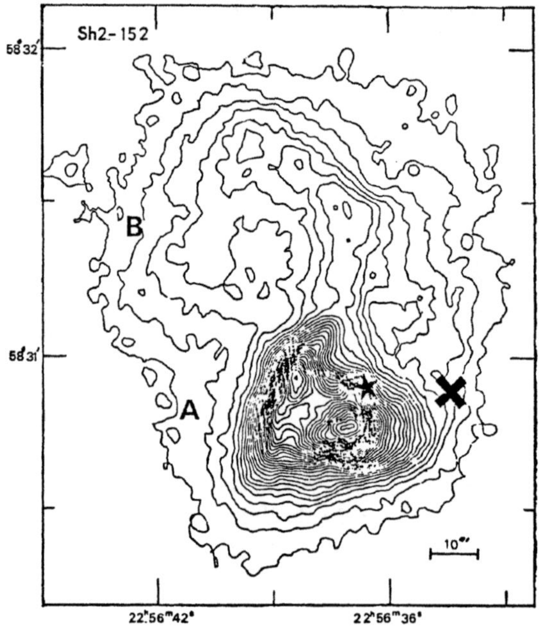

Fig. 3 Hα photograph of Sh2-152 (bandpass 12 Å, exposure time 15 min). The asterisk indicates the exciting star of spectral type O9V, and the cross the position of the dust cloud lying at 15" west of the star. The components A and B are also shown.

We have evaluated ^{13}CO column densities, using model calculations by Lucas (1974). In the direction of Sh2-152 we find $N(^{13}CO) = 3.6 \times 10^{16} cm^{-2}$ for $n(H_2) = 10^3 cm^{-3}$, and $4.6 \times 10^{16} cm^{-2}$ for $n(H_2) = 10^4 cm^{-3}$. Throughout most of the cloud $N(^{13}CO)$ ranges from 2×10^{15} to $10^{16} cm^{-2}$.

Assuming a ^{13}CO to H_2 abundance ratio of 2×10^{-6} (Dickman, 1978), we deduce molecular hydrogen column densities ranging from 10^{21} to $5 \times 10^{21} cm^{-2}$ ($\sim 2 \times 10^{22} cm^{-2}$ in the direction of Sh2-152). The total mass of the cloud is then found to be between 2×10^4 to 3×10^4 $(D/3.5\ kpc)^2 M_\odot$.

From the similarity of velocity determinations (Table 1) one sees that the HII regions and other objects marked in Figure 1 probably belong to this complex.

The SNR G109.1-1.0 (Hughes et al., 1981), which is also an X-ray source (Gregory and Fahlman, 1980), may be associated with the molecular cloud (see Heydari-Malayeri et al., 1981). This view is supported by the apparent compression of the 0.610 Ghz radio structure at its western border adjacent to the molecular cloud, by the apparent connection between the ^{13}CO peak and the NE radio concentration, and by rather good agreement between different estimations of distance for the SNR, the X-ray source, the HII region Sh2-152, and the molecular cloud.

CHARACTERISTICS OF INDIVIDUAL OBJECTS

The HII regions of the complex have been observed at different wavelengths by several authors. However, relatively little is known about most of them, except for Sh2-152 which was extensively studied by Heydari-Malayeri and Testor (1981) and Heydari-Malayeri (1981). The main characteristics of the nebula are as follows:

The champagne model applies to this HII region. The exciting star (of spectral type O9V following Crampton et al., 1978) lies now outside the cavity A (Figure 3) created by the Lyman continuum photons of the star. The champagne flow (the faint component B) has an unusual receding motion of 13.5 km s^{-1} (Pişmiş and Hasse, 1980) with respect to component A.

There is an absorbing cloud of mass greater than 0.1 $M_\odot$ mixed with gas lying at 15" west of the exciting star (Figure 3, see also Figure 3d in Heydari-Malayeri and Testor, 1981). IR observations by Frogel and Persson (1972) show a 3.5 μm limb-brightening in this area.

Another interesting feature in Sh2-152 is the apparent association of dust with the ionization front at the eastern border of the dense component A.

Table 1

RADIAL VELOCITIES IN THE DIRECTION OF THE COMPLEX

	ν (GHz)	V_{LSR} (km s^{-1})	ΔV (km s^{-1})	Resolution (arc min)	References
Ionized Gas					
—Hα					
Sh2-147		-51			
148		-51			1
149		-51			
152 A		-54.8			
152 B		-41.3			2
153		-54.3			
—H109α	5.009				
Sh2-152		-50.4		2.3	3
Molecular Gas					
—CO	115.3				
Sh2-147					
148		-54.0			
149			2.6	8	4
152		-51.0			
153					
—H_2O	22.2				
AS 501/		-51.7	1.2	1.5	5
CRL 2999		-51.0	0.4	2.4	6
		-52	0.4	0.7	7
		-52	0.4	0.7	8
—OH					
AS 501/	1.665/1.667	-53 TO -65	1	3.5 X 18	9
CRL 2999	1.612	-30 TO -70	0.7	7.9	10
SH2-152/153		-48	1	3.5 X 18	11
		-43	1	3.5 X 18	
—^{13}CO	110.20				
Sh2-147		-54.4			
148		-52.9			
149			0.27	5	this work
152		-50.5			
153		-51.3			
—HCO^+	89.2				
148		-52.9			
149			0.34	5	this work
152		-50.3			

References

1) Recillas-Cruz & Pişmiş (1979)
2) Pişmiş & Hasse (1980)
3) Kazès et al. (1977)
4) Israel (1980)
5) Lo et al. (1975)
6) White & Macdonald (1979)
7) Genzel & Downes (1979)
8) Thum et al. (1981)
9) Baudry et al. (1977)
10) Gahm et al. (1980)
11) Le Squeren (1980) Between Sh2-152 and 153

The highest observed electron densities derived from the ratio of [SII]λλ6717/6731 line intensities (about 2000 cm^{-3}) correspond to the compact component A, while the smallest values (<500 cm^{-3}) are systematically found in the northern parts of component B.

Heydari-Malayeri (1981) has shown that the line ratios [OIII] λ5007/Hβ and HeI λ5876/Hβ are enhanced unexpectedly in the peripheral zones of the nebula, probably due to a lack of matter in the outer zones of the champagne flow.

A detailed study of the physical characteristics of the other HII regions of the complex is at hand. An Hα photograph of Sh2-148 is presented in Figure 4. Sh2-148 is more extended and structured while Sh2-149 is fainter, smaller, and circular. Following Crampton et al. (1978), the stars apparently responsible for the excitation of Sh2-148 and 149 are of spectral types O8V and B0V respectively. Wramdemark (1981) gives O7 for the exciting star of Sh2-148. This HII region is less excited than Sh2-152, as its [OIII]/Hβ ratio reaches values about 0.5 (2.2 for component A of Sh2-152). Sh2-149 should be much less excited, as no [OIII]/λ5007 was detected in it (Chopinet and Lortet-Zuckermann, 1976). The WSRT observations at 6 cm by Felli and Harten (1981) show that only Sh2-148 has associated radio emission.

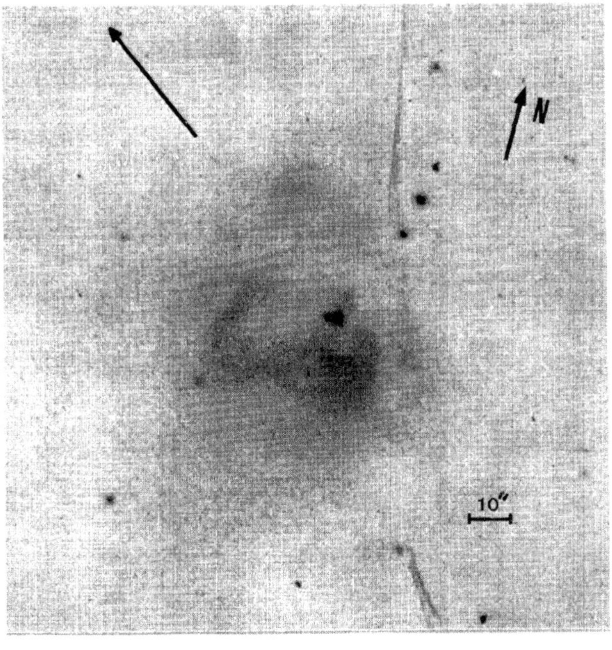

Fig. 4 Hα photograph of Sh2-148. Sh2-149 lies at nearly 1'.5 from the central star, in the direction indicated by an arrow.

Sh2-147 and Sh2-153 are apparently the most evolved HII regions of the complex. Seemingly, Sh2-147 has completely disrupted its cavity. From Sh2-153 remains a spherical structureless nebula showing no discernable ionization front. It seems that star formation has taken place at the two ends of the molecular cloud.

On the northern rim of Sh2-149 Bergeat et al. (1975) detected a very bright IR source at 2.2 μm. Optical spectroscopy and IR 2-4 μm observations reported by Russell (1978) show that this source is rather a background M supergiant and not a highly reddened O star as suggested by Bergeat et al.

AS 501/CRL 2999 is one of the best studied objects lying in the direction of the complex. At this position an H_2O maser was reported by Lo et al. (1975), Genzel and Downes (1979), White and Macdonald (1979) and Thum et al. (1981). An OH source was also detected there by Baudry et al. (1977) and Gahm et al. (1980). The 1612-MHz emission extends from -30 to -70 km s^{-1}, while the emission at 1665 MHz is strongest near $V_{LSR} \sim -70$ km s^{-1}, whereas up to now the H_2O has been seen only near -52 km s^{-1}.

Cohen (1974), Lebofsky et al. (1976) and Gehrz and Hackwell (1976) identified AS 501 as CRL 2999, an IR source from the AFGL survey (Price and Walker, 1976). Cohen and Kuhi (1977) derived a spectral type of M2 II and a visual magnitude $m_V = 13.2$ from their scanner observations, which yields a distance of ~3.3 kpc. This is in good agreement with Eiroa (1981).

Gahm et al. (1980) concluded from the "type I" characteristics of the OH emission that AS 501 might be a visible part of a compact HII region. However, Thum et al. (1981) suggest that AS 501 is in fact a long-period variable. The argument supporting this view is as follows: AS 501 shows variable luminosity and variable Balmer emission, possesses a late-type spectrum, is a strong IR source, does not emit in the continuum at 1.3 cm and 6 cm, and OH and H_2O masers are found in its immediate vicinity.

REFERENCES

Baudry, A., Brillet, J., Desbats, J.M., Lacroix, J., Montignac, G., Encrenaz, P., Lucas, R., Beaudin, G., Dierich, P., Germont, A., Landry, P., Rerat, G.: 1980, J. Astrophys. Astr. 1, 193

Baudry, A., Le Squeren, A.M., Brillet, J.: 1977, 21e Colloque International Astrophys. Liège

Bergeat, J., Sibille, F., Lunel, M.: 1975, Astron. Astrophys 40, 347

Bodenheimer, B., Tenorio-Tagle, G., Yorke, H.W.: 1979, Astrophys. J. 25, 179

Chopinet, M., Lortet-Zuckermann, M.C.: 1976, Astron. Astrophys. Suppl. 25, 179

Cohen, M.: 1974, Monthly Notices Roy. Astron. Soc. 169, 257
Cohen, M., Kuhi, L.V.: 1977, Publ. Astron. Soc. Pacific 89, 829
Crampton, D., Georgelin, Y.M., Georgelin, Y.P.: 1978, Astron. Astrophys. 66, 1
Dickman, R.L.: 1978, Astrophys. J. Suppl. 37, 407
Eiroa, C.: 1981, Astron. Astrophys. Suppl. Ser. 44, 77
Elmegreen, B.G., Lada, C.J.: 1977, Astrophys. J. 214, 725
Felli, M., Harten, R.H.: 1981, Astron. Astrophys. 100, 42
Frogel, J.A., Persson, S.E.: 1972, Astrophys. J. 178, 667
Gahm, G.F., Lindroos, K.P., Sherwood, W.A., Winnberg, A.: 1980, Astron. Astrophys. 83, 263
Gehrz, R.D., Hackwell, J.A.: 1976, Astrophys. J. 206, L161
Genzel, R., Downes, D.: 1979, Astron. Astrophys. 72, 234
Gregory, P.C., Fahlman, G.G.: 1980, Nature 287, 805
Heydari-Malayeri, M.: 1981, Astron. Astrophys. (in press)
Heydari-Malayeri, M., Kahane, C., Lucas, R.: 1981, Nature (submitted)
Heydari-Malayeri, M., Testor, G.: 1981, Astron. Astrophys. 96, 219
Hughes, V.A., Harten, R.H., Van den Bergh, S.: 1981, Astrophys. J. (Letters) 246, L127
Israel, F.P.: 1980, Astron. J. 85, 1612
Israel, F.P.: 1977, Astron. Astrophys. 61, 377
Kazès, I., Walmsley, C.M., Churchwell, E.: 1977, Astron. Astrophys. 60, 293
Lebofsky, M.J., Kleinmann, S.G., Rieke, G.H., Low, F.J.: 1976, Astrophys. J. 206, L157
Le Squeren, A.M.: 1980, (private communication)
Lo, K.Y., Burke, B.F., Haschick, A.D.: 1975, Astrophys. J. 202, 81
Lucas, R.: 1974, Astron. Astrophys. 36, 465
Pişmiş, P., Hasse, I.: 1980, Rev. Mexicana Astron. Astrof. 5, 39
Price, S.D., Walker, R.G.: 1976, Air Force Geophysics Laboratory, AFGL-TR-76-0208, the AFGL Four Colour Infrared Sky Survey: Catalog of Observations at 4.2, 11.0, 19.8 and 27.4 μm.
Recillas-Cruz, E., Pişmiş, P.: 1979, Rev. Mexicana Astron. Astrof. 4, 337
Russell, R.W.: 1978, Astron. Astrophys. 67, 273
Tenorio-Tagle, G.: 1979, Astron. Astrophys. 71, 59
Thum, C., Bertout, C., Downes, D.: 1981, Astron. Astrophys. 94, 80
White, G.J., Macdonald, G.H.: 1979, Mon. Not. R. Astr. Soc. 188, 745
Wramdemark, S.: 1981, Astron. Astrophys. Suppl. Ser. 43, 103

THE HII REGION W40 AND ITS MOLECULAR CLOUD

Richard M. Crutcher and You-Hua Chu
University of Illinois and University of California, Berkeley

ABSTRACT: Extensive new and published older observations of the HII region W40 and its associated molecular cloud have been analyzed. The HII region is near the center of an extensive dark cloud which is contracting at 3 km/s along the line of sight. The HII region is adjacent to the high-density, warm core of the molecular cloud and is responsible for heating it. The ionized hydrogen is expanding away from the high-density molecular gas at velocities up to 8 km/s and is just beginning to break through holes in the near side of the dark cloud.

I. INTRODUCTION

W40 provides an excellent opportunity to study the interaction with the surrounding neutral medium of a more modest HII region than such well-studied examples as Orion A. A variety of data has been published for W40, including interferometric radio continuum maps, an infrared map at 2.2μ and observations out as far as 140μ, and radio recombination line and a few radio molecular line observations. We have performed extensive mapping of lines of ^{12}CO, ^{13}CO, HCO^+, HCN, and Hα, observations of lines of OH and the 166α carbon recombination line, and B,V photometry of the exciting stars. We have used both the published and the new data to study in detail the HII region, the surrounding molecular clouds, and their interaction.

II. OPTICAL RESULTS

W40 is a thermal radio continuum source about five arcminutes in diameter centered at $\alpha(1950) = 18^h28^m51^s$, $\delta(1950) = -2°07.'5$, or ℓ = 28.°8, b = +3.°5. We shall subsequently refer to this as the center position. W40 is near the center of an obvious (on the Palomar Sky Survey prints) dark cloud which has a diameter of just over one degree. On the red print, faint and very patchy Hα emission is visible near the center position. The Hα in the northwest quadrant is all in a cone with an opening angle of about 45° and an apex very near the center position. To

R. S. Roger and P. E. Dewdney (eds.), Regions of Recent Star Formation, 53–60.
Copyright © 1982 by D. Reidel Publishing Company.

the southeast the emission is generally fainter, with a prominent cone defined by the absence of Hα which also points toward the center position. Emission in the other quadrants is very much fainter. The Hα emission has the appearance of being heavily obscured by patchy foreground dust.

Several faint red stars are visible within about one arcminute of the center. We have measured the positions and B,V magnitudes of the four stars closest to the center. The brightest of these is only 15 arcseconds from the center; we find V = 13.4 and B−V = 2.5. The other stars are fainter by one to two magnitudes, but all have B−V ≈ 2.5. If these are the exciting stars of the HII region, they are O or early B type, and $(B-V)_o \approx -0.3$; thus, E(B−V) ≈ +2.8, and $A_V \approx 9$ for a normal extinction law. Hence, $V_o \approx 4.4$ for the brightest star.

The three brightest of the four stars coincide within 4 to 6 arcseconds with the three infrared sources noted by Zeilick and Lada (1978) on their 2.2μ map, which had 44 arcsecond resolution. The K magnitude of the brightest star derived from their data is 5.5, which with the small extinction correction at 2.2μ gives $K_o \approx 4.7$. Since V−K ≈ −0.9 for O and early B stars, we find $V_o \approx 3.8$. This agrees within uncertainties with our other estimate. Clearly the three infrared sources observed at 2.2μ are primarily stellar radiation from the probable exciting stars of the HII region, although at longer wavelengths emission from hot dust dominates.

If we assume that the spectral type of the brightest star is BOV (this will be justified below), $M_V \approx -4.0$, m−M ≈ 8.0, and the distance to W40 is 400 pc.

III. THE HII REGION

Goss and Shaver (1970) have produced maps of the radio continuum from W40 at 408 and 5000 MHz. A representative of their lower frequency map is shown in Figure 2a. For a distance of 400 pc, the parameters of the HII region are: r = 0.44 pc, EM = 1.2×10^5 cm^{-6} pc, n_e = 200 cm^{-3}, M = 3.4 $M_\odot$, U = 23 pc cm^{-2}, and T_e = 8500 K. The excitation parameter is that of a BOV star, which is the basis for that assumption above. (Since the derived values of U and M_V depend quite differently on distance, one can find both uniquely.)

Pankonin, Thomasson, and Barsuhn (1977) find that the velocities of the hydrogen and carbon 166α recombination lines from W40 are 0.7 ± 0.2 and 6.4 ± 0.5 (1−σ errors), respectively.

We have mapped the Hα emission line from W40 with a Fabry-Perot spectrometer using a one or two arcminute aperture. For 16 positions with a sufficiently strong signal, we fit a gaussian profile to the data in order to derive the Hα velocity at each position. In two quadrants the line was too weak. In the northwest quadrant all velocities were negative with a mean of -2.7 ± 0.4 km/s; in the southeast quadrant all velocities were positive with a mean of +1.0 ± 0.4 km/s. Clearly, there is a significant velocity difference between the two quadrants.

IV. MOLECULAR OBSERVATIONS

Toward the continuum source the 1667 MHz line of OH is in absorption (see Figure 1), with V = 7 km/s. (The OH data are from Crutcher (1977), who performed limited OH and ^{12}CO mapping of the region; Zeilik and Lada (1978) also carried out ^{12}CO mapping.)

We have mapped an extensive area around W40 in the ^{12}CO and ^{13}CO J = 1-0 lines with the Aerospace 4.6-m antenna. The profiles toward $(\Delta\alpha, \Delta\delta) = (6', 24')$ shown in Figure 1 are typical of positions more than 15 arcminutes from the center. There is a single ^{13}CO peak at about 8 km/s, but the ^{12}CO line profile is flat or depressed at the center and extends from about 3 to 11 km/s. The ^{13}CO peak is often as strong or even slightly stronger than the ^{12}CO line. Hence, the ^{13}CO optical depth appears to be > 1, and ^{12}CO may be somewhat self-absorbed.

As one moves toward the center, a second component appears in ^{13}CO at about 5 km/s. The total width of the ^{12}CO emission does not change, but the extreme low-velocity wing acquires a strong spike which typically has a velocity of 4 km/s. Hence, there is an offset of 1 to 1.5 km/s between the velocities of the ^{12}CO and ^{13}CO peaks. The ^{13}CO line is often stronger than the ^{12}CO line at velocities between 5 and 8 km/s. Clearly, strong self-absorption of ^{12}CO emission by a relatively cool foreground cloud is indicated. Observations were made of the $^{12}C^{18}O$ line at (0,3), where self-absorption appeared particularly strong. Two components were seen, and no evidence for ^{13}CO self-absorption was found. The line-strength ratios $^{13}CO/^{12}C^{18}O$ were about 12 and 5 for the 5 and 8 km/s components, respectively. For the latter component, the small ratio again suggests that $\tau(^{13}CO) > 1$.

The strength of the ^{12}CO spike at 4 km/s peaked at about $T_A^* \approx 23$ K at (-3,-1). Ten positions near this peak were observed with the NRAO 11-m telescope in order to achieve higher angular and velocity resolution. The data for (-3,-2) are shown in Figure 1. The velocity displacement between ^{12}CO and ^{13}CO is clearly present for the 5 km/s component; the higher velocity component is also obvious in ^{13}CO at about 7.5 km/s.

Finally, at some positions there seemed to be a third component in ^{12}CO at 1 to 2 km/s. Usually this component had $T_A^* \approx 1$ to 2 K and was only marginally detected, but at (3,-3) it is quite strong (see Figure 1).

We also mapped the J = 1-0 lines of HCO^+ and HCN with the NRAO 11-m telescope near the CO peak position. Both lines were strongest at (-3,-1); the HCO^+ line at that position is shown in Figure 1. The velocities of these lines were +4.8 km/s with only a single component.

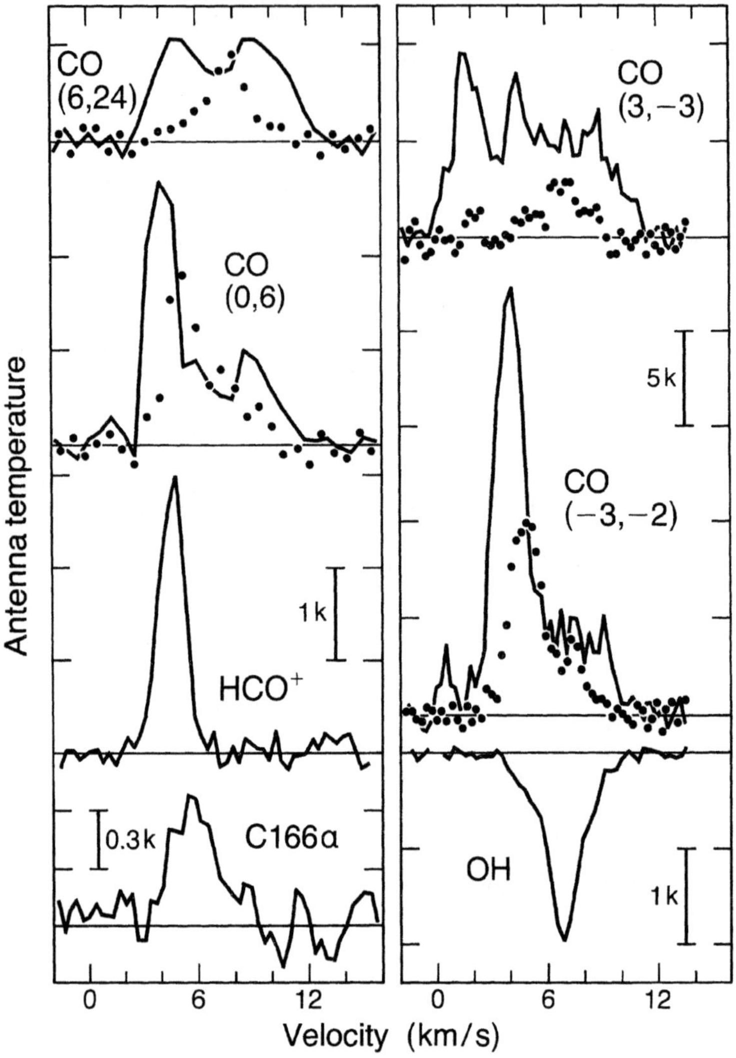

Figure 1. Selected line profiles observed toward W40. The top four are the ^{12}CO (solid lines) and ^{13}CO (dots) observed at the listed positions relative to the center. The $T_A{}^*$ scales for all CO data are the same. At bottom left is the HCO^+ data measured toward the molecular peak position (-3,-1) and the carbon 166α line observed toward the center position; at bottom right is the 1667 MHz OH line observed toward the center.

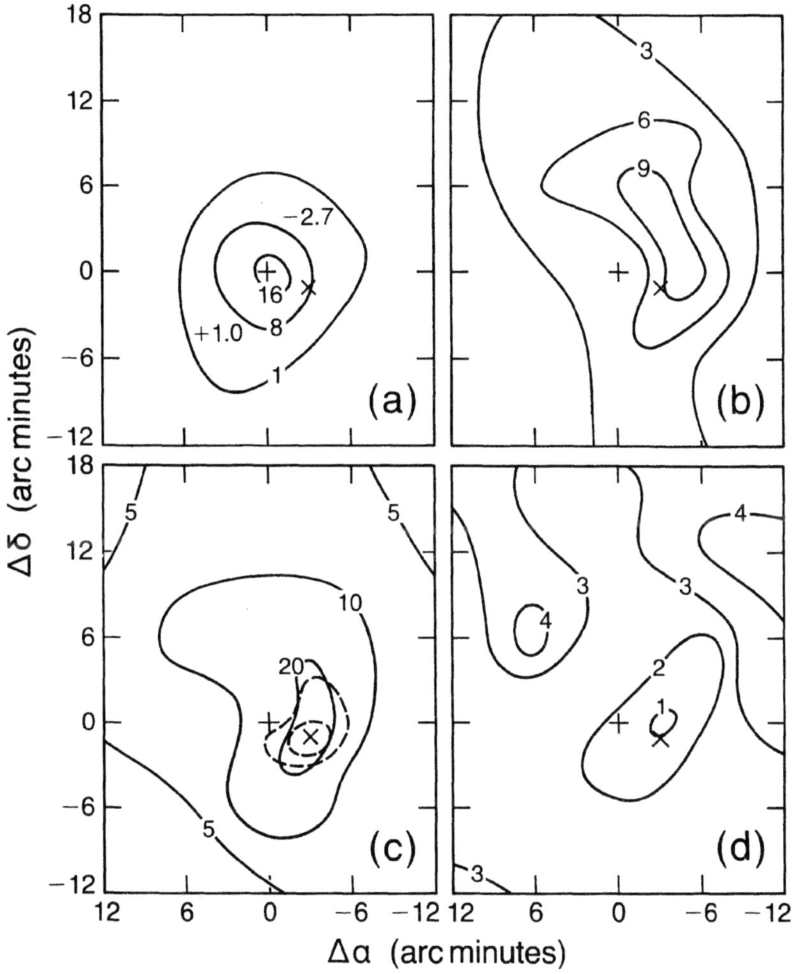

Figure 2. (a) The 408 MHz continuum brightness temperature contours (in 100's of degrees). The cross marks the center position. Also shown are the averaged Hα velocities measured in the northwest and southeast quadrants. (b) Peak $T_A^*(^{13}CO)$ for the 5 km/s component. (c) Peak T_A^* (^{12}CO) for the 5 km/s component (solid contours), and peak T_A^* (HCN) (dashed contours; 1 K contour interval). The X marks the center of the dense molecular core. (d) Mean T_A^* (^{13}CO) for V = 8 ± 1 km/s.

In Figure 1 we also show the carbon 166α recombination line which we observed with the NRAO 43-m telescope pointed to the center position. The centroid velocity is +5.7 km/s, with emission from about +4 to +8 km/s.

In Figure 2b we display the spatial distributions of the peak ^{13}CO emission for V = 4 to 5 km/s. Figure 2c shows the peak ^{12}CO and HCN distribution for the same velocity component. The velocity of the 8 km/s component measured at various positions varied from 7 to almost 9 km/s; Figure 2d shows the mean T_A^* between V = 7 and 9 km/s for ^{13}CO. The mean ^{12}CO line strength for this component was about $T_A^* \approx$ 4-5 K everywhere, so no map was drawn.

V. DISCUSSION

There are two main molecular components, with velocities of about 5 and 8 km/s. It is clear that the 8 km/s component is closer to us than is the HII region, since it produces strong absorption lines of OH. This component appears throughout the dark cloud area and is evidently associated with the visible dust. The column densities of OH and ^{13}CO for the 8 km/s component are consistent with the measured $A_V \approx 9$ and calibrations of N(OH)/A_V (Crutcher, 1979) and N(^{13}CO)/A_V (Dickman, 1978). If we take the ~ 10 pc diameter of the dark cloud in the plane of the sky to be a measure of its thickness along the line of sight and compute the hydrogen column density from $A_V \approx 9$ and the usual N_H/A_V ratio, we find $n(H_2) \approx 10^3$ cm^{-3}. On the other hand, the detection of strong emission lines of HCO^+ and HCN from the core of the 5 km/s cloud implies $n(H_2) \sim 10^5$ cm^{-3}. This cloud has a significant extension to the north especially (as delineated by ^{13}CO) which is probably of lower density, since emission from the high-excitation molecules HCN and HCO^+ is not present. The 5 km/s component is clearly slightly <u>more</u> distant than the HII region, since it produces little OH absorption. Also, the dense core area of this component is precisely where the strongest Hα is seen. With this geometry, the self-absorption of ^{12}CO emission from the 5 km/s component may be understood as being due to the low velocity wing of the obviously very optically thick and cool ^{12}CO of the foreground 8 km/s component. Finally, we point out that comparison of Figures 2b,c, and d shows that the two molecular components are spatially anti-correlated, as though the 5 km/s gas is simply 8 km/s gas which has been displaced toward us in velocity. There is no evidence that there is spherical collapse of the dark cloud, for the CO lines at the edges of the cloud ~ 30 arcminutes from the center are no narrower than at the cloud center. The size of the dark cloud and the fact that the back and front are coming together at about 3 km/s sets an upper limit of about three million years to the duration of the present kinematics.

The core of the 5 km/s cloud has a diameter of about 0.5 pc and is centered on the sky about 0.4 pc west of the center of the HII region, which has a diameter (to half-power) of about 0.9 pc. The high temperature of this molecular core ($T_K \approx 30$ K is estimated from ^{12}CO) is probably due to the heating of dust by the exciting stars of the HII region. Infrared measurements out to 140μ imply a dust temperature of 35-40 K. The heating rate necessary to maintain this temperature is about $4 \times 10^4 L_\odot$, or slightly less than the luminosity of a BOV star. If $n(H_2) = 10^5 cm^{-3}$, collisions of the gas and dust will maintain a gas kinetic temperature of 30K (see Goldsmith and Langer, 1978). On the other hand, for $n(H_2) = 10^3 cm^{-3}$, the gas and dust are not coupled and $T_K \approx 10$ K due primarily to cosmic-ray heating; hence, one would not expect significant heating of the 8 km/s component.

The masses of the ionized gas, the stars, and the molecular gas are roughly 3, 30, and $10^4 M_\odot$, respectively. Only about 10% of the molecular gas is in the dense molecular core (as defined above).

We believe that the following picture adequately accounts for the observations. Perhaps two million years ago or so the dark cloud had a velocity of 8 km/s, $n(H_2) \approx 10^3 cm^{-3}$, $T_K \approx 10$ K, and no star formation. A one-dimensional compression (supernova shock wave, spiral-arm shock,?) started on the far side of the cloud. This compression shifted gas to a velocity of 5 km/s and increased its density. The density became highest near the projected center of the cloud, and several B stars formed from the eastern part of this dense clump. These stars began ionizing the molecular gas. However, while lognT ≈ 6.5 for the background 5 km/s gas, it is only about 4 for the foreground 8 km/s gas. Hence, the ionized hydrogen expands primarily away from the dense gas and toward us. The observed carbon recombination line is at a velocity greater than that of the background cloud but less than that of the foreground material, and may arise in either or both. The ionized hydrogen is just beginning to break through holes in the foreground gas and may be seen as patchy Hα emission. The HII velocity with respect to the background dense cloud is 4 to 8 km/s toward us; the variation is probably due to clumpiness in the neutral gas surrounding the HII region. The fact that the greatest velocity difference occurs toward the densest molecular gas supports the idea that the HII region is most prohibited from expansion away from us where the surrounding pressure is greatest.

This work was partially supported by NSF grant AST 76-20810. We especially thank R.L. Dickman for his assistance with the Aerospace CO observations.

REFERENCES

Crutcher, R.M.: 1977, Astrophys. J. 216, 308.
Crutcher, R.M.: 1979, Astrophys. J. 234, 881.
Dickman, R.L.: 1978, Astrophys. J. Suppl. 37, 407.
Goldsmith, P.F. and Langer, W.D.: 1978, Astrophys. J. 222, 881.
Goss, W.M. and Shaver, P.A.: 1970, Australian J. Phys. Astrophys. Suppl. 14, 1.
Pankonin, V., Thomasson, P. and Barsuhn, J.: 1977, Astron. Astrophys. 54, 335.
Zeilik, M. and Lada, C.J.: 1978, Astrophys. J. 222, 896.

DISCUSSION FOLLOWING PAPER BY CRUTCHER AND CHU

BALLY: In your discussion of the heating and cooling of the molecular cloud you state that the dust temperature, from infrared measurements, is 35 - 40 K. How do you know that this dust co-exists with the molecular gas rather than with the ionized gas? If the hot dust were associated with the HII region, then the heating of the molecular cloud would require a mechanism other than gas-dust coupling. There is evidence that some other mechanism has to be the dominant source in some molecular clouds associated with HII regions (e.g. S155/Cepheus B and S87).

CRUTCHER: In this particular case everything in the model agrees very well if the hot dust is associated with the dense molecular gas.

HII BUBBLES AND SHOCKS IN MOLECULAR CLOUDS

T.J. Mazurek
Department of Physics
S.U.N.Y. at Stony Brook
Stony Brook, New York 11794

Supersonic gas velocities and large density gradients are observed in compact HII regions. Possible causes of such flows are examined analytically. It is shown that density gradients in either the HII region or the exterior molecular cloud can naturally give supersonic gas speeds. Cases discussed include the swelling of uniform HII bubbles into clouds with decreasing densities, isothermal shock propagation within the ionized region, and homologous expansion of non-uniform HII bubbles. Supersonic flows are readily generated under reasonable conditions. Thus the observed supersonic motions and density gradients in compact HII regions may be intimately related.

1. INTRODUCTION

Supersonic velocities of about twice that of sound speed are observed in the compact HII regions near K 3-50 and large density gradients are also present within the ionized gas (van Gorkom, et al. 1981). For the compact HII regions of DR 21, density profiles that vary over at least one order of magnitude are observed (Harris 1973). It is well known that uniform expansion of HII regions into a medium of constant density results in subsonic expansion velocities (Spitzer 1968). Hence the observed supersonic motions have been interpreted in terms of a "champagne" model (cf., Tenorio-Tagle, et al. 1979, and references therein), where the expanding HII bubble bursts through the surface of the molecular cloud. When this occurs the large density discontinuity between the HII bubble and the tenuous gas exterior to the cloud generates a shock/rarefaction wave that accelerates the gas to supersonic speeds (ibid.). In this model it is a sharp density contrast that is responsible for the supersonic motion.

In this communication, it is suggested that density gradients and/or sharp discontinuities may generate supersonic motion in HII bubbles prior to their "blistering" through the surface of the cloud. The traditional approach to the expansion of spherical and uniform HII bubbles (Spitzer 1968, Mazurek 1980) does not result in supersonic

R. S. Roger and P. E. Dewdney (eds.), Regions of Recent Star Formation, 61–66.
Copyright © 1982 by D. Reidel Publishing Company.

expansion due to the assumption of uniform density for the clouds. For a realistically decreasing density profile in the molecular cloud, the expansion velocity of a dense shell around an HII region can become supersonic. In expansion, the shell material can be eroded by and absorbed into the HII region. The shell may disappear during expansion, producing large discontinuities in both density and velocity. Such discontinuities could then result in isothermal shocks within HII bubbles that produce supersonic motion. Finally density gradients present at inception within the HII bubble will affect gas flows. It is shown that homologous expansion of a non-uniform HII region gives supersonic motion for most of the mass in the HII bubble. While the Gaussian density profile in this case may be somewhat idealistic, it is noteworthy that supersonic motion occurs at the mean radius of such a configuration. These considerations show that in the presence of density variations within HII bubbles, supersonic gas velocities may be generated in compact HII regions before they "blister" through the surface of the molecular cloud.

2. HII BUBBLES IN MOLECULAR CLOUDS WITH DENSITY GRADIENTS

Current work on the evolution of protostellar envelopes (Yorke 1979) indicates that compact HII regions around massive stars are initially formed with a thin dense shell that contains most of the mass. The density gradient exterior to this shell is that of free-fall under steady-state conditions: $n = n_o(r_o/r)^{3/2}$, where n_o is the proton density, r the radius, and the subscript o indicates the region just outside the dense shell. The density gradient within the expanding portion of the HII region is relatively flat with maximum contrast below five; the mass of the shell is 4 to 5 that of the HII region; and the infall velocities exterior to the shell are negligible in comparison to those of shell expansion (ibid., Fig. 2). Although these results were obtained for the formation of a roughly ten solar mass star, let us assume then to apply generally to more massive stars. Note that the density profile outside the shell has been established quite generally (Larson and Starfield 1971, Shu 1977).

To examine the subsequent evolution of such HII bubbles analytically, let's assume a uniform density within the HII bubbles and that the thickness of the dense shell is negligible. One can then make the traditional approximations for HII bubble expansion to derive the momentum equation for the motion of the shell (cf., Mazurek 1980, eq. 6)

$$\frac{d}{dt}(M_s u_s) = 4\pi r_o^2 a^2 \frac{n_{IIo}}{N_o} \left(\frac{r_s}{r_o}\right)^{1/2} - \frac{u_s}{2} \frac{d}{dt}(M_{II}), \qquad (1)$$

where M_s is the mass of the shell, u_s and r_s are its radius and velocity, respectively, a is the isothermal sound speed in the HII region, n_{IIo} and M_{II} are its initial proton density (pure hydrogen assumed) and mass, respectively, r_o is the initial radius of the shell, and N_o is Avogadro's number. The effects of cloud infall and self-gravity on the shell have been ignored. Let M_i be the initial mass of the shell and $\theta \equiv n_{II}/n_o$. The masses of the bubble and shell at arbitrary radius are given by

$$M_{II} = \theta M_o \left(\frac{r_s}{r_o}\right)^{1/2} \; ; \; M_s = M_i(1+\beta)\{1-\gamma\left(\frac{r_s}{r_o}\right)^{3/2}\}$$

$$M_o = \frac{4\pi}{3}\frac{n_o}{N_o} r_o^3 , \tag{2}$$

with $\beta = (\theta - 2)M_o/M_i$ and $\gamma = \beta/(1+\beta)$. Substituting eqs.(2) into eq. (1) and introducing dimensionless variables $x = r_s/r_o$ and $y = u_s^2/a^2$, one can reduce eq. (1) to the following dimensionless form

$$\frac{dy}{dx} + 3\frac{M_o}{M_i}x^{1/2}\, y = 12\frac{M_o}{M_i}\, x^{1/2} \qquad (\theta = 2) \tag{3a}$$

$$\frac{dy}{dx} + \left(\frac{\theta-4}{\theta-2}\right)\left\{\frac{d}{dx}\ln(1-\gamma x^{3/2})\right\} y = \frac{6\theta}{\theta-2}\,\frac{\gamma x^{1/2}}{1-\gamma x^{3/2}} \qquad (\theta \neq 2) \tag{3b}$$

The solutions of this equation are given by

$$y = 4\{1-\exp[\frac{2M_o}{M_i}(1-x^{3/2})]\} \qquad (\theta = 2), \tag{4a}$$

$$y = 8\ln[(1-\gamma)/(1-\gamma x^{3/2})] \qquad (\theta = 4) \tag{4b}$$

$$y = \frac{4\theta}{4-\theta}\{1-[\frac{1-\gamma x^{3/2}}{1-\gamma}]^{\left(\frac{4-\theta}{\theta-2}\right)}\} \qquad (\theta \neq 2,4) \tag{4c}$$

Note that $\gamma^{-1} = (\theta-2)^{-1}M_i/M_o + 1$ is typically positive for $\theta>2$ and $M_i/M_o >> 1$. Thus in this case, the shell velocities must become supersonic. For the case $\theta=2$, the mass accumulated from the cloud by the shell just equals the mass eroded into the HII region. The shell's mass is constant and its velocity approaches an asymptotic value of 2 a. For $\theta>4$, the rate of erosion exceeds that of accumulation and the shell velocity gets arbitrarily high as its mass decreases to zero. Of course the true shell velocity will be limited to around a few times sound speed since the assumptions leading to equation (1) will break down. If $\theta<2$, then β and γ are negative. In this case the accumulation dominates erosion, and the shell mass grows. However to the extent that the shell's self gravity continues to be negligible, the shell velocity approaches supersonic speeds asymptotically. Thus it is seen that the decreasing density gradient expected in the collapsing cloud gives rise to supersonic shell expansion with traditional assumptions for HII bubble expansion.

2. SHOCKS WITHIN HII BUBBLES

The considerations of the previous section lead to an interesting possibility for the generation of shocks within an HII region. If the initial formation of a shell bounded HII bubble results in a density contrast between the cloud and ionized material such that $\theta \geq 2$, then the shell vanishes due to erosion into the HII region as $x \to \gamma^{-3/2}$. In the process it achieves supersonic velocities that should continue into the HII region. The original density contrast θ will be maintained since $n_{II} \propto n \propto n^{-3/2}$. Thus when the shell vanishes, it produces sharp discontinuities in both velocity and density. Subsequently, an ionization front should move into the lower density material. Thus two isothermal waves may be generated: a rarefaction into the high density material and a shock into the more tenuous matter.

An estimate of the resulting flow is obtained via planer shock-tube theory (cf., Zel'dovich and Raizer 1966). Assuming that the post-shock conditions are identical to the post-rarefaction ones, one can write the equations for the rarefaction and shock, respectively, as

$$m_s - m_o = \ln \frac{n_o}{n_s} \quad \text{and} \quad m_s^2 = \left(\frac{n_s}{n_1} - 1\right)\left(1 - \frac{n_1}{n_s}\right), \tag{5}$$

where $m = u/a$, the subscript s denotes post-shock conditions, and the subscripts o and 1 denote original HII bubble and post-ionization cloud, respectively. These equations can be manipulated to yield the initial density contrast in terms of the fluid flow Mach numbers,

$$\theta \equiv \frac{n_o}{n_1} = \{(\frac{m_s}{2}) + [(\frac{m_s}{2})^2 + 1]^{1/2}\}^2 \, e^{m_s - m_o}. \tag{6}$$

This agrees with published results for $m_o = 0$ (Tenorio-Tagle 1979).

The observed velocities of $\sim$ 25 km/s in K 3-50 (van Gorkom et al., 1981) imply Mach numbers of $\sim$2 for standard HII temperatures of $\sim 10^4$K. Using $m_s \sim m_o \sim 2$ in equation (6) one obtains a density contrast $\theta \sim 6$. This value is of the order of the density contrast observed in the HII condensations of DR 21 (Harris 1973). Thus the observed Mach numbers and density contrasts are consistent with the propagation of isothermal shock/rarefaction waves.

3. HOMOLOGOUS EXPANSION OF NONUNIFORM HII BUBBLES

The considerations of the previous two sections assumed a roughly flat density gradient in the original HII bubble. The current work on protostellar envelopes (Yorke 1979) indicates a mild density gradient for the case of an envelope of about half of a solar mass. The situation may be different for larger envelope masses. Thus the possibility exists that the density gradients of HII regions may be steeper then presently indicated. This could have significant consequences for the subsequent expansion. To examine potential effects of density gradients let us consider the somewhat idealized case of homologous expansion.

Assuming that the gas remains isothermal throughout, one needs to solve only the fluid equations for mass and momentum, respectively,

$$n_{II}^{-1} \frac{dn_{II}}{dt} = - \frac{2u}{r} - \frac{\partial u}{\partial r} \quad \text{and} \quad \frac{du}{dt} = - \frac{a^2}{n_{II}} \frac{\partial n_{II}}{\partial r} . \qquad (7)$$

We seek a solution that is a function of one scale parameter λ that is independent of radius. The Lagrangian coordinate r of an arbitrary mass point at time t is related to its initial value r_o by $r = \lambda r_o$ (where $\lambda_o \equiv 1$), and its velocity is $u = (\dot{\lambda}/\lambda)r$. Substitution of the latter into the first of equations (7) yields $n_{II}/n_{II}^{o} = \lambda^{-3}$. Using these results in the second of equations (7), one obtains an equation that is separable in its spatial or time variation. The result can be put in the form

$$2\nu = \lambda \ddot{\lambda} = - \frac{a^2}{n_{II}^{o} r_o} \frac{\partial n_{II}^{o}}{\partial r_o} , \qquad (8)$$

where ν is a constant. Assuming that $\dot{\lambda}_o = 0$, the temporal equation can be integrated to give

$$g(\ell n^{1/2} \lambda) = \nu^{1/2} \frac{t}{\lambda}, \text{ with } g(x) = e^{-x^2} \int_0^x e^{y^2} dy. \qquad (9)$$

The spatial integration yields a Gaussian distribution,

$$n_{II}^{o}(r_o) = n_{II}^{o}(o) \exp\left(- \frac{\nu r_o^2}{a^2}\right). \qquad (10)$$

The function g(x) is roughly constant over the interesting range: $0.42 \leq g(x) \leq .55$ for $0.5 \leq x \leq 1.5$ (cf. Karpov 1965). Thus one has $g(x) \sim .5$ for $1.3 \leq \lambda \leq 10$. In this range $\lambda \sim 2\nu^{1/2} t$ and hence $u \sim 2\nu^{1/2} r_o$. Thus a roughly constant velocity distribution is established after an expansion of $\sim$30% in radius. The mean radius of the HII bubble is $R = 2a\lambda/(\pi\nu)^{1/2}$. In terms of the mean radius the expansion velocity at r is $u = 4ar/(\pi^{1/2} R)$. Hence the expansion velocity at the mean radius of the HII bubble exceeds twice that of sound. The fractions of total mass, momentum, and energy, above a given flow Mach number are given in the following table. The last column gives totals for an HII bubble with the following parameters: R = 0.05 pc, $n_{II}(o) = 10^5 cm^{-3}$, a = 13 km/s.

$m = u/a \geq$	1	2	3	4	Total(c.g.s.)
mass fraction:	.92	.57	.21	.05	2.4×10^{33}
momentum fraction:	.97	.74	.34	.09	2.2×10^{40}
kinetic energy fraction:	.99	.85	.48	.16	1.2×10^{47}

The table shows that most of the HII bubble is characterized by flow mach numbers exceeding two, that an appreciable fraction of the matter is at Mach numbers of three, and that essentially all of the matter flows supersonically. This indicates that density gradients can induce supersonic flows if they are steep enough.

4. SUMMARY

Observations of compact HII regions indicate the presence of supersonic gas flows and large density gradients. Density variations arise naturally in both HII bubbles and the exterior cloud during star formation. It is shown above that expansion of HII bubbles into clouds with density gradients generally results in supersonic flows. Under favorable conditions, the expansion may form an HII region with large internal discontinuities in both velocity and density, and thus produce internal shocks. In addition, large density gradients within the ionized region may produce supersonic flows. For the case of homologous expansion essentially all of the flow is supersonic. Thus on these theoretical grounds, the observed supersonic velocities and density gradient should be intimately connected.

REFERENCES

Gorkom, J.H. van, Shaver, P.A., Pottasch, S.R., Blair, G.N. and Matthews, H.E. 1981, Astron. Ap. 94, 259.

Harris, S. 1973, M.N.R.A.S., 162, 5 p.

Karpov, K.A. 1965, Tables of the Function $W(Z) = e^{-Z^2} \int_0^Z e^{X^2} dX$ in the Complex Domaine, Macmillian, New York.

Larson, L.B. and Starrfield, S. 1971, Astron. Ap., 13, 190.

Mazurek, T.J. 1980, Astron. Ap., 90, 65.

Shu, F.H. 1977, Ap.J., 214, 488.

Spitzer, L. 1968, Diffuse Matter in Space, Interscience Pub., N.Y.

Tenorio-Tagle, G. 1979, Astron. Ap., 71, 59.

Tenorio-Tagle, G. Yorke, H.W. and Bodenheimer, P. 1979, Astron. Ap., 80, 110.

Yorke, H.W. 1979, Astron. Ap., 80, 308.

Zel'dovich, Ya.B. and Raizer, Yu.P. 1966, Physics of Shock Waves and High-Temperature Hydrodynamic Phenomena, Vol. I, eds. W.D. Hayes and R.F. Probstein, Academic Press, New York.

A MULTI-PURPOSE SCANNING FABRY-PEROT INTERFEROMETER SYSTEM

J. R. Roy, R. Arsenault and G. Joncas
Départment de physique et Observatoire astronomique du mont Mégantic, Université Laval, Quebec, QC.

ABSTRACT

We have built a self contained two-mode Fabry-Pérot interferometer system to obtain (a) radial velocity field maps and (b) photoelectric line profiles of extended emission line sources. The servo-controlled Fabry-Pérot is piezo-scanned by using capacitance micrometry to detect deviations from parallelism or changes in the absolute spacing. The imaging mode uses a 183mm focal length f/8 collimator which images the pupil on the Fabry-Pérot; the field with the characteristic fringes is reimaged by a high speed Angénieux 25mm focal length f/0.95 refocussing lens. Stripped from its Fabry-Pérot, this mode acts as a focal reducer to obtain filtergrams of extended sources. The photoelectric configuration uses two Vivitar 200mm f/3.5 lenses as collimator/camera optics. A Fabry lens forms an image of the primary on the photocathode of a Ga-As PM tube. Scanning of the Fabry-Pérot and data acquisition is done by a dedicated Rockwell AIM 65 microcomputer soon to be replaced by a HP-85 equipped with the HP-IB IEEE 488 interface bus.

1. INTRODUCTION

The efficient spectral study of extended sources such as galactic and extragalactic HII regions requires the optimization of all optical and detector components. If one is concerned mainly with radial velocity field information in a few spectral elements, the ideal approach is to utilize a focal reducer, a scanning Fabry-Pérot interferometer and a photon counting imaging device. However, very few such sophisticated devices are presently in operation (Taylor and Atherton, 1980).

In a focal reducer system (Courtès, 1960; 1972), the light from the telescope is recollimated (to be passed parallel through the

R. S. Roger and P. E. Dewdney (eds.), Regions of Recent Star Formation, 67–72.
Copyright © 1982 by D. Reidel Publishing Company.

Fabry-Pérot etalon) and imaged through a high speed refocussing camera lens. The luminosity of the system is increased as the square of the telescope focal ratio divided by the square of the camera focal ratio. Unfortunately the size, complexity and cost of the collimator increases rapidly with field size and a compromise must be found. Nevertheless, the nature of most astronomical sources observed with such a system lessens the importance of this impediment. The use of a Fabry-Pérot interferometer permits us to exploit its high $R \times \mathcal{L}$ value. Provision for scanning allows a stepping of the Fabry-Pérot etalon in wavelength and, at each step, a recording of the image of the sky with the characteristic ring pattern superposed. One obtains for each point in the field a spectral scan having the range of the interorder length for the etalon.

If only the axial fringe is illuminated, the use of a photomultiplier as a detector allows the system to act as a line spectrometer when the gap of the etalon or its spectral index is varied (Jacquinot, 1960; Cruvellier, 1967; Vaughan, 1967; Smith and Weedman, 1970, and Meaburn, 1976). When numerous fringes over a region of the sky are observed, photographic methods can be used with or without an image tube (Courtès, 1960; Monnet, 1970; Georgelin, 1970a,b; de Vaucouleurs et al. 1974, de Vaucouleurs and Pence, 1980; Tully, 1974), but they are severely limited because of the weakness of most sources. A high quantum efficiency detector is particularly well adapted to this problem. However, the implementation of a photon-counting imaging system requires major investments and puts the ideal device out of range for small observatories.

We describe a less ambitious scanning Fabry-Pérot system which, in its first stage, achieves many of the possibilities of the ideal cronfiguration described above and which can be developed to a high level of sophistication as new resources become available.

2. INSTRUMENTATION

Our "poor man's" Fabry-Pérot system is modular in design to allow for the efficient use of various detectors. An important constraint has been to design a self-contained, reliable and portable system which can travel with its dedicated control and data acquisition unit. There are two configurations which use the same interferometer and the same "bonette" for pointing, narrow-band interference filters, reticles, knife-edge focussing, off-set guiding and diaphragms, but different collimator/camera optics and detectors.

In the imaging mode, the system acts as a focal reducer and provides monochromatic images of extended sources when used with the appropriate filter. When the Fabry-Pérot interferometer is introduced in the collimated beam, one has the classical high resolution premonochromator filter and scanning Fabry-Pérot camera. The second mode turns the scanning Fabry-Pérot into a spectrometer in exploiting the advantages of a photomultiplier. An interference filter is again necessary to reject any other emission line which could be picked by another order of the etalon. In the photoelectric mode, accurate profiles of nebular emission lines can be obtained from extended sources in a number of relatively large spatial elements, but small spectral elements.

2.1 The imaging configuration

The overall layout and design of the instruments is shown in Figure 1. The 80.6mm diameter collimating optics were designed at Marseille Observatory and built by Sud-Optique (Société Bertin & Cie). The field lens is 47.54mm behind the focal plane of the telescope. The f/8 collimator has a 183mm focal length. The 23mm exit pupil is located after the collimating achromat; when the exit pupil coincides with the median plane of the Fabry-Pérot, the light reaching the etalon has been treated equally by the whole primary mirror of the telescope. The optics is coated for the red spectral region. The refocusing camera lens is a commercially available Angénieux 25mm focal length f/0.95 lens; it has a 25μm circle of least confusion and sets a resolution limit of 3" when used with a III a F photographic plate as a detector on the Mont Mégantic 1.60m f/8 telescope. The f/0.95 focal reducer decreases the input f ratio by a factor of 8.42, thus giving a scale of 126 arcsec mm^{-1} on the detector when observing on the same telescope. With the image tube, the camera lens is changed to a Takumar 85mm focal length f/1.8 objective which forms an image on the entrance face of a 40mm magnetically focused one-stage ITT-4089 S-20 fiber optics image intensifier; IIaD plates pushed against the exit face of the image tube are then used.

The scanning Fabry-Pérot interferometer was designed and built by Queensgate Instruments Ltd. (London). The etalon has a 50mm working aperture out of an 81mm overall diameter. Plate spacing is 348 μm±5 μm with gap matched flat to better than λ/50 at 5461Å after coating. The reflectivity of the mirrors is 0.82±0.01 over at least a 1000Å bandwidth centred on 65000Å.

The etalon is piezo-scanned and servo-controlled using a technique of capacitance micrometry to detect deviation from parallelism or change in the absolute spacing. This servo system was developed at Imperial College of Science and Technology by Hicks and his colleagues (Hicks <u>et al.</u>, 1974; Atherton <u>et al.</u>, 1978). In this type of etalon, piezoelectric ceramics are incorporated into the three spacers separating the plates, thus permitting external fine adjustment and scanning. "Pads" of chromium overlaid with gold are evaporated around the peripheries of the plate to serve as capacitance error detectors.

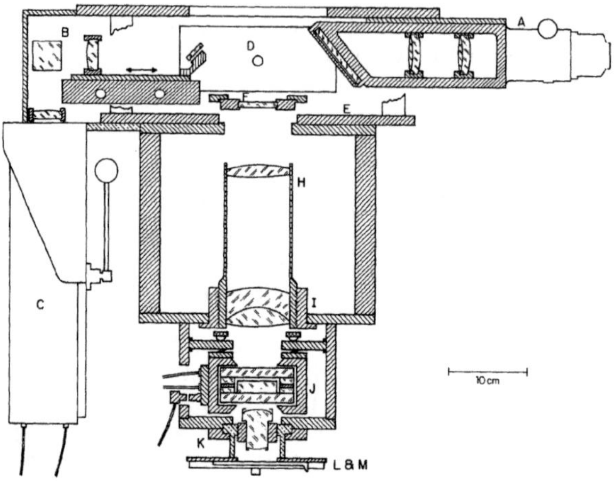

FIGURE 1. Optical layout of the focal reducer and Fabry-Perot interferometer (imaging mode)

A - Field finding optics
B - Guiding
C - ISIT TV camera
D - Calibration lamp
E - Knife-edge, focal plane
F - Interference filter (tiltable)
H - Field lens
I - Collimating lens f/8, 183mm
J - Etalon : 30mm effective diameter
K - Camera lens f/0.95, 25mm
L - Plate-holder
M - Photographic plate
(CCD, photon counting or image tube)

2.2 The Fabry-Pérot spectrometer configuration

To allow more flexibility and versatility in the use of the Fabry-Pérot interferometer, we have built an independent unit for use of the etalon as a photoelectric spectrometer. The collimator/camera optics are two Vivitar 200mm focal length f/3.5 TX lenses which are commercially available and which satisfy the less severe optical constraint of working with the field corresponding to the central

Fabry-Pérot fringe. A remote control filter wheel is located in the parallel beam between the collimator and the interferometer which is at the exit pupil. The cameral lens images the interference fringes on an iris diaphram which isolates the central spot. A Fabry lens re-images the entrance pupil (i.e. the primary mirror of the telescope) on the red sensitive photocathode of a RCA C31034A photomultiplier.

3. MICROCOMPUTER CONTROL

The closed loop servo-stabilization unit, also built by Queensgate, maintains parallelism and spacing of the etalon to an accuracy better than $\lambda/200$ et 5000 Å. The controller is provided with a digital-analog converter which transforms any input binary number into a potential difference applied to the piezo stacks. This 12 bit number can be given by any computer provided with a TTL compatible interface module.

We have temporarily interfaced the Fabry-Pérot interferometer with a Rockwell AIM 65 microcomputer, soon to be replaced with the more powerful and versatile HP-85 microcomputer equipped with the HP-IB IEEE-488 interface bus.

The scanning program on the AIM 65 introduces the index of a loop into the memory space allocated to the output part. Completing this loop simply increases the binary number sent to the Fabry-Pérot controller, producing a wider or narrower gap whether N increases or decreases. The detector readout which is stored in the memory may be synchronized with the scans of the etalon.

4. DATA ACQUISITION AND REDUCTION

At present, data is acquired on photographic plates which can be digitized and analyzed with the normal techniques of image processing. Because of the large plate scale, it is important to use a high S/N emulsion such as a IIIaF photographic plate which is sensitized by baking in N_2 and N_2H_2. The image size is less than 7mm in diameter and as many as eight successive observations can be taken on each 6cm x 9cm plate. When the image tube is used, only one observation per plate is taken on a different type of emulsion to match the green phosphor window.

In order to obtain the complete line profile $I(\lambda)$ of an emission line at each spatial position (x,y) in the sky, the etalon gap z is scanned in a way to obtain N_R independent Fabry-Pérot frames of $I_z(x,y)$ to achieve a complete velocity map (where the finesse N_R is ~15). Obviously, this requirement puts a severe constraint on the detector and dramatizes the need for a high quantum efficiency imaging detector. Velocity information is obtained by measuring the diameter of the Fabry-Pérot rings (Monnet, 1970; Georgelin, 1970a,b; Taylor and Atherton, 1980) and velocity dispersion can be deduced by deconvolving the instrumental profile and the temperature broadening from the observed profile.

ACKNOWLEDGEMENTS

The technical skills of R. Bisson, A. Bouffard, R. Labbé and G. Pigeon of Laval University insured the success of the project. We are most grateful to Y.P. Georgelin, J. Boulesteix and J. Caplan of the Marseille Observatory (France) for their encouragement and numerous suggestions in the early phase of the project; the imaging collimator optics has been kindly loaned to us by Y.P. Georgelin. The assistance of T. Hicks, M. Wells and K. Reay was most helpful.

We wish to acknowledge financial support of the Natural Science and Engineering Research Council of Canada and of the Ministère de l'Education du Québec.

REFERENCES

Atherton, P.D., Hicks, T.R., Reay, N.K. and Wells, M.: 1978, Proc. of the 4th International Colloquim on Astrophysics, Trieste, ed. Hack, M.
Courtès, G.: 1960, Ann. Astrophys. 23, 115
Courtès, G.: 1972, in Vistas in Astronomy 14, 81
Cruvellier, P.: 1967, Ann. Astrophys. 30, 1059
Georgelin, Y.P.: 1970a, Astron. Astrophys. 9, 436
Georgelin, Y.P.: 1970b, Astron. Astrophys. 9, 441
Hicks, T.R., Reay, N.K. and Scaddan, R.J.: 1974, J. Phys. E. Sci. Instrum. 7, 27
Jacquinot, P.: 1960, Rep. Progr. Phys. 24, 267
Meaburn, J.: 1976, "Detection and Spectrometry of Faint Light," Reidel, Dordrecht
Monnet, G.: 1970, Astron. Astrophys. 9, 420
Smith, M.G. and Weedman, D.W.: 1970, Astrophys. J. 161, 33
Taylor, K. and Atherton, P.D.: 1980; Monthly Notices Roy. Astron. Soc. 191, 675
Tully, R.B.: 1974, Astrophys. J. Suppl. 27, 415
de Vaucouleurs, G., de Vaucouleurs, A. and Pence, W.: 1974, Astrophys. J. 194, L119
de Vaucouleurs, G. and Pence, W.D.: 1980, Astrophys. J. 242, 18
Vaughan, A.H.: 1967, Ann. Rev. Astron. and Astrophys. 5, 139

FINE STRUCTURE LINES IN HII REGIONS INTERACTING WITH MOLECULAR CLOUDS

D.A. Naylor, R. Emery, B. Fitton
Astronomy Division, Space Science Department of ESA, ESTEC, Noordwijk, The Netherlands
I. Furniss, R.E. Jennings, K.J. King
University College, London, U.K.

This paper presents observations of far infrared fine structure emission lines of OIII at 52 and 88 μm, OI at 63 μm and NIII at 57 μm, obtained using a balloon borne telescope with a Michelson interferometer system. The sources M42 and M17 have been mapped with a beam of 1.6' FWHM. Physical quantities derived from these measurements are presented and some of the results are discussed.

INTRODUCTION

The measurement of collisionally excited infrared fine structure emission lines provides a useful means of studying the physical processes occurring in gaseous nebulae. From measurements of these lines it is possible to determine the electron density, ionic abundance and ionization structure of the emitting region. Reviews of the infrared observational work and the theoretical aspects have been given recently by Lacy[1] and by Watson and Storey.[2]

The interpretation of infrared emission lines is often easier than that of their optical counterparts. Firstly, since the excitation temperatures of most infrared fine structure levels is small compared to the ambient nebular temperature, the quantities derived from infrared line measurements are virtually independent of the electron temperature assumed for the nebula. Secondly, the infrared lines are relatively unaffected by extinction due to interstellar material and this provides a means of probing regions which are heavily obscured in the visible.

Many of the infrared lines are unobservable at ground level due to low atmospheric transmission; this is especially true in the far infrared spectral region 40-100 μm. This paper presents results obtained on a balloon flight from N.S.B.F. Palestine, Texas in September 1979 using the University College London 60 cm diameter

R. S. Roger and P. E. Dewdney (eds.), Regions of Recent Star Formation, 73–80.
Copyright © 1982 by D. Reidel Publishing Company.

Table 1
M42 Line Measurement Data

Pos.	RA/Dec (1950)	OIII 52μm	OIII 88μm	NIII 57μm	n_e cm^{-3} ($T_e=10^3K$)	E $10^6 cm^{-6} pc$ (Ref 14)	$\frac{n(OIII)}{n_e}$ 10^{-4}	$\frac{n(NIII)}{n_e}$ 10^{-5}
		10^{-9}W cm^{-2} sr^{-1}						
1	$5^h32^m49^s$ −5°25'16"	5.9	0.82	<1.9	10700	5.3	3	<15
2	$5^h32^m46.3$ −5°24'0"	5.9	1.24	<1.9	3800	3.9	1.9	< 9
3	$5^h32^m45.1$ −5°26'18"	2.5	0.82	<1.9	1750	1.8	1.1	<10
4	$5^h32^m39.8$ −5°24'16"	2.6	<1.5	1.5	>660			
7	5^h32^m45 −5°28'2"	1.1	0.76	1.7	450	1.1	0.6	4.4
8	$5^h32^m50.6$ −5°27'13"	2.5	<1.5	1.8	>600	2.2	>0.7	> 4
9	5^h32^m55 −5°26'15"	2.3	<1.5	1.8	>600	3.1	>0.5	> 4
10	$5^h33^m0.5$ −5°25'5"	<2.3	<1.5	<1.9				
11	$5^h32^m53.3$ −5°23'50"	5.0	1.2	1.1	3100	3.6	1.5	2.6
12	5^h32^m50 −5°22'16"	3.6	1.1	<1.9	2000	1.8	1.5	<13
13	5^h32^m43 −5°22'0"	1.4	1.1	<1.9	390	0.7	1.2	<10

Table 2
M17 Line Measurement Data

Pos.	RA/Dec (1950)	OIII 52μm	OIII 88μm	NIII 57μm	n_e cm^{-3} ($T_e=10^3K$)	$\frac{n(OIII)}{n_e}$ 10^{-4}	$\frac{n(NIII)}{n_e}$ 10^{-4}	$\frac{NIII}{OIII}$
		10^{-9}W cm^{-2} sr^{-1}						
1	$18^h17^m32.5$ −16°14'30"	6.8	2.3	1.7	1440	2.6	.72	.28
2	18^h17^m26 −16°15'45"	<1.1	< 7	< .9				
3	18^h17^m39 −16°15'17"	3.8	1.6	1.1	1040	1.6	.44	.28
4	18^h17^m41 −16°13'27"	2.6	1.2	< .9	890	1.75		
5	18^h17^m43 −16°11'42"	3.1	1.1	<1.1	1320	2.35		
6	18^h17^m50 −16°12'13"	2.7	1.3	< .7	820	3.3		
7	18^h17^m58 −16°12'48"	1.7	< .6	.6		3.4	1.1	.32
8	18^h17^m55 −16°11'2"	1.5	2.2		27	1.4		
9	18^h17^m48 −16°10'29"	2.4	1.5		480	1.3		
10	18^h17^m32 −16° 8'39"	2.5	1.6	<1.0	500	1.7		

telescope system and the ESTEC high resolution Michelson interferometer.[3] The interferometer was set up to measure the wavelength range 45–100 μm which included the two OIII lines at 52 and 88 μm, the OI line at 63 μm and the NIII line at 57 μm. The resolving power was ~4000 at 50 μm and the effective beam size was 1.6' FWHM. The objects chosen for study were M42 and M17; M42 being a region in which the electron temperature distribution is not well known, M17 being a region of high visible extinction. In addition, these objects present different geometrical configurations to the observer as regards the position of the molecular cloud and the HII region.

M42

The mapping of M42 was chosen to complement the observations of Moorwood et al.[4] and Storey et al.[5] and covered a region about 6 arc minutes diameter centered on the Trapezium; the values obtained are given in Table 1. Contour maps of the OIII and OI lines are shown in Figures 1 through 3. Since NIII was detected in only 4 positions a contour map could not be drawn. The electron densities and ionic abundances were derived using the method outlined by Simpson[6], and these values are also shown in Table 1.

The peak of the 52 μm map of OIII lies a short distance to the NE of the Trapezium whereas the radio continuum peak and the brightest optical region lie just to the SW. This difference is within the positional uncertainty, which stems mainly from the large beam sizes involved and the pointing accuracy of the telescope.

From Table 1 it is seen that the O^{++} abundance drops off in all directions away from the Trapezium, as would be expected from the falling level of ionization. The values of electron density show a similar trend and are shown in Figure 4.

The most striking feature of the observations is that at three points around the edge of the nebula the NIII line becomes as strong as the OIII 52 μm line. This might be considered questionable were it not that a similar phenomenon is seen in M17 (Emery et al.[7] and Moorwood et al.[8]). We must accept that it is real.

Simple models of the density and ionization structure of the region indicate that the observed intensities are consistent with the following conclusions:

a) The emitting region in Orion is not homogeneous. Variations in electron density between several hundred and in excess of ten thousand are indicated. Such a feature has been previously deduced in the radio-recombination models of Brocklehurst and Seaton[9] and Lockman and Brown[10] and is also necessary to explain the SIII IR line intensities measured by Moorwood et al.[4].

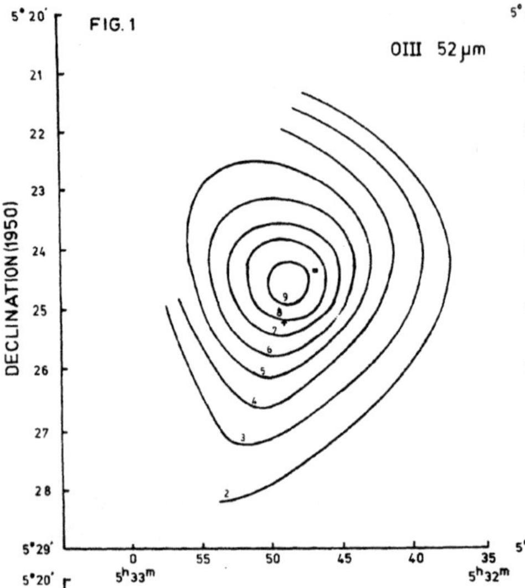

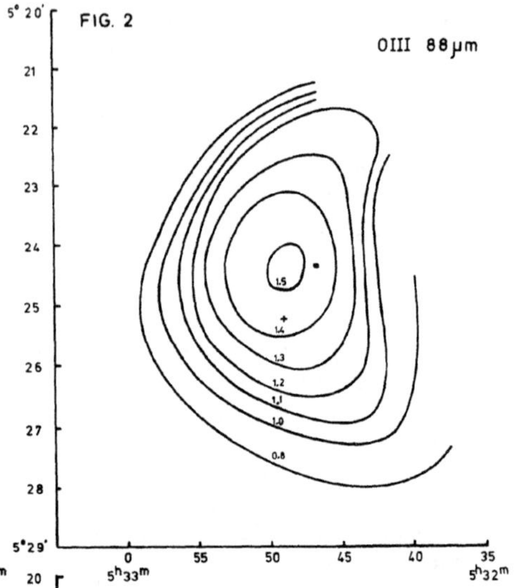

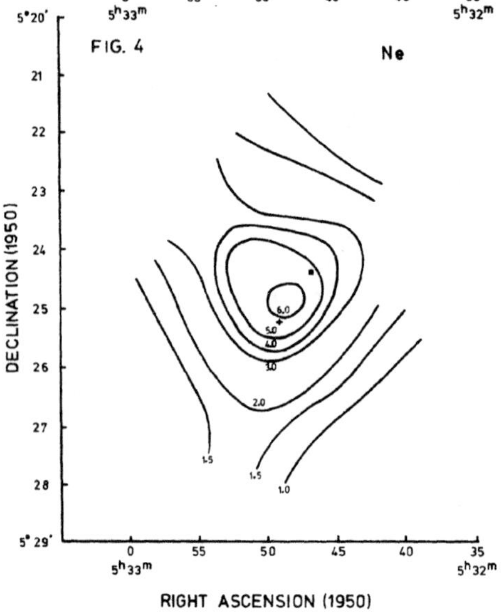

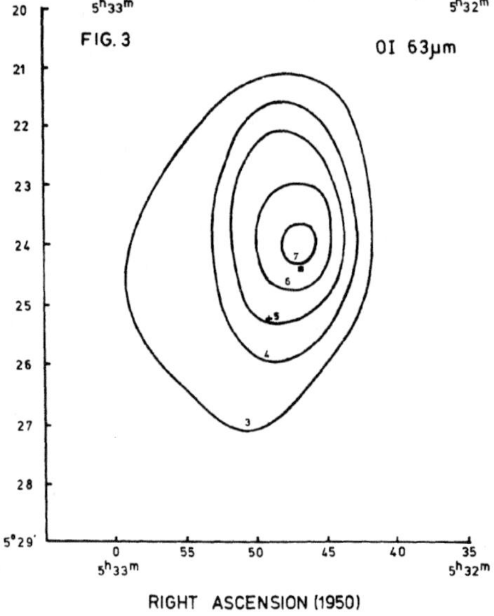

FIG. 1 M42, OIII 52 μm map, contour level = 7.68 x 10^{-10} W $cm^{-2}str^{-1}$.
FIG. 2 M42, OIII 88 μm map.
FIG. 3 M42, OI 63 μm map.
FIG. 4 M42, electron density map, contour level = 10^3 cm^{-3}

b) The ionic abundance of OIII has a maximum, over the Trapezium, of 3.3*10**(-4). Given the uncertainties of deriving the ionic abundance from the emission measure, our models indicate that a solar abundance of oxygen is present.

The abundance of nitrogen is more uncertain, but an overabundance by a factor of two seems necessary to explain the observed line intensities at the edge of the nebula.

c) The variation of derived electron density and emitting depth across the nebula is consistent with the cup-shape for M42 proposed by Pankonin et al. . This structure causes us to see the ionization fronts at the edge of the nebula almost edge on. This, together with a separation of the outer edges of the ionization spheres of OIII and NIII and the proposed overabundance of nitrogen can explain the bright NIII 57 μm line seen.

d) The OI peak is displaced to the NW with respect to the OIII peak indicating that some of the radiation originates from the shock region as indicated by the strong emission in the vibrational lines of H_2 (Hollenbach and Shull[12]). An extension to the SE of the OI 63 μm line towards the bright bar seen in the OI 6300 Å line is also evident. Hill and Hollenbach[13] have modelled various stages in the growth of an HII region with the special reference to the propagation of shocks, dissociation waves and ionization fronts and show the OI 63 μm emission line as a major source of cooling. With improved sensitivity and spatial resolution the OI far infrared emission lines will be extremely powerful probes of shock regions and the HI/HII boundary.

M17

The mapping of M17 was also chosen to complement the observations of Storey et al.[5] and Moorwood et al.[8], and the results are given in Table 2. Our measurements have been combined with the published data, adjusting for beam size differences where necessary, and cover the region including M17a and b as defined by the radio map of Wilson et al.[15]. Contour maps of the two OIII lines are shown in Figures 5 and 6. There was insufficient data to construct a contour map of the NIII line and the OI line was not detected at the positions observed.

The two OIII maps (Figures 5 and 6) show general similarities as expected. The lower critical density of 670 electrons cm^{-3} for the 88 μm OIII line compared with 4,900 electrons cm^{-3} for the 52 μm OIII line (Osterbrock[16]) indicates the 88 μm emission is probably collisionally saturated over most of the region. The 52 μm emission should be more peaked in the higher electron density region, as appears to be the case.

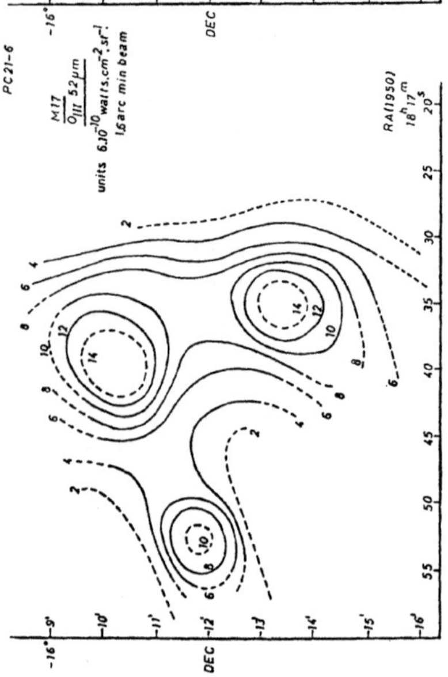

FIG. 5 M17, OIII 52 μm map, contour level = 6 x 10^{-10} W cm^{-2} str^{-1}

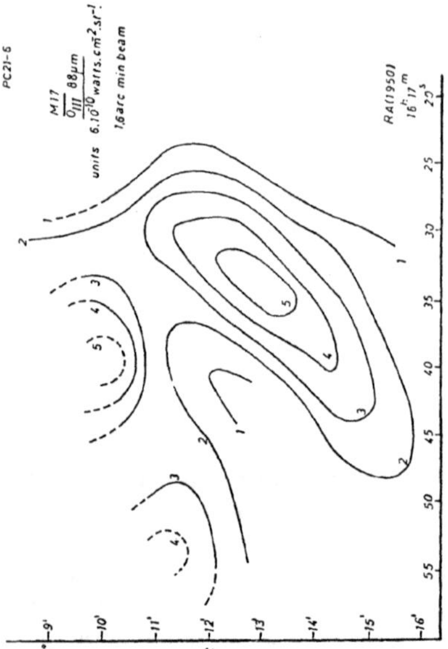

FIG. 6 M17, OIII 88 μm map

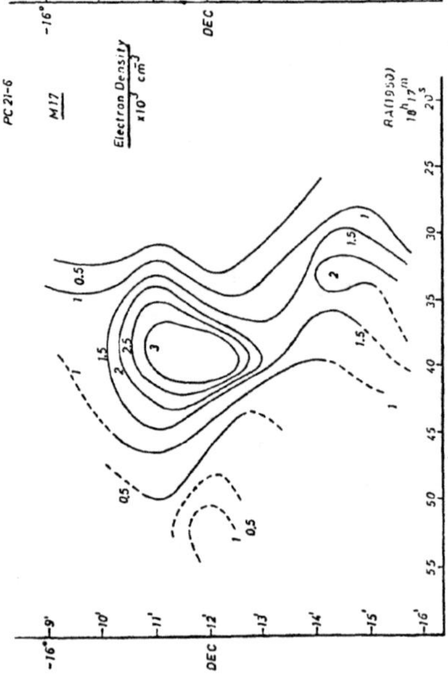

FIG. 7 M17, electron density map, contour level = 10^3 cm^{-3}

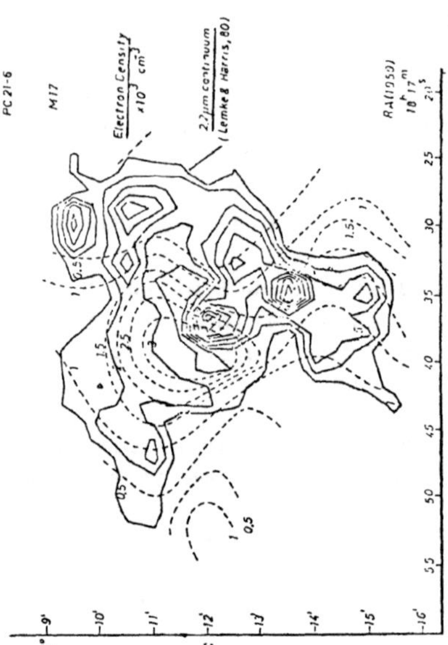

FIG. 8 2.2 μm continuum map of Lemke and Harris[19]

The location of the two peaks in the OIII maps associated with M17a and b are essentially coincident with the continuum radio emission peak of Wilson et al.[15]. Likewise the IR emission maps at 30, 50 and 100 μm reported by Gatley et al.[17] are strongest in the region of the M17a position. The 21 μm continuum map of Lemke and Low[18] shows three peaks which coincide with those indicated in the OIII maps. A 2.2 μm emission peak (Lemke and Harris[19]) is coincident with the M17a position but this 2.2 μm map also shows that there is a complex distribution of sources through M17 contributing to the ionization. At least some of these appear to be associated with regions of high OIII emission.

The electron density map derived from the ratio of the two OIII line intensities is shown in Figure 7. The peak n_e values are close to the 2.2 μm peak in the region of Kleinmann's star as shown in the 2.2 μm map of Lemke and Harris[19] in Figure 8, and are displaced from the radio continuum contours. Since the n_e determined from the ionic line intensity ratios refers to those electrons which are involved in the actual excitation/de-excitation processes, these values will be peaked at particular density ranges and ionization conditions, depending upon the transitions observed. Moorwood et al.[8] have already shown that the n_e determined from the SIII 18.7 μm and 33 μm line ratios leads to apparent electron densities substantially higher than those obtained with the two OIII lines, due to the higher critical densities for the SIII states. Hence the line of sight rms electron density which is determined from the radio continuum measurements can differ appreciably from the n_e values derived from ionic line measurements.

Although the NIII values generally had the lowest signal/noise of the fine structure line data some comments of the NIII distribution can be made. Extended emission comes from the region of the bright bar around M17b. Clear NIII emission is also seen to the SW of M17a where the nebula is ionization bounded by the molecular cloud. Since trhe n_e dependence of the NIII transition is similar to that of the OIII transitions, this agreement is to be expected assuming that the intrinsic abundance ratio N/O is constant.

CONCLUSIONS

The measurement of far infrared fine structure lines is proving to be a useful tool in understanding the physical processes occurring in gaseous nebula. With improved spectral resolution, it will be possible to examine in detail the dynamics of expanding HII regions.

Data from a second flight of a modified interferometer system with resolving power of 2 x 10^4 at 50 μm is presently undergoing analysis. With the recent detections of molecular rotational lines in the far infrared [20,21] which can be measured with the present interferometer system, we should be able to observe simultaneously, molecular, ionic and atomic transitions in gaseous nebulae. The simultaneous measurement of these transitions will remove the major uncertainties when comparing far infrared maps, namely calibration of the intensity scale and absolute pointing errors.

REFERENCES

1. Lacy, J.H.: 1981,"Infrared Astronomy",I.A.U. Symposium 96, 237
2. Watson, D.M., Storey, J.W.V.: 1980, Int. J. Infrared and Millimeter Waves 1, 609
3. Anderegg, M., Moorwood, A.F.M., Salinari, P., Furniss, I., Jennings, R.E., King, K.J., Towlson, W.A., Venis, T.E.: 1980, Astron. Astrophys. 82, 86
4. Moorwood, A.F.M., Baluteau, J.-P., Anderegg, M., Coron, N., Biraud, Y.: 1978, Ap. J. 224, 101
5. Storey, J.W.V., Watson, D.M., Townes, C.M.: 1979, Ap. J. 233, 109
6. Simpson, J.P.: 1975, Astron. Astrophys. 39, 43
7. Emery, R.J., Naylor, D.A., Fitton, B., Furniss, I., Jennings, R.E., King, K.J.: in preparation
8. Moorwood, A.F.M., Baluteau, J.-P., Anderegg, M., Coron, N., Biraud, Y., Fitton, B.: 1980, Ap. J. 238, 565
9. Brocklehurst, M., Seaton, M.J.: 1972, M.N.R.A.S. 157, 179
10. Lockman, F.J., Brown, R.L.: 1975, Ap. J. 201, 134
11. Pankonin, V., Walmsley, C.M., Harwit, M.: 1979, Astron. Astrophys. 75, 34
12. Hollenbach, D.J., Shull, J.M.: 1977, Ap. J. 216, 419
13. Hill, J.K., Hollenbach, D.J.: 1978, Ap. J. 225, 390
14. Johnston, K.J., Hobbs, R.W.: 1969, Ap. J. 158, 145
15. Wilson, T.L., Fazio, G.G., Jaffe, D., Kleinmann, D., Wright, E.L., Low, F.J.: 1979, Astron. Astrophys. 76, 86
16. Osterbrock, D.E.: 1974, "Astrophysics of Gaseous Nebulae," Freeman and Company, San Francisco
17. Gatley, I., Becklin, E.E., Sellgren, K., Werner, M.W.: 1979, Ap. J. 233, 575
18. Lemke, D., Low, F.J.: 1972, Ap. J. 177, L53
19. Lemke, D., Harris, A.W.: 1980, preprint
20. Watson, D.M., Storey, J.W.V., Townes, C.H., Haller, E.E., Hansen, W.L.: 1980, Ap. J. 239, L129
21. Storey, J.W.V., Watson, D.M., Townes, C.M.: 1981, Ap. J. 244, L27

EXCITED OH IN ABSORPTION TOWARDS W3(OH)

C. M. Walmsley,[1,2] A. Winnberg,[2] T. L. Wilson,[2] A. Baudry[3]
[1]Physics Dept., University of Illinois at Urbana-Champaign
[2]Max Planck Institut für Radioastronomie, Bonn
[3]Observatoire de l'Université de Bordeaux

There is considerable evidence for clumps or condensations of approximately stellar mass in the neutral gas surrounding compact HII regions. There are a variety of observational approaches to studying such gas and some of these have been discussed at this Conference (see e.g. Bastien et al. (this volume), Forster et al. (this volume)). In this contribution, we discuss recent absorption line observations of excited (300 to 500 K above ground) Λ-doublet transitions of OH towards the compact HII region W3(OH). Such observations are interesting due to the large spontaneous decay rates of the far infrared transitions which depopulate these excited levels ($\sim 1\ s^{-1}$). This implies either a very large density ($> 10^8\ cm^{-3}$) or a very strong far-infrared radiation field in the OH absorption region. In either case, it is hard to avoid the conclusion that what one is observing is a very high density clump close to the HII region ionization front.

We have observed the $^2\Pi_{3/2}$, J = 7/2 and J = 9/2 Λ doublet transitions towards W3(OH). In both cases, we have been able to observe the two $\Delta F = 0$ lines ("main lines") and the respective line ratios are consistent with optically thin L.T.E. Observation of the J = 7/2, F = 4 → 4 line is hampered by a maser emission line (Turner et al. (1970)) at somewhat higher velocity. The maser itself is of some interest because it appears to show Zeeman splitting with a frequency separation corresponding to a magnetic field of 6 milligauss. In the absorbing region, we estimate from our observations that the OH rotational temperature is 160 ± 30 K. This seems to imply a kinetic temperature in the region of at least this order of magnitude and hence that the absorbing OH is very close to the ionization front of the compact HII region. The size of the region is probably of order 0.01 to 0.02 pc. It is interesting that the OH absorption velocity (-45.1) is intermediate between that of 6 cm H_2CO (-46.7, Forster et al. in this volume) and NH_3 (-44.5, Pauls and Wilson (1980)). The OH 18 cm masers have an average velocity close to that of NH_3 (Reid et al. (1980)) and all absorption features are red-shifted relative to the ionized gas. The velocity differences between the various

R. S. Roger and P. E. Dewdney (eds.), Regions of Recent Star Formation, 81–82.
Copyright © 1982 by D. Reidel Publishing Company.

molecules may not be significant but it is tempting to speculate that bright infrared lines from the OH absorption clump may be capable of pumping the OH in the presumably colder OH maser cloudlet and thus of causing the maser phenomenon. The Λ-doublet would be inverted by a mechanism analogous to that discussed by Lucas (1980) which relies upon the velocity shift between two clouds to introduce an asymmetry into the problem. In any event, the exact relationship between OH-maser and OH-absorption regions could clearly be a key piece of information in unravelling the maser problem.

A more detailed account of this work is given in an article soon to be published in Astronomy and Astrophysics (Baudry et al. (1981)). C. M. W. would like to acknowledge the partial support of NSF grant AST 80-23230. He would also like to thank the University of Illinois Physics Department for its hospitality during his stay in Champaign-Urbana.

REFERENCES

Bastien P., Bieging J., Henkel C., Martin R. N., Pauls T., Walmsley C. M., Wilson T. L., Ziurys L. M. (1982) this volume, p. 335.

Baudry A., Walmsley C. M., Winnberg A., Wilson T. L. (1981) Astron. Astrophys. (in press).

Forster J. R., Goss W. M., Dickel H. R. (1982) this volume, p. 409.

Lucas R. (1980) Astron. Astrophys. 84, 36.

Pauls T., Wilson T. L. (1980) Astron. Astrophys. 91, L11.

Reid M. J., Haschick A. D., Burke B. F., Moran J. M., Johnston K. J., Swenson G. W. (1980) Astrophys. J. 239, 89.

Turner B. E., Palmer P., Zuckerman B. (1970) Astrophys J. 160, L125.

ULTRA-COMPACT HII REGIONS EMBEDDED IN INFRARED SOURCES

Sun Kwok
Herzberg Institute of Astrophysics
National Research Council of Canada
Ottawa, Canada K1A 0R6

The formation of an ionized region and its subsequent interaction with surrounding neutral material can occur in either the very early or very late stages of stellar evolution. Both newborn and old stars are often surrounded by thick dust clouds and the interaction of a newly formed compact HII region with this circumstellar material can take place either: (1) in a site of star formation where a compact HII region develops in a dark molecular cloud or (2) when a star is in transition from a red-giant to a planetary nebula and an ionized nebular shell interacts with the remnant circumstellar envelope of the red-giant progenitor. Although both types of objects are likely to be first identified as infrared or molecular line sources, yet they can be distinguished in the following ways:

(1) Infrared emission: dust grains surrounding young stars are likely to be primordial and not ejected by the stars themselves. The infrared spectrum is often broad (extending to the far infrared) and spatially extended (Harvey *et al.* 1979). On the other hand, circumstellar envelopes formed by mass loss from red-giants have strong density gradients ($n \propto r^{-2}$) and emit strongly only in the near and mid-infrared.

(2) Molecular lines (e.g. CO) associated with young stars have line widths between 5 and 10 km s^{-1}, much narrower than the Doppler-broadened lines observed in circumstellar envelopes ($\Delta V \sim 20-40$ km s^{-1}, Zuckerman *et al.* 1978). Since the total mass in star-forming clouds is large ($10-10^3$ $M_\odot$) compared to circumstellar envelopes ($0.1-1$ $M_\odot$), molecular line strengths are higher in proto-stellar objects. Maser lines, when present, are also often more energetic in young objects.

Compact HII regions embedded in dense neutral clouds can be most effectively probed by observing: (1) radio free-free emission and (2) infrared recombination line emission. Table 1 lists 4 examples of compact HII regions which are believed to be star formation sites and have been observed to possess infrared recombination line emission.

R. S. Roger and P. E. Dewdney (eds.), Regions of Recent Star Formation, 83–89.
Copyright © 1982 by D. Reidel Publishing Company.

TABLE 1

	PROTO-STELLAR OBJECTS				PROTO-PLANETARY NEBULA	SLOW NOVA
	LkHα101	V645 Cyg	NGC 2264	GL 490	GL 618	HM Sge
D(pc)	800	6000	800	900	2000	1000?
$L(L_\odot)$	$1.2x10^4$ (a)	$9x10^4$ (b)	$3.5x10^3$ (c)	$1.5x10^3$ (a)	$3.2x10^4$ (d)	$3x10^3$ (e)
	➔ B 0.5 ZAMS	➔ 07.5 ZAMS	➔ B1 ZAMS	➔ B2-B3 ZAMS	➔ $M_* = 1.06\ M_\odot$ (Paczynski core mass-luminosity relationship)	
Bγ(erg $cm^{-2}s^{-1}$)	$4.2x10^{-11}$ (f)	$1.92{\pm}0.3x10^{-12}$ (g)	$1.7{\pm}0.2x10^{-12}$ (h)	$7{\pm}1x10^{-13}$ (i)	$< 2.5x10^{-14}$ (j)	$1.4{\pm}0.15x10^{-11}$ (k)
A_v(mag)	14.2±0.4 (a)	4	12-72 (h)	12±3 (i)	> 45 (j)	∿ 12
$N_e^2V(cm^{-3})$	$2.45x10^{60}$	$3x10^{60}$	$7.5x10^{58}$-$6x10^{60}$	$4.8x10^{58}$	$2x10^{59}$	$1.3x10^{60}$
	➔ 09.5 ZAMS	➔ 09 ZAMS	➔ B 0.5 - 08.5 ZAMS	➔ B 0.5 ZAMS	(from radio)	
S_{ff}(Jy) (from B_γ)	10	0.2	4.2-340	1.75	< 0.0024	4
S_{ff}(Jy)	∿ 0.2 (l)	$S_{5\ GHz}$ < 0.0005 (m)	$S_{10.6\ GHz}$ ∿ 0.015±0.004	$S_{5\ GHz}$ < 0.001 (n)	> 0.09 (o)	> 0.1
v(km s^{-1})	> 800 (p)	> 700 (q)				1700 (r)
CO : ΔV(km s^{-1})	3 (b)	5 (g)			40 (t)	
T_A(K)	17 (s)	9 (g)			0.35 (t)	
shock excited H_2					yes (j)	

(a) Harvey, Thronson and Gatley 1979
(b) Harvey, Thronson and Gatley 1981
(c) Harvey, Campbell and Hoffman 1977
(d) Kleinman et al. 1978
(e) Davidson, Humphreys and Merrill 1978
(f) Thompson et al. 1977
(g) Harvey and Lada 1980
(h) Thompson and Tokunaga 1978
(i) Thompson and Tokunaga 1979
(j) Thronson 1981
(k) Thronson and Harvey 1981
(l) Purton et al. 1981
(m) Kwok 1981
(n) Simon et al. 1981
(o) Kwok and Feldman 1981
(p) Thompson et al. 1976
(q) Cohen 1977
(r) Wallerstein 1978
(s) Knapp et al. 1976
(t) Lo and Bechis 1976

The emission measure (n_e^2V) can be estimated from the reddening-corrected line strenghts, and the free-free flux density can be predicted from the Brackett γ flux by the formula $S_{\lambda 6\ cm}(Jy) = 1.14 \times 10^{14}\ S_{B\gamma}(erg\ cm^{-2}\ s^{-1})$ (Wynn-Williams *et al.* 1978). In all cases, the observed radio flux is orders of magnitudes below the value inferred from the infrared data. For example, the Bγ line observed in V 645 Cygni suggests an embedded O-type star, and this is supported by the high luminosity observed in the far infrared. Both the CO and H_2O maser line give a velocity of $-50\ km\ s^{-1}$, implying a kinematic distance of 6 kpc. The Strömgren sphere of a 07.5 ZAMS star is expected to have n_e^2V exceeding $10^{60}\ cm^{-3}$, which should lead to an optically-thin radio flux density of ≳0.2 Jy. Figure 1 shows a λ6-cm VLA map of V 645 Cygni. No continuum source is detected to the limit of 0.5 mJy. Similarly, recombination line strengths in LkHα 101 predict a free-free flux density of 10 Jy but the observed optically-thin flux is only ∿0.2 Jy (Brown, Broderick and Knapp 1976, Purton *et al.* 1981). Recent VLA observations of GL 490 by Simon *et al.* (1981) show a λ6-cm upper limit of 1 mJy, far less than expected. Although no radio interferometric data is available for NGC 2264, single-dish measurements made at the Algonquin Radio Observatory indicate that the observed radio fluxes are again hundreds of times weaker than the IR-inferred value.

There are several possible explanations to such discrepancies:

(1) The ionized HII region is extremely dense and compact, and the radio spectrum is optically thick to very high frequencies (>100 GHz). As an example, the low λ2-cm flux density of V 645 Cygni implies $\theta_{\lambda 2\ cm} < 0\overset{''}{.}06$, or $<5\times10^{15}$ cm. It is more likely, however, that such high turn-over frequencies result from the presence of a stellar wind (e.g. MWC 349 has a turn-over frequency of 600 GHz, Harvey *et al.* 1979). The radio structure of LkHα 101 indicates

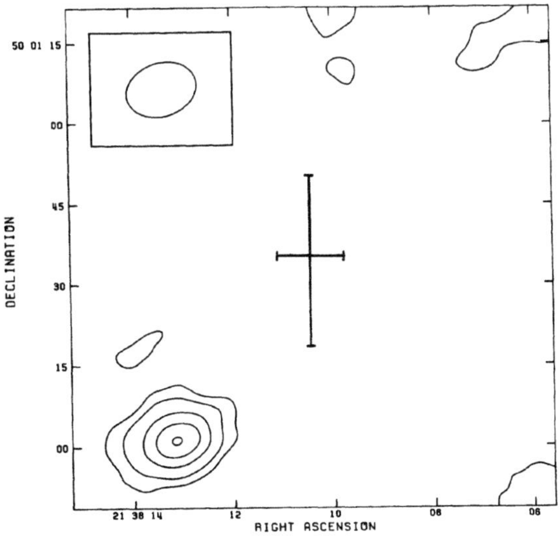

Fig. 1. The λ6-cm map of the 1!5×1!5 field around V 645 Cygni. The infrared position is marked with a "+".

a stellar wind operating inside an ionized halo of ∿1' in size (Brown *et al.* 1976). The turn-over frequency of the wind is ∿50 GHz but the observed optically thin flux is still inadequate to completely remove the above discrepancy.

(2) The underlying star is not a ZAMS object, but a high luminosity B or A supergiant. For example, Humphreys *et al.* (1980) have suggested that the spectrum of V 645 Cygni resembles that of an A5e shell star.

(3) The density in the HII region is so high that collisional processes begin to compete with radiative transitions and the B_n coefficients calculated under hydrogen recombination theory are underestimated. Since the Einstein A coefficients are of the order 10^4-10^8 s^{-1}, the electron density has to be extremely high ($>10^{10}$ cm^{-3}) to have an effect. However, in a stellar wind situation, the electron density can be very high near the star. For example, a stellar wind of mass loss rate 10^{-5} $M_\odot$ yr^{-1} and velocity 1000 km s^{-1} has a density $>10^{11}$ cm^{-3} at $r = 10^{12}$ cm from the star. If the observed recombination lines arise from such a wind, n_e^2V is quite likely to have been overestimated.

While the relative merit of these explanations is difficult to assess at this time, it is clear that the conventional picture of a uniform density Strömgren sphere surrounding a young star inadequately describes the compact HII regions discussed above. Stellar winds and their interaction with the circumstellar neutral cloud may be a significant factor. Recent VLA observations (Matthews *et al.* 1981) show ring structure in a number of compact HII regions, probably formed by stellar winds sweeping up the circumstellar material. Realistic recombination line calculations applied to a stellar wind situation may help to clarify the situation.

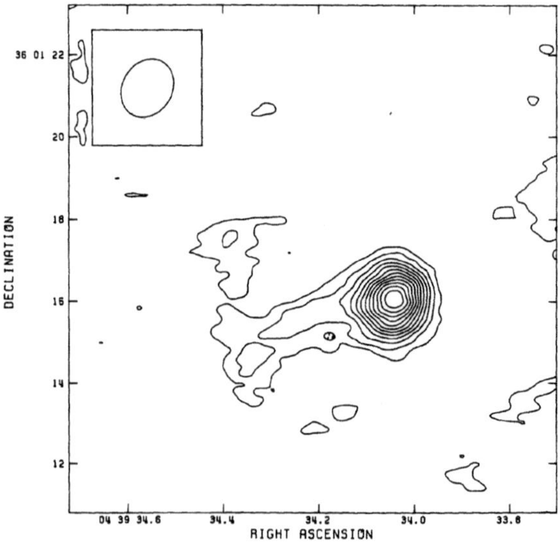

Fig. 2. λ2 cm map of GL 618.

The case of a proto-planetary nebula, where a star is in the process of moving from the red to the blue side of the HR diagram, is quite different. GL 618 is an infrared object which has a molecular envelope similar to the molecular envelopes of late-type stars (Lo and Bechis 1976). However it has a B-type central star, and an ultra-compact HII region has been found to be coincident with the infrared source (optically thick to 10 μm) of 0$''$.4 in size. Figure 2 shows the λ2-cm VLA map of GL 618. The HII region has an angular size of ∿0$''$.2 and has a tail extending to a reflection nebula seen in optical photographs. This tail is probably due to ionizing photons leaking through the dust envelope. Comparison of spectra taken at epochs 1977 and 1980 shows that radio emission from GL 618 has brightened by a factor of 2 over a period of 3 years (Kwok and Feldman 1981), suggesting that the HII region has expanded, probably due to an increasing effective temperature of the central star and an advancing ionization front. It is interesting to note, however, that no Brackett lines are observable (Thronson 1981). In order to reconcile this with the observed radio flux, at least 45 magnitudes of visual extinction (A_V) is necessary. This is exactly the opposite situation to that found in protostellar objects, where there is an excess of recombination line flux. Shock-excited emission from the H_2 molecule has been seen in GL 618, which could be the result of the collision between the planetary-nebular shell and the remnant of the red-giant wind (Thronson 1981).

Another example of an HII region interacting with neutral matter is HM Sge. It had an optical outburst in 1975 and is believed to be a very slow nova (Paczyński and Rudak 1980, Kwok 1981b). It has a very strong silicate feature which suggests the presence of an M-giant wind. A dense HII region is evident from a very rich emission line spectrum both in the optical and in the ultraviolet. HM Sge is resolved at the VLA and is found to have an angular size of ∿0$''$.2 at λ2-cm. Observed infrared recombination line fluxes suggest an optically-thin radio flux of ∿4 Jy (Thronson and Harvey 1981). Figure 3 shows the radio spectrum of HM Sge. Its 1980 optically-thin flux, although not precisely determined, is probably no more than 0.3 Jy. Therefore, in this late-type object, the recombination lines are again enhanced.

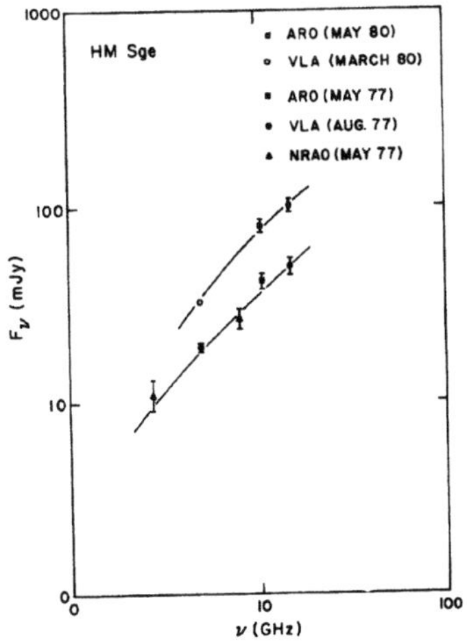

g. 3. Radio spectrum of HM Sge at epochs 1977 and 1980.

Before we conclude our discussion on late-type objects, I would like to note that interactions with the remnant red-giant envelope in a planetary nebula system can lead to interesting dynamical consequences. Many planetary nebula nuclei are now found to have fast winds with mass loss rates of 10^{-7} $M_\odot$ yr^{-1} and velocities

$\sim$2000 km s^{-1} (Benvenuti and Perinotto 1980; Castor, Lutz and Seaton 1981). Since the transition time from red-giants to planetary nebulae is very short (10^3-10^4 years, Renzini 1981), interactions between these central-star winds with the remnant red-giant envelopes will cause a snow-plow effect, leading to the formation of a dense shell at the interface of the two winds. It can be shown that, in several thousand years time, such shells will develop all the characteristics of planetary nebulae and a sudden violent ejection may not be necessary for the formation of planetary nebulae (Kwok, Purton and FitzGerald 1978; Kwok 1981c).

In summary, we are just beginning to realize the importance of stellar winds in the evolution of very young and very old stars. The energy and momentum input provided by such winds can profoundly alter the dynamical picture of circumstellar clouds during the early and late evolutionary stages.

REFERENCES

Benvenuti, P. and Perinotto, M. 1980, in *Proceedings of Second European IUE Conference,* p. 187.
Brown, R.L., Broderick, J.J. and Knapp, G.R. 1976. *Mon. Not. Roy. Astron. Sco.,* 175, 87P.
Castor, J.I., Lutz, J.H. and Seaton, M.J. 1981. *Mon. Not. Roy. Astron. Soc.,* 194, 547.
Cohen, M. 1977. *Astrophys. J.,* 215, 533.
Davidson, K., Humphreys, R. and Merrill, K.M. 1978. *Astrophys. J.,* 220, 239.
Harvey, P.M. and Lada, C.J. 1980. *Astrophys. J.,* 237, 61.
Harvey, P.M., Campbell, M.F. and Hoffmann, W.F. 1977. *Astrophys. J.,* 215, 151.
Harvey, P.M., Thronson, H.A. and Gatley, I. 1979. *Astrophys. J.,* 231, 115.
Harvey, P.M., Thronson, H.A. and Gately, I. 1981, private communication.
Humphreys, R.M., Merrill, K.M. and Black, J.H. 1980. *Astrophys. J. (Letters),* 237, L17.
Kleinmann, S.G., Sargent, D.G., Moseley, H., Harper, D.A., Loewenstein, R.F., Telesco, C.M. and Thronson, H.A. 1978. *Astron. Astrophys.,* 65, 139.
Knapp, G.R., Kuiper, T.B.H., Knapp, S.L. and Brown, R.L. 1976. *Astrophys. J.,* 206, 443.
Kwok, S. 1981a. *Publ. Astron. Soc. Pacific,* in press.
Kwok, S. 1981b. in *Effects of Mass Loss on Stellar Evolution,* ed. C. Choisi and R. Stalio, D. Reidel, Holland, p. 347, 499.
Kwok, S. 1981c. *Astrophys. J.,* submitted.
Kwok, S. and Feldman, P.A. 1981. *Astrophys. J. (Letters),* in press.
Kwok, S., Purton, C.R. and FitzGerald, P.M. 1978. *Astrophys. J. (Letters),* 219, L125.
Lo, K.Y. and Bechis, K.P. 1976. *Astrophys. J. (Letters),* 205, L21.
Matthews, H.E., Kwok, S., Turner, B.E. and Winnberg, A. 1981, in preparation.

Paczyński, B. and Rudak, B. 1980. *Astron. Astrophys.*, 82, 349.
Purton, C.R., Feldman, P.A., Marsh, K.A., Wright, A.E. and Allen, D.A. 1981. *Mon. Not. Roy. Astron. Soc.*, in press.
Renzini, A. 1981. in *Physical Processes in Red Giants*, ed. I. Iben and A. Renzini, Reidel, p. 421.
Simon, M., Righini-Cohen, G., Felli, M. and Fischer, J. 1981. *Astrophys. J.*, 245, 552.
Thompson, R.I. and Tokunaga, A.T. 1978. *Astrophys. J.*, 226, 119.
Thompson, R.I. and Tokunaga, A.T. 1979. *Astrophys. J.*, 231, 736.
Thompson, R.I., Erickson, E.F., Witteborn, F.C. and Strecker, D.W. 1976. *Astrophys. J. (Letters)*, 210, L31.
Thompson, R.I., Strittmatter, P.A., Erickson, E.F., Witteborn, F.C. and Strecker, D.W. 1977. *Astrophys. J.*, 218, 170.
Thronson, H.A. 1981, preprint.
Thronson, H.A. and Harvey, P.M. 1981. *Astrophys. J.*, in press.
Wallerstein, G. 1978. *Publ. Astron. Soc. Pacific*, 90, 36
Wynn-Williams, C.G., Becklin, E.E., Matthews, K. and Neugebauer, G. 1978. *Mon. Not. Roy. Astron. Soc.*, 183, 237.
Zuckerman, B., Palmer, P., Gilra, D.P., Turner, B.E. and Morris, M. 1978. Astrophys. J. (Letters), 220, L53.

DISCUSSION FOLLOWING PAPER BY KWOK

BEICHMAN: From the change in the spectrum of GL618 what "expansion velocity" do you derive?

KWOK: About 200 to 220 km s^{-1} depending on the distance. This is rather high if interpreted as a physical expansion. It may represent only the advancement of the ionization front, which has a time scale of a few years, consistent with the observations.

DYNAMICAL EFFECTS OF STELLAR WINDS AND SHOCKED GAS ON INTERSTELLAR CLOUDS

J. Michael Shull
Joint Institute for Laboratory Astrophysics, University of Colorado and National Bureau of Standards, Boulder, Colorado 80309

Abstract

Stellar winds are a ubiquitous feature of nearly all stars. Their energy input may affect the dynamics of interstellar gas on scales from 10^{15} to 10^{21} cm. New observations suggest that powerful winds are also present during the pre-main sequence phase of massive stars. This review summarizes recent observations and theories of winds and high velocity gas, with special emphasis on inhomogeneities in O-star winds, shocked molecular clouds, H_2O masers, and Herbig-Haro objects. The difficulties of current models for these phenomena are discussed, and several consequences of protostellar ejection of planet-sized condensations are explored.

I. INTRODUCTION

Among the most exciting results to emerge from X-ray, ultraviolet (UV), infrared (IR), and radio astronomy over the past 15 years is the demonstration of the importance of stellar winds. In addition to their intrinsic value in tying together observations at such disparate wavelengths, these data have forced astrophysicists to re-examine the influence of stellar mass loss on star formation, stellar evolution, and interstellar gas dynamics. Rarely have so many sub-disciplines of astronomy been studied in such close connection.

In this review I wish to discuss stellar winds and high velocity interstellar gas, dealing with such subjects as their morphology, origin, and interaction. Beginning in Sec. II with a historical summary of the evidence for winds, I shall proceed to some recent observations which furnish new insight and raise new questions about the details of the phenomenon. A recurrent theme of this discussion will be an attempt to bring some measure of unity to a variety of astronomical objects and observations. In Sec. III, I will delve somewhat deeper into the mechanisms by which stellar winds may clear cavities in

R. S. Roger and P. E. Dewdney (eds.), Regions of Recent Star Formation, 91–105.
Copyright © 1982 by D. Reidel Publishing Company.

interstellar clouds, accelerate clumps of gas, and excite atomic or molecular spectral features. Finally, in Sec. IV, I will speculate about several intriguing observations: the Orion molecular outflow, X-ray emission from O-star winds, and Herbig-Haro object proper motions. The dynamical effects of stellar winds may not only explain the features of such observations, but may also provide information about pre-main sequence evolution, the stellar multiplicity and initial mass functions, and the evolution of young H II regions.

II. OBSERVATIONS

Mass loss is a common phenomenon among stars, ranging from the 10^{-14} $M_\odot$ yr^{-1} in the solar wind to 10^{-4} $M_\odot$ yr^{-1} in some Wolf-Rayet stars (Conti and McCray 1980). A variety of observations (circumstellar absorption lines, infrared emission from dust, Ca II/Mg II line asymmetries) are attributed to slow winds in cool giants and supergiants (Castor 1980). Related support for the notion of cool star winds comes from the empirical division of the cool half of the Hertzsprung-Russell diagram into two distinct regions, separating stars with prominent chromospheres and transition regions from stars with weak chromospheric activity and strong winds (Linsky and Haisch 1979; Stencel and Mullan 1980; Hartmann, Dupree, and Raymond 1980). T-Tauri stars are believed to lose mass at a rate between 10^{-9} and 10^{-8} $M_\odot$ yr^{-1} (Kuhi 1964, 1966), and the rapidly rotating Be stars probably have similar mass loss rates at somewhat greater velocities (Snow and Marlborough 1976).

However, the most violent main sequence stellar winds arise in the O-type stars (see Fig. 1), with mass loss rates ranging from 10^{-8} to 10^{-5} $M_\odot$ yr^{-1} at velocities V_w between 1000 and 3500 km s^{-1} (Snow and Morton 1976; Conti and Garmany 1980). A recent analysis of UV P-Cygni lines from 31 O-type stars observed by the International Ultraviolet Explorer (IUE) satellite (Garmany <u>et al.</u> 1981) and 6 O-stars observed with the <u>Copernicus</u> satellite (Olson and Castor 1981) yields a relationship between mass loss rate $\dot{M}$ and luminosity L,

$$\dot{M} = (3.8 \times 10^{-6} \; M_\odot \; yr^{-1})(L/10^6 \; L_\odot)^{1.73} \quad . \tag{1}$$

This relationship is in fair agreement with that found by VLA radio observations (Abbott, Bieging, and Churchwell 1981),

$$\dot{M} = (9.1 \times 10^{-6} \; M_\odot \; yr^{-1})(L/10^6 \; L_\odot)^{1.56} \quad , \tag{2}$$

when one takes into account the fact that current radio techniques can only detect the largest mass loss rates while UV methods are only reliable for low $\dot{M}$. The former study finds a large spread in $\dot{M}$ at a given L, suggesting evolutionary effects along the main sequence, wind variability, or a dependence of the mass loss rate on some additional parameter, such as effective gravity. However, this interpretation is still controversial.

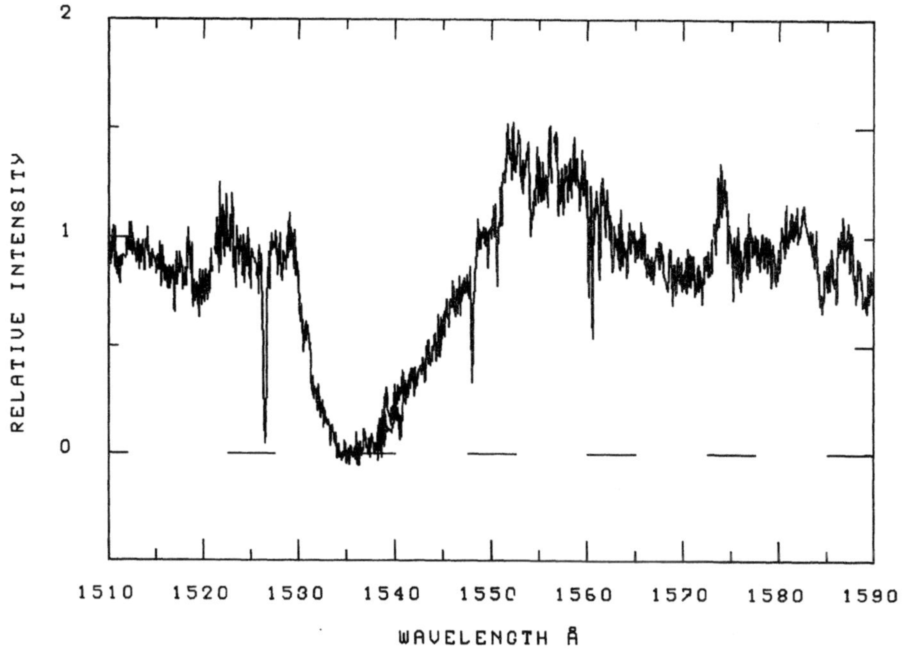

Fig. 1. A stellar wind P-Cygni resonance line of C IV λλ1549 toward HD 164794 (9 Sgr), an O4 V star, taken with the IUE satellite (Shull, unpublished). The wind terminal velocity (~3700 km s^{-1}) and mass-loss rate (~3 × 10^{-6} $M_\odot$ yr^{-1}) are derived by fitting the violet absorption component of this line.

Even larger mass loss rates have been inferred for several objects embedded in molecular clouds. A wide range of observations in the core of the Orion molecular cloud is best understood in terms of a molecular outflow from a central source, either the BN object or the infrared source IRc2. This outflow is observed in the vibrational lines of H_2 (Gautier et al. 1976; Beckwith et al. 1978) which show broad 40-100 km s^{-1} wings (Nadeau and Geballe 1979); in millimeter CO (1-0)emission (Zuckerman, Kuiper, and Rodriguez-Kuiper 1976; Kwan and Scoville 1976); in the broad wings of higher-J CO lines (Phillips et al. 1977); and in other molecules such as SO, SiO, SO_2, OCS and HCO^+. Recent studies of high velocity CO (2-1) and SO_2 emission in Orion (Knapp et al. 1981, see Fig. 2) suggest a roughly spherical outflow with velocity V ~ r and a rapid density falloff. The mass of high velocity gas is estimated as ≳5 $M_\odot$, the energy ~2.5 × 10^{46} ergs, and the mass loss rate 10^{-3} to 10^{-2} $M_\odot$ yr^{-1} at 100 km s^{-1}. The dynamical age of this flow is ~10^3 yr; the "Hubble-like" velocity law is indicative that the flow began suddenly from a central source. Proper

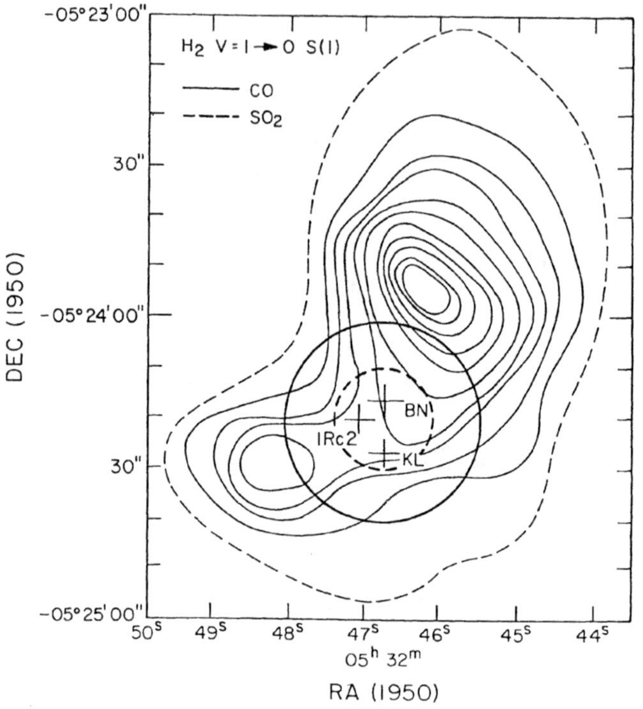

Fig. 2. A comparison of the sizes of the CO, SO_2, and H_2 emitting regions in Orion (Knapp et al. 1981). The H_2 map is from Beckwith et al. (1978); the centering of the CO and SO_2 maps is schematic.

motion studies of nearby high velocity H_2O masers (Genzel et al. 1981) are also consistent with radial motion away from the center of CO emission. Likewise, the H_2 quadrupole emission at the periphery of the CO emission is suggestive that it shares a common source. A major difficulty in the interpretation is the fact that the flow momentum is ~100 times the single-scattering momentum, L/c, in the radiation field of either BN or IRc2.

Such high velocity outflows have also been detected in other giant molecular clouds, with dynamical lifetimes of order 10^4 yr. Two amazing objects, the cometary nebulae NCG 2261 surrounding R Mon (Cantó et al. 1981) and the source in the molecular cloud L1551 (Snell, Loren, and Plambeck 1980), exhibit CO emission contours and velocities with a double-lobed structure emanating from a central compact source. The former source is associated with the Herbig-Haro object HH 39, while the latter is associated with HH 28, 29, 30, and

102. Adopting 10^4 yr as the dynamical age and assuming ~4000 giant molecular clouds in the galaxy, Solomon, Huguenin, and Scoville (1981) estimate that the overall galactic birthrate for such sources may be as large as 0.1 yr^{-1}. This rate is similar to the stellar birthrate in OB associations (assuming several thousand OB associations of age 3×10^7 yr each containing 100-1000 stars). Because there are only about 10^4 O-stars in the galaxy (Conti and McCray 1980), the most likely sources for the outflows are B stars. While nearly all B supergiants have winds, Lamers and Snow (1978) have found that on the main sequence only stars earlier than B5 (T_{eff} > 12,000 K) have detectable winds. However, the evidence is compelling that pre-main sequence stellar evolution is accompanied by much stronger winds, perhaps at lower velocity.

Evidence is also accumulating to suggest that pre-main sequence winds are collimated, either by a circumstellar accretion disk (Snell et al. 1980) or by density gradients in the molecular cloud (Cantó and Rodriguez 1980; Meaburn and Walsh 1980). Proper motion studies of Herbig-Haro objects (Cudworth and Herbig 1979; Herbig and Jones 1981 -- see Fig. 3) indicate that the objects are moving away from a central source (Cohen and Schwartz 1979) at 100 to 350 km s^{-1}. Such motion is similar to that seen in the Orion H_2O masers, although on a somewhat larger scale.

Stellar winds may have dramatic effects on interstellar gas. Over the main sequence life of an O-star, the wind may inject 10^{49}-10^{50} ergs of mechanical energy into the interstellar medium -- this is comparable to the dynamical effects of a supernova, although it is spread over 10^6 yr. The thermal pressure from the cavity created by this wind as it impacts on the interstellar gas may drive a "bubble" of 10^6 K low density gas, surrounded by a dense shell of swept up interstellar gas (Castor, McCray, and Weaver 1975; Weaver et al. 1977). The theory of these bubbles, scaled to parameters appropriate to dense molecular clouds, is discussed in Sec. III.

On even larger scales, the combined action of stellar winds and supernovae may create "superbubbles" hundreds of parsecs in size (McCray and Snow 1979). Evidence for high velocity ($\gtrsim$100 km s^{-1}) exhausts from OB associations has been reported in UV absorption features of Si III, C III, and other ions seen by the *Copernicus* satellite (Shull and York 1977; Shull 1977; Cohn and York 1977; Cowie, Songaila, and York 1979) toward stars in the Orion and Eridanus regions. Reynolds (1976) and Reynolds and Ogden (1979) have obtained optical Fabry-Perot observations of similar shells of high velocity gas toward the Gum Nebula and Orion, respectively. Perhaps the most spectacular examples of superbubbles are the large 21-cm galactic "supershells" seen by Heiles (1979), the Magellanic cloud superbubbles seen in optical lines by Meaburn (1979, 1980), and the X-ray Cygnus superbubble found by Cash et al. (1980). An expanding region of hot gas roughly 450 pc in diameter, the Cygnus superbubble is driven by

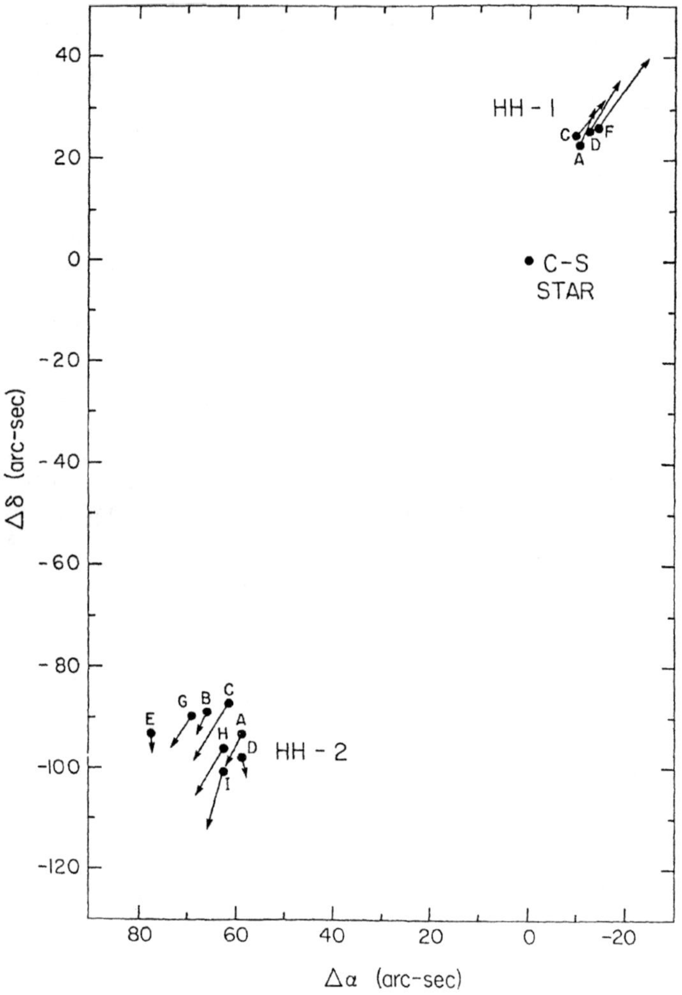

Fig. 3. The positions of HH-1, HH-2, and the Cohen-Schwartz star for epoch 1968.0 (Herbig and Jones 1981). The arrows indicate the shift in 100 years due to proper motion. The motion of the CS star is too small to show on this scale.

2×10^6 K plasma containing 10^{52} ergs. Although both winds and supernovae are implicated in the origin of this cavity (Cash et al. 1980; Bruhweiler et al. 1980), recent work by Abbott et al. (1981) has determined that five O-stars in Cyg OB2 are injecting energy at a rate (5×10^{38} ergs s^{-1}) sufficient to explain the shell kinetics if the association is 2 million years old and if the ambient medium has a particle density 0.35 cm^{-3}. It seems that the most massive OB associations are capable of moving interstellar gas across distances exceeding several hundred parsecs.

III. THEORIES

While many observations suggest that the interactions of a star's ionizing radiation and winds with the surrounding medium deviate markedly from spherical symmetry, the basic effects may be modeled by a simple analysis. In the absence of a wind, a compact H II region begins to expand after the time

$$t_{II} = (788\ \text{yr}) y_o\, S_{48}^{1/3}\, n_5^{-2/3} \qquad (3)$$

for a pressure disturbance to cross the Strömgren sphere (Spitzer 1978). Here, S_{48} is the star's Lyman continuum photon luminosity in units 10^{48} s^{-1}, n_5 is the ambient density in units 10^5 cm^{-3}, and $y_o \lesssim 1$ is a factor accounting for Lyman continuum dust opacity (Petrosian, Field, and Silk 1972). As shown by Shull (1980), the mechanical input from the wind will dominate the subsequent dynamics if

$$L_{36} > (0.068)\, S_{48}^{2/3}\, y_o^2\, n_5^{-1/3} \quad , \qquad (4)$$

where

$$L_{36} = (1.27) \left[\frac{\dot{M}}{10^{-6}\ M_\odot\ \text{yr}^{-1}}\right] \left[\frac{V_w}{2000\ \text{km s}^{-1}}\right]^2 \qquad (5)$$

is the mechanical luminosity of the wind in units 10^{36} ergs s^{-1}. Note that both the slower molecular outflows and the fast O-star winds have $L_{36} \sim 0.1$ to 10.

Castor, McCray, and Weaver (1975) showed that after an early stage in which the stellar wind impacts directly on the ambient medium (Steigman, Strittmatter, and Williams 1975), the system evolves into a thermal-pressure-driven bubble. Applying this theory to the case of an O-star embedded in a molecular cloud, Shull (1980) found the radius, velocity, and X-ray free-free luminosity of the bubble to be

$$R_s = (0.165\ \text{pc})\, L_{36}^{1/8}\, n_5^{-7/8} (t/t_{cr})^{3/5} \qquad (6)$$

$$V_s = (9.64\ \text{km s}^{-1})\, L_{36}^{1/4}\, n_5^{1/4} (t/t_{cr})^{-2/5} \qquad (7)$$

$$L_{ff} = (4.4 \times 10^{35}\ \text{ergs s}^{-1})\, L_{36} (t/t_{cr})^{16/35} \qquad (8)$$

where

$$t_{cr} = (10^4\ \text{yr})\, L_{36}^{-1/8}\, n_5^{-9/8} \qquad (9)$$

is the time at which the shell expansion stalls due to interior cooling. The bubble interior has a density and temperature

$$n_b = (159\ \mathrm{cm}^{-3})\ L_{36}^{1/4}\ n_5^{5/4}(t/t_{cr})^{-22/35} \qquad (10)$$

$$kT_b = (677\ \mathrm{eV})\ L_{36}^{1/4}\ n_5^{1/4}(t/t_{cr})^{-6/35} \quad . \qquad (11)$$

After the time t_{cr}, the shell will again be driven by the wind momentum. Once it encounters a region of lower density, the shell will likely fragment via the Rayleigh-Taylor instability. The resulting fragments could form the masers or HH-objects (cf. Kahn 1974).

The gas outside the bubble will be fully ionized in a thin shell, preceded by an extended "jacket," partially ionized by the more penetrating X-rays. Further ahead, the hydrogen will become molecular in a broad "dissociation front" (Hill and Hollenbach 1978; London 1978). Because the H_2 may be dissociated by Lyman and Werner band absorption (912-1120 Å), the details of this dissociation front depend on the dust opacity and line self-shielding (Hollenbach, Werner, and Salpeter 1971; Shull 1978). The gas near the edge of this front may be heated by conversion of the UV absorption in the Lyman-Werner bands and by the translational energy of the H_2 newly formed on grains (Black and Dalgarno 1976; Hunter and Watson 1978). The effects of the stellar wind on this structure are difficult to distinguish. The best indicators are probably shell structure in radio recombination lines or radio continuum emission of compact H II regions (Harris and Wynn-Williams 1976; Dreher and Welch 1981), or detection of the X-rays from the hot interior (either directly or by a search for IR fine structure lines of highly ionized ions, such as [Ar V] or [S IV]).

We now consider deviations from spherical symmetry, which are obviously required to explain bipolar nebulae, Herbig-Haro objects, high velocity masers, and Orion-type molecular outflows. The H_2 vibrational emission near 2.2 μm was modeled initially by 10-15 km s^{-1} shocks into molecular clouds (Hollenbach and Shull 1977; Kwan 1977; London, McCray, and Chu 1977). (For an alternative view involving magnetic precursors, see Draine [1980].) The relationship of the H_2 to the 100 km s^{-1} CO poses a theoretical problem, since H_2 should be destroyed in dense shocks with $V \gtrsim 35$ km s^{-1} (Hollenbach and McKee 1980). Perhaps, the 40-80 km s^{-1} wings of the H_2 (1-0) S(1) line are formed in dense blobs in a general outflow colliding with the ambient cloud (see Fig. 2). The H_2 line width would then characterize the bulk flow, whereas individual blobs would contain shocks of lower velocity. This hybrid shock model would also explain the fact that the centroid of the high velocity emission has the same velocity as the ambient cloud (Knapp et al. 1981).

The X-rays observed from O-star winds (Harnden et al. 1979) are also best understood by a model in which the wind breaks up into shocked blobs, created perhaps by a "drift instability" connected with radiation pressure on lines (Lucy and Solomon 1970; Lucy and White 1980). There is a wide range in scales for these wind inhomogeneities. The blobs in O-star winds are estimated at 10^{15} g (Lucy and White 1980), based on the pressure scale height at the sonic point. The H_2 clumps probably have masses $\sim 10^{23}$ g, if one assumes that their shape is spherical and determined by the H_2 cooling length at the appropriate density (10^6 cm^{-3}). Herbig-Haro objects and H_2O masers are evidently much larger clumps; estimates for their mass range from 10^{26} to 10^{29} g for the line-emitting regions (Genzel and Downes 1977; Brugel, Böhm, and Mannery 1981b).

Norman and Silk (1978) and Cantó and Rodríguez (1980) suggest that HH objects and H_2O masers are compact clouds ejected at high velocity by the supersonic winds of pre-main sequence stars (see also Strel'nitskii and Sunyaev 1973). Observational evidence for the maser-HH object connection has been supplied by Rodríguez et al. (1980) and Haschick et al. (1980). In the "interstellar bullet" model of Norman and Silk, the ejection results from the Rayleigh-Taylor breakup of an accelerating, wind-driven shell, and the heat source for the masers arises from magnetic dissipation and reconnection. The Cantó-Rodríguez model relies on a pressure-density gradient near the edge of a molecular cloud to channel the wind through a nozzle at the far end of an ovoid cavity. The latter model has the advantage of providing a more efficient use of the wind pressure. A third model (Schwartz and Dopita 1980) envisages HH objects as condensations in the ambient molecular cloud, swept up and shocked by the wind.

Several recent observations are relevant here. First, the location of several HH objects inside the L1551 cavity (Snell et al. 1980) and the 100 to 350 km s^{-1} proper motions of HH 1, 2, 28, 29, and 39 (Cudworth and Herbig 1979; Herbig and Jones 1981) suggest that the HH objects are ejected blobs rather than cloud condensations or regions of focused flow. There also seems to be substantial dispersion in velocity among the various sub-condensations. Second, Ortolani and d'Odorico (1980), Böhm, Böhm-Vitense, and Brugel (1981), and Brugel, Böhm, and Mannery (1981a) have found that the continua of several HH objects rise steeply into the blue and UV. The total UV luminosity of HH 1 was found to be ~ 1 $L_\odot$, compared with 0.05 $L_\odot$ in optical emission! Third, Pravdo and Marshall (1981) have identified possible soft X-ray emission near HH 1 and 2. Traditionally, the optical emission line spectra of HH objects have been interpreted as radiative shocks of velocity 50-70 km s^{-1} (Schwartz 1978; Dopita 1978; Raymond 1979). However, the UV lines in HH 1 (C IV λ1549, C III] λ1909, Si IV λ1400) are indicative of $V_s \sim$ 100 km s^{-1} (Böhm et al. 1981; Shull and McKee 1979). The X-ray emission would require $V_s > 300$ km s^{-1}, while the H_2 infrared lines seen by Elias (1980) are characteristic of 15 km s^{-1} shocks. Clearly, a complex situation is involved. Even more disturbing is the fact that there is no simple way to explain the steeply rising UV continuum as emission from radiative shocks. The presence of an early-type star is doubtful, owing to the weakness of the optical continuum, Hβ and He II λ4686 and to the absence of a strong IR source. Non-thermal synchrotron radiation also seems unlikely, since its extension into the EUV is excluded by the same reasoning. Schmidt and Miller (1979) suggest that the optical continuum of HH 24A is formed by dust reflection, but this source is atypical (Brugel et al. 1981a). Furthermore, a steeply rising UV continuum is in conflict with the spectral type of any nearby exciting stars and with the dust albedo in the far UV. Conceivably, the variety of shock velocities required to explain the emission lines could result from density variations in the blob or from obliquities in the bow shock flow. The UV continuum could result from collisionally excited two-photon emission from H I, for which F_λ peaks near 1520 Å, and from He I and II. Non-steady flow or a fast, ionizing shock (Chevalier, Kirshner, and Raymond 1980) may be required to model this strong continuum (see Sec. IV for a more speculative model).

IV. SUMMARY AND SPECULATION

Through the variety of observational and theoretical evidence discussed in the preceding sections, it seems apparent that inhomogeneities are a ubiquitous feature of stellar winds. The range of masses is enormous, from 10^{15} g X-ray emitting blobs in O-star winds to 10^{29} g condensations seen as H_2O masers and HH objects. Most theoretical models tend to be *ad hoc*, relying primarily on analogy with other situations. Therefore, in the spirit of one Wall Street investment firm's philosophy that "what everyone knows is not worth knowing," I would like to conclude this review with a list of questions and with some wild-eyed speculations of my own.

(1) What is the driving source for pre-main sequence winds and molecular outflows?

a) Is the outflow driven by a pulsational instability of an over-massive pre-main-sequence star (an η Carinae phenomenon)?

b) If the source is radiation pressure, how does one understand the ratio of flow momentum to L/c?

c) Alternatively, do protostars rid themselves of angular momentum and magnetic field by driving a wind containing 1 to 5% of the star's luminosity?

(2) Do stellar wind inhomogeneities result from a common mechanism (Rayleigh-Taylor instability or drift instability of line-driven winds)?

(3) Can one understand the excitation mechanism of the lines and continua of HH objects with a single, locally-produced model such as radiative shocks?

(4) Are the blobs and outflows collimated intrinsically by the wind or by surrounding gas (e.g. accretion disks or ovoids)?

(5) Are the blobs ejected in the plane of a circumstellar disk or perpendicular to it?

(6) Is a substantial amount of wind momentum carried by the blobs? How do these blobs affect the spectral and dynamical interactions with ambient interstellar gas?

Some of these questions have already been addressed in the literature. For example, Castor (1981) has constructed models of Wolf-Rayet mass loss in which the wind momentum exceeds L/c by the observed factor of 5 to 10. Each photon is scattered several times off resonance lines distributed stochastically in frequency. Solomon et al. (1981) suggest that the molecular outflows are driven by optically thick IR radiation pressure on dust. Elmegreen and Morris (1979) suggest that masers are created by the interaction of a strong protostellar wind with an accretion disk. MacGregor, Hartmann, and Raymond (1979) and Nelson and Hearn (1978) have considered the amplification of sound waves in stellar winds by drift and Rayleigh-Taylor instabilities as a means of obtaining non-radiative heating from inhomogeneities. However, the 10^{15} range in mass between O-star blobs and HH objects suggests that several mechanisms may be operating. Main-sequence winds are probably driven by radiation pressure on UV resonance lines (Castor, Abbott, and Klein 1975), but there may well be a non-radiative source for pre-main sequence mass ejection.

In view of the large (planet-sized) masses inferred for masers and HH objects and the uncertain source of their high velocities and collimation, I would like to suggest an alternative exotic origin. Suppose that these objects really are planets or cometary objects, ejected from the fragmenting protostar or protocluster in its attempt to shed excess mass and angular momentum. (This ejection could be a result of pulsational instability, magnetic torques, centrifugal forces, or a gravitational slingshot.) The stellar wind would help accelerate these loosely bound condensations, and the collimation could result either from channeling of the flow or from a preferential ejection plane. The large dispersion in velocity among the maser or HH sub-condensations would result from the breakup of these cometary systems. Pursuing this connection one step further, one might attribute the puzzling UV continuum emission in some HH objects to the excitation of molecular bands in cometary material. Note that a 2000 Å wide emission bump near 6500 Å seen in the bipolar nebula near HD 44179 ("The Red Rectangle") was attributed to such effects (Schmidt, Cohen, and Margon 1980).

Admittedly, this is a speculative model, which may prove unnecessary if the bursting wind-driven shell picture proves adequate to explain masers and HH objects. However, in either case, the deposition of substantial wind momentum in dense condensations has important consequences for interstellar gas dynamics and stellar mass distributions. For instance, the discrepancy between shell momentum and wind input seen in several Wolf-Rayet star ring nebulae (Treffers and Chu 1981) could result from such "leaks." Also, the ratio of primary to secondary masses in O-star systems could be affected by the winds and molecular outflows of the pre-main sequence phase (cf. Garmany, Conti, and Massey 1980). Future spectral studies of ejecta and a correlation of proper motions with the intensity and orientation of IR/radio emission contours may help unravel some of the mysteries of these protostellar phenomena.

I am grateful to Ed Brugel, Bruce Elmegreen, Gill Knapp, and Dick McCray for their valuable help in preparing this review.

REFERENCES

Abbott, D.C., Bieging, J.H. and Churchwell, E.: 1981, Astrophys. J., 250, in press.
Beckwith, S.E., Persson, S.E., Neugebauer, G. and Becklin, E.E.: 1978, Astrophys. J., 223, p. 464.
Black, J.H. and Dalgarno, A.: 1976, Astrophys. J., 203, p. 132.
Böhm, K.H., Böhm-Vitense, E. and Brugel, E.W.: 1981, Astrophys. J. (Letters), 245, p. L113.
Brugel, E.W., Böhm, K.H. and Mannery, E.: 1981a, Astrophys. J., 243, p. 874.
Brugel, E.W., Böhm, K.H. and Mannery, E.: 1981b, Astrophys. J. Suppl., 51, in press.
Bruhweiler, F.C., Gull, T.R., Kafatos, M. and Sofia, S.: 1980, Astrophys. J. (Letters), 238, p. L27.
Cantó, J. and Rodríguez, L.F.: 1980, Astrophys. J., 239, p. 982.
Cantó, J., Rodríguez, L.F., Barral, J.F. and Carral, P.: 1981, Astrophys. J., 244, p. 102.
Cash, W., Charles, P., Bowyer, S., Walter, F., Garmire, G. and Riegler, G.: 1980, Astrophys. J. (Letters), 238, p. L71.
Castor, J.I.: 1981, in I. Iben and A. Renzini (eds.), Physical Processes in Red Giants, D. Reidel Publ. Co., Dordrecht, Holland, p. 285.
Castor, J.I.: 1981, unpublished work.
Castor, J.I., Abbott, D.C. and Klein, R.I.: 1975, Astrophys. J., 195, p. 157.
Castor, J.I., McCray, R.A. and Weaver, R.: 1975, Astrophys. J. (Letters), 200, p. L107.
Chevalier, R.A., Kirshner, R.P. and Raymond, J.C.: 1980, Astrophys. J., 235, p. 186.
Cohen, M. and Schwartz, R.D.: 1979, Astrophys. J. (Letters), 233, p. L77.
Cohn, H. and York, D.G.: 1977, Astrophys. J., 216, p. 408.
Conti, P.S. and Garmany, C.D.: 1980, Astrophys. J., 238, p. 190.
Conti, P.S. and McCray, R.: 1980, Science, 208, p. 9.
Cowie, L.L., Songaila, A. and York, D.G.: 1979, Astrophys. J., 230, p. 469.

Cudworth, K.M. and Herbig G.H.: 1979, Astron. J., 84, p. 548.
Dopita, M.A.: 1978, Astrophys. J. Suppl., 37, p.117.
Draine, B.T.: 1980, Astrophys. J., 241, p. 1021.
Dreher, J.W. and Welch, W.J.: 1981, preprint.
Elias, J.H.: 1980, Astrophys. J., 241, p. 728.
Elmegreen, B.G. and Morris, M.: 1979, Astrophys. J., 229, p. 593.
Garmany, C.D., Conti, P.S. and Massey, P.: 1980, Astrophys. J., 242, p. 1063.
Garmany, C.D., Olson, G.L., Conti, P.S. and Van Steenberg, M.: 1981, Astrophys. J., 250, in press.
Gautier, T.N., Fink, U., Treffers, R.R. and Larson, H.P.: 1976, Astrophys. J. (Letters), 207, p. L129.
Genzel, R. and Downes, D.: 1977, Astron. Astrophys., 61, p. 117.
Genzel, R., Reid, M.J., Moran, J.M. and Downes, D.: 1981, Astrophys. J., 244, p. 884.
Harnden, F.R., Branduardi, G., Elvis, M., Gorenstein, P., Grindlay, J., Pye, J.P., Rosner, R., Topka, K. and Vaiana, G.S.: 1979, Astrophys. J. (Letters), 234, p. L51.
Harris, S. and Wynn-Williams, C.G.: 1976, Mon. Not. Roy. Astron. Soc., 174, p. 649.
Hartmann, L., Dupree, A.K. and Raymond, J.C.: 1980, Astrophys. J. (Letters), 236, p. L143.
Haschick, A.D., Moran, J.M., Rodriguez, L.F., Burke, B.F., Greenfield, P. and Garcia-Barreto, J.A.: 1980, Astrophys. J,. 237, p. 26.
Heiles, C.: 1979, Astrophys. J., 229, p.533.
Herbig, G.H. and Jones, B.F.: 1981, Astron. J., in press.
Hill, J.K. and Hollenbach, D.J.: 1978, Astrophys. J., 225, p. 390.
Hollenbach, D.J. and McKee, C.F.: 1980, Astrophys. J. (Letters), 241, p. L47.
Hollenbach, D.J. and Shull, J.M.: 1977, Astrophys. J., 216, p. 419.
Hollenbach, D.J., Werner, M.W. and Salpeter, E.E.: 1971, Astrophys. J., 163, p. 165.
Hunter, D.A. and Watson, W.D.: 1978, Astrophys. J., 226, p. 477.
Kahn, F.: 1974, Astron. Astrophys., 37, p. 149.
Knapp, G.R., Phillips, T.G., Hugggins, P.J. and Redman, R.O.: 1981, Astrophys. J., in press.
Kuhi, L.V.: 1964, Astrophys. J., 140, p. 1409.
Kuhi, L.V.: 1966, Astrophys. J., 143, p. 991.
Kwan, J.: 1977, Astrophys. J., 216, p. 713.
Kwan, J. and Scoville, N.: 1976, Astrophys. J. (Letters), 210, p. L39.
Lamers, H.J. and Snow, T.P.: 1978, Astrophys. J., 219, p. 504.
Linsky, J.L. and Haisch, B.M.: 1979, Astrophys. J. (Letters), 229, p. L27.
London, R.: 1978, Astrophys. J., 224, p. 405.
London, R., McCray, R. and Chu, S.-I.: 1977, Astrophys. J., 217, p. 442.
Lucy, L.B. and Solomon, P.M.: 1970, Astrophys. J., 159, p. 879.
Lucy, L.B. and White, R.L.: 1980, Astrophys. J., 241, p. 300.
MacGregor, K.B., Hartmann, L. and Raymond, J.C.: 1979, Astrophys. J., 231, p. 514.

McCray, R.A. and Snow, T.P.: 1979, Ann. Rev. Astron. Astrophys., 17, p. 213.
Meaburn, J.: 1979, Astron. Astrophys., 75, p. 127.
Meaburn, J.: 1980, Mon. Not. Roy. Astron. Soc., 192, p. 365.
Meaburn, J. and Walsh, J.R.: 1980, Mon. Not. Roy. Astron. Soc., 191, p. 5P
Nadeau, D. and Geballe, T.R.: 1979, Astrophys. J. (Letters), 230, p. L69.
Nelson, G.D. and Hearn, A.G.: 1978, Astron. Astrophys., 65, p. 223.
Norman, C. and Silk, J.I.: 1979, Astrophys. J., 228, p.197.
Olson, G.L. and Castor, J.I.: 1981, Astrophys. J., 244, p. 179.
Ortolani, S. and d'Odorico, S.: 1980, Astron. Astrophys., 83, p. L8.
Petrosian, V., Silk, J. and Field, G.B.: 1972, Astrophys. J. (Letters), 177, p. L69.
Phillips, T.G., Huggins, P.J., Neugebauer, G. and Werner, M.W.: 1977, Astrophys. J. (Letters), 217, p. L161.
Pravdo, S.H. and Marshall, F.E.: 1981, Astrophys. J., in press.
Raymond, J.C.: 1979, Astrophys. J. Suppl., 39, p. 1.
Reynolds, R.J.: 1976, Astrophys. J., 206, p. 679.
Reynolds, R.J.: and Ogden, P.M.: 1979, Astrophys. J., 229, p. 942.
Rodriguez, L.F., Moran, J.M., Ho, P.T.P. and Gottlieb, E.W.: 1980, Astrophys. J., 235, p. 845.
Schmidt, G.D., Cohen, M., and Margon, B.: 1980, Astrophys. J. (Letters), 239, p. L133.
Schmidt, G. and Miller, J.S.: 1979, Astrophys. J. (Letters), 234, p. L191.
Schwartz, R.D.: 1978, Astrophys. J., 223, p. 884.
Schwartz, R.d. and Dopita, M.A.: 1980, Astrophys. J. 236, p. 543.
Shull, J.M.: 1977, Astrophys. J., 212, p. 102.
Shull, J.M.: 1978, Astrophys. J., 219, p. 877.
Shull, J.M.: 1980, Astrophys. J., 238, p. 860.
Shull, J.M. and McKee, C.F.: 1979, Astrophys. J., 227, p. 131.
Shull, J.M. and York, D.G.: 1977, Astrophys J., 211, p. 803.
Snell, R.L., Loren, R.B. and Plambeck, R.L.: 1980, Astrophys. J. (Letters), 239, p. L17
Snow, T.P. and Marlborough, J.M.: 1976, Astrophys. J. (Letters), 203, p. L87
Snow, T.P. and Morton, D.C.: 1976, Astrophys. J. Suppl., 32, p. 429.
Solomon, P.M., Huguenin, G.R. and Scoville, N.Z.: 1981, Astrophys. J. (Letters), in press.
Spitzer, L.: 1978, "Physical Processes in the Interstellar Medium" (New York: Wiley-Interscience), Chap. 12.
Steigman, G., Strittmatter, P.A. and Williams, R.E.: 1975, Astrophys. J., 198, p. 575.
Stencel, R.E. and Mullan, D.J.: 1980, Astrophys. J., 238, p. 221.
Strel'nitskii, V.S. and Syunyaev, R.A.: 1973, Sov. Astron., 16, p. 579.
Treffers, R.R. and Chu, Y.-H.: 1981, private communication
Weaver, R., McCray, R., Castor, J., Shapiro, P. and Moore, R.: 1977, Astrophys. J., 218, p. 377.
Zuckerman, B., Kuiper, T.B.H. and Rodriguez-Kuiper, E.N.: 1976, Astrophys. J. (Letters), 209, p. L137.

DISCUSSION FOLLOWING REVIEW BY SHULL

LOREN: There are two clouds (Mon R2 and NGC 2071) which have bipolar stellar wind outflows where the velocity gradient of optically thin molecular emission shows that the channelled outflow lies along the rotational axis of the system.

UNDERHILL: When one is considering the ejection of blobs from stars, I think it is wise to note that the Sun is known to eject small amounts of material impulsively as a result of magneto-hydrodynamic events. Similar forces may be active, much more active, in the condensed peculiar objects which are being considered for powering flows in HII regions. The Sun is a good example of what types of force are active in nature; it is, however, a very feeble star so far as level of activity is considered. It is not obvious that one may rule out interactions between differential motion and magnetic fields as sources for outward motion of material in and around a star.

SHULL: It is a question of scale. What you have to explain is the ejection of planetary size objects. The Sun doesn't kick out Jupiters!

UNDERHILL: We know that the Sun is a very weak little star, but when you see certain forces in action on the Sun, I think you should seriously consider them in other stars.

SANDELL: I thought that the continuum in Herbig-Haro objects could be explained by reflection from hidden stars while the lines would be produced in situ, i.e. by shocks. Is this idea now totally abandoned?

SHULL: The explanation of Herbig-Haro emission as due to reflection was generally used before 1975. However, the exciting star does not have sufficient continuum and, secondly, the albedo of the grains varies in the wrong way.

A CALCULATION OF MOLECULE ABUNDANCES BEHIND SLOW SHOCKS

George F. Mitchell and Terry J. Deveau
Saint Mary's University, Halifax, Nova Scotia

ABSTRACT

Post-shock abundances of 105 chemical species are followed after the passage of a 10 kms^{-1} shock through an interstellar cloud of initial density 10^4 cm^{-3}. We find significant enhancement in the column densities of H, H_2O, NH_3, and HS. The column densities of most ions decrease in abundance by rather large factors.

1. INTRODUCTION

Numerous processes are known which must produce shocks in the interstellar gas and there is, indeed, considerable observational evidence of high temperature, high velocity post-shocked gas. At the high temperature of the shocked gas, endothermic chemical reactions can become rapid and will alter molecular abundances in the gas. The effects of interstellar shocks on molecular abundances have been studied by a number of authors (Elitzur and de Jong 1978; Elitzur and Watson 1978; Iglesias and Silk 1978; Elitzur 1979; Hollenbach and McKee 1979; Hartquist, Oppenheimer and Dalgarno 1980; Elitzur and Watson 1980). A major achievement of shock models has been an explanation of the ion CH^+ in diffuse clouds. Apart from this, however, the models have had rather limited success in accounting for observed molecular abundances in regions which are believed to be shocked.

HCO^+ is observed to be enhanced in abundance by a factor of 10 to 100 in IC 443 (Dickinson et al. 1980; de Noyer and Frerking 1981) but is unchanged in five other regions which show evidence of shock waves (Rodriguez-Kuiper and Dickinson 1981). The model of Iglesias and Silk (1978), on the other hand, predicts a decrease in the number density of HCO^+ of at least two orders of magnitude due to the shock. Elitzur (1979) has invoked a shock to account for the apparent high abundance of H_2O in Orion (Phillips et al. 1978). There are, however, cold gas-phase pathways which lead to a high water abundance (e.g. Mitchell,

R. S. Roger and P. E. Dewdney (eds.), Regions of Recent Star Formation, 107–116.
Copyright © 1982 by D. Reidel Publishing Company.

Ginsburg, and Kuntz 1978). In IC 443, OH is some 100 times more abundant than normal (de Noyer and Frerking 1981). In the shock models of Iglesias and Silk (1978) and of Hartquist, Oppenheimer and Dalgarno (1980), the OH abundance increases in number density for a brief time after the shock but then plummets as OH is converted to H_2O. In neither paper is the shocked column density of OH given, but it is likely smaller than the unshocked column density. In the wide velocity ("plateau") source in Orion, the CS abundance is very close to that in the unshocked cloud, $n(CS)/n \simeq 5 \times 10^{-10}$ (Goldsmith et al. 1980). Hartquist, Oppenheimer and Dalgarno (1980) predict an enhancement of CS to $n(CS)/n \simeq 5 \times 10^{-8}$.

In this paper we present results of a calculation in which a shock of 10 kms^{-1} passes through a cloud whose initial density is 10^4 cm^{-3}. We use as starting values the molecular abundances of Prasad and Huntress (1980) and follow the post-shocked abundances for a few times 10^4 years. Details of the shock model and the chemical reaction network are given in section 2 and the results are discussed in section 3.

2. THE SHOCK MODEL AND CHEMICAL SYSTEM

We use the equations of a one-dimensional shock adapted from Field et al. (1968) and Hartquist, Oppenheimer and Dalgarno (1980). With an appropriate cooling rate, these equations permit the determination of the post-shock temperature and density as a function of time.

H_2 rotational-vibrational cooling rates are taken from Hartquist, Oppenheimer, and Dalgarno (1980). We use the rotational cooling expression of Hollenbach and McKee (1979) which includes the effects of self absorption and dust opacity, for cooling due to H_2O, CO, OH, and CH. Cooling by atomic forbidden lines is included but never becomes significant. Two heating mechanisms were incorporated: cosmic ray heating and grain formation of H_2.

Our unshocked cloud is given the chemical abundances of the equilibrium dark cloud (cloud 4) model of Prasad and Huntress (1980). Our chemical reaction library consists essentially of the 1423 reactions listed by Prasad and Huntress (1980) augmented by 160 high temperature reactions taken from various sources. Selected chemical species (105 in this case) are passed as input to our computer program which extracts all reactions involving these species from the library. The system of differential equations represented by these rates is integrated numerically using the GEARS package. Each time step in the integration uses the abundances from the previous step with a new temperature and density calculated from the shock equations.

3. RESULTS AND DISCUSSION

3.1 Shock profile and cooling rates

The density and temperature profiles are shown in Figure 1. The immediate post-shock temperature is 4160 K and the immediate post-shock particle density is 5.6×10^4 cm^{-3}. In about 30 years the shocked gas has cooled to 30 K. Although the cooling processes, as formulated here, would continue to cool the gas, we arbitrarily keep the temperature at 30 K from this time on. This has the additional effect of freezing the density. After 30 years, the density has reached $\simeq 10^7$ cm^{-3}. Such large compression will not occur in the presence of magnetic fields. A useful project for the future will be the inclusion of a magnetic field in the shock calculation.

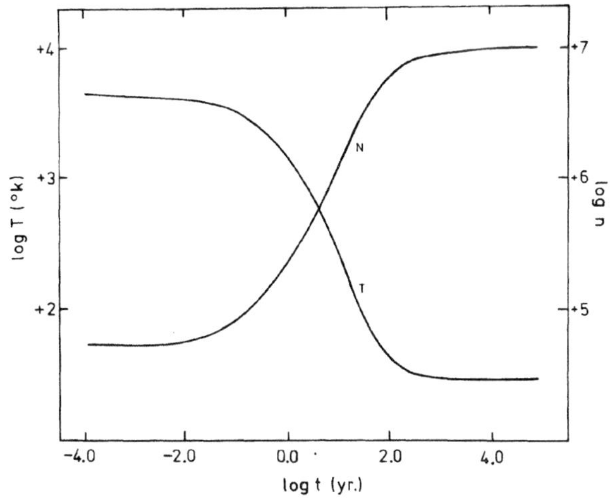

Figure 1. Temperature and density profiles

Molecular cooling rates are shown in Figure 2. At high temperatures, H_2 is the dominant coolant. About one year after the shock, H_2O becomes the most important coolant.

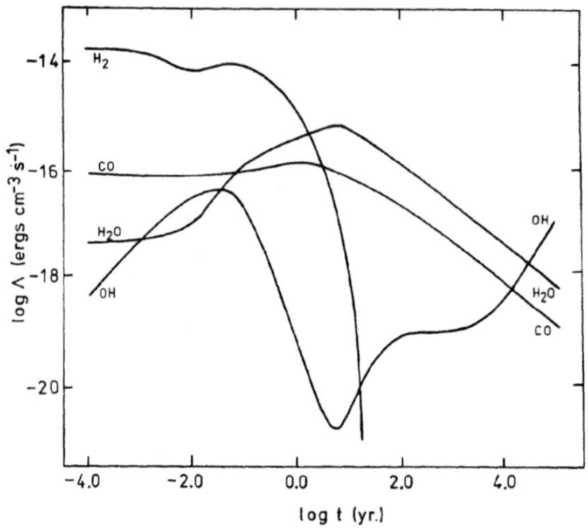

Figure 2. Molecular cooling rates

3.2 Molecular abundances behind the shock

Figures 3-6 display molecular abundances behind the shock. The major oxygen-bearing species are shown in Figure 3. For these species our results are in qualitative agreement with previous work (Iglesias and Silk 1978; Hartquist, Oppenheimer and Dalgarno 1980). CO/n remains constant, H_2O/n increases to a relative abundance of $\approx 10^{-4}$, while O_2/n and O/n both decrease. OH/n initially increases but drops off quickly as it is converted into H_2O.

The atomic hydrogen abundance increases rapidly to high value (H/n $\simeq 7 \times 10^{-4}$) due to collisional dissociation of H_2. Below about 3000 K, H is formed primarily by

$$H_2 + OH \rightarrow H_2O + H$$

and is destroyed by

$$NH + H \rightarrow N + H_2$$

and

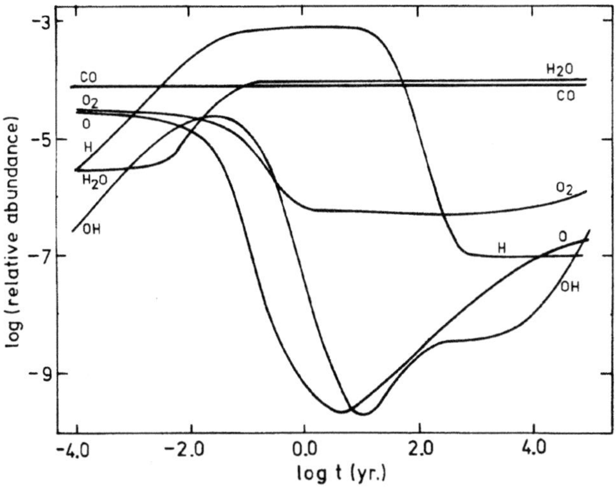

Figure 3. Post-shock fractional abundances of atomic hydrogen and the major oxygen-bearing species.

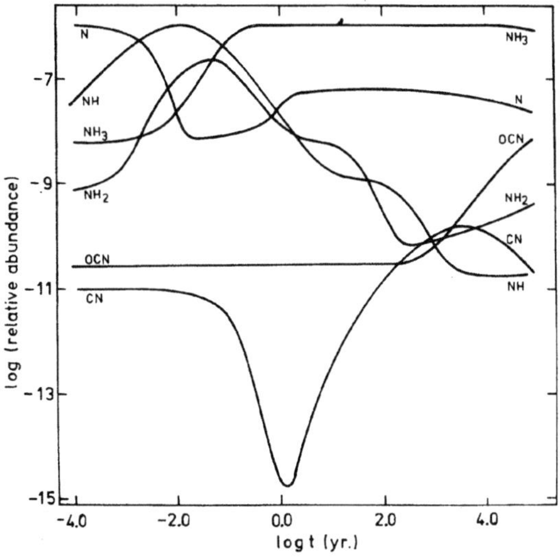

Figure 4. Fractional abundances of nitrogen species.

$$HS + H \rightarrow S + H_2$$

The rapid decrease in H/n after $\simeq$ 100 years is due to the switching off of endothermic reactions and the subsequent grain formation of H_2. Within less than 1000 years, a new equilibrium is established between formation processes, such as

$$H_3O^+ + e \rightarrow H_2O + H$$

and

$$H_2^+ + H_2 \rightarrow H_3^+ + H,$$

and destruction by conversion into H_2 on grains.

Several nitrogen species are shown in Figure 4. Perhaps the most interesting process illustrated by Figure 4 is the conversion of much of the available atomic nitrogen into ammonia. The following reaction sequence occurs:

$$N + H_2 \rightarrow NH + H,$$

$$NH + H_2 \rightarrow NH_2 + H$$

$$NH_2 + H_2 \rightarrow NH_3 + H.$$

While NH and NH_2 briefly achieve a high abundance, by several hundred years after the shock they have declined in abundance to $NH/n \simeq NH_2/n \simeq 10^{-10}$. Ammonia, on the other hand, increases from $NH_3/n = 7 \times 10^{-9}$ to $NH_3/n \simeq 10^{-6}$ and maintains the new high abundance for the 30,000 years of our calculation.

The abundance of CN drops to a very low level in the first year due to the reaction

$$CN + H_2 \rightarrow HCN + H.$$

The formation mechanisms, $NO + C \rightarrow CN + O$ and $H_2CN^+ + e \rightarrow CN + 2H$, cannot compete with the destruction (by H_2) until the temperature has dropped to about 300 K. At low post-shock temperatures the main destruction process is

$$CN + O_2 \rightarrow OCN + O.$$

Figure 5 displays abundances of several sulphur species. Particularly interesting is the behaviour of S, HS, and H_2S. Their abundances at high temperature are controlled by the reactions:

$$S + H_2 \rightarrow HS + H,$$

$$HS + H \rightarrow S + H_2,$$

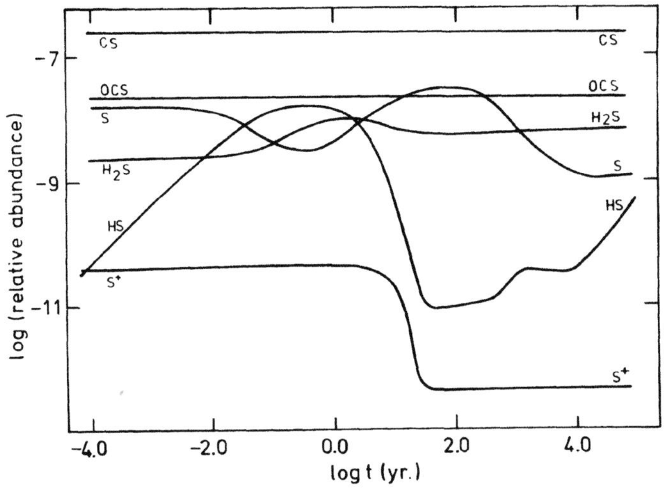

Figure 5. Fractional abundances of sulphur species.

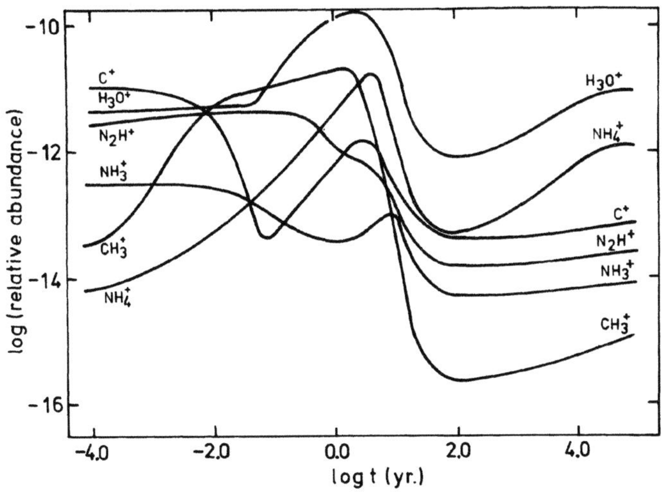

Figure 6. Fractional abundances of some major ions.

$$HS + H_2 \rightarrow H_2S + H,$$

$$H_2S + H \rightarrow HS + H_2.$$

The abundance of atomic sulphur first decreases as HS and H_2S increase in abundance. As the temperature falls, this trend reverses, with atomic sulphur increasing again in abundance as HS/n sharply drops off. In contrast to HS, the abundance of H_2S does not change significantly, increasing from $H_2S/n \simeq 2\ 10^{-9}$ to $\simeq 5 \times 10^{-9}$. At low post-shock temperatures the abundance of HS and H_2S is determined by other reactions, such as:

$$H_2S^+ + H_2O \rightarrow H_3O^+ + HS,$$

$$HS + O \rightarrow SO + H,$$

$$H_3^+ + H_2S \rightarrow H_3S^+ + H_2,$$

$$H_3S^+ + e \rightarrow H_2S + H,$$

$$H_2S^+ + Mg \rightarrow Mg^+ + H_2S.$$

Many of the major ions in the model decrease in abundance as the compression procedes and stabilizes at a relative abundance roughly 100 times lower than their initial abundance. Examples are HCO^+, H^+, and He^+ which decrease in relative abundance from $\simeq 10^{-10}$ to 10^{-12}. This behaviour can be attributed to the density increase alone, since, in cold cloud models, the abundance of most ions is inversely proportional to the gas density (e.g. Table 7 of Mitchell, Ginsburg, and Kuntz 1978). In Figure 6 we show a number of ions which behave in a more complex fashion. Particularly anomalous are H_3O^+ and NH_4^+ which reach an abundance maximum near one year and appear to stabilize at post-shock abundances higher than their pre-shock values. The high H_3O^+ abundance is attributable to the increase in H_2O/n, since one of the major formation mechanisms for H_3O^+ is

$$H_3^+ + H_2O \rightarrow H_3O^+ + H.$$

The higher NH_4^+ abundance has two causes. Most important is the reaction

$$NH_3^+ + H_2 \rightarrow NH_4^+ + H,$$

which proceeds quite rapidly at high temperatures. Secondly, the higher NH_3 abundance will increase NH_4^+/n via the reaction

$$H_3O^+ + NH_3 \rightarrow NH_4^+ + H_2O.$$

We find that formaldehyde is affected little by the shock, its relative abundance declining from 5.4×10^{-10} initially to 4.6×10^{-10} after 10^4 years. This result is in disagreement with Iglesias and Silk (1978) who found that formaldehyde was almost totally converted to formyl (HCO) by the reaction

$$H_2CO + H \rightarrow HCO + H_2$$

Since we include this reaction, the reason for the difference is not clear. Iglesias and Silk also predicted an enhancement in the abundance of HCO. We find, instead, that HCO/n decreases to a very low value ($\simeq 10^{-13.7}$) during the high temperature post-shock phase, due to the reaction

$$HCO + H \rightarrow CO + H_2$$

(Prasad and Huntress 1980). After the gas cools, however, the abundance of HCO quickly increases to somewhat more than its pre-shock value due to the charge exchange

$$HCO^+ + Mg \rightarrow Mg^+ + HCO$$

so that the column density is not strongly affected (Table 1 below).

3.3 Column densities

While the abundance profiles are of considerable interest, they are not directly observable. The observed quantity is the column density of molecules along the line of sight. To facilitate comparison with observations we present in Table 1 column densities through a cloud of initial thickness 0.1 pc and of initial density 10^4 cm^{-3}. Such a cloud consists of a thin slab viewed face-on. The shocked column densities of Table 1 represent the column densities through the cloud after a 10 kms^{-1} shock has just left, having entered about 10^4 years before.

We see from Table 1 that the column densities of O_2 and O are much reduced by the shock, while $N(H_2O)$ is correspondingly increased and N(OH) is unchanged to within a factor of 2. The column density of atomic hydrogen is significantly increased.

Ammonia is predicted to have its column density increased by more than two orders of magnitude. In contrast, atomic nitrogen and the hydrides NH and NH_2 should have slightly reduced column densities.

The shock has increased N(HS) by an order of magnitude, but $N(H_2S)$ has increased by only a factor of two.

Ions tend to have smaller column abundances, often by two orders of magnitude.

Table 1: Column abundances before and after the shock

Species	N(pre-shock) (cm^{-2})	N(post-shock) (cm^{-2})	Species	N(pre-shock) (cm^{-2})	N(post-shock) (cm^{-2})
O_2	1.6(17)	3.0(15)	OCN	1.5(11)	2.0(12)
O	1.6(17)	2.6(14)	S	7.4(13)	1.7(13)
H_2O	1.4(16)	4.7(17)	HS	2.1(10)	1.7(11)
OH	1.7(13)	3.2(13)	H_2S	1.2(13)	2.8(13)
H	6.8(15)	4.1(16)	OCS	8.3(13)	1.7(14)
N	6.4(15)	2.9(14)	Mg^+	3.3(14)	1.1(14)
NH	1.2(12)	4.7(11)	H_3^+	2.6(11)	2.3(09)
NH_2	3.0(12)	8.3(11)	HCO^+	2.4(11)	1.4(09)
NH_3	3.4(13)	6.1(15)	C^+	6.7(10)	3.0(08)
HCO	3.5(11)	5.8(11)	He^+	3.5(10)	6.6(08)
CN	4.6(10)	6.2(11)	N_2H^+	2.3(10)	1.4(08)

Acknowledgement: This work was partially supported by an operating grant from the Natural Sciences and Engineering Research Council.

REFERENCES

De Noyer,L.K. and Frerking,M.A.:1981,Astrophys.J.(Letters)246,p.L37.
Dickinson, D.F., Rodriguez-Kuiper, E.N., St. Clair Dinger, A. and Kuiper, T.B.H.: 1980,Astrophys.J.(Letters)237,p.L43.
Elitzur,M.:1979,Astrophys.J.229,p.560.
Elitzur,M. and de Jong,T.:1978,Astron.Astrophys.67,p.323.
Elitzur,M. and Watson,W.D.:1978,Astrophys.J.(Letters)222,p.L141
Elitzur,M. and Watson,W.D.;1980,Astrophys.J.236,p.172.
Field,G.B.,Rather,J.D.G.,Aannestad,P.A. and Orszag,S.A.:1968, Astrophys. J.151,p.953.
Goldsmith,P.F.,Langer,W.D.,Schloerb,F.P. and Scoville,N.Z.: 1980, Astrophys.J.240,p.524.
Hartquist,T.W.,Oppenheimer,M. and Dalgarno,A.:1980,Astrophys.J.236, p.182.
Hollenbach,D. and McKee,C.F.:1979,Astrophys.J.(Suppl.)41,p.555.
Iglesias,E.R. and Silk,J.:1978,Astrophys.J.226,p.851.
Mitchell,G.F.,Ginsburg,J.L. and Kuntz,P.J.:1978,Astrophys.J. (Suppl.) 38,p.39.
Phillips,T.G.,Scoville,N.Z.,Kwan,J.,Huggins,P.J. and Wannier,P.G.: 1978,Astrophys.J.(Letters)222,p.L59.
Prasad,S.S. and Huntress,W.T.,Jr.:1980,Astrophys.J.(Suppl.)43,p.1.
Rodriguez-Kuiper,E.N. and Dickinson,D.F.:1981,paper presented at 157th meeting of Am.Astron.Soc., Albuquerque, New Mexico.

GRAVITATIONAL INSTABILITIES IN SHOCK COMPRESSED GAS LAYERS

Gary L. Welter
Sonderforschungsbereich Radioastronomie
Auf dem Hügel 71
5300 Bonn 1, Federal Republic of Germany

ABSTRACT

In previous studies of the stability properties of shock compressed gas layers, Elmegreen and Elmegreen (1978) and Welter and Schmid-Burgk (1981) made the computational simplification of treating the layer as stationary and pressure bounded rather than moving and shock bounded. In the present paper we reconsider the problem taking proper account of the layer's motion. Our results verify that the simpler proceedure yields fairly accurate results.

I. INTRODUCTION

Observational evidence for the occurance of shock induced star formation has been reviewed by Elmegreen and Lada (1977) and Elmegreen and Elmegreen (1978, hereafter EE). Based on the observational evidence, which does seem fairly convincing, Elmegreen and Lada discussed the qualitative features expected for such a process occurring in the shocked layer of gas preceeding an H II region as the region expands into a molecular cloud. EE extended the theory by modeling layers of shocked gas as self-gravitating, pressure-bounded, isothermal, plane-paralled gas sheets, and then studied the dispersion relation for unstable perturbations of such sheets. Welter and Schmid-Burgk (1981, hereafter WSB) performed similar calculations for the case of curved sheet geometry.

The applicability of the calculations presented by EE and WSB to the phenomenon of shock induced star formation is dependent on two assumptions: a) that the growth time of a given perturbation mode should be much shorter than 1) the time scale for the evolution of the shocked layer that is being modeled, and 2) the time required for the layer to expand by a perturbation scale length; and b) that the use of boundary conditions which do not allow for the passage of material through the simulated front does not introduce significant effects on the dispersion relation being studied. One can easily verify that the

R. S. Roger and P. E. Dewdney (eds.), Regions of Recent Star Formation, 117–122.
Copyright © 1982 by D. Reidel Publishing Company.

solutions obtained by EE and WSB do satisfy assumption (a) for typical model parameters in the modes of principal interest. To test assumption (b), however, one must repeat the calculations using the appropriate shock front boundary conditions. In this paper we present the dispersion relation which results from such a calculation, and a brief description of the model to which it applies. A fuller description will be made available later in Welter (1981).

II. THE MODEL - AN OVERVIEW

The calculation assumes plane-stratification for the unperturbed model. We imagine a hot, tenuous gas expanding against a cold cloud, driving a shocked layer ahead of it. The interface between the hot medium and the cloud is assumed to move at constant velocity, V. The gas cools rapidly behind the shock, forming a layer whose density is substantially greater than the ambient cloud density. For such compression, we require $V \gg c$, c being the speed of sound in the shocked layer. For the results discussed in this paper we have taken $V=10c$. We assume that the compression is great enough that the layer feels essentially only its own gravitational field, with negligible contribution from the cloud. Given the above conditions, the layer will be essentially in gravitational equilibrium at all times (see EE or Welter). The density distribution within the layer is given by the Ledoux (1951) stratification:

$$\rho = \rho_c(1-\mu^2), \quad -A<\mu<A, \tag{1}$$

$$\mu = \tanh(\tilde{z}/H), \tag{2}$$

$$H = c/(2\pi G\rho_c)^{1/2}, \tag{3}$$

$$A^2 = t^2/((\pi\rho_0 G/2)^{-1} + t^2), \tag{4}$$

where ρ_c is the density at the layer's center, $\tilde{z}$ measures the distance from that center, t is the time since the shock was first driven into the cloud, ρ_0 is the cloud's ambient density, and G is Newton's constant. See Welter (1981) for a derivation of equation (4). For normalization purposes, we define a length scale H_0 given by

$$H_0 = c/(2\pi G\rho_0)^{1/2}. \tag{5}$$

We apply a perturbation to the above model. The linearized hydrodynamic perturbation equations are

$$\partial\vec{v}/\partial t + \vec{u}\cdot\nabla\vec{v} + \vec{v}\cdot\nabla\vec{u} = -\nabla\psi - c^2\nabla(q/\rho), \tag{6}$$

$$\partial q/\partial t + \vec{u}\cdot\nabla q + \vec{v}\cdot\nabla\rho = -q\nabla\vec{u} - \rho\nabla\cdot\vec{v}, \tag{7}$$

$$\nabla^2\psi = 4\pi G q, \tag{8}$$

where $\vec{v}$, q, and ψ represent the perturbation of velocity, density, and potential; and $\vec{u}$ represents the unperturbed velocity, which is of order c^2/V. We assume a solution for the perturbation quantities, Q, of the form

$$Q = Q(z,t)\exp(ikx), \tag{9}$$

where z is measured from the hot-gas/shocked-layer interface, and x is a coordinate measured parallel to the layer. We solve equations (6)-(8) for $\vec{v}$, q, and ψ using a 1-D hydrodynamics computer code, starting with the initial conditions

$$A^2(t) = .2, \tag{10}$$

$$q = q_0(1 - |\tilde{z}|/2W), \tag{11}$$

$$\vec{v} = 0, \tag{12}$$

where W, the layer width at the beginning of the calculation, equals $.089H_0$. $A^2=.2$ corresponds to a time when $W \approx H$.

The boundary conditions at the shock front are 1) conservation of mass, 2) conservation of momentum, and 3) smooth and continuous variation of the full potential through the shock. The trailing edge boundary conditions are exactly as described by EE for an impenetrable interface. We further require that the perturbation tends to zero at $z \to \pm\infty$. The solution for the perturbation in the unshocked cloud will be described in Welter (1981).

We find that the perturbation grows essentially exponentially, with a mild oscillation superimposed as a consequence of the particular form of the arbitrarily chosen initial perturbation. We plot the exponential growth rate, Ω, vs. k as curve (c) in Figure 1. For comparison we include the dispersion curves for stationary, pressure bounded layers which correspond with the moving layer's density distribution at the beginning of the calculation (curve(a), $t=H_0/c$, A=.45) and near the end (curve(b), $t=1.55H_0/c$, A=.61). We see that the effect of using the proper boundary conditions at the shock front is to suppress slightly the instability; the critical wavenumber above which modes are stable is reduced by about 30%, while the peak growth rate remains approximately the same. The changes in the curve are small when compared with the idealizations used in setting up the model initially. We conclude that using a stationary, pressure bounded gas layer as a model for a moving, shock bounded layer does not introduce significant error in calculating dispersion relations for graviationally driven instabilities. The calculations by Elmegreen and Elmegreen (1978) and Welter and Schmid-Burgk(1981) which used such a procedure may thus be taken more reliably as accurate representations of the evolution of shocked gas layers.

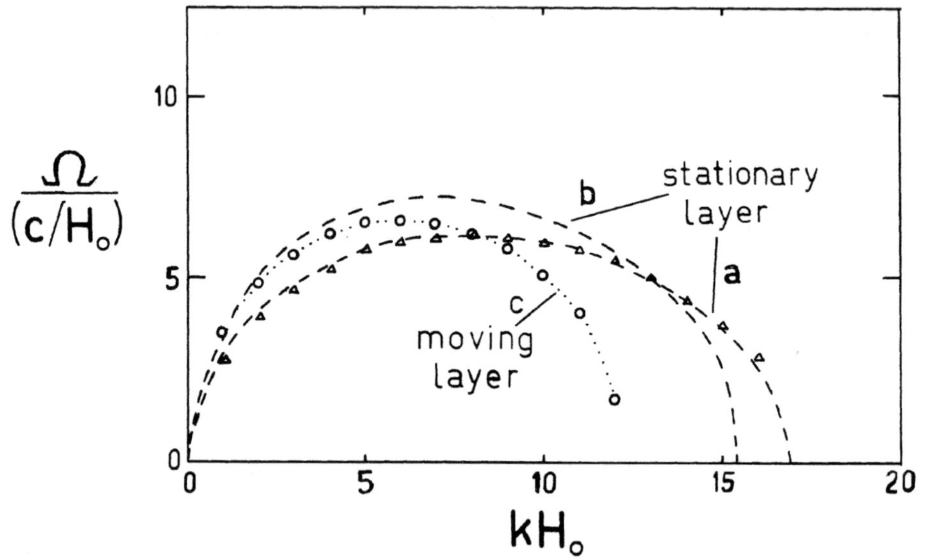

Figure 1: Dispersion relations for self-gravitating gas layers. The dashed curves are interpolations taken from Elmegreen and Elmegreen (1978) for stationary, pressure bounded layers with a) A=.45, and b) A=.61. The triangles and circles represent points calculated by the methods described in the text. The triangles are for a stationary layer with A=.45; their close fit to curve (a) confirms the accuracy of the numerical methods. The circles were calculated for the case of a moving, shock bounded layer. During the time of the calculation, the layer evolves from t=1 (A=.45) to t=1.65 (A=.64), units of time being H_0/c.

REFERENCES

Elmegreen, B.G. and Elmegreen, D.M.: 1978, Astrophys. J. 220, 1051 (EE).
Elmegreen, B.G. and Lada, C.J.: 1977, Astrophys. J. 214, 725.
Ledoux, P.: 1951, Ann.d'ap, 14, 438.
Welter, G. and Schmid-Burgk, J: 1981, Astrophys. J. 245, 927 (WSB).
Welter, G.L.: 1981, in preparation, to be submitted to Astron. Astrophys.

This work was supported by the Deutsche Forschungs Gemeinschaft through the Sonderforschungsbereich Radioastronomie.

DISCUSSION FOLLOWING PAPER BY WELTER

LARSON: We have heard a great deal about the Elmegreen-Lada mechanism of successive star formation. Could the evidence that has been put forth in support of this theory equally well be interpreted as evidence that the propagation of strong shocks into molecular clouds, driven for example by the expansion of HII regions, actually acts to terminate rather than to initiate star formation? There is evidence that in many groups of young stars, star formation has not occurred in a single brief burst but has continued for a period of at least 10^7 years. For example, this is true in the Orion Nebula cluster. The age of such a region as determined from the most massive stars on the main sequence is then just the age of the youngest stars present, and really measures the time when star formation stopped rather than when it started. Whatever theoretical arguments one may put forth that shock compression can trigger star formation, it is equally plausible that a violent disturbance such as a strong shock would disrupt a molecular cloud and disperse the gas rather than compress any significant fraction of it into stars. A variety of disruptive hydrodynamical instabilities exist which would tend to have this effect, and it is not clear that, in a real situation, idealized gravitational instability analyses are ever very relevant. Thus I wonder whether the Elmegreen-Lada theory is actually a good theory to explain the termination, rather than the initiation, of star formation in regions where star formation has previously been going on for an extended period of time.

WELTER: Personally, I am somewhat prejudiced in favour of the picture presented by Elmegreen and Lada. It seems to me that if the passage of a shock stops star formation, one would expect to observe O stars of all ages laced throughout molecular clouds, rather than concentrated towards the edges.

HARTEN: There is some evidence, in the Cepheus region for example, for a progression of age spatially - from older to younger aggregates with the remaining neutral aggregate at one end of the chain.

LARSON: Could this evidence for a progressive sequence of bursts of star formation equally well be interpreted as evidence for the progressive cessation of star formation in those regions?

HUGHES: When we consider star formation we must remember the very large range of mass and luminosity and also the large range of time of formation. When we talk of star formation by the Elmegreen-Lada process, I assume we are talking about O-type stars because they are the ones that produce ionization fronts. How does one consider the later type stars whose formation will take much longer than it takes for the shocks to move through the cloud?

SANDFORD: The formation of low-mass objects is not addressed by the E-L mechanism, but can be considered as a natural consequence of O-B star formation which subsequently produces ionization-shock compressions of cloud clumps. (See paper by Sandford, Whitaker and Klein, this volume) In this view, star formation can be regarded as occurring continuously and as terminating when the O-B stars that drive ionization-shock compressions no longer exist.

MAGNETIC FIELDS AND THE EVOLUTION OF SHOCKED GAS CLOUDS

Johann Nittmann
Department of Applied Mathematical Studies
University of Leeds, Leeds, LS2 9JT, England

ABSTRACT

We are studying the influence of large scale interstellar magnetic fields on the early evolution of a high density gas cloud which is hit by a strong shock wave. The incident shock is assumed, a priori, to be driven by a spiral density wave. Results are presented for the flows which develop in the interstellar gas with magnetic field strength of $1\mu G$ and $3\mu G$, respectively.

1. INTRODUCTION

In contrast to magnetic fields in stars, where, in general, it is known that magnetic fields have energies which are small compared with gravitational energy, magnetic fields in the interstellar medium could easily be associated with energies which are in equipartition or larger than the kinetic or thermal energies of shocked interstellar gas flows (Heiles 1976). In such a case one expects, and indeed as we shall later see, the magnetic field to determine the flow field.

There are several mechanisms which are able to set up shocks in the interstellar medium (e.g. see McKee and Hollenbach 1980). We are particularly interested in the large scale shocks associated with spiral density waves (Shu et al 1972, Roberts et al 1975). Interstellar gas clouds with densities close to the critical density for collapse could be further compressed by these shocks initiating star formation. As such, we are investigating the problem of a shock wave hitting a cold dense interstellar gas cloud which is in pressure equilibrium with the surrounding hot interstellar gas. We have introduced large scale magnetic fields into the system and are primarily concerned with the influence of such fields on the early evolution of shocked gas clouds. The set of non-linear time dependent partial differential equations used to model the magnetohydrodynamic flow field makes it necessary for us to use numerical techniques in order to obtain a solution.

R. S. Roger and P. E. Dewdney (eds.), Regions of Recent Star Formation, 123–128.
Copyright © 1982 by D. Reidel Publishing Company.

2. EQUATIONS OF MOTION AND INITIAL CONDITIONS

The following set of partial differential equations is used to model the flow in the absence of heat conduction and viscosity.

$$\frac{\partial \underline{v}}{\partial t} + (\underline{v}\underline{\nabla})\underline{v} = -\frac{1}{\rho}\underline{\nabla}\left(p + \frac{B^2}{8\pi}\right) + \frac{1}{4\pi\rho}(\underline{B}\underline{\nabla})\underline{B} \tag{1}$$

$$\frac{\partial \rho}{\partial t} + \underline{\nabla}(\rho\underline{v}) = 0 \tag{2}$$

$$\frac{\partial I}{\partial t} + (\underline{v\nabla}) I = -\frac{p}{\rho}\underline{\nabla v} \tag{3}$$

$$\frac{\partial B}{\partial t} + (\underline{v\nabla})\underline{B} = (\underline{B\nabla})\underline{v} - \underline{B}(\underline{\nabla v}) \tag{4}$$

where $\underline{v}$, p, ρ, $\underline{B}$ and I represent the velocity, pressure, density, magnetic field strength and internal energy respectively, of the system. The gas is assumed to be adiabatic with an indefinite electric conductivity. The internal energy is related to pressure and density by the following perfect gas law

$$I = \frac{p}{(\gamma-1)\rho}$$

where $\gamma = \frac{5}{3}$. The system is taken as being axisymmetric and as such a two-dimensional computation is sufficient to describe the flow. Neglecting any θ-dependence, we assume a magnetic field of the form $\underline{B} = (B_r(r,z), 0, B_z(r,z))$ parallel to the propagation of the shock and post-shock velocities of the form $\underline{v} = (v_r(r,z), 0, v_z(r,z))$. Under these restrictions the Lorentz force in Equation 1 and the coupled magneto-velocity terms in Equation 4 remain in the r,z plane. The initial system under investigation is illustrated diagramatically in Fig.1. We assume the interstellar gas to have a density of $\rho_o = 0.019$ atoms/cm^3 and a temperature of 11000K and to be in pressure equilibrium with a spherical gas cloud of uniform density $100\rho_o$ and radius 15 parsec. (This represents a gas cloud with a mass of approx. 700 $M_\odot$.) In addition the incident shock is taken as having a pressure jump of 8 through its discontinuity. The shock velocity is found to be 20.1 km/sec.

In order to solve the non-linear equations 1 - 4 the numerical technique FLIC (Gentry et al 1966) has been extended to treat magneto-hydrodynamic problems. The differential equations 1 - 4 are approximated by finite difference equations which are, in principle, first order in both space and time. An artificial viscous pressure term is introduced into the model equations to allow the treatment of shocks. The properties of this technique, in particular stability conditions, the conservation of mass, momentum and energy and the problem of numerical diffusion are discussed by Nittmann (1981).

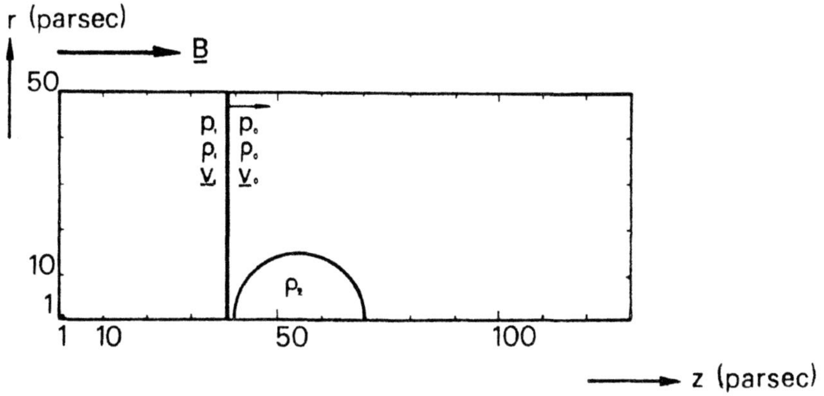

Fig.1 : Diagramatic representation of the system at time t = 0

3. RESULTS AND DISCUSSION

In Fig.2 we have plotted the velocity vectors of the postshock flow field and two density contours (which indicate the surface of the gas cloud) for the flow fields without a magnetic field (a) and with a 3μG field (b) at 1.5×10^6 years after the start of the motions. Woodward (1976) has argued for the zero magnetic field case that large velocity gradients between the nearly static gas cloud and the fast moving postshock gas will produce Kelvin-Helmholtz (KH) instabilities at the interface. In our calculations the slip motions are not sufficient to set up KH instabilities over the time interval considered. We think that numerical diffusion could be responsible for retarding the growth of these instabilities. However, in very recent computations (Nittmann and Gaskell 1981) we have employed a low, diffusive fully second-order flux-corrected algorithm (Book et al. 1975) for the same problem. Preliminary results indicate only a mild onset of KH instabilities, at the cloud surface.

Regardless of whether or not KH instabilities will develop at the surface of the cloud for the field free model, Fig.2b indicates that, for a magnetic field of 3μG parallel to the shock flow, no slip motions at either the front or rear of the cloud will develop because the post-shock gas flow is forced to follow the magnetic field lines. The post-shock flow field has a kinetic energy which is in equipartition with the magnetic energy of a 2.1μG field and therefore the 3μG field is associated with a magnetic energy larger than the kinetic energy of the flow field and thus determines the physical evolution of the system. For the 1μG field case we found that the kinetic energy of the post-shock gas exceeds the magnetic energy of the field and the slip motions are only fractionally influenced at the front of the cloud. However, they are very much reduced at the rear of the cloud. Fig.3 illustrates the evolution of the magnetic field for the 1μG and 3μG field at

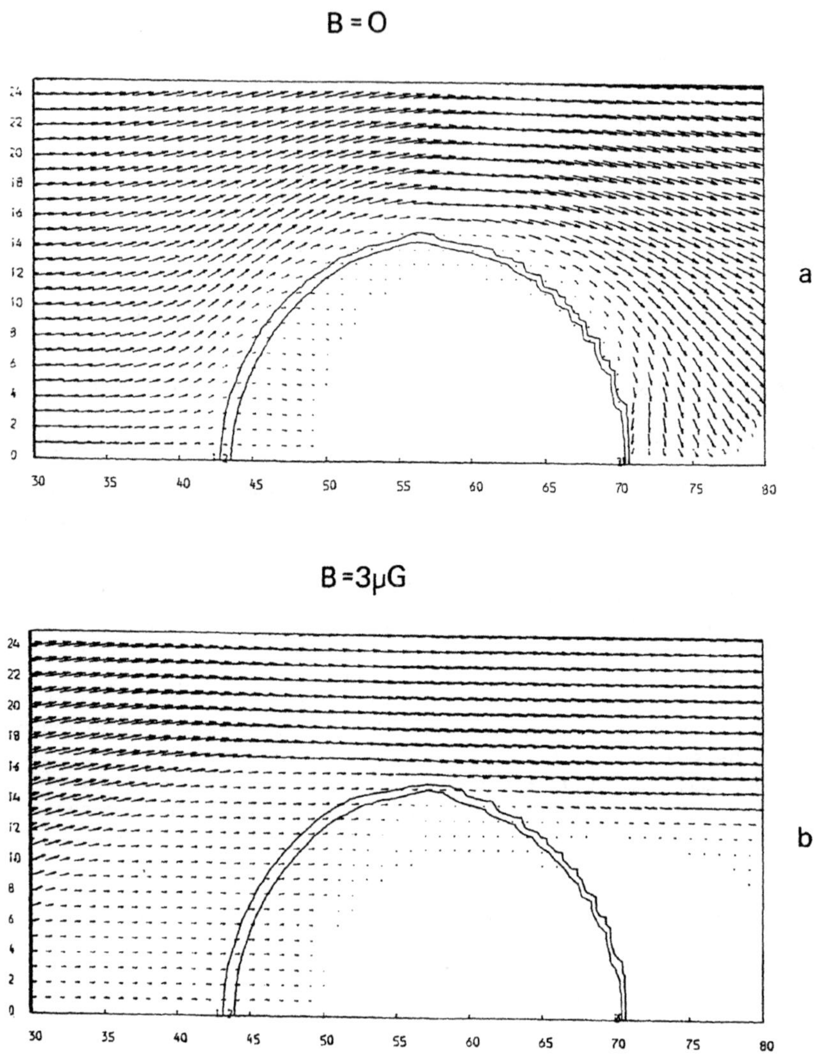

Fig.2 : Vector field representation of the gas flow around the high density gascloud for a magnetic field of 0μG (a) and 3μG (b).

intervals of 1.0, 2.0, 2.5 and 3.0 × 10^6 years. As indicated above in the 1μG case, the flow field determines the evolution of the system and the magnetic field is dragged heavily by the motions of the gas in which it is frozen. In contrast the magnetic field lines are only slightly perturbed in the 3μG case. The density contours indicate that the compression of the gas cloud is somewhat lower for the larger fields because part of the kinetic energy of the shock is converted into kinetic energy of the gas cloud, whereas, in the 1μG case, and in particular in the zero field case, a shock enters the cloud from the rear as indicated by Woodward (1976) and Nittmann (1981) and effectively holds the cloud back. A full discussion of the particulars of the post-shock flow fields for the assumed magnetic field strengths is given in Nittmann (1981).

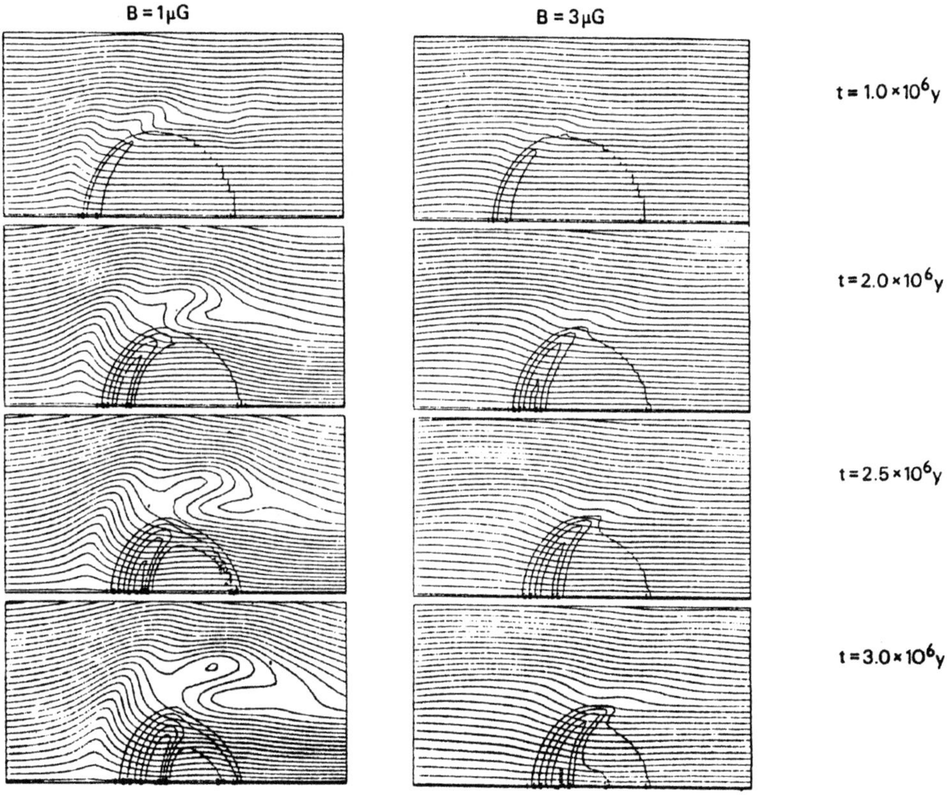

Fig.3 : Magnetic Field lines and density contours for 1μG and 3μG field

4. CONCLUSION

Our results show that interstellar magnetic fields of a few μG have considerable influence on the evolution of shocked gas flows. Concerning the observational scale of interstellar magnetohydrodynamic phenomena, we think that the orientation as well as the amplitude of the magnetic field around gas clouds should give some indication as to whether spiral density waves have traversed through them.

REFERENCES

Book, D.L., Boris, J.P., Hain, K.: 1975, J.Comput.Phys., 18, 248.
Gentry, R.A., Martin, R.E., Daly, B.J.: 1966, J.Comput.Phys., 1, 87.
Heiles, C.: 1976, Annual Review of Astron. & Astrophys., 14, 1.
McKee, C.F., Hollenbach, D.J.: 1980, Annual Review of Astron. & Astrophys., 18, 219.
Nittmann, J.: 1981, M.N.R.A.S., accepted for publication.
Nittmann, J., Gaskell, P.H.: 1981, in preparation.
Roberts, W.W.Jr., Roberts, M.S., Shu, F.H.: 1975, Astrophys.J., 196, 381.
Shu, F.H., Milione, V., Gebel, W., Yuan, C., Goldsmith, D.W., Roberts, W.W.: 1972, Astrophys.J., 173, 557.
Woodward, P.R.: 1976, Astrophys.J., 207, 484.

DISCUSSION FOLLOWING PAPER BY NITTMANN

GULL: I studied this problem some years ago and was interested in how the clouds broke up. My calculations were done in Cartesian coordinates assuming a cylindrical sort of model with flow on both sides, so I could study the effects of introducing slightly asymmetrical shapes, slightly skew to the flow. I found they did break up due to Rayleigh-Taylor instabilities, not Kelvin-Helmholtz. I would predict that by the time the shock has reached the centre of your blob, the R-T perturbations will dominate. How much do you think your modelling in r and z coordinates influences the growth of R-T perturbations near your "stagnation" point?

NITTMANN: It is difficult to say. I think that K-H instabilities must be considered first.

GULL: Can you run the calculations to see if the magnetic field stabilizes the perturbations?

NITTMANN: Yes. I have considered here only the first 3×10^6 years. I would have to go further in time.

GULL: I suspect that you may have to abandon axial symmetry and model it in three dimensions.

RADIATION-HYDRODYNAMICS OF HII REGIONS AND MOLECULAR CLOUDS

Maxwell T. Sandford II and Rodney W. Whitaker
University of California, Los Alamos National Laboratory
Richard I. Klein
Dept. of Astronomy, U. of California, Berkeley

ABSTRACT

Two-dimensional calculations of ionization-shock fronts surrounding neutral cloud clumps reveal that a radiation-driven implosion of the clump can occur. The implosion of a cloud clump results in the formation of density enhancements that may eventually form low mass stars. The smaller globules produced may become Herbig-Haro objects, or maser sources.

INTRODUCTION

The details of two-dimensional radiation-hydrodynamics calculations leading to our conclusion that radiative implosion is a possible mechanism for star formation are given in Klein et al. (1980) and in Sandford et al. (1981). Briefly, a consequence of the interaction of O-B star radiation with nearby neutral cloud gas can be the compression of cloud clumps to densities significantly greater than is achieved by planar compression as suggested by Elmegreen and Lada (1977). Two-dimensional compressions occur on a timescale of 10^4 years and produce central densities 10^4 - 10^5 cm^{-3}. Calculations to date have demonstrated that radiative implosion is capable of forming objects of at least $1M_\odot$ from cloud clumps initially only about twice as massive. This paper presents a pictorial summary of a star formation hypothesis that includes the radiative implosion mechanism, and which requires further observational and theoretical study.

STAR FORMATION HYPOTHESIS

The sequence of events leading to the radiation implosion of cloud clumps is shown schematically in Figures 1-3. Star formation of a few massive objects in a molecular cloud is initiated by passage of a

R. S. Roger and P. E. Dewdney (eds.), Regions of Recent Star Formation, 129–132.
Copyright © 1982 by D. Reidel Publishing Company.

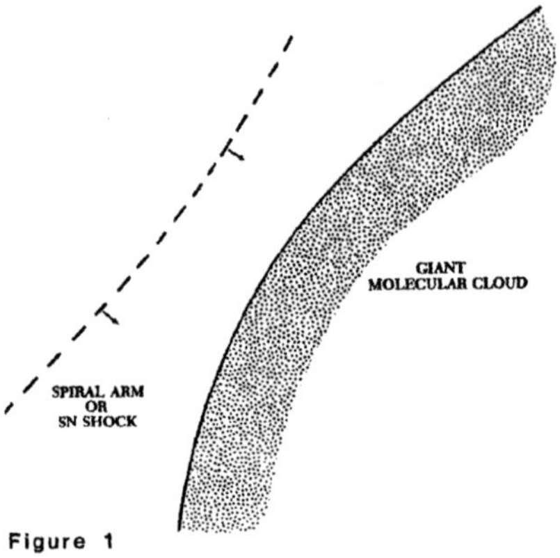

Figure 1

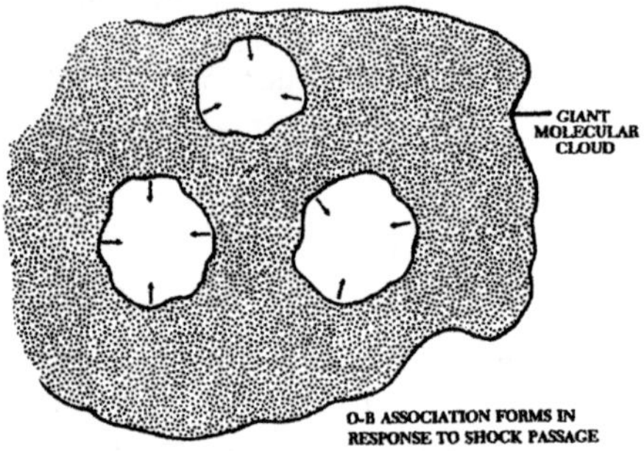

Figure 2

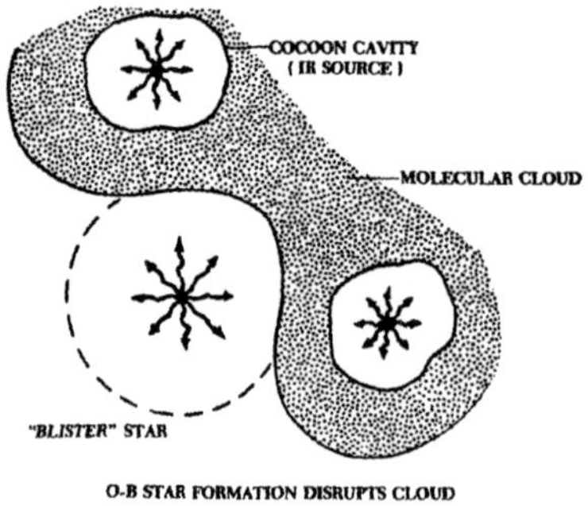

Figure 3

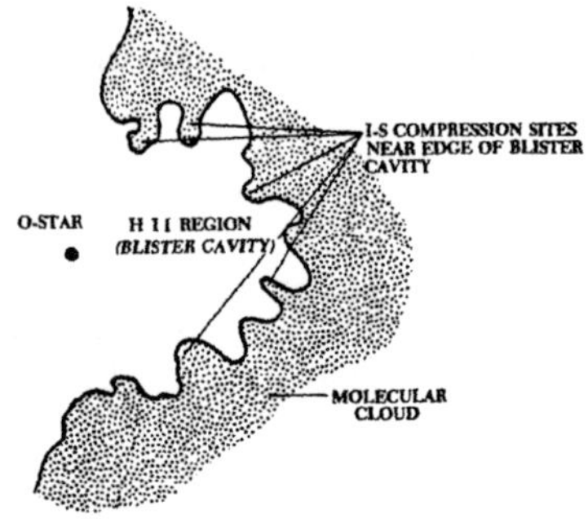

Figure 4

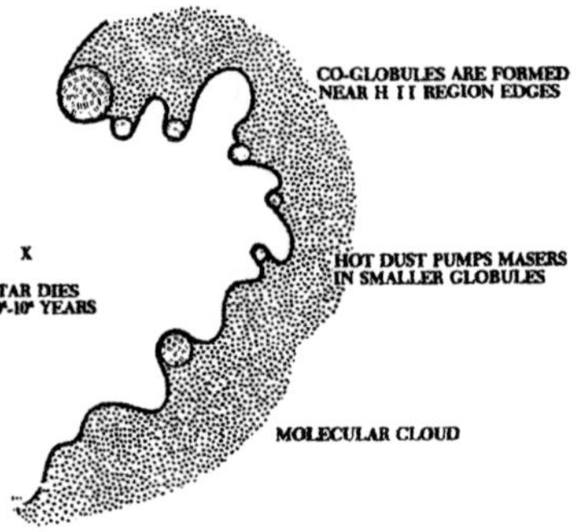

Figure 5

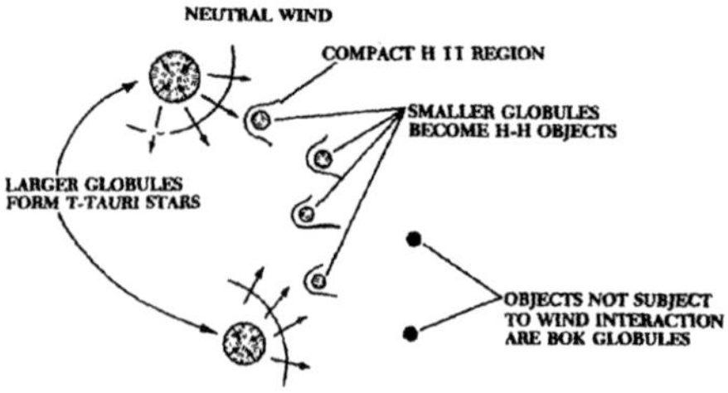

Figure 6

shockwave (Woodward, 1976), by the collision of clouds, or by the Elmegreen-Lada mechanism. Once formed, O-B stars disrupt the cloud by forming "blisters" of HII near the cloud edges (Bodenheimer et al. 1979, and Whitworth 1979). As a result of this process molecular cloud-HII complexes form which we postulate to be characterized by an irregular interface such as is observed in many regions, and is schematically illustrated in Figure 4. The irregular interface can result from the interaction of stellar winds with the neutral material (Schneps et al. 1980), the propagation of the ionization front into the cloud (Giuliani, 1980), or the existence of internal cloud structure (Larson, 1981).

During the 10^5 year lifetime of the exciting star, radiation-driven implosions form neutral condensations of various masses near the edge of the HII region, within the molecular cloud material. These are postulated to be observable in CO and HI and would be surrounded by ionized material observable in the radio continuum (Figure 5). The larger globules are expected to self-gravitate to form stars of BO and later classes. In particular, we propose that smaller globules (not self-gravitating) may be visible as H-H objects when subject to the winds of T-Tauri stars, and as dark globules when viewed in projection (Figure 6).

Finally, the last stage of star formation is viewed as the dissipation or disruption of the smaller globules and the formation of a few main sequence stars (Figure 7). This picture of the star formation process is qualitative and necessarily simplistic, but is presented to indicate the potentially important role of radiation-driven implosion.

SUMMARY

The principle features of the radiation implosion model for star formation can be stated as follows:

- O-B stars are necessary to form condensations on short timescales,
- parent clouds in which O-B stars form must be disrupted to leave cloud inhomogeneities exposed to ionizing radiation,
- radiation implosion compresses irregularities to n/n_o 100-1000,
- Jeans' limit is reduced by the presence of a velocity field in the neutral gas (Hunter, 1979),
- low mass objects are formed by this mechanism,
- globules are formed near the edges of HII regions,
- H-H objects and dark globules may result from radiation implosion of cloud clumps.

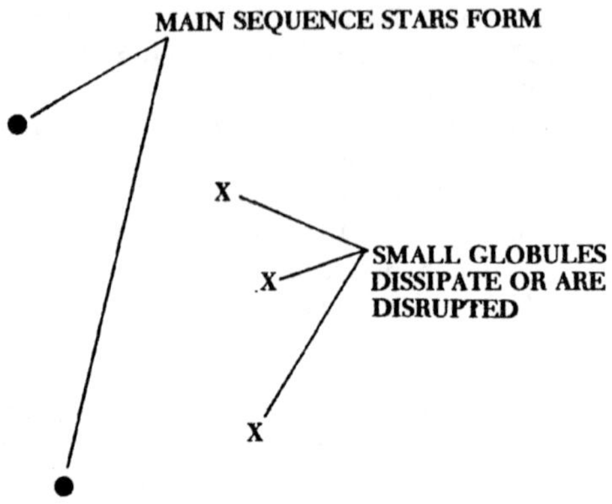

FIGURE 7

REFERENCES

Klein, R. I., Sandford II, M. T. and Whitaker, R. W., 1980, Space Sci. Rev. 27, 275.

Sandford II, M. T., Whitaker, R. W. and Klein, R. I., 1981, submitted to Astrophys. J.

Elmegreen, B. G. and Lada, C. J., 1977, Astrophys. J., 214, 725.

Woodward, P. R., 1976, Astrophys, J. 207, 484.

Bodenheimer, P., Tenorio-Tagle, G. and Yorke, H. W., 1979, Astrophys. J., 233, 85.

Whitworth, A. 1979, Monthly Not. R. A. S., 186, 59.

Schneps, M. H., Ho, P. T. P. and Barrett, A. H., 1980, Astrophys. J., 240, 84.

Giuliani, J. L., 1980, Astrophys. J., 242, 219.

Larson, R. B., 1981, Monthly Not. R. A. S., 194, 809.

Hunter, J. H., 1979, Astrophys. J., 233, 946.

HIGH VELOCITY GAS IN MOLECULAR CLOUDS

Ronald L. Snell
Five College Radio Astronomy Observatory, University of Massachusetts, Amherst, Massachusetts
Suzan Edwards
Five College Astronomy Department, Smith College, Northhampton, Massachusetts

ABSTRACT

A new source of high velocity molecular gas to the south of the HII region NGC 2068(M78) near the Herbig-Haro objects HH 25-26 is reported. The redshifted and blueshifted CO wings seen in this region are spatially separated indicating that this is another region of bi-polar mass outflow from a young stellar object. We compare this region to the other known regions of bi-polar mass outflow and discuss the implications of these energetic outflows.

INTRODUCTION

The recent discoveries of high velocity molecular gas associated with young stellar objects have been attributed to energetic mass outflow from these young objects interacting with the surrounding ambient cloud (Kwan and Scoville 1976; Snell, Loren and Plambeck 1980; Rodriguez, Ho and Moran 1980; Lada and Harvey 1981; Snell and Edwards 1981a). The high velocity gas observed in many of these regions has spatially separated asymmetrical wings suggesting that the outflow is bi-polar. Several of these regions of bi-polar mass outflow are associated with Herbig-Haro objects or H_2O masers. The kinetic energy transferred to the molecular cloud during this evolutionary phase may have important consequences on the future evolution of the cloud.

We report results of an ongoing program to search for high velocity molecular gas associated with Herbig-Haro objects. To date we have discovered two regions of high velocity gas which show evidence for bi-polar mass outflow from an embedded infrared source. These regions are associated with the Herbig-Haro objects HH 7-11 which lie south of the reflection nebula NGC 1333 and the Herbig-Haro objects HH 25-26 which lie south of the HII region NGC 2068. We

R. S. Roger and P. E. Dewdney (eds.), Regions of Recent Star Formation, 133–139.
Copyright © 1982 by D. Reidel Publishing Company.

present a summary of our observational results for the HH 25-26 region and briefly describe our results for the HH 7-11 region which have been presented in Snell and Edwards (1981a). The properties of these two regions are compared with the properties of other known bi-polar mass outflow regions.

OBSERVATIONS AND RESULTS

Observations of the J=1→0 transitions of ^{12}CO and ^{13}CO have been obtained at the 14 m telescope of the Five College Radio Astronomy Observatory (FCRAO) located in New Salem, Massachusetts. Observations of the J=2→1 ^{12}CO transition were obtained at the 4.9 m telescope of the Millimeter Wave Observatory in Ft. Davis, Texas using a receiver developed at FCRAO. Details of the observational procedure are described in Snell and Edwards (1981a).

Our first detection of high velocity molecular gas was in the vicinity of HH 7-11. The CO observations of this region show evidence for an energetic bi-polar outflow of gas from a low luminosity infrared source, HH 7-11 IR. A spatial-velocity map of the region around HH 7-11 is shown in Figure 1, this diagram shows the large spatial separation between the redshifted and blueshifted high velocity gas. The asymmetry is centered about HH 7-11 IR where the high velocity gas has a velocity extent of >40 km s^{-1}. These observations are presented in Snell and Edwards (1981a).

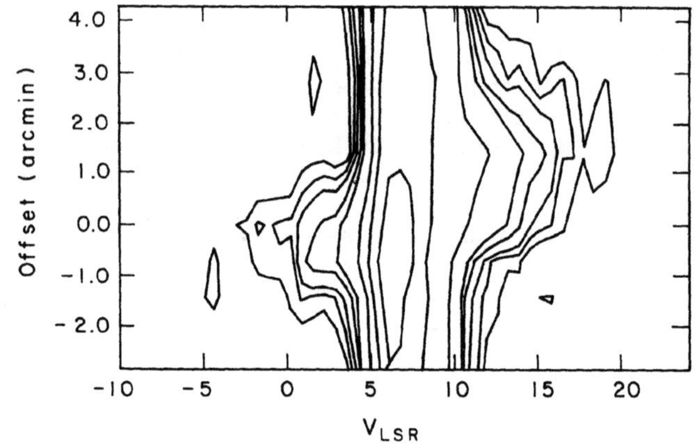

Figure 1. Spatial-velocity map of the region near HH 7-11. The offsets are relative to the infared source HH 7-11 IR along an axis from the SE to the NW.

The high velocity gas detected in the HH 25-26 region is similar to that found near HH 7-11. It is centered about an infrared source, #59 in the list of Strom, Strom, and Vrba (1976), which we have labelled HH 26 IR. A complete description of this region is given in Snell and Edwards (1981b). Spectra of ^{12}CO and ^{13}CO J=1→0 taken at two positions, one toward HH 26 IR and the other 1.4 arcmins to the north-west, are shown in Figure 2. These spectra show both the high velocity redshifted and blueshifted ^{12}CO wings and also the spatial asymmetry of these wings. The ratio of the integrated intensities of $^{12}CO/^{13}CO$ in the high velocity wings at both positions shown in Figure 2 indicates that the high velocity ^{12}CO gas is optically thin. The region around HH 26 IR has been mapped in both the J=1→0 and J=2→1 transitions of ^{12}CO. These maps show that the high velocity gas has a limited spatial extent and that the red and blue wings are distributed asymmetrically about HH 26 IR. Using the relation of Goldsmith, Plambeck and Chiao (1976) and the intensities of the J=2→1 and J=1→0 ^{12}CO lines, we have estimated the kinetic temperature of the high velocity gas to be roughly 10 - 25 K. The kinetic temperature of the high velocity gas may be slightly higher than the kinetic temperature of the ambient cloud.

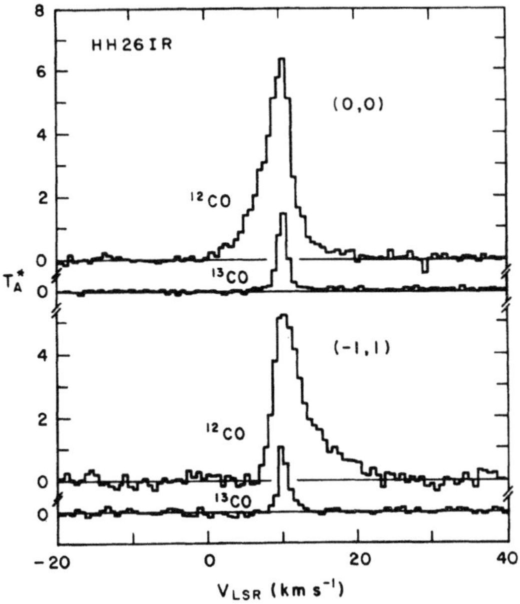

Figure 2. Spectra of the ^{12}CO and ^{13}CO J=1→0 transition of CO toward HH 26 IR and at a position 1 arcmin west and 1 arcmin north of HH 26 IR.

Using the expression in Snell and Edwards (1981a), we have estimated the mass of high velocity gas to be 3.1 $M_\odot$. The mass loss rate necessary to produce the observed high velocity gas can be estimated assuming conservation of momentum between the wind and the ambient gas. Assuming a stellar wind velocity of 200 km s^{-1}, we estimate the mass loss rate to be 2×10^{-6} $M_\odot$ yr^{-1}.

DISCUSSION

A summary of the known regions of bi-polar mass outflow is presented in Table 1. In addition to these sources, the star R Mon (Cantó et al. 1981) may also be a source of bi-polar mass outflow. Many of the lower luminosity bi-polar mass outflow regions are associated with Herbig-Haro objects (the first three in Table 1 and R Mon).

TABLE 1
SOURCES OF BI-POLAR MASS OUTFLOW

Source	Luminosity ($L_\odot$)	Mass of High Velocity Gas ($M_\odot$)	Outflow Velocity (km s^{-1})	Mass Loss Rate(a) ($M_\odot$ yr^{-1})	Kinetic Energy (ergs)	Ref.
L1551 IRS-5	25	0.3	15	8×10^{-7}	3×10^{44}	1,2
HH 26 IR	>3 50-300(b)	3.1	15	2×10^{-6}	2×10^{45}	3,4
HH 7-11 IR	70	4.2	20	8×10^{-6}	7×10^{45}	5
AFGL 490	1.4×10^{3}	30	25	3×10^{-4}	2×10^{47}	6
Ceph A	2.5×10^{4}	$\geqslant 10$	25	$\geqslant 2 \times 10^{-4}$(c)	$\geqslant 6 \times 10^{46}$(c)	7

Notes to Table:

(a) Mass loss rate calculated assuming a stellar wind velocity of 200 km s^{-1}.
(b) The near infrared luminosity is 3 $L_\odot$. The total luminosity of 50-300 $L_\odot$ was estimated by comparing the near infrared flux of HH 26 IR with that of L1551 IRS-5 and HH 7-11 IR and assuming a distance of 500 pc.
(c) Calculated using data given in paper.

1 Snell, Loren and Plambeck (1980)
2 Fridlund et al (1980)
3 Snell and Edwards (1981b)
4 Strom, Strom and Vrba (1976)
5 Snell and Edwards (1981a)
6 Lada and Harvey (1981)
7 Rodriguez, Ho and Moran (1980)

The close physical association of high velocity molecular gas and HH objects have suggested to some that there is a common origin of both via the interaction of a stellar wind with the surrounding ambient cloud (Norman and Silk 1979); Rodriguez et al 1980). In L1551 IRS-5 the proper motion measurements and radial velocity measurements of the compact Herbig-Haro objects (Cudworth and Herbig 1980; Strom, Grasdalen and Strom 1974) and the CO observations (Snell, Loren, and Plambeck 1979) strongly support this hypothesis. Similarly the radial velocity measurements of HH 7-11 (Strom, Grasdalen and Strom 1974) and the CO observations (Snell and Edwards 1981a) in the vicinity are also consistent with this model. Unfortunately there is no information on the motions of the Herbig-Haro objects around HH 26 IR to further test this model.

But not all high velocity Herbig-Haro objects are associated with high velocity molecular gas. We were unable to detect high velocity CO gas around the Herbig-Haro objects HH 1-2. The proper motion measurements of these HH objects (Herbig and Jones 1981) have shown them to be diverging from the position of a faint red star, thought by Cohen and Schwartz (1979) to be the exciting star of HH 1. Herbig and Jones suggest a model in which the HH objects are dense knots of material which are ejected with the wind from the Cohen-Schwartz star. For the source L1551 IRS-5 the dynamical time scale of the HH-objects is an order of magnitude shorter than the dynamical time scale of the high velocity molecular gas. This discrepancy suggests that the HH-objects may have a much shorter lifetime than the wind and may be continually produced throughout the lifetime of the wind. The absence of high velocity gas around the Cohen-Schwartz star may indicate that the wind has not had sufficient time to sweep up an observable amount of molecular gas.

The mass loss rate, the mass of high velocity gas, and the outflow velocity that have been derived for the bi-polar outflow regions all increase with increasing luminosity. The increased mass loss rate cannot depend solely on luminosity since the inferred momentum of the wind greatly exceeds the momentum avalable in photons. Therefore the winds are not radiatively driven. The conversion of rotational or magnetic energy of the source into mechanical energy may be required to explain the mass outflow.

The confinement mechanism for the bi-polar mass outflow is not understood. The bi-polar flows in L1551 IRS-5, HH 7-11 IR and HH 26 IR are roughly aligned with the direction of polarization in these regions. This alignment may indicate that the confinement mechanism may be related to the magnetic fields.

The large variations in luminosity among the sources in Table 1 suggest that both high and low mass stars undergo an early phase of mass loss. The more luminous stars may lose a substantial fraction of their mass during this phase which may have important implications on pre-main sequence evolution. In addition the observed high mass loss rates may also have important consequences on the evolution of the surrounding molecular cloud. If the kinetic energy of the high velocity gas in AFGL 490 were distributed uniformly over 5000 $M_{\odot}$, this gas would have a mean velocity of 2 km s^{-1}. These velocities would be sufficient to explain the broad CO lines (much greater than the thermal linewidth) seen in most clouds. The formation of low mass stars, such as L1551 IRS-5 or HH 7-11 IR, may also be important in determining the dynamics of the surrounding cloud. Norman and Silk (1980) have suggested that young low mass stars are the dominant source of turbulant energy in molecular clouds. Winds from low mass stars may provide the continuous energy input necessary to support molecular clouds for greater than their free-fall lifetime without.

The Five College Radio Astronomy Observatory is operated with support from the National Science Foundation under grant AST 80-26702 and with permission of the Metropolitan District Commission, Commonwealth of Massachusetts.

REFERENCES

Cantó , J., Rodriguez, L.F., Barrel, J.F. and Carrel, P.: 1981, Ap. J. 244, 102.
Cohen, M. and Schwartz, R.D.: 1980, M.N.R.A.S. 191, 165.
Cudworth, K.M. and Herbig, G.: 1979, Astron. J. 84, 548.
Fridlund, C.V., Nordh, H.L., van Duinen, R.J., Aalders, T.W.G. and Sargent, A.I.: 1980, Astron. Ap. 91, L1.
Goldsmith, P.F., Plambeck, R.L. and Chiao, R.Y.: 1975, Ap. J. Lett. 196, L39.
Herbig, G.H. and Jones, B.R.: 1981, preprint.
Kwan, J. and Scoville, N.Z.: 1976, Ap. J. Lett. 210, L39.
Lada, C.J. and Harvey, P.M.: 1981, Ap. J. 245, 58.
Norman, C. and Silk, J.: 1979, Ap. J. 228, 197.
Norman, C. and Silk, J.: 1980, Ap. J. 238, 158.
Rodriquez, L.F., Ho, P.T.P. and Moran, J.M.: 1980. Ap. J. Lett. 240, L149.
Rodriquez, L.F., Moran, J.M., Ho., P.T.P. and Gottlieb, E.W.: 1980, Ap. J. 235, 845.

Snell, R.L., Loren, R.B. and Plambeck, R.L.: 1980, Ap. J. Lett. 239, L17.
Snell, R.L. and Edwards, S.: 1981a, Ap. J. in press.
Snell, R.L. and Edwards, S.: 1981b, in preparation.
Strom, S.E., Grasdalen, G.L. and Strom, K.M.: 1974, Ap. J. 191, 111.
Strom, K.M., Strom, S.E. and Vrba, F.J.: 1976, Astron. J. 81, 308.

DISCUSSION FOLLOWING PAPER BY SNELL AND EDWARDS

HARTEN: What are the typical scale sizes of the bipolar CO regions seen?

SNELL: The largest is the L1551 region. It shows very elongated bipolar flow with a full extent of almost 1 pc. In the other two cases described here (HH 7-11 and HH 25C) the scales are more than 0.2 pc.

UNDERHILL: Could you clarify how you determine the luminosity of the underlying infrared star?

SNELL: The luminosities come from the near-infrared and far-infrared continuum measurements. The observations indicate a point source of emission which we interpret as radiation from dust around a star, radiation converted by the dust grains to the infrared.

HIGH VELOCITY CO EMISSION AROUND T TAURI STARS

Suzan Edwards
Five College Astronomy Department, Smith College
Northampton, Massachusetts
Ronald L. Snell
Five College Radio Astronomy Observatory, University of Massachusetts, Amherst, Massachusetts

ABSTRACT

T Tauri winds are detected via their interactions with ambient cloud material. The magnitude of the mass flows are calculated and dynamical effects on molecular clouds are discussed.

INTRODUCTION

Recent discoveries of high velocity molecular gas around low luminosity IR sources indicate that strong bipolar mass flows are associated with these young, presumably low mass objects. (Snell, Loren and Plambeck 1980; Beichman and Harris 1981; Snell and Edwards 1981). Optically visible T Tauri stars may represent the subsequent evolutionary phase where these low mass objects have recently emerged from their placental cloud material. A stellar wind from T Tauri stars has long been recognized on the basis of the P Cygni structure often found at Hα (Herbig 1962). Quantitative analysis of the T Tauri winds, based on Hα emission intensity or line profile structure, has been frought with ambiguity, and yields uncertainties of $\gg$3 orders of magnitude in the determination of $\dot{M}$ (DeCampli 1981). Determination of the magnitude, duration and geometry of the T Tauri wind will allow the evolutionary transition from an active embedded IR source to an optically visible pre-main sequence star to be examined. It will also clarify the dynamical effects of T Tauri stars on molecular clouds.

In an effort to study T Tauri winds via the interactions with ambient cloud material, we have searched for high velocity molecular gas at ^{12}CO around 22 T Tauri stars. Two stars, T Tau and AS 353A, show spatially extended blueshifted and redshifted wings. The implications for mass ejection from these and other T Tauri stars are examined, and the effect on molecular cloud dynamics is discussed.

R. S. Roger and P. E. Dewdney (eds.), Regions of Recent Star Formation, 141–146.
Copyright © 1982 by D. Reidel Publishing Company.

OBSERVATIONS

The J=1-0 ^{12}CO and ^{13}CO observations were obtained with the Five College Radio Astronomy Observatory 14m telescope located in New Salem, Ma. The J=2-1 ^{12}CO and ^{13}CO data were taken at the Millimeter Wave Observatory with the 4.9m telescope at Ft. Davis, Tx. The observational procedures are described in Snell an Edwards (1981).

A sample of 22 T Tauri stars that exhibited the full range of observed values of A_v, Hα emission strength, spectral type and luminosity characteristic of T Tauri stars was selected from the list of Cohen and Kuhi (1979). All stars were observed at J=1-0 ^{12}CO; only T Tau and AS 353A showed extended wings at $T \geqslant 0.1$ K. (See Edwards and Snell 1981 for a complete description of the observational material.) These two stars are distinguished from the remainder of the sample in that they are the only two T Tauri stars known to be accompanied by nearby Herbig-Haro Objects (separation < 1').

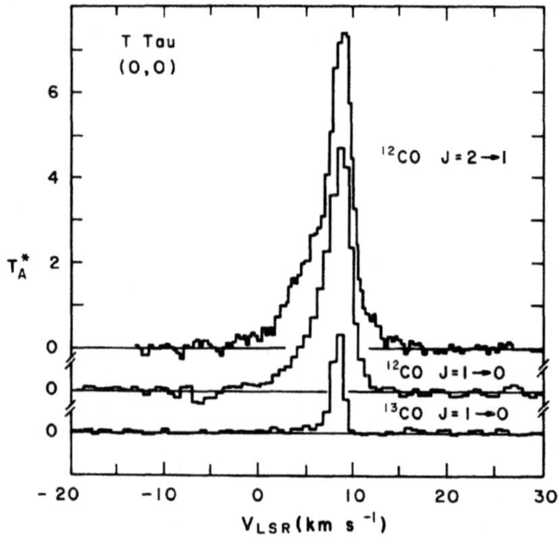

Figure 1. The J=2-1 and J=1-0 ^{12}CO spectra and the J=1-0 ^{13}CO spectrum obtained toward T Tau.

Figure 1 shows the J=1-0 and J=2-1 ^{12}CO spectra and the J=1-0 ^{13}CO spectrum towards T Tau. The ^{12}CO spectra of both T Tau and AS 353A are characterized by a narrow peak at the velocity of the surrounding molecular cloud, with broad wings exhibiting a maximum blueward extent of ~13 km s^{-1} and a redward extent of only ~4 km s^{-1}. The 12/13 CO ratio of the integrated intensities in the extended wings indicates τ~6 towards T Tau and an optically thin high velocity gas towards AS 353A. The ^{12}CO J=2-1/1-0 ratio in the high velocity wings toward both stars is close to unity, indicating excitation temperatures of 8 to 15 K. This represents a lower limit to the kinetic temperature of the high velocity gas, which may not be thermalized; it does not differ significantly from that of the ambient cloud.

Maps of the spatial distribution of the ^{12}CO J=1-0 emission towards T Tau and AS 353A reveal that the broad wings have a limited spatial extent and are roughly centered about the optical objects. The 36 point map towards T Tau is shown in Figure 2. For both stars

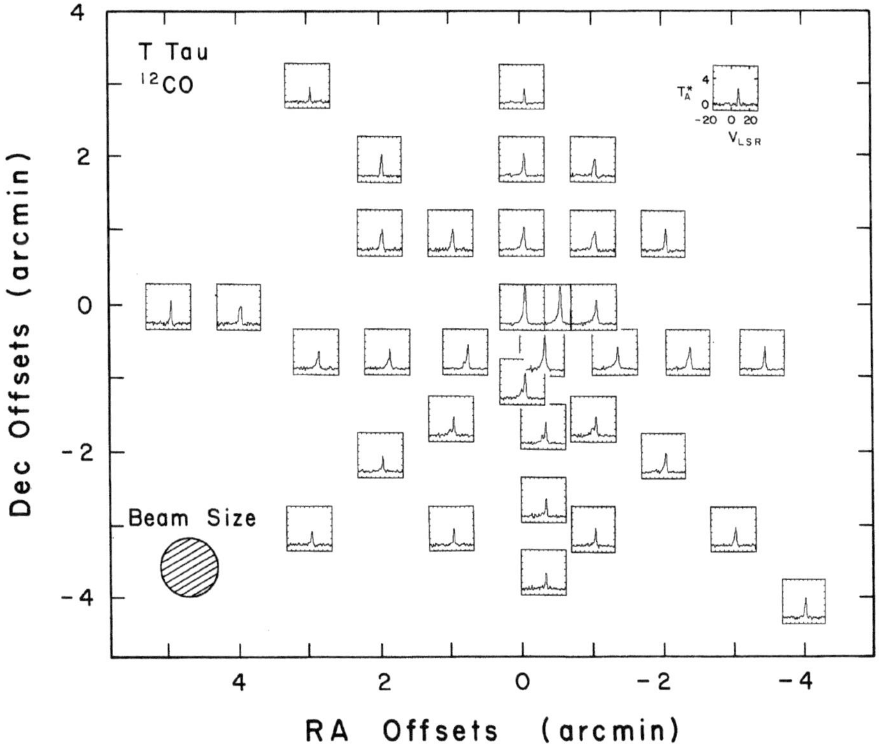

Figure 2. The 36 point spatial map of the J=1-0 ^{12}CO transition in the vicinity of T Tau.

the blue high velocity component is more prominent and considerably more spatially extended than the redward component. The distinct spatial separation of the peaks of the integrated intensities of the blueward and redward wings that characterizes the CO distribution around the embedded, presumably younger, objects (Snell and Edwards 1982) is not clearly present for the two T Tauri stars studied here.

Column densities of high velocity gas appropriate to an optically thin material, with an H_2/CO abundance ratio of 2×10^4, are 2×10^{20} cm^{-2} towards T Tau and AS 353A. The total mass of high velocity gas, assuming a distance of 140 pc to T Tau (Elias 1978) and 150 pc to AS 353A (Edwards and Snell 1981), is 1×10^{-1} $M_\odot$ around T Tau and 4×10^{-2} $M_\odot$ around AS 353A. Ninety five percent of the mass of the high velocity material associated with T Tau is from blueshifted gas. In AS 353A the blueshifted gas represents 55% of the high velocity material.

DISCUSSION

We attribute the high velocity wings around T Tau and AS 353A to expanding material driven by a stellar wind from these T Tauri stars. Radial velocities from optical spectra of the emission nebulae and reflection nebula (NGC 1555) associated with T Tau also suggest the presence of an extended, supersonic mass outflow from T Tau (Schwartz, 1975). As with the low luminosity embedded IR sources, the maximum of the integrated intensity of blueshifted material coincides with the positions of the optical HH objects near T Tau and AS 353A. Mass loss rates can be estimated by requiring conservation of momentum between the stellar wind and the swept up molecular gas. Assuming a stellar wind of 200 km s^{-1} and dynamical timescales of a few times 10^4 yr, estimated from the observed extent and velocity of the molecular gas, implies $\dot{M}$ of 1×10^{-7} $M_\odot$ yr^{-1} for T Tau and 7×10^{-8} $M_\odot$ yr^{-1} for AS 353A. These implied mass loss rates are about an order of magnitude smaller than those inferred from the embedded low luminosity objects discussed in Snell and Edwards (1982).

The total kinetic energy in the high velocity gas associated with T Tau is $\sim 3 \times 10^{43}$ erg and with AS 353A is $\sim 1 \times 10^{43}$ erg. Again, this is more than an order of magnitude lower than the energy in the high velocity gas from the embedded low luminosity objects. The impact of this kinetic energy on the dynamics of molecular clouds will be dependent on the duration of such mass flows and on the frequency of the phenomenon among all T Tauri stars.

The lack of high velocity molecular gas around 90% of our sample of T Tauri stars could be either the result of insufficient molecular gas ambient to the T Tauri stars or from the lack of a stellar wind of sufficient magnitude to drive the gas. The former argument is favored, since the P Cygni profiles in T Tau and AS 353A at Hα and Ca II K (Kuhi 1964; Ulrich and Knapp 1979; Boesgaard 1981), are indicative of mass outflow and are found in $\sim$70% of all T Tauri stars (Kuhi 1978).

If continuous flows are assumed, the observed dynamical timescales in both the T Tauri stars and the embedded low luminosity objects are on the order of 10^4 yr. The T Tauri phase probably lasts $\sim 10^6$ yr for a 1 $M_\odot$ star, although T Tauri 'activity' is believed to decline during this time (Cohen and Kuhi 1979). Sporadic outbursts from T Tauri stars are known to occur (Herbig 1977), and a discontinuous mass ejection cannot be discounted. It is interesting to note that the protype of this group, FU Ori, shows no high velocity wings (Edwards and Snell 1981). The duration and continuity of the mass flows thus remain uncertain.

Continuous mass flows of the observed magnitude from T Tauri stars have been shown by Norman and Silk (1980) to be sufficient to account for the energetics, longevity and clumpy structure of molecular clouds if T Tauri densities are ≥ 10 pc^{-3}, although several generations of T Tauri stars would be required to account for cloud lifetimes as great as 10^7 yr. However, many of the optically visible T Tauri stars studied here seem to be sufficiently removed from the parent cloud that dynamical effects are minimal. The more energetic mass flows observed in the (younger) embedded low luminosity objects might then have a larger impact on cloud dynamics.

CONCLUSIONS

T Tauri mass flows of 10^{-7} to 10^{-8} $M_\odot$ yr^{-1} are inferred from the high velocity molecular emission centered on two T Tauri stars. In both cases H-H Objects are found to be spatially coincident with the maximum intensity of the blueshifted material. Dynamical impact on the ambient clouds may be small for most optically visible T Tauri stars. If the low luminosity IR stars discussed in Snell and Edwards (1982) are predecessors to optically visible T Tauri stars, then the transition is probably accompanied by a decrease in the momentum and energy of the stellar winds.

The Five College Radio Astronomy Observatory is operated with support from the National Science Foundation under grant AST 80-26702 and with permission of the Metropolitan District Commission, Commonwealth of Massachusetts.

REFERENCES

Beichman C. and Harris S.: 1981 Ap. J. 245, 589.
Boesgaard A.M.: 1981 private communication.
Cohen M. and Kuhi L.V.: 1979 Ap. J. Suppl. 41, 743.
DeCampli W.: 1981 Ap. J. 244, 124.
Edwards S. and Snell R.L.: 1981 in preparation.
Elias J.: 1978 Ap. J. 224, 857.

Herbig G.H.: 1962 Adv. Ast. Ap. 1, 47.
Herbig G.H.: 1977 Ap. J. 217, 693.
Kuhi L.V.: 1964 Ap. J. 140, 1409.
Kuhi L.V.: 1978 "Protostars and Planets", Univ. of Az. Press, 708.
Norman C. and Silk J.: 1980 Ap. J. 238, 158.
Schwartz R.D.: 1975 Ap. J. 195, 631.
Snell R.L. and Edwards S.: 1981 Ap. J. submitted.
Snell R.L. and Edwards S.: 1982, this volume, p. 133.
Snell R.L., Loren R.B. and Plambeck R.L.: 1980 Ap. J. 239, L17.
Ulrich R.K. and Knapp G.: 1979 Ap. J. 230, L99.

DISCUSSION FOLLOWING PAPER BY EDWARDS AND SNELL

WELTER: Why do the ^{13}CO profiles (Fig. 1) not show the extended wings seen in the ^{12}CO profiles? The noise level seems to be low enough that such wings would be detected.

SNELL: The narrow component is very optically thick in ^{12}CO. However, the emission in the wings is not, and the (terrestial) ratio ^{12}CO/^{13}CO of 90 will apply.

J. DICKEL: Could the association be just with the Herbig-Haro object and not the T Tauri star?

EDWARDS: These are the only two known T Tauri stars to be accompanied by Herbig-Haro objects and they both have high velocity flow.

J. DICKEL: Are there any Herbig-Haro objects which don't show the double velocity effect?

EDWARDS: Yes, there are Herbig-Haro objects which do not have high velocity gas associated with them.

TURNER: The OH emission around Herbig-Haro objects, as observed with the 3' beam of the Arecibo telescope, shows a unique form of excitation anomaly. Although weak ($T_A \approx 0.4$ K), the ratio of 1667 to 1665 MHz intensities is systematically greater than the LTE value of 1.8 - the ratios vary from 2.2 to 5.1. This is suggestive of pumping by streams of charged particles, which may be related to the gas flows discussed in this paper.

OBSERVATIONS AND INTERPRETATION OF THE LINE PROFILES OF EXCITED H_2 IN ORION

T.R. Geballe
Mount Wilson and Las Campanas Observatories,
Carnegie Institution of Washington

D. Nadeau
Palomar Observatory,
California Institute of Technology

ABSTRACT

The profiles of the V=1-0 S(1) line of H_2 in the Orion molecular cloud show a variation from the center of the emission region where the profiles are broad and have extended blue wings, to the periphery where the profiles are narrow and have a peak at the velocity of the molecular cloud. These observations can be understood by a model of a radially expanding supersonic flow of gas colliding with the surrounding molecular cloud. It is concluded that the high velocity emission originates in the flow and the low velocity emission originates in the molecular cloud. Density variations in the flow and the surrounding cloud are necessary to explain the detailed line profiles and morphology.

INTRODUCTION

The present paper contains a summary of the spectroscopic observations of the excited H_2 in the Orion molecular cloud (OMC-1), followed by a discussion which it is hoped will serve as a guide in understanding the origin of the excited H_2. Although a general model for understanding the line profiles is advocated here, a detailed explanation is still lacking; the ball remains in the theorists' court.

OBSERVATIONS

The observations discussed here were made with a Fabry-Perot spectrometer, and consist mainly of spectra of the V=1→0 S(1) line at 4713 cm^{-1}. This line was observed over the region mapped by Beckwith et al. (1978), although not with complete spatial coverage. The observations were made with both 5" and 10" diameter apertures, and usually at a spectral resolution corresponding to $\Delta V \cong 20$ km s^{-1}.

R. S. Roger and P. E. Dewdney (eds.), Regions of Recent Star Formation, 147–153.
Copyright © 1982 by D. Reidel Publishing Company.

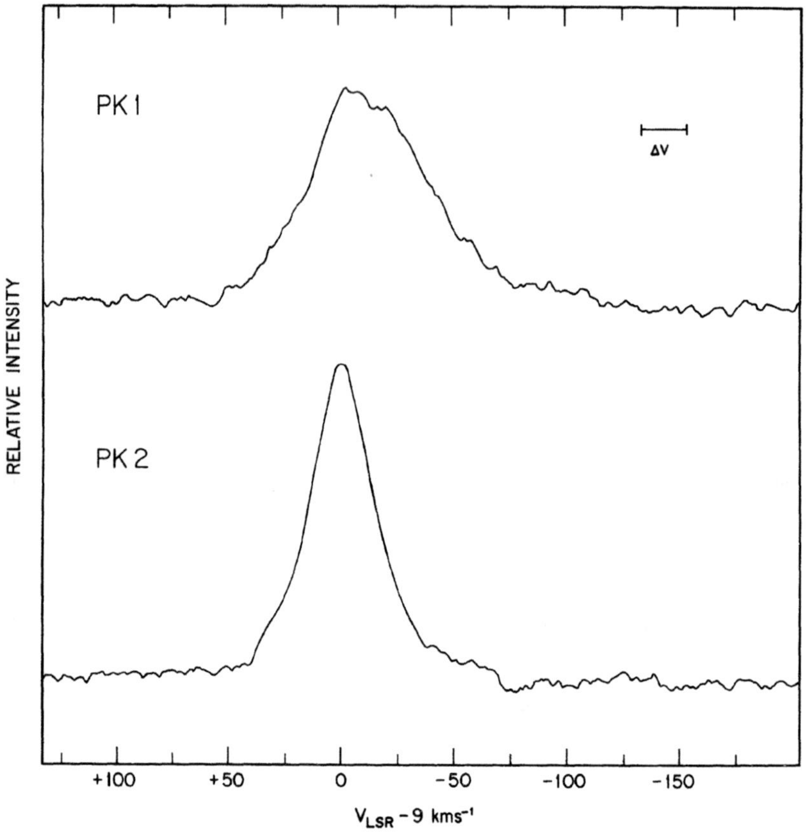

Fig. 1 - Spectra of the V=1→0 S(1) line of H_2 at the positions of Pk 1 and Pk 2. These spectra illustrate the range of H_2 line profiles seen in OMC-1 at 2μm.

Many details of these and other spectra are discussed by Nadeau et al. (1981); there are, however, for the purposes of this paper, two principal types of observed line profile, of which examples are shown in Fig. 1. One of these, definitely the more unusual of the two, is an asymmetric line having a peak which is slightly blueshifted from the velocity of OMC-1 and a pronounced blue wing which extends to velocities as extreme as 100 km s^{-1} from the velocity of the peak. The second type is a much more symmetric and narrower line, which peaks at the velocity of OMC-1. The broad lines are found in the center of the line emission region, the narrow lines in the periphery. Although from the center to the periphery line intensities vary somewhat unpredictably with radial direction, the transition of the line profile from broad

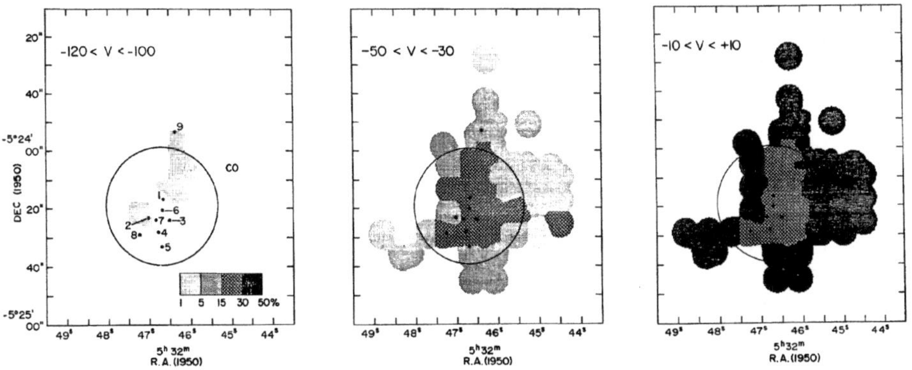

Fig. 2 - Maps of the fraction of the total energy in the S(1) line emitted in a given velocity interval. Shading of different densities indicates the percentage of the total intensity; the scale is shown in the left hand map. On each map the velocity interval is indicated in the top left corner ($V=V_{LSR}$ - 9 km s^{-1}), a circle of diameter 40" shows the location of the high velocity CO emission, and the dots refer to the locations of the infrared continuum sources, using the nomenclature of Downes et al. (1981).

to narrow is fairly smooth in all directions. Fig. 2 illustrates the symmetry of the distribution of line shapes. The peaks of total line intensity found by Beckwith et al. (1978) have different spectral characteristics and have little significance for the present discussion. Some peaks, such as 2 and 5, are at the positions of narrow lines and indicate maxima of spectral intensity. Others, most notably Peak 1, are not peaks in spectral intensity but are due to line emission over a wide range of velocities at their positions.

DISCUSSION

a) Kinematics

The S(1) line profiles apparently are due to real motion of the emitting molecules, rather than scattering off fast moving grains, since the 12.3μm V=0→0 S(2) line, observed by Beck (1981) and her collaborators, shows comparably broad profiles but is much less susceptible to scattering. The above together with the symmetric spatial distribution of the S(1) line profiles imply a large scale organized motion of the gas. One can reasonably conclude that this motion is a radial expansion, most readily by comparing the S(1) line profiles to the broad lines of CO emitted in the millimeter wave region. The symmetry of the latter line profiles and the apparent asymmetry of the H_2 line can be reconciled if there is dust present within the emitting volume, which attenuates the near infrared H_2 line radiation from the back of the region. The amount of dust needed is in satisfactory agreement with the column density of gas in the expanding region (Knapp et al. 1981) and a normal dust-to-gas ratio.

For a radial expansion, the source of the gas must lie near the center of the broad-lined region. This places the source near the cluster of infrared sources in OMC-1, in agreement with the millimeter line measurements, and suggests that the cause of the expansion is an outflow of material from one or more members of the cluster. The velocity of the source(s) of the flow must be very nearly that of the extended molecular cloud, since the observed velocity of the S(1) line at the periphery is that of the molecular cloud. Velocities in the flow must therefore reach or exceed 100 km s^{-1}. It is apparent that the flow must be very energetic, because it has affected a volume in the core of OMC-1 with a characteristic radius of 0.1 pc in a short time ($\sim 10^3$ years). If this volume originally contained molecular gas at a density of $\sim 10^5$ cm^{-3}, the mass which has been accelerated by the flow to date is ~ 20 $M_\odot$.

b) Shock Models

A natural consequence of the collision of a supersonic flow of gas with a quiescent molecular cloud is the generation of shock waves, and it is by now generally accepted that shock waves are the most plausible mechanism for exciting the observed H_2 in OMC-1. An important constraint on shock models of H_2 line emission, however, is that, in the absence of magnetic precursors and for preshock densities $>10^4$ cm^{-3}, shocks which propagate at velocities higher than 24 km s^{-1} will dissociate all of the H_2 which passes through the shock (see e. g. Hollenbach and McKee 1980). The effects of magnetic precursors to shock waves (Draine 1980) have not yet been calculated extensively, and while they may raise the dissociation velocity above 24 km s^{-1}, the amount of the increase is uncertain. It remains a strong possibility, then, that the highest velocities at which H_2 is observed are substantially higher than the dissociation velocity for H_2 passing through a shock, with or without a precursor.

Simple models of excitation behind a single shock front propagating into the molecular cloud thus may not be able to explain a simultaneous vibrational excitation of the H_2 and its acceleration to the observed high velocities, and it is useful to consider other, somewhat more complex possibilities. If the high velocities are caused by shock-acceleration of the cloud then either (1) the emitting molecules may have reformed behind a dissociative shock or (2) multiple low velocity and non-dissociative shocks may be required. Another possibility is that (3) the observed high velocity line emission occurs in gas in the flow which already has a high velocity prior to its excitation. The third possibility appears to us to have fewer difficulties than the first two possibilities. There are several plausible mechanisms by which H_2 in the flow can be excited. Small fluctuation in the flow velocity would result in the propagation of non-dissociative shocks through parts of the flow, resulting in H_2 line emission at high velocities. Dense clumps in the flow (perhaps created in the above shocks), which encounter less dense stationary cloud material, will result in non-dissociative shocks being driven into them, as suggested by Chevalier (1980). Magnetic precursors may mediate the effects of shocks in the flow at the cloud boundary, which otherwise would be dissociative. The above mechanisms can result in a continuous range of emission line velocities, similar to what is seen toward the center of the emitting region. A final point in support of this picture is that the enormous momentum needed to accelerate perhaps $\sim 20\ M_{\odot}$ of cloud material to ~ 100 km s^{-1} is required by the first two possibilities, but not by the third.

It remains to account for the emission at low velocities. It is noteworthy that even in the central region, where the measured radial velocities probably approximately represent the actual gas motions, the peak of the line occurs at low velocities which are only slightly blueshifted from that of the ambient cloud. There is also a limb brightening effect at low velocities, seen in almost every direction from the center. These phenomena suggest that the bulk of the low velocity line emission occurs near the outside boundary of the expansion region. The most reasonable guesses as to the average density of the flow at the boundary ($\sim 10^4$ cm^{-3}) (Knapp et al. 1981) and of the molecular cloud (10^5 cm^{-3}), (Evans et al. 1975) suggest that their collision on the average would produce a low velocity (non-dissociative) shock propagating into the molecular cloud.

Fig. 3 summarizes our conclusions concerning the location of the 2μm line emitting H_2 in OMC-1. We suggest that the high velocity emission occurs in shocked gas in the flow and the low velocity emission occurs in shocked gas in the molecular cloud. The figure is a gross simplification; closer examination of the line profiles and the morphology of the region show evidence of both large scale and small scale inhomogeneities. Some of these may be due to peculiarities in the flow, others may be directly related to inhomogeneities in the surrounding molecular cloud (see e.g. Bastien et al. 1982).

Further details of this work can be found in Nadeau, Geballe, and Neugebauer (1981). This work was supported by NASA and NSF grants. D. N. was a fellow of the Natural Sciences and Engineering Research Council of Canada.

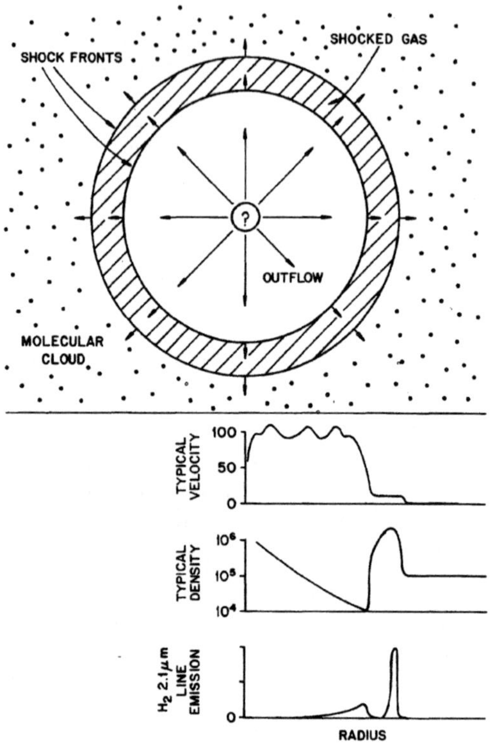

Fig. 3 - Simplified suggested model of the H_2 line emission region. Typical variations of velocity, density, and line emissivity with radius are given in the lower part of the figure.

REFERENCES

Bastien, P., Bieging, J., Henkel, C., Martin, R. N., Pauls, T., Walmsley, D. M., Wilson, T. L. and Ziurys, L. M.: 1982, this volume, pp. 335-336.

Beck, S. C.: 1981, in preparation.

Beckwith, S., Persson, S. E., Neugebauer, G. and Becklin, E. E.: 1978, Astrophys. J. 223, pp. 464-470.

Chevalier, R. A.: 1980, Astrophys Lett. 21, pp. 57-62.

Downes, D., Genzel, R., Becklin, E. E. and Wynn-Williams, C. G.: 1981, Astrophys. J. 244, pp. 869-883.

Draine, B. T.: 1980, Astrophys. J. 241, pp. 1021-1038.

Evans II, N. J., Zuckerman, B., Sato, T. and Morris, G.: 1975, Astrophys J. 199, pp. 383-397.

Hollenbach, D. J. and McKee, C. F.: 1980, Astrophys. J. (Letters) 241, pp. L47-L50.
Knapp, G. R., Phillips, T. G., Huggins, P. J. and Redman, R. O.: 1981, Astrophys. J., in press.
Nadeau, D., Geballe, T. R. and Neugebauer, G.: 1981, Astrophys. J. submitted.

DISCUSSION FOLLOWING PAPER BY GEBALLE AND NADEAU

PHILLIPS: Are the low velocity ($\sim$10 km s^{-1}) shocks that you envisage consistent with the excitation temperature of 2000 K for the H_2?

GEBALLE: Yes, they are.

PHILLIPS: Is there any evidence for a variation of excitation temperature with velocity in the line emission?

GEBALLE: We would have been able to detect differences of 500 K but we saw none.

HIGH VELOCITY H_2 LINE EMISSION IN THE NGC 2071 REGION

S. E. Persson, T. R. Geballe
Mount Wilson and Las Campanas Observatories
Carnegie Institution of Washington

Theodore Simon, Carol J. Lonsdale
Institute for Astronomy, University of Hawaii

F. Baas
Sterrenwacht Huygens Laboratorium
Rijksuniversiteit te Leiden

Abstract The line profile of the v=1→0 S(1) line of H_2 in NGC 2071 is $\sim$100 km s^{-1} wide (FWZI), and asymmetric; it resembles S(1) profiles seen toward Orion KL. The NGC 2071 region represents the second detection of high velocity H_2 emission in a region showing signs of ongoing star formation.

Several molecular clouds have now been found to contain regions which emit radiation in the infrared vibration-rotation lines of the hydrogen molecule (see Beckwith 1981 for a review). By far the best studied of these regions is the core of the Orion Molecular Cloud (OMC-1). The intensities of the various infrared lines are most easily explained as arising from a collisional process, probably in a strong shock wave which is being driven into the molecular cloud. The H_2 lines are up to 150 km s^{-1} wide, and asymmetric, with an excess of emission at negative velocities. The most extreme profiles are found toward the center of the emission region, near KL (Nadeau and Geballe 1979 [NG], and 1982; Scoville et al. 1981). These high velocities present serious problems for a simple shock picture (Kwan 1977; Hollenbach and McKee 1980).

In order to extend the study of H_2 gas motions to other sources, we have obtained a 20 km s^{-1} resolution profile of the v=1→0 S(1) line in a source recently discovered by Simon and Joyce (1981) and Lane and Bally (1982) in their low spectral resolution searches for new H_2 sources in molecular clouds. The particular region of interest, located about 4' North of the NGC 2071 reflection nebula, shows several signs of ongoing star forming activity. These include the presence of OH and H_2O masers (Johansson et al. 1974; Genzel and Downes 1979 and references therein), a bright, heavily reddened 2μm source (No. 41; Strom et al. 1976), a cluster of 10μm sources (Persson et al. 1981; Evans et al.

R. S. Roger and P. E. Dewdney (eds.), Regions of Recent Star Formation, 155–159.
Copyright © 1982 by D. Reidel Publishing Company.

1979), a far infrared source (Harvey et al. 1979), and probably most importantly, a J=1→0 ^{12}CO line profile which is 70 km s^{-1} wide, (full width at zero intensity; Bally 1982).

The observations were made on the NASA 3-m infrared telescope (IRTF) of Mauna Kea Observatory. We used a cold grating infrared spectrometer in conjunction with a Fabry-Perot interferometer. (The instrument is described in detail by Persson, Geballe, and Baas 1981). Our spectrum, taken with 20 km s^{-1} resolution at the position of a local maximum in the extended H_2 distribution, is shown in Figure 1. This position, at RA=$05^h\ 44^m\ 31^s\!.2 \pm 0^s\!.2$, DEC=+00° 20' 48" ± 2 (1950.0) happens to correspond, to within the uncertainties, with that of the H_2O maser (Genzel and Downes 1979), and also one of the fainter 10μm sources (Persson et al. 1981). The integrated S(1) line intensity is $1.8 \pm 0.3 \times 10^{-20}$ W cm^{-2} into a 7" diameter beam; this is roughly 40 times fainter than that seen at Peak 1 of the Orion H_2 source (Beckwith et al. 1978).

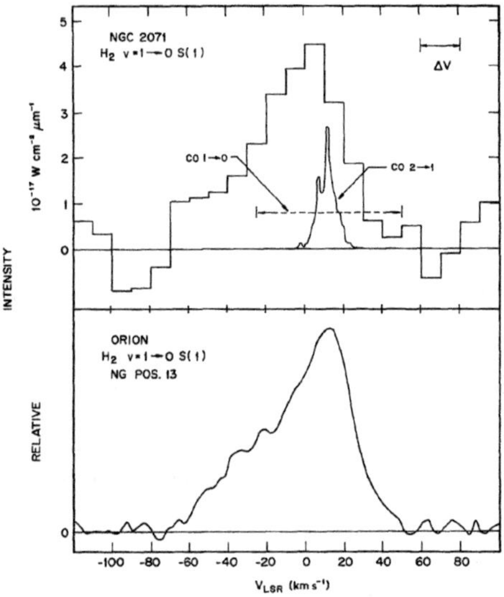

Fig. 1 - Spectrum of the v=1→0 S(1) 2.12 micron line of H_2 towards the infrared source region north of NGC 2071. The line observed by Nadeau and Geballe (1979) in the KL nebula in OMC-1, at their position 13, is shown for comparison. The FWHM velocity resolution (ΔV=20 km s^{-1} for both spectra) is also shown. For NGC 2071 the data were binned into 10 km s^{-1} intervals. The baseline was determined by averaging the data in 50 km s^{-1} intervals at each end of the spectrum. The CO J=2-1 profile is adapted from the work of Loren et al. (1981). The full extent of the emission in the CO J=1-0 line as measured by Bally (1982) is also shown.

The NGC 2071 H_2 line profile is wide and asymmetric. S(1) line emission extends over a range of at least 100 km s^{-1}, and the blueshifted side of the line is much stronger than the redshifted side. The peak of the S(1) line occurs at $V_{LSR}=+5\pm10$ km s^{-1}, which is close to the +10 km s^{-1} velocity of the molecular cloud (Bally 1982). NGC 2071 is thus the second example, after OMC-1, of a source showing high velocity emission in the H_2 line. The fact that the velocity extent of the H_2 is larger than that of the CO line (Bally 1982) may be due to a difference in sensitivity of the two techniques, or could simply reflect the fact that the H_2 emission arises in a different hotter part of the cloud.

The S(1) line profile observed in the NGC 2071 region is remarkably similar to those observed by NG in the center of the H_2 line emission region in OMC-1. The lower panel in Fig. 1 is NG's S(1) spectrum at their position 13, obtained at essentially the same spectral resolution as the present measurement. Position 13 lies close to the center of the KL nebula, where a number of luminous young objects are found. Although this particular OMC-1 spectrum was selected as being the most similar to the one seen in the NGC 2071 region, all of the spectra near KL show the same general characteristics as position 13.

In OMC-1 the broad radio molecular line profiles, the pattern of proper motions of the high velocity masers, and the broad, asymmetric H_2 emission lines have all been attributed to an outward flow of material from a central object, presumably one of the infrared cluster sources (Kwan and Scoville 1976; Zuckerman et al. 1976; NG, Downes et al. 1981; Genzel et al. 1981; Welch et al. 1981; Reid and Moran 1981). The absence of red wings in the near infrared emission lines of H_2 is thought to be due to extinction within the expanding volume (NG; Beck 1981).

The present H_2 line spectroscopic data together with the existence nearby of several 10μm sources show that similar phenomena may be taking place in the core of the NGC 2071 source. The amounts of extinction within the high velocity flows in NGC 2071 and OMC-1 may be quite different however, because the intensities of the ^{12}CO high velocity wings in NGC 2071 (Bally 1982) and OMC-1 (e.g. Scoville 1980) differ by at least a factor of 10. The asymmetries seen in the H_2 line profiles cannot then be attributed simply to extinction within the flows without invoking additional mechanisms.

The luminosities of the brighter objects within the infrared cluster of OMC-1 are in the range 10^4-10^5 $L_\odot$ (Downes et al. 1981) while Harvey et al. (1979) have detected a total of only 750 $L_\odot$ in the vicinity of the infrared sources north of NGC 2071. It may be concluded therefore that the luminosity of a source responsible for an outflow is not directly related to the velocity of that outflow. The similarity of the maximum velocity extents of the H_2 emission in NGC 2071 and OMC-1 is apparently due to similar ratios of flow momentum to density of the quiescent molecular clouds which surround the two sources. Both the flow momentum as estimated from the CO broad line intensity, and the density of the surrounding cloud in NGC 2071 (Bally 1982) are one to two orders of magnitude lower than in the core of OMC-1 (Zuckerman et al. 1976; Goldsmith et al. 1980; Bastien et al. 1981). The observed H_2 S(1) line intensity in the NGC 2071 region is also lower than in OMC-1 by a similar factor, although the amount of extinction toward the NGC 2071 H_2 source is not determined, and could be very different from that toward the Orion source. In these respects, then, the NGC 2071 region seems to be simply a scaled down version of th core of OMC-1. Whether a common physical mechanism underlies the apparent similarity of the ratio of luminosity to flow momentum in OMC-1 and NGC 2071 is not clear. Radiatively driven winds are a possibility, but have been considered inadequate to explain the high velocity flows; the momentum in photons is typically a factor of at least 50 too small (Zuckerman 1981; Lada and Harvey 1981). For the case of OMC-1, Solomon, Huguenin and Scoville (1981) and Phillips and Beckman (1980) have suggested that a large infrared grain opacity close to the central power source strongly traps the infrared radiation and a large multiplication of the photon momentum imparted to the gas can be achieved. It seems doubtful, however, that such a picture can work for NGC 2071, because the total luminosity of the region is roughly two orders of magnitude smaller than that of OMC-1 (see equation (3) of Solomon et al. and their ensuing discussion).

A more detailed account of this work will appear in Astrophysical Journal Letters.

REFERENCES

Bally, J.: 1982, this volume, p. 287.
Bastien, P., Bieging, J., Henkel, C., Martin, R.N., Pauls, T., Walmsley, C.M., Wilson, T.L. and Ziurys, L.M.: 1981, Astr. Ap. in press.
Beck, S.C.: 1981, private communication.
Beckwith, S.: 1981, IAU Symposium No. 96,"Infrared Astronomy",ed. C.G. Wynn-Williams and D.P. Cruikshank (Dordrecht; Reidel) p. 167.
Beckwith, S., Persson, S.E., Neugebauer, G. and Becklin, E.E. 1978: Ap. J. 223, 464.
Downes, D., Genzel, R., Becklin, E.E. and Wynn-Williams, C.G.: 1981, Ap. J. 244, 869.
Evans II, N.J., Beckwith, S., Brown, R.L. and Gilmore, W.: 1979, Ap. J. 227, 450.
Genzel, R. and Downes, D.: 1979, Astr. Ap. 72, 234.
Genzel, R., Reid, M.J., Moran, J.M. and Downes, D.: 1981, Ap. J. 244, 884.

Goldsmith, P.F., Langer, W.D., Schloerb, F.P. and Scoville, N.Z.: 1980, Ap. J. 240, 524.
Harvey, P.M., Campbell, M.F., Hoffmann, W.F., Thronson Jr., H.A. and Gatley, I.: 1979, Ap. J. 229, 990.
Hollenbach, D. and McKee, C.F.: 1980, Ap. J. (Letters) 241, L47.
Johansson, L.E.B., Höglund, B., Winnberg, A., Nguyen-Q-Rieu and Goss, W.M.: 1974, Ap. J. 189, 455.
Kwan, J.: 1977, Ap. J. 216, 713.
Kwan, J. and Scoville, N.Z.: 1976, Ap. J. 210, L39.
Lada, C.J. and Harvey, P.M.: 1981: Ap. J. 245, 58.
Lane, A.P. and Bally, J.: 1982, this volume, p. 301.
Loren, R.B., Plambeck, R.L., Davis, J.H. and Snell, R.L.: 1981, Ap. J. in press.
Nadeau, D. and Geballe, T.R.: 1979, Ap. J. (Letters) 230, L169.
Nadeau, D., Geballe, T.R.: 1982, preprint, this volume, p. 147.
Persson, S.E., Geballe, T.R. and Baas. F.: 1981, in preparation.
Persson, S.E., Gaballe, T.R., Simon, T., Lonsdale, C.J. and Baas, F.: 1981, Ap. J. (Letters), in press.
Phillips, J.P. and Beckman, J.E.: 1980, MNRAS 193, 245.
Reid, M.J. and Moran, J.M.: 1981, article to appear in Ann. Rev. Ast. Ap.
Scoville, N.Z.: 1980, IAU Symposium 87, "Interstellar Molecules", ed. B.H. Andrew (Reidel, Dordrecht), p. 33
Scoville, N.Z., Kleinmann, S.G., Hall, D.N.B. and Ridgway, S.T.: 1981, Ap. J., in press.
Simon, T. and Joyce, R.R.: 1981, in preparation.
Solomon, P.M., Huguenin, G.R. and Scoville, N.Z.: 1981, Ap. J. (Letters) 245, L19.
Strom, K.M., Strom, S.E. and Vrba, F.J.: 1976, Ap. J. 81, 308.
Welch, W.J., Wright, M.C.H., Plambeck, R.L., Beiging, J.H. and Baud, B.: 1981, Ap. J. (Letters) 245, L87.
Zuckerman, B.: 1981, IAU Symposium No. 96, "Infrared Astronomy", (ed. C.G. Wynn-Williams and D.P. Cruikshank) p. 275.
Zuckerman, B., Kuiper, T.B.H. and Rodriguez-Kuiper, E.N.: 1976, Ap. J. 209, L137.

INFRARED ATOMIC HYDROGEN LINE FORMATION IN LUMINOUS STARS

Julian H. Krolik
Department of Physics, Center for Space Research, and Center for Theoretical Physics, Massachusetts Institute of Technology
and
Howard A. Smith
Steward Observatory, University of Arizona and E.O. Hulburt Center for Space Research, Naval Research Laboratory

ABSTRACT

Infrared atomic hydrogen lines observed in luminous stars, generally attributed to compact circumstellar HII regions, can also be formed in the winds likely to emanate from these stars. We show how the mass-loss rate may be estimated from line fluxes, and also how constraints on the velocity profile may be deduced. Winds capable of producing lines of the observed flux should also be weak, but detectable, sources of radio emission.

A number of bright ($10^{4\pm1}$ $L_\odot$), but dust-obscured, stars exhibit strong Brα, Brγ, and occasionally Pfund β and γ emission lines.[1,2,5] It has been suggested that these lines are formed in a compact HII region surrounding the star, but it was immediately recognized that no star of the measured luminosity following a standard evolutionary track could generate enough ionizing photons to produce lines of the observed strength by recombination.[3] A variety of solutions to the problem have been proposed, including UV radiation from viscous circumstellar disks.[4] Instead, we suggest that these lines are produced thermally in an optically thick wind surrounding the star. If our suggestion is correct, the lines will be a new diagnostic in the study of stellar mass-loss - and, of course, the roster of compact HII regions will be diminished. In this paper we will first demonstrate the plausibility of our picture, and then show how relatively crude measurements of line fluxes can be used to estimate physically interesting quantities like the mass-loss rate, the flow speed, and the velocity profile of the outflow (though the last is only weakly determined).

R. S. Roger and P. E. Dewdney (eds.), Regions of Recent Star Formation, 161–166.
Copyright © 1982 by D. Reidel Publishing Company.

Perhaps the best reason for investigating the line-generating abilities of winds from these stars is that the lines are observed to be broad. Hall, et al.[5] found that the Brα wings of BN extend out to ±100 km s^{-1}. More recently, Simon, et al.[6] measured Brγ widths of 150 km s^{-1} and 135 km s^{-1} (FWHM) in GL 490, and M17 IRS1, respectively. From a more theoretical point of view, we may link these stars with unobscured bright stars of comparable luminosity, which we can unambiguously observe shedding mass at a rate of 10^{-6} $M_\odot$ yr^{-1}.[7] Alternatively, stars buried deep in dusty clouds, as these are, may be young, and, like T Tauri's, possess substantial winds.

Futhermore, there are physical reasons why it is easy for stellar winds to make strong IR lines. In an HII region ∿15 Lyman continuum photons, or ∿200 eV, must be absorbed in order to generate one Brα photon by recombination cascade.[8] On the other hand, an exposed surface with a moderately well-defined temperature radiates most strongly in those transitions which are optically thick and fall near the peak in the Planck spectrum for that temperature. Near infrared lines are near enough the emissivity peak of typical stellar atmospheric gas (∿1/10 the energy of the Planck maximum when the temperature is ∿1 eV), that they are emitted with fair efficiency. If we make the comparison between thermal radiation and recombination for Brα, we find the ratio:

$$L^{(th)}_{Br\alpha}/L^{(rec)}_{Br\alpha} \simeq 0.5\ \Delta v_{100}\ R^2_{\alpha,13}\ N^{-1}_{47} \qquad (1)$$

where Δv_{100} is the equivalent full width that is optically thick in units of 100 km s^{-1}, $R_{\alpha,13}$ is the areal equivalent radius of the Brα photosphere in units of 10^{13} cm, and N_{47} is the number of Lyman continuum photons absorbed, in units of 10^{47} s^{-1}. We have assumed in Eq. 1 that the excitation temperature of Brα is 10^4 K. A line is visible against the continuum because its greater opacity expands its radiating surface area; we shall see shortly that $R_{\alpha,13}$ ∿1 is a fair estimate. We chose to normalize to 10^{47} photons s^{-1} because that is the ionizing flux from a ZAMS B0.5 star, whose luminosity is ∿2x10^4 $L_\odot$.[9] Both before and after zero-age, stars are redder, so 10^{47} s^{-1} is an upper limit on the number of ionizing photons from a star of this luminosity. Thus we see that thermal processes can easily compete with recombination in the generation of infrared photons.

To quantitatively analyze the situation we employ a simplified version of the standard Sobolev theory.[10] Here we merely sketch our reasoning; for a proper derivation of our results, we refer readers to our original paper.[11] In a dense wind, the populations of the excited states which define these infrared lines are roughly in thermal equilibrium with the free electrons.[11,12] These populations are then proportional to the square of the electron density, hence to $[r^2v(r)]^{-2}$, where v(r) is the speed at radius r. Therefore, the opacity drops very rapidly with radius, and the radius is well-defined at which rays passing through the wind at a given offset from the line of sight to the center of the star are marginally optically thick to photons of a given frequency offset from line-center. The characteristic scale for all such rays and line frequencies is:

$$R_\ell = v^{-1}(R_\ell)\left\{\frac{\pi e^2}{m_e}\frac{f_{nn'}}{\nu_{nn'}}\left(\frac{\dot{M}}{4\pi\mu_e}\right)\frac{N_n^{th}\left[T_{ex}(R_\ell)\right]}{n_e^2}\left(1-\frac{1}{Y_\ell}\right)\right\}^{1/3}$$

$$= 3.10 \times 10^{13}\, v_{100}^{-1}(R_\ell)\left[f_{nn'}\,\nu_{13}^{-1}\,\dot{M}_5^2\left(\frac{m_p}{\mu_e}\right)^2\right]^{1/3} T_{ex,4}^{-1/2}\ \times$$

$$\times \left[n^2 \exp\left(\frac{I_H}{kT_{ex}n^2}\right)\left(1-\frac{1}{Y_\ell}\right)\right]^{1/3} \text{cm} \qquad (2)$$

where $f_{nn'}$ is the oscillator strength of the transition n'-n with frequency $\nu_{nn'}$ (measured in units of 10^{13} Hz) and having excitation temperature T_{ex} at R_ℓ ($Y_\ell = \exp(\frac{h\nu_{nn'}}{kT_{ex}})$); $\dot{M}$ is the mass-loss rate (the units are $10^{-5} M_\odot\ yr^{-1}$), μ_e is the mass per electron; and N_n^{th} is the population level n would have in thermal equilibrium. We immediately see that the radiating surface area for a hydrogen infrared line can easily exceed the continuum photospheric surface area.

The total luminosity in the line is found by multiplying the actual radiating surface area for each frequency in the optically thick portion of the line by the Planckian at that frequency and integrating over the optically thick bandwidth:

$$L_\ell = 8\pi^2 R_\ell^2 \frac{v(R_\ell)}{c} B_\nu\left[T_{ex}(R_\ell)\right]\nu_\ell\ W_\ell \qquad (3)$$

where $W_\ell \sim O(1)$ hides the details of the calculation. To actually evaluate W_ℓ requires a knowledge of $T_{ex}(R_\ell)$, the velocity profile of the wind, and the size of the stellar disk capable of occulting the line. W_ℓ is only weakly dependent on variations in T_{ex}, but decreases slowly with increasing magnitude of the velocity gradient. Alternatively, if we measure the line luminosity, we can use Eq. 3 to place constraints on these quantities:

$$W_\ell = 2.58\ L_{\ell 30}\ R_{\ell 13}^{-2}\ \nu_{13}^{-4}\ v_{100}^{-1}(R_\ell)\ (Y_\ell - 1) \qquad (4)$$

where the unit of line luminosity is 10^{30} erg s^{-1}. Unfortunately, the luminosity is always subject to two intrinsic uncertainties: the distance to the source, and the fraction of the flux in the line which is due to radiation from dust along the line of sight.

It is sometimes more useful to reverse the reasoning and obtain the mass-loss rate which would be necessary to support the line:

Table 1- Line and Radio Emission from Obscured Objects

Object	Brα flux observed (10^{-12} erg $cm^{-2}s^{-1}$)	estimated extinction (A_V: mag)	intrinsic Brα flux (10^{-12} erg $cm^{-2}s^{-1}$)
BN	20 [5]	35	78
GL 490	3.9 [13]	12	6.2
LkHα101	200 [13]	14	340
R Mon	3.5 [13]	40	16
NGC 2264	5.3 [13]	35	21
M8E	1.6 [14]	>10	>2
M17/IRS1	3.6 [14]	15	6
S106 IR	62 [15]	5	73

Object	Distance (kpc)	mass-flux ($10^{-6} M_{\odot}$ yr^{-1})	6cm flux predicted (mJy)	6cm flux observed (mJy)
BN	0.48	1.1	3.7	<3 [14]
GL 490	0.90	0.44	0.30	<3 [14]
LkHα101	0.80	7.4	16.	30 [16]
R Mon	0.70	0.61	0.77	<5 [16]
NGC 2264	0.80	0.92	1.0	<12 [17]
M8E	1.50	>0.41	>0.096	4 [14]
M17/IRS1	2.0	1.4	0.29	4 [18]
S106 IR	2.3	11.	3.5	<5 [19]

$$\dot{M} = 1.3\times10^{-4} \frac{\mu_e}{m_p} \nu_{13}^{-5/2} D_{kpc}^{3/2} \left[v_{100} (R_\ell) F_{12} T_{ex,4} (R_\ell) W_\ell^{-1}\right]^{3/4} \times$$

$$\times f_{nn'}^{-1/2} n^{-1} \exp\left(\frac{-7.89}{n^2 T_{ex,4}}\right) (Y_\ell - 1)^{1/4} M_\odot \, yr^{-1} \qquad (5)$$

where D_{kpc} is the distance to the star in kpc and F_{12} is the measured line flux in units of 10^{-12} erg cm^{-2} s^{-1}. In Table 1 we show the predicted mass-loss for a variety of objects with Brα lines. The numerical values all assume $v_{100}(R_\ell) = W_\ell = \frac{\mu_e}{m_p} = T_{ex,4} = 1$.

In order to maintain the excited-state population that keeps these infrared lines optically thick, a sizable electron density is necessary. These electrons, and their associated ions, will participate in free-free interactions with radio-frequency photons. Due to the rapid absorption and emission of low-frequency photons, the wind will present a sizable thermally radiating surface throughout the radio band. Its radio flux scales in proportion to the infrared line flux:

$$\frac{L_\nu(RF)}{L_\ell} = \frac{1}{2} \frac{B_{\nu RF}(T_{RF})}{B_{\nu\ell}(T_{ex,\ell})} c v(R_\ell) \left[\frac{m_e}{3\pi e^2} \frac{(\kappa_{ff}/n_e^2)}{f_{nn'} \nu_\ell^{1/2} v^2(R_{RF})} \times \right.$$

$$\left. \times \left(\frac{n_e^2}{N_n^{th}}\right) \frac{Y_\ell}{Y_\ell - 1}\right]^{2/3} \qquad (6)$$

where κ_{ff} is the free-free opacity. Table 1 shows our predictions for the radio flux from a number of Brα-emitting objects. In fact, the radiating surface area in some of the nearer objects is big enough to be resolved by the VLA. The half-angle is:

$$\theta_{RF} = 0.17 \left(\dot{M}_5 \frac{m_p}{\mu_e}\right)^{2/3} v_{100}^{-2/3} (R_{RF}) T_{RF,4}^{-1/2} \nu_{5GHz}^{-2/3} D_{kpc}^{-1} \text{ arc sec} \qquad (7)$$

We have now shown that the strong infrared atomic hydrogen lines observed in a number of luminous, but highly obscured, stars may be easily explained if these stars possess strong ($\dot{M} \sim 10^{-5} M_\odot$ yr^{-1}), moderately speedy ($v \sim 100$ km s^{-1}) winds. The existence of these winds is both suggested by the breadth of the lines and conforms to some of our theoretical prejudices about the nature of these stars. To be sure, in a sense we have merely substituted implicitly postulated unknown heating agents to supply the lines' power in order to replace their implausibly large demand for ionizing photons in the recombination model. However, it is much easier to imagine diverting a mere 10^{-7} of the stellar luminosity, which is all the power the lines require, than it is to amplify the production of hard UV photons by one or two orders of magnitude.

Our picture also possesses several corollaries of relevance to observation. First, not all atomic hydrogen lines signal the existence of photoionized HII regions. Second, the infrared line ratios are not necessarily those given by recombination cascade, and therefore cannot be used for extinction analyses. Third, radio emission, perhaps resolvable, should accompany infrared atomic hydrogen line emission.

ACKNOWLEDGMENTS

J.H.K. was partially supported by NASA Grant NSG-7643.

REFERENCES

[1] Thompson, R.I. 1980, preprint.
[2] Smith, H.A., Larsen, H.P. and Fink, U. 1979, *Astrophs. J.*, *233*, 132.
[3] Thompson, R.I. and Tokunaga, A.T. 1978, *Astrophys. J.*, *226*, 119.
[4] Thompson, R.I., Strittmatter, P.A., Erickson, E.F., Witteborn, F.C. and Strecker, D.W. 1977, *Astrophys. J.*, *218*, 170.
[5] Hall, D.N.B., Kleinmann, S.G., Ridgway, S.T. and Gillett, F.C. 1978, *Astrophys. J. (Letters)*, *223*, L47.
[6] Simon, M., Righini-Cohen, G., Fischer, J. and Cassar, L. 1981, submitted to *Astrophys. J. (Letters)*.
[7] Cassinelli, J.P. 1979, *Ann. Rev. Astron. Astrophys.*, *17*, 275.
[8] Seaton, M.J. 1959, *M.N.R.A.S.*, *119*, 90.
[9] Spitzer, L., Jr. 1978, "Physical Processes in the Interstellar Medium" Wiley-Interscience, New York.
[10] Castor, J.I. 1970, *M.N.R.A.S.*, *149*, 111.
[11] Krolik, J.H. and Smith, H.A. 1981, *Astrophys. J.*, in press.
[12] Kunasz, P.B. 1980, *Astrophys. J.*, *237*, 819.
[13] Simon, T., Simon, M. and Joyce, R.R. 1979, *Astrophys. J.*, *230*, 127.
[14] Simon, M., Righini-Cohen, G., Felli, M. and Fischer, J. 1980, preprint.
[15] Pipher, J.L., Sharpless, S., Savedoff, M.P., Kerridge, S.J., Krassner, J., Schurmann, S., Soifer, B.T. and Merrill, K.M. 1976, *Astron. Astrop.*, *51*, 255.
[16] Altenhoff, W.J., Braese, L.L.E., Olnon, F.M. and Wendker, H.J. 1976, *Astron. Astrop.*, *46*, 11.
[17] Harris, S. 1976, *M.N.R.A.S.*, *174*, 601.
[18] Felli, M., Johnston, K.J. and Churchwell, E. 1980, preprint.
[19] Israel, F.P. and Felli, M. 1978, *Astron. Astrop.*, *63*, 325.

ATOMIC HYDROGEN ZONES ASSOCIATED WITH HII REGIONS

R.S. Roger
Dominion Radio Astrophysical Observatory
Herzberg Institute of Astrophysics
Penticton, B.C., Canada

ABSTRACT

The broad HI zones detected in 21-cm emission near three HII regions, NGC 281, IC 5146 and NGC 1579, are described and compared. The formation of such zones in low and medium density gas by dissociation of H_2 with UV radiation from the exciting star, is discussed. The HI spin temperatures are probably in the range 100K - 300K and for the two most evolved sources the zones are expanding. Neither of these shows evidence of the thin dense shell of shocked gas predicted to lie just ahead of an ionization front.

INTRODUCTION

The detection of atomic hydrogen directly associated with HII Regions was first convincingly demonstrated by Riegel (1967). These observations, however, lacked sufficient resolution to show any details of the associations. Theoretical and numerical treatments of the expansion of an HII region (e.g. Savedoff and Greene 1955; Lasker 1966; Tenorio-Tagle 1976) have long predicted a thin dense shell of shocked gas just ahead of the advancing ionization front. The details of the atomic hydrogen component of the shocked shell and of the region ahead of the shock front, however, have not been put to observational test.

Recently the environs of a number of HII regions have been mapped in the 21-cm spectral line of HI with synthesis radiotelescopes. In this paper I will compare the atomic hydrogen associated with three HII regions at quite different stages of evolution. The individual observations are each reported in more detail elsewhere.

OBSERVATIONS

NGC 1579 (Dewdney and Roger 1981), IC5146 (Roger and Irwin 1982), and NGC 281 (Roger and Pedlar 1981) were observed in the 21-cm

R. S. Roger and P. E. Dewdney (eds.), Regions of Recent Star Formation, 167–174.
Copyright © 1982 by D. Reidel Publishing Company.

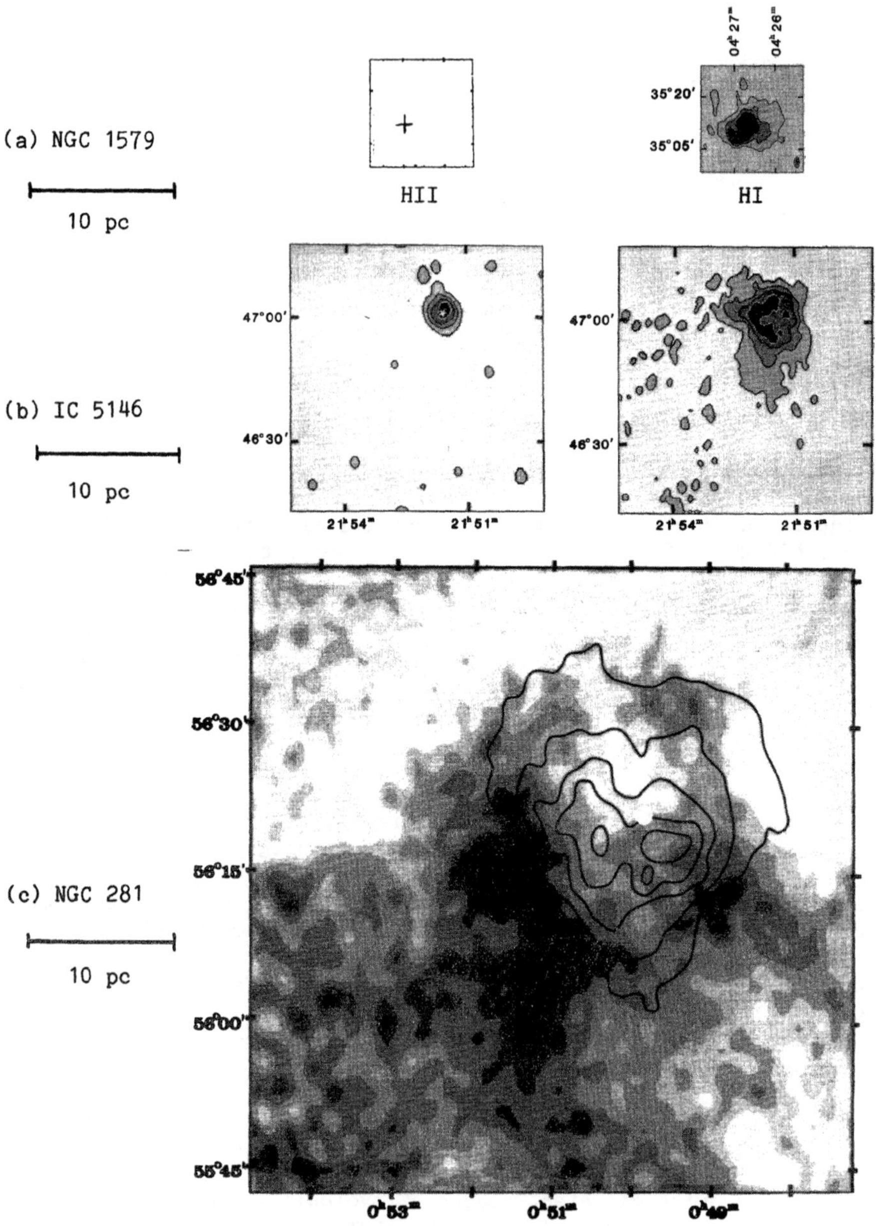

Figure 1. Maps of the integrated HI emission and of the thermal continuum emission for the three regions to the same scale; (a) NGC 1579 - left field shows only position of star and HII region, (b) IC 5146 - position of exciting star shown (✱) on left map of continuum emission, (c) NGC 281 - contours depict the thermal emission, grey levels the HI emission and the white dot the position of the exciting star.

continuum and spectral line of HI using the Synthesis Radiotelescope at the Dominion Radio Astrophysical Observatory (Roger et al. 1973) equipped with a 125 channel digital cross-correlation spectrometer. Table 1 lists the telescope parameters relevant to the three sets of observations.

Table 1. Telescope Parameters

Primary Field (FWHM)	106 arcmin
Synthesized beam (FWHM)	2.0' EW X 2.0' csc(δ) NS
Continuum bandwidth	15 MHz
Range in radial velocity	53 km s^{-1} (125 channels)
Radial velocity resolution	0.67 km s^{-1}

Fine scale structure for each source was measured for 12 hours at each of 68 interferometer spacings from 61λ to 1421λ. Broad structure, equivalent to spacings <61λ, was measured with the DRAO 26-m paraboloid (and the same digital spectrometer) and added to the synthesized maps.

Each of the three HII regions is excited mainly by a single star. The following is a short description of each region, considered in order of increasing size and age.

NGC 1579 is optically apparent as a reflection nebula (S222) whose source of illumination is the emission-line star Lk Hα101 (Herbig 1971). The star and the nebula, at ~800 pc distance, are associated with a large diffuse dark cloud embedded in which are dense concentrations on a smaller scale. One of these dense dust bands obscures the star by ~11^m. Lk Hα101, with a luminosity equivalent to a B0.5 ZAMS, is surrounded by an HII region of diameter ~0.15pc, density ~800 cm^{-3} and total mass ~0.08 $M_\odot$ (Altenhoff et al. 1976; Dewdney and Roger 1981).

A dense core component of ionization (Altenhoff et al. 1976), and infrared sources detected at .84 and 2.2 μm (Neugebauer and Leighton 1969) and at 100 μm (Harvey at al. 1977) are possibly due to gas and dust left from the star's formation.

We have mapped an isolated cloud of atomic hydrogen surrounding the HII region, shown in Figure 1(a). The cloud comprises ~85 $M_\odot$ and extends over a projected area ~3.5 pc across. The measured column densities and extent suggest densities in the range 100 - 200 cm^{-3}, probably similar to that in the diffuse dark cloud as a whole. The distribution shown in Figure 1(a) is an integration of the range in radial velocity of -6 to +6 km s^{-1}, uncorrected for apparent foreground narrow band absorption. This absorption is probably diffuse cold atomic gas associated with the near side of the dark cloud in which the nebula is embedded. When corrected for this absorption the central profiles show a peak brightness temperature of ~55K.

The gas is most extended towards the northwest, with a similar asymmetry to that of the reflection nebula. Column densities of distributed CO, on the other hand, indicate higher average densities of molecular gas to the southeast of Lk Hα101 (Christie et al., this volume). No systematic motions of the atomic gas (e.g. expansion, contraction or rotation) are apparent.

IC 5146 is a circularly symmetric HII region surrounding a BOV star (HD 46°3474) situated on the near side of the bulbous end of a lane of obscuration. The ionized region, ~2.5 pc in diameter is at a distance of 960 pc (Crampton and Fisher 1974) and comprises about 10 $M_\odot$.

Surrounding the ionized gas is a zone of atomic hydrogen amounting to ~440 $M_\odot$ (Roger and Irwin 1982). The HI shows peak brightness temperatures of ~90K. The distributions of both the HII thermal continuum emission and the HI emission, integrated over all radial velocities, are shown in Figure 1(b). Gas at "approaching" velocities, relative to the mean velocity, is brightest in a ring of emission surrounding the HII region. Gas at receding velocities, on the other hand, is centrally bright and clearly represents HI on the far side of the HII region. The distribution in position and velocity is consistent with a general expansion of ~2 km s^{-1} outwards from the central nebula. All evidence points to the star and its HII region being on the near edge of a roughly spherical molecular cloud. The modelled expansion of the HII region indicates an age of ~10^5 years for the atomic and ionized gas. At about half this age the HII region would have broken out from the parent cloud in the manner of a Champagne flow (Bodenheimer et al 1979). Hα and H142α measurements (Kuiper et al. 1976; Williamson 1970) indicate that much of the ionized gas has an approach velocity of ~6 km s^{-1} relative to the bulk of the HI.

It is particularly interesting to note that although the bulbous dark cloud encompassing IC5146 extends to a dark lane to the northwest, the HI cloud shows no such extension. Thus the warm atomic gas is clearly associated with the nebula and its exciting star.

IC5146 has three dense concentrations of molecular gas on the periphery of the HII region, as indicated by CO observations (Lada and Elmegreen 1979; McCutcheon, Roger and Dickman 1982). At least one of these possesses an embedded infrared source (Sargent et al. 1981).

NGC 281 is an extended HII region surrounding an O6.5 star at a distance of 2.3 kpc. The ionized nebula, 20 pc across and comprising ~2200 $M_\odot$, appears to have expanded until it is now only partially ionization-limited and has a mean density of <20 cm^{-3}.

The main concentration of associated HI is a broad zone on the east, southeast, and southwest sides of the nebula as illustrated in

Figure 1(c), which shows contours of thermal continuum emission superposed on a grey-level representation of the HI emission integrated over all channels. The densest concentrations of HI coincide with the steepest gradients in HII emission. The mass of atomic hydrogen directly associated with the nebula is ~3500 $M_\odot$. From the excess line width of the HI profile on the southeast side, the atomic gas appears to be expanding outwards at ~6 km s^{-1}.

Maps of CO emission (Elmegreen and Lada 1978; Elmegreen and Moran 1979) show molecular gas within both the southwest and southeast HI zones. The southwest molecular cloud coincident with foreground obscuration almost certainly represents the most dense and massive concentration in the region and contains within it a source of H_2O maser emission (Elmegreen and Moran 1979). Much of the ionized gas of NGC281 appears to be receding with respect to this massive foreground cloud (Johnson et al. 1981) which may represent erosion of it by the exciting star.

Table 2. Summary of the HII Regions and HI Zones

	NGC1579 (Dewdney & Roger, 1981)	IC5146 (Roger & Irwin, 1982)	NGC281 (Roger & Pedlar 1981)
Distance	800pc	960pc	2300pc
Exciting star	B0.5 ZAMS	B0 V	O6.5V((f))
HII Density	800 cm^{-3}	100 cm^{-3}	20 cm^{-3}
Mass	0.08 $M_\odot$	10 $M_\odot$	2200 $M_\odot$
Mean diameter	0.15 pc	2.5 pc	20 pc
HI Density	150 cm^{-3}	200 cm^{-3}	20 cm^{-3}
Mass	85 $M_\odot$	440 $M_\odot$	3500 $M_\odot$
Mean diameter	3.5 pc	6 pc	-
Expansion velocity	-	2 km s^{-1}	6 km s^{-1}
Age	< 10^4 yrs	8 x 10^4 yrs	10^6 yrs

THE ATOMIC ZONES

The properties of the three HII regions and their HI zones are summarized in Table 2. The existence of the zones and their radial extent are satisfactorily explained in each case (see the individual papers cited above) as due to the two-step dissociation of H_2 proposed by Stecher and Williams (1969). The molecules are excited to the first few electronic levels by absorption in the Lyman-Werner bands in the range 91 to 112 nm and, in cascading down, occasionally (~11% of the time) gain sufficient vibrational energy to dissociate. The equilibrium between dissociation and reformation at any radius from the central star has been described by Hill and Hollenbach (1978) and is represented by the following relation for the density of molecular hydrogen:

$$n_{H_2} = \frac{R\, n_t^2}{2\, Rn_t + I_o \left(\frac{r_o}{r}\right)^2 \exp\,(-KN_t)\, \beta N_{H_2}^{-\frac{1}{2}}}$$

where $n_t \equiv n_{HI} + 2n_{H_2}$ relates total, atomic, and molecular gas densities at radius r, and N_t and N_{H_2} are the corresponding column densities out to this radius. R is the molecular formation rate coefficient, I_o the dissociation rate at distance r_o and K is the dust absorption coefficient. The term $\beta N_{H_2}^{-\frac{1}{2}}$ accounts for self-shielding of the Lyman band absorption in the square-root portion of the curve of growth where $\beta = 4.2 \times 10^5$ cm^{-1} (Jura 1974). The relation can be numerically integrated from the ionization boundary outwards until the dissociation term is insufficient to overcome the reformation.

The HI zones of the three HII regions have been successfully modelled using the above relation (Roger and Pedlar 1981; Dewdney and Roger 1981; Roger and Irwin 1982) with very little deviation from accepted values of the rate coefficients or the dust absorption coefficient. Figure 2 illustrates the HI radial profiles expected for dissociation equilibrium for three stellar luminosities in gas of uniform density 100 atoms cm^{-3}. The total mass which would be initially dissociated by various stars completely embedded in a uniform molecular cloud of this density is shown in Figure 3, using photo-dissociation rates given by Hollenbach, Chu and McCray (1976). Figure 4 illustrates the dependence of dissociated mass on the gas density. This inverse dependence is due to the dissociation rate varying directly as density and reformation on dust grains varying as the square of the density.

Thus the HI zones will be found preferentially in lower density gas. Not only is the total mass of dissociated HI greater in lower density gas but a longer time will elapse before the extensive zone is overtaken by the shock front of the expanding HII region. This is a major point of difference between the atomic zones we have observed and those of density 10^3 - 10^4 cm^{-3} considered in the models of Hill and Hollenbach (1978).

A second point of difference is the elevated temperature we observe for the HI. Peak brightness temperatures for IC 5146, for example, are at least 90K which indicates gas spin temperatures $\gtrsim$100K. A statistical study of 250 individual spectral profiles for this source reveals a significant number whose line widths, if due entirely to thermal broadening, imply temperatures lower than 300K.

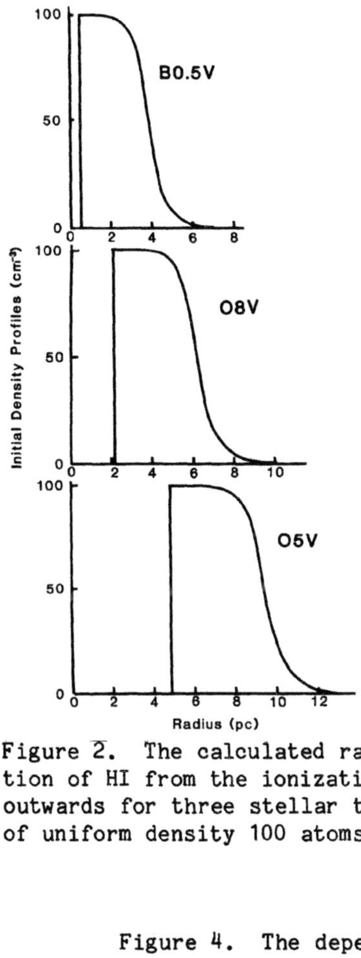

Figure 2. The calculated radial distribution of HI from the ionization boundary outwards for three stellar types in gas of uniform density 100 atoms cm^{-3}.

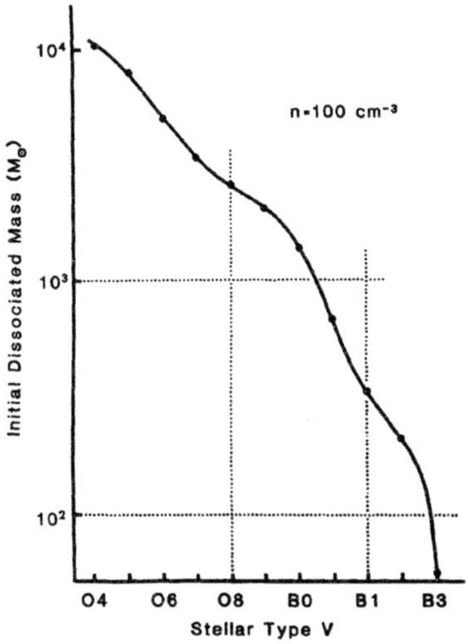

Figure 3. The total mass that can be initially dissociated by stars of various types totally embedded in gas of uniform density 100 atoms cm^{-3}.

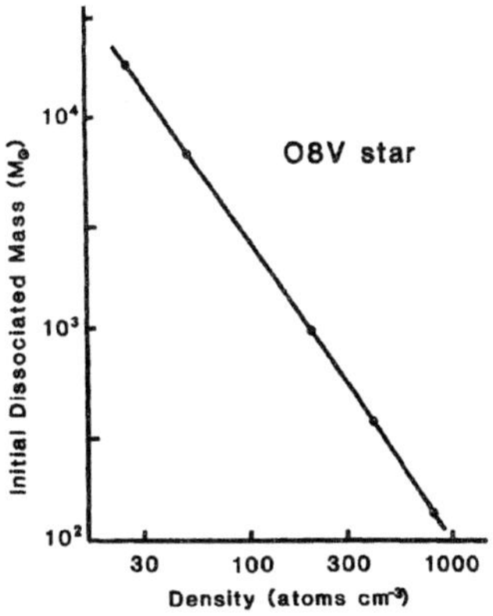

Figure 4. The dependence of initial dissociated mass on gas density for an O8V star.

Temperatures in this range (100K - 300K) are considerably in excess of that of the molecular gas from which the HI formed and imply that heating has taken place. London (1978) has suggested that energy left over from the dissociation process (0.4 eV per dissociation) is capable of heating the gas. He further predicts that the warm HI zone may expand in a way similar to HII region expansion. Our observations of expansion of the zones of IC 5146 and NGC 281, the two most evolved regions, lend weight to this hypothesis. Increased pressure in such dissociated warm HI zones when newly formed could decrease the critical mass for gravitational collapse of embedded dense concentrations, such as those apparent from CO observations of IC 5146 (Lada and Elmegreen 1979).

The absence of any indication of atomic hydrogen in dense shocked shells for NGC 281 and IC 5146 has yet to be accounted for. The relatively compact HII region in NGC 1579 was not resolved in our observations, but it is of particular interest to see if higher resolution observations can reveal the predicted dense shell of HI in this relatively young object.

I am indebted to my co-workers Peter Dewdney, Alan Pedlar and Judy Irwin for numerous discussions relating to these regions.

REFERENCES

Altenhoff, W.J., Braes, L.L.E., Olnon, F.M., Wendker, H.J. 1976, Astr. Ap. 46, 11.
Bodenheimer, P., Tenorio-Tagle, G., Yorke, H.W. 1979, Ap. J. 233, 85.
Crampton, D., Fisher, W.A. 1974, Publ. Dom. Ap. Obs. XIV, 283.
Dewdney, P.E., Roger, R.S. 1981, Ap. J. (in press).
Elmegreen, B.G., Lada, C.J. 1978, Ap. J. 219, 467.
Elmegren, B.G., Moran, J.M. 1979, Ap. J. Lett. 227, L93.
Harvey, P.M., Thronson, H.A., Gatley, I. 1979, Ap. J. 231, 115.
Herbig, G.H. 1971, Ap. J. 169, 537.
Hill, J.K., Hollenbach, D.J. 1978, Ap. J. 225, 390.
Hollenbach, D.J., Chu, S.I., McCray, R. 1976, Ap. J. 208, 458.
Johnson, P.G., White, N.J., Pedlar, A. 1981, Mon. Not. R.A.S. 196, 995.
Jura, M. 1974, Ap. J. 191, 375.
Kuiper, T.B.H., Knapp, G.R., Rodriguez Kuiper, E.N. 1976, Astr. Ap. 48, 475.
Lada, C.J., Elmegreen, B.G. 1979, Astr. J. 84, 336.
Lasker, B.M. 1966, Ap. J. 143, 700.
London, R. 1978, Ap. J. 225, 405.
McCutcheon, W.H., Roger, R.S., Dickman, R.L. 1982,(submitted to Ap. J.).
Riegel, K.W. 1967, Ap. J. 148, 87.
Roger, R.S., Costain, C.H., Lacey, J.D., Landecker, T.L., Bowers, F.K. 1973, Proc. IEEE 61, 1270.
Roger, R.S., Irwin, J.A. 1982, (submitted to Ap. J.).
Roger, R.S., Pedlar, A. 1981, Astr. Ap. 94, 238.
Sargent, A.I., van Duinen, R.J., Fridlund, C.V.M., Nordh, H.I., Aalders, J.W.G. 1981 (preprint).
Savedoff, M.P., Greene, J. 1955, Ap. J. 122, 477.
Stecher, T.P., Williams, D.A. 1967, Ap. J. (Letters) 149, L29.
Tenorio-Tagle, G. 1976, Astr. Ap. 53, 411.
Williamson, R.A. 1970, Ap. & Sp. Sci. 6, 45.

THE NGC 7538 REGION: THE DISTRIBUTION AND DYNAMICS OF MOLECULES COMPARED WITH THOSE OF HI AND H^+

Hélène R. Dickel and John R. Dickel
Astronomy Department, University of Illinois, Urbana, IL
William J. Wilson
The Jet Propulsion Laboratory, Pasadena, CA

SUMMARY

CO maps and preliminary H_2S and H_2CO data for the molecular cloud associated with the HII region NGC 7538 are compared with the distributions of ionized and neutral hydrogen. South of the optical HII region is a ridge of high ^{13}CO column density with cold, self-absorbed HI gas just beyond it. A dense clump within the ridge is found adjacent to the HII region in the southeast. The percentage of the hydrogen in atomic form varies from ~ 0.1% in the dense region to ~ 0.8% in the outskirts. The lower-density region of expanding gas seen next to the HII region in the southwest is attributed to the passage of a molecular dissociation wave.

INTRODUCTION

The nebular complex associated with the HII region NGC 7538[1] exhibits a remarkable assortment of young objects ranging over main sequence O stars and compact HII regions (Israel, Habing and de Jong 1973), infrared sources (Wynn-Williams, Becklin and Neugebauer 1974), the first (and at present only) known formaldehyde maser (Forster et al. 1980), and CO hot-spots (Wilson et al. 1974). The tremendous variety makes this region a key candidate for studying the process of star formation and the evolution of molecular clouds. Very recently, a number of high resolution observations have been made of the emission from ionized hydrogen (Israel 1977; Deharveng, Lortet and Tester 1979), heated dust (Werner et al. 1979), vibrationally-excited molecular hydrogen (Fischer et al. 1980) and neutral hydrogen (Read 1980) in the NGC 7538 complex. Here we present some molecular distributions and compare them with the other constituents. (Full details will appear in a paper being submitted to the Astrophysical Journal 1981).

R. S. Roger and P. E. Dewdney (eds.), Regions of Recent Star Formation, 175–180.
Copyright © 1982 by D. Reidel Publishing Company.

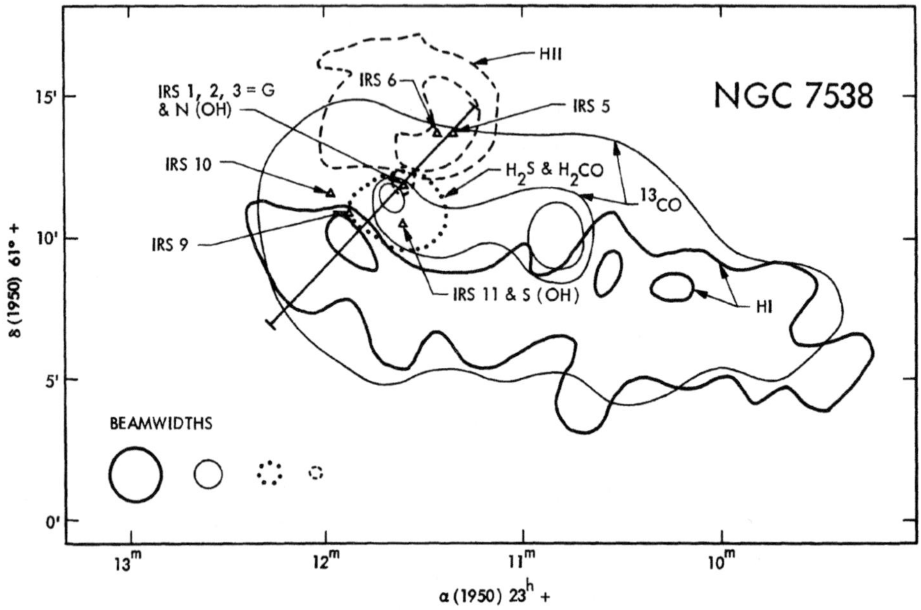

FIG. 1. -Comparison of the distributions of molecules, neutral hydrogen and ionized hydrogen in NGC 7538. Infrared sources (some with associated OH masers) are shown as open triangles (see Werner et al. 1979). The N(^{13}CO) contours (—) are 2 x 10^{16}, 8 x 10^{16} and 9 x 10^{16} cm^{-2}. The 50% level of the equivalent width of the emission lines of H_2S and H_2CO at 2mm is indicated by the dotted contour. The contours of equivalent width of the HI absorption (—) are -300 and -500 K km s^{-1} (adapted from Figure 9 of Read 1980). The contours of the radio continuum intensity at 20 cm (--) are 5 and 150 mJy per synthesized beam (taken from Fig. 6b of Israel, Habing and de Jong 1973).

MOLECULAR OBSERVATIONS

The CO emission stretches from the optical HII region NGC 7538 in the north to the Cas A supernova remnant, 3° away to the south (Wilson et al. 1974; Thaddeus, private communication). The present results shown in Figure 1 are from our CO mapping program with the NRAO 11-meter telescope and Aerospace 4.6-meter telescope (for details refer to Dickel, Dickel and Wilson 1977; Dickel et al. 1980). In April 1980 we mapped the emission from higher-excitation molecules in the region around NGC 7538 G (= IRS 1+2+3, see Figure 1). The 50% level of the integrated brightness of the H_2S emission is shown as a dotted contour, the 50% level for H_2CO is virtually identical to this. Similar distributions are found for the emission from HCN, CS and SO (observed with a beam about twice that used for H_2S).

Selected (light) contours of the ^{13}CO (LTE) column densities are shown in Figure 1 for the inner part of the NGC 7538 molecular cloud. These values were calculated as described in our earlier papers (Dickel, Dickel and Wilson 1977). The most significant feature of the CO emission is the ridge of high column density south of the optical HII region. There are local maxima at each end; the eastern maximum is adjacent to source G (see Figure 1). The mass within the outer $N^{13}CO$ contour is between 5×10^4 and 5×10^5 $M_\odot$ (depending on the choice of the abundance ratio and the distance) and the mean molecular hydrogen density is between 10^3 and 5×10^3 cm^{-3}. However, the presence of the high excitation emission from CS, HCN, H_2CO, and H_2S in the region just south of NGC 7538 G implies much higher densities, on the order of 10^5 cm^{-3} which are probably associated with clumping as seen in the high resolution VLA maps of DR 21 (observed by Dickel, Lubenow, Forster, Rots and Goss in May 1981).

SPATIAL STRUCTURE AND DYNAMICS

The structure and dynamics of the molecular cloud are related to the optical HII region as described below:

The radio continuum emission from NGC 7538 at 20 cm is outlined by the two dashed contours in the upper part of Figure 1. The nebula is excited by two O stars, which are coincident with IRS 5 and 6 shown as triangles. The compact continuum component G, located on the southern edge of the extended HII region, is made up of a cluster of late O and B stars which are surrounded by compact HII regions (Martin 1973; Wynn-Willams, Becklin and Neugebauer 1974). OH, H_2O and H_2CO masers are associated with IRS 1 (refer to Forster et al. 1978, Dickel et al. 1981). There is a steep gradient in the visual extinction from $A_V \sim 4.5$ magnitudes towards IRS 6 in the north through $A_V \sim 16$ magnitudes further south at IRS 2 to $A_V > 37$ magnitudes deeper into the molecular cloud at IRS 1 where we encounter the ridge of high ^{13}CO column density (Glushkov, Denisyuk and Karyagina 1975, Fischer et al. 1980). Emission from high-excitation molecules (dotted contours) and embedded infrared sources (triangles) are found at the eastern end of the CO ridge. South of the ridge we see self-absorbed HI gas (heavy, solid contours) with the maximum absorption occurring just outside the ridge towards the southeast and southwest.

Following the diagonal line in Figure 1, we find that the temperature decreases and the density increases as one goes from the edge of the optical HII region ($\sim$ 1' NW of G) into the molecular cloud ($\sim$ 1' SE of G). Then as the molecular emission (and density) decline outward, the HI gas becomes cold and self-absorbing. Read (1980) estimated that 0.5% of the hydrogen in the whole cloud is in atomic form. From the observed gradients in the HI absorption and ^{13}CO emission, we find the following variation: $\sim$ 0.1% near G where the total gas density is highest, $\sim$ 0.6% where the column density of cold HI is greatest, and $\sim$ 0.8% in the remaining outskirts.

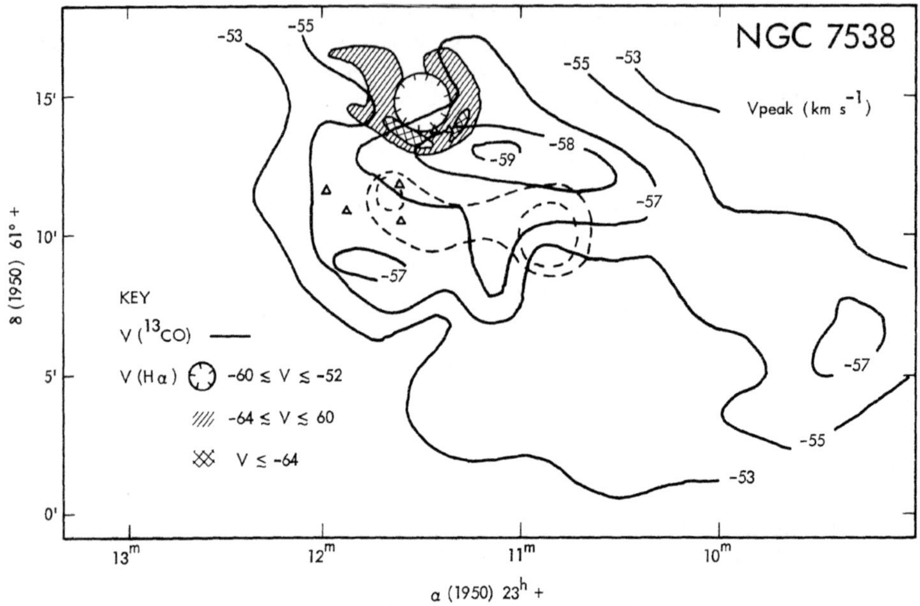

FIG. 2. -The large scale velocity structure in NGC 7538. The lsr velocity (in km s^{-1}) of the maximum ^{13}CO emission line is shown by the solid contours. The region of the most negative velocities of the optical Hα emission (from Deharveng, Lortet, and Testor 1979) is hatched. For comparison, the highest contours of the ^{13}CO column density have been dashed in and the positions of the infrared sources have been marked by triangles.

The large scale velocity structure in NGC 7538 is shown in Figure 2. The systemic (lsr) radial velocity of the NGC 7538 complex is found from several different data to be around -54 km s^{-1}. The ionized gas has measured radial velocities between -44 and -79 km s^{-1} (Deharveng, Lortet and Testor 1979, Baluteau et al. 1981). The most negative values (approach) occur in the region of highest emission measure (cross-hatched area in Figure 2) where the HII region is nestled into the molecular cloud and the ionized gas is streaming off the edge of the cloud toward the observer; such a situation is reminiscent of the "Champagne models" of Tenario-Tagle (1979) and Bodenheimer, Tenario-Tagle and Yorke (1979). The most negative velocities of the CO gas fall along a band south and west of the HII region and just north of the N(^{13}CO) ridge as shown in Figure 2. Similar velocities are found there for the HI emission. In the southern part of the molecular cloud, the velocities of the cold HI and ^{13}CO gas are between -53 and -57 km s^{-1}.

MOLECULAR DISSOCIATION WAVE

We suggest that the expansion of the gas in the region of low column density seen just southwest of the HII region might be related to the passage of a dissociation wave into this more tenuous gas -- that is, where the hydrogen molecules in the region outside the ionization front of the HII region are dissociated after they absorb the stellar photons in the Lyman-Werner bands longward of the Lyman limit. The pressure of the HI, heated by the dissociation of H_2, may lead to dynamical expansion of the HI zone, analogous to the expansion of a hot HII region (London 1978). The thickness of the HI zone will depend on the density of the gas. From the calculations of Hill and Hollenbach (1978) one sees that after $\sim 10^5$ years (the age estimated for the HII region), the dissociation front (where the gas is $\sim$ 90% molecular) would be coincident with a shock front for ambient densities $\gtrsim 10^4$ cm^{-3} (i.e. the region SE of the HII region). However, for ambient densities $\sim 10^3$ cm^{-3}, appropriate for the region to the SW, the thickness of the dissociation front would be $\sim$ 0.5 pc which is comparable to the -59 km s^{-1} contour in Figure 2. (Note that the edge of the HII region defined by the radio continuum contours in Figure 1 is beyond the optical edge shown in Figure 2). For the kinetic temperatures of several hundred K derived by London, the gas in the transition zone could be expanding at up to $\sim$ 2 km s^{-1} (1/2 the sound speed). Such a difference in velocity is observed between the SW region and the region near source G.

Thus the passage of a dissociation wave is compatible with the observations. We expect that the HII region will soon burst out of the molecular cloud in the southwest as it has already done in the northeast where a plume of HI emission is attributed by Read (1980) to the passage of a dissociation wave there. The high density gas in the southeast is presumably the remains of the densest part of the cloud from which IRS 5 and 6 formed and in which other lower mass stars have just recently formed.

ACKNOWLEDGEMENTS

This research of HRD was initially partially-supported by NSF grants AST 75-22208 and AST 77-21021. The National Radio Astronomy Observatory is operated by Associated Universities, Inc. under contract with the National Science Foundation. The Aerospace spectral line radio astronomy program was supported jointly by the National Science Foundation grant MPS-73-04554 and the Aerospace Corporation Program for Research and Investigation.

NOTE

[1]NGC 7538 (Sulentic and Tifft 1973) = Sh2 158 (Sharpless 1959) is located at l $\sim$ 111.5°, b $\sim$ 0.8°.

REFERENCES

Baluteau, J.P., Moorwood, A.F.M., Biraud, Y., Coron, N., Anderegg, M. and Filton, B. 1981, Ap. J., 244, 66.

Bodenheimer, P., Tenorio-Tagle, G. and Yorke, H.W. 1979, Ap. J., 233, 85.

Deharveng, L., Lortet, M.C. and Testor, G. 1979, Astr. Ap., 71 151.

Dickel, H.R., Dickel, J.R. and Wilson, W.J. 1977, Ap. J., 217, 56.

Dickel, H.R., Dickel, J.R., Wilson, W.J. and Werner, M.W. 1980, Ap. J., 237, 711.

Dickel, H.R., Rots, A.H., Goss, W.M. and Forster, J.R. 1981, M.N.R.A.S., submitted.

Fischer, J., Righini-Cohen, G., Simon, M., Joyce, R.R. and Simon, T. 1980, Ap. J. (Letters), 240, L95.

Forster, J.R., Goss, W.M., Wilson, T.L., Downes, D. and Dickel, H.R. 1980, Astr. Ap., 84, L1.

Forster, J.R., Welch, W.J., Wright, M.C.H. and Baudry, A. 1978, Ap. J., 221, 137.

Glushkov, Yu. I., Denisyuk, E.K. and Karyagina, Z.V. 1975, Astr. Ap. 39, 481.

Hill, J.K. and Hollenbach, D.J. 1978, Ap. J., 225, 390.

Israel, F.P. 1977, Astr. Ap., 59, 27.

Israel, F.P., Habing, H.J. and de Jong, T. 1973, Astr. Ap., 27, 143.

London, R. 1978, Ap. J., 225, 405.

Martin, A.H.M. 1973, M.N.R.A.S., 163, 141.

Read, P.L. 1980, M.N.R.A.S., 192, 11.

Sharpless, S. 1959, Ap. J. Suppl., 4, 257.

Sulentic, J.W. and Tifft, W.G. 1973, "The Revised New General Catalogue of Non-Stellar Objects", (U. Ariz. Press, Tucson, AZ).

Tenorio-Tagle, G. 1979, Astr. Ap., 71, 59.

Werner, M.W., Becklin, E.E., Gatley, I., Matthews, K., Neugebauer, G. and Wynn-Williams, C.G. 1979, M.N.R.A.S., 188, 463.

Wilson, W.J., Dickel, H.R., Dickel, J.R. and McKenna, D.L. 1974, Pub. Astr. Soc. Pacific, 86, 602.

Wynn-Williams, C.G., Becklin, E.E. and Neugebauer, C. 1974, Ap. J., 187, 473.

ATOMIC AND IONIZED HYDROGEN IN CEPHEUS OB3

P.E. Dewdney
Dominion Radio Astrophysical Observatory,
Herzberg Institute of Astrophysics,
Penticton, B.C., Canada

The Cepheus OB3 association ($\ell=110^{\circ}$, $b=2.5^{\circ}$) appears to be a region of sequential star formation and for this reason has received much recent interest. The association has two subgroups, one older than the other, and also has indications of continuing star formation.

The region of star formation is the giant molecular cloud detected by Sargent (1977) in ^{12}CO and ^{13}CO lines. Several peaks of CO emission (Cep A and Cep B) also have IR sources in them (Beichman et al., 1979; Evans et al., 1981). H_2O maser sources (Blitz and Lada, 1979), and an OH maser source (Rodriguez et al., 1980) are also present in Cep A.

Cep B is coincident with an HII region (S155) which appears to be ionized by the O7n star HD 217086 (Felli et al., 1978). Observations were made with the Penticton synthesis telescope of HI and 21cm continuum in a 2 x 2 degree field centered between Cep A and Cep B with a resolution of 2 x 2 arcmin by .67 km s^{-1} in velocity. Figure 1 shows our 21cm continuum map with the positions of the features mentioned above.

Simonson and van Someren Greve (1976) have mapped a large region including the one covered here. They detected a positive velocity feature coincident in position with the Cepheus OB3 association. (Positive velocities are forbidden in differential Galactic rotation in this longitude range.)

They associate the positive velocity gas with a shell surrounding S155. The diameter of the shell is about 5 degrees corresponding to 70 pc at a distance of 800 pc (Felli et al., 1978). Our 2 degree field is insufficient to delineate a shell of this size, but we do detect a depression in the emission at the position of S155 at similar positive velocities.

Figure 2 illustrates this depression in a greyscale map of HI integrated in velocity over a 2-16 km s^{-1} range representing 35

R. S. Roger and P. E. Dewdney (eds.), Regions of Recent Star Formation, 181–184.
Copyright © 1982 by D. Reidel Publishing Company.

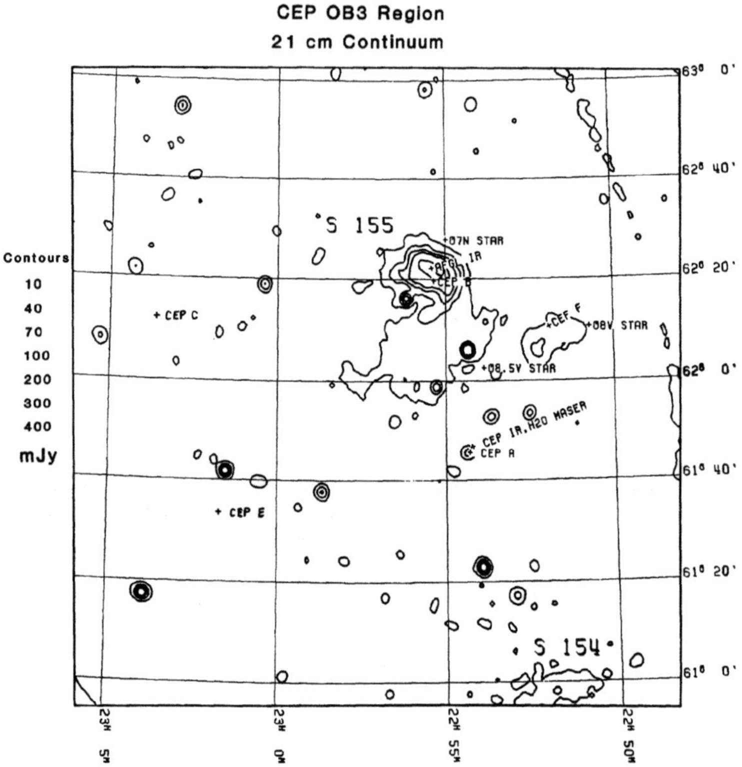

Figure 1: 21-cm continuum map of S155. The three O stars belonging to the Cepheus OB3 association are shown along with the position of 5 peaks of ^{12}CO emission (Sargent, 1977). Star formation is likely taking place near the Cepheus A IR source (Evans et al., 1981), and probably near Cepheus F as well. The IR source near Cepheus B may be the result of the heating of a concentration of dust from the outside (Evans et al., 1981). This particular continuum map is uncorrected for the polar diagram of the synthesis telescope antennas and does not contain spacings less than 13m.

velocity channels in our observations. The depression is visible in most of the individual channels throughout this range.

These maps were derived by subtracting out-of-band continuum from line-plus continuum so that absorption of continuum emission would appear as a depression. To determine whether the continuum is absorbed by HI we integrated scans in a strip (Figure 2) across the depression in the HI maps and across the S155 continuum map. (Both types of maps included low-order spatial frequencies taken with the Penticton 26-m telescope.) However, there is insufficient continuum brightness to "fill in" the depression in the HI emission thereby

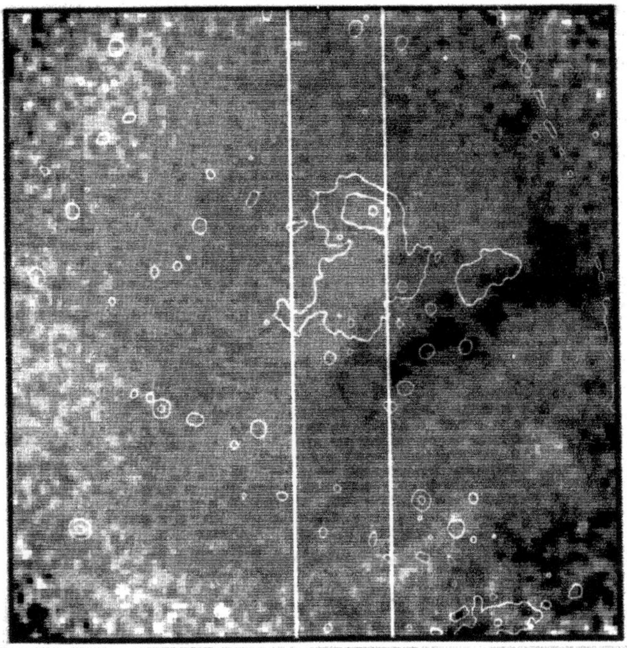

Figure 2: A greyscale map of HI integrated over a velocity range of 2-16 km s^{-1} (35 channels) and covering the same field as in Figure 1. The HI map contains the low-order spacings and is corrected for the polar diagram of the synthesis telescope antennas. The overlaid contours of 21-cm continuum emission illustrate the position of S155 on the HI map. Strip scans integrated over the North-South strip outlined here are shown in Figure 3 (see text).

eliminating absorption as the sole cause of the depression. Representative strip scans are shown in Figure 3 for various velocities in the range 2-16 km s^{-1}.

The other possibility is that there is a dearth of HI at the position of S155, which may be due to ionization of HI around S155. We estimated the total mass of HI represented by the depression by interpolating a smooth surface between four reference areas around the depression at each velocity. The depth of the depression from this surface was treated as a drop in column density assuming optically thin conditions. The total mass integrated over the depression and over velocity was found to be 70 ± 10 $M_\odot$ assuming a distance of 800 pc (Felli et al., 1978). Equivalent column density of HI in the depression is about 2 x 10^{20} cm^{-2}. In comparison, the total mass of ionized gas in S155 is about 170 $M_\odot$ (Felli et al., 1978). Although the mass of ionized gas is larger than the mass of HI, some of the ionized gas may have originally been in molecular form.

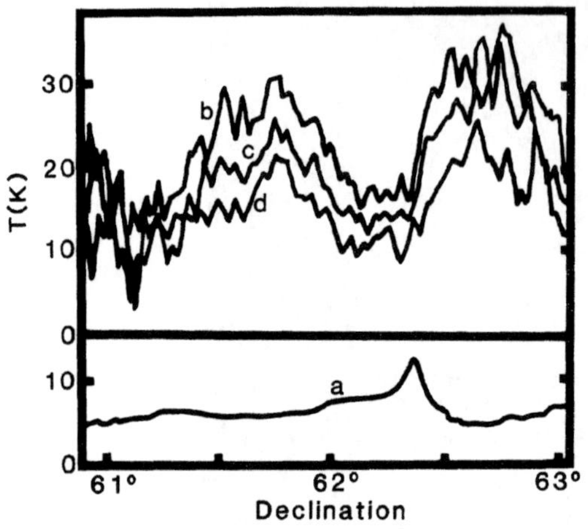

Figure 3: Strip scans integrated over the strip illustrated in Figure 2 for continuum (a) and 3 velocity channels (b=8.5 km s^{-1}; c=9.4 km s^{-1}; d=10.2 km s^{-1}). Both continuum and HI emission are shown to the same scale.

The spatial coincidence of this depression with the HII region, S155, gives credence to the conclusion of Simonson and van Someren Greve (1976) that the positive velocity gas is associated with the Cepheus OB3 region. The depression does not fit well, however, with their dynamical picture which includes the positive velocity gas as the redshifted part of an expanding HI shell around S155. It is more likely that this gas is at the same location along the line-of-sight rather than 35 pc farther away as expected in the shell hypothesis.

The velocity of the molecular cloud is about −11 km s^{-1} (Sargent, 1977). Preliminary analysis of our observations indicate that there is "disturbed" HI near −23 km s^{-1}. This gas may also be associated with the Cepheus OB3 association since this velocity is about the same magnitude of displacement with respect to the rest velocity of the cloud as is the positive velocity gas.

I gratefully acknowledge discussions with my colleague R.S. Roger on the interpretation of these observations.

REFERENCES

Beichman, C.A., Becklin, E.E. and Wynn-Williams, C.G. 1979, Ap. J., (Letters), 232, L47
Blitz, L. and Lada, C.J. 1979, Ap. J., 227, 152
Evans II, N.J., Becklin, E.E., Beichman, C.A., Gatley, I., Hildebrand, R.N., Keene, J., Slovak, M.H., Werner, M.W., Whitcomb, S.E. 1981, Ap. J., 244, 115
Felli, M., Tofani, G., Harten, R.H. and Panagia, N. 1978, Astr. Ap., 69, 199
Rodriguez, L.F., Morgan, J.M., Ho, P.T.P., Gottlieb, E.W. 1980b, Ap. J., 235, 845
Sargent, A.I. 1977, Ap. J., 218, 736
Simonson, S.C. and van Someren Greve, H.W. 1976, Astr. Ap., 49, 343

NEUTRAL HYDROGEN OBSERVATIONS OF THE PUPPIS WINDOW

J. Gregory Stacy and Peter D. Jackson
Astronomy Program, University of Maryland
College Park, Maryland 20742 USA

1. INTRODUCTION

A 21-cm neutral hydrogen survey has been carried out in a region of the galactic disk known as the "Puppis Window." This unique area of the Milky Way contains much less absorbing interstellar dust than is found in more typical regions of the Galaxy. In fact, galaxies have been observed near the galactic equator (Dodd and Brand 1976). It has been estimated that the total visual extinction out to the edge of the Galaxy in the direction l=245° is only four to five magnitudes (FitzGerald and Moffat 1976). Hence, this region of the Galaxy is ideally suited for the comparison of optical and radio observations, extending out to much greater distances than is normally feasible in other portions of the galactic disk.

An extensive optical study of this region of the Galaxy is currently being undertaken by a group of Canadian astronomers (FitzGerald, Innanen, Moffat, Turner and van den Bergh). The radio observations described here are intended to complement this study. In particular, we have attempted to establish correlations between distinct HI features and those optical spiral tracers which have been previously observed, with a view towards delineating the spiral structure of the Milky Way in this direction. An obvious additional benefit of such a study, of course, is a better understanding of the relationship between the neutral gas component of the interstellar medium and sites of recent and ongoing star formation.

2. OBSERVATIONS AND DATA PRESENTATION

Observations were conducted on the 43-m (140-ft) NRAO radio telescope at Green Bank, West Virginia[1] over the course of two observing sessions: 1979 June 22-28 and 1980 May 12-19. The HPBW of the telescope at the 21-cm wavelength (=1420 MHz) is 21 arc-minutes. The area

[1]The National Radio Astronomy Observatory is operated by Associated Universities, Inc., under contract with the National Science Foundation.

R. S. Roger and P. E. Dewdney (eds.), Regions of Recent Star Formation, 185–191.
Copyright © 1982 by D. Reidel Publishing Company.

observed (see Figure 1) consists of a 61 x 61 grid of points in the sky, extending from l=239° to l=251° and from b = -9° to b = +3°, every 0°.2 in l and b (i.e., every 0.57 beamwidth). Thus, over a 12° x 12° grid, a total of 3,721 HI profiles were obtained. A 40 second integration time per point gave an rms channel-to-channel noise of approximately 0.5 K at 0.69 km/sec resolution. The standard position S8 (Williams 1973) and the baseline regions WWH2 and WWH7 (Wrixon and Heiles 1972) were observed for the purposes of temperature and baseline calibration, respectively. After data reduction, a 421 point velocity-temperature line profile was constructed for each observed grid point, extending from -45 to +165 km/sec, with brightness temperature values interpolated at half-integral velocity values.

The final reduced data are presented in the form of constant-velocity HI maps, which were generated using the COMTAL image-processing system of the Laboratory for Astronomy and Solar Physics of the NASA Goddard Space Flight Center in Greenbelt, Maryland. Each of the accompanying maps is actually a 121 x 121 point array of the observed region of the sky (i.e., observations are interpolated to every 0°.1 in l and b). Tick marks are included in each map at 2° intervals at odd values of galactic longitude and latitude. The LSR velocity values (in km/sec) corresponding to each individual HI map appear in the lower right corner of each map. Note that an intensity scale is presented with each set of maps. The 256-level intensity scale was chosen so as to enhance the lower levels of 21-cm emission (see Figure 2 for calibration scale).

3. RESULTS

Examination of the photographs representing the neutral hydrogen distribution in Puppis reveals several striking features. The intricate filamentary structure evident throughout on virtually all scales is highly suggestive of a hot, turbulent interstellar medium, even in this relatively quiescent corner of the Galaxy. Extremely large "holes" or gaps in the HI gas are apparent, particularly in the velocity ranges -5 to +10 km/sec (i.e., the local solar neighborhood) and +35 to +50 km/sec. Whether these holes represent static or expanding HI shells and/or supershells, as Heiles (1979) has proposed, or simply interarm regions naturally devoid of gas, is still subject to some debate. In addition to these very large-scale features, one can also readily discern several prominent rings and shell-like structures in the HI distribution (most notably those in the vicinity of +15 and +29 km/sec), indicative of past energetic explosive events in this region of the Galaxy. A preliminary examination of such optical surveys as the Palomar Observatory Sky Survey, and the emission line survey of Parker et al. (1979), has so far failed to uncover any obvious optical counterparts of these very extended HI features. Still closer inspection of our HI maps reveals that such shell-like features persist down to the lower resolution limit of the telescope. Several of these small rings appear to be related to particular hot spots of neutral hydrogen emission, which gas in turn is found to be coincident with the positions of previously detected optical spiral tracers (OB associations, HII regions or young stellar clusters).

In the following tables we have attempted to quantify the physical parameters describing a selection of these HI clouds. Due to space limitations, we have restricted ourselves in the present report to only the more prominent features. A fuller and more detailed description of our results will be presented elsewhere. It is worth noting that several of the more promising of these regions have been targeted for CO observations in the near future.

The following is a description of the column headings in Tables I and II: Column 1: object name/type, followed by LSR velocity at which the HI emission is seen to be most intense or prominent (followed by the names of any optical objects believed to be associated with the HI feature); Columns 2 and 3: galactic longitude and latitude of the center of the radio feature; Columns 4 and 5: approximate angular extent of the radio feature in l and b; Columns 6 and 7: minimum and maximum LSR velocities between which the HI feature is detected; Column 8: distance to the feature from the Sun, calculated by applying a flat rotation curve (R_o = 10 kpc, Θ_o = 250 km/sec) and using the LSR velocity of peak emission (note that, for those HI features we believe to be associated with particular optical objects, we have also carried out on a second line calculations based on the photometrically-determined distances to these objects); Column 9: linear diameter of the HI feature in parsecs; Column 10: HI column density within the shell or cloud (cm^{-2}); Column 11: HI number density within the shell or cloud (cm^{-3}); Column 12: total HI mass contained within the shell or cloud expressed in solar masses; Column 13: the number of the accompanying figure which contains the feature in question.

4. CONCLUSION

Our observations in Puppis strongly confirm the existence of a turbulent interstellar medium, as evidenced by the presence of numerous HI shells and other detailed filamentary structure throughout the region of our survey. Several HI features appear to be physically related to sites of recent star formation. The correlation of such neutral gas features with young optical spiral tracers (resulting from the present study and future planned observations), in addition to providing valuable information on galactic dynamics, should tell us much about star formation processes in the more distant parts of our Galaxy.

We would like to acknowledge partial support of this research by the National Science Foundation through grant AST 77-26898 to Prof. F. J. Kerr of the University of Maryland. We also thank the University of Maryland Computer Science Center for a grant of computer time, and D. Klinglesmith of the NASA Goddard Space Flight Center for arranging use of the Interactive Astronomical Data Analysis Facility at the Goddard Space Flight Center.

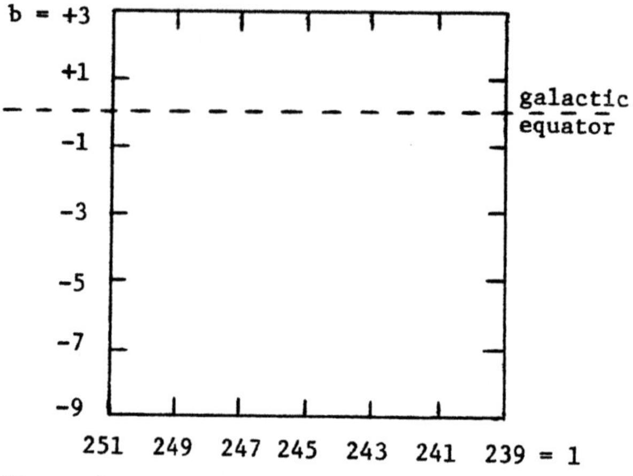

Figure 1. Schematic of Observed Region in Puppis.

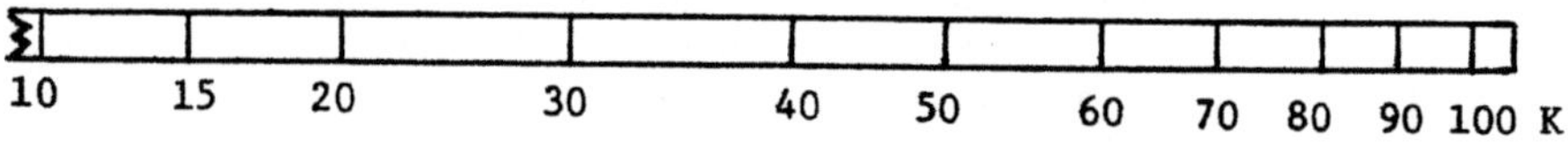

Figure 2. Brightness Temperature vs. Intensity Calibration Scale (right-hand edge coincides with right-hand edge of intensity scales given with maps).

REFERENCES

Dodd, R.J. and Brand, P.W.: 1976 Astron. Astrophys. Suppl. 25, 519.
FitzGerald, M.P. and Moffat, A.F.J.: 1976, Sky and Telescope 52, 104.
Havlen, R.J.: 1976, Astron. Astrophys. 47, 193.
Heiles, C.: 1979, Astrophys. J. 229, 533.
Parker, R.A.R., Gull, T.R., Kirschner, R.P.: 1979, "An Emission Line Survey of the Milky Way", NASA, Washington, DC.
Pişmiş, P. and Moreno, M.A.: 1976, Revista Mexicana de Astronomia y Astrophysica 1, 373.
Williams, D.R.W.: 1973, Astron. Astrophys. Suppl. 8, 505.
Wrixon, G.T. and Heiles, C.: 1972, Astron. Astrophys. 18, 444.

TABLE I: HI SHELLS

Object	l_{ctr}	b_{ctr}	Δl	Δb	V_{min} (km/sec)	V_{max} (km/sec)	d (kpc)	LD (pc)	N_{sh} (cm^{-2})	n_{sh} (cm^{-3})	M_{sh} $(M_\odot)$	Related Figure
Shell +15 =GS242-01+11 (Heiles 1979)	241°.	-1°.	6°	8°	+7	+17	1.4	170	$3.2x10^{20}$	0.8	$4.3x10^{4}$	3
Shell +29	249°.	0°.0	4°	5°	+24	+38	3.	240	$7.1x10^{20}$	0.7	$1.2x10^{5}$	4
Shell +54 (NGC2467/Haf 18-19?)	244°.	+0°.5	1°.75	2°	+47	+61	5.2	170	$5x10^{20}$	0.8	$5x10^{4}$	5
							4.1[1]	130	$5x10^{20}$	1.0	$3.1x10^{4}$	5
Shell +57 (Pup OB2/Rup 44?)	245°.	+2°.	2°.5	2°.5	+51	+65	5.6	240	$3.9x10^{20}$	0.6	$1x10^{5}$	5
							4.3[2]	190	$3.9x10^{20}$	0.7	$5.9x10^{4}$	5

TABLE II: HI CLOUDS

Object	l_{ctr}	b_{ctr}	Δl	Δb	V_{min} (km/sec)	V_{max} (km/sec)	d (kpc)	LD (pc)	N_{cl} (cm^{-2})	n_{cl} (cm^{-3})	M_{cl} $(M_\odot)$	Related Figure
Cloud A +53 NGC2467/ Haffner 18-19	243°.	+0°.4	2°.	2°.	+38	+60	5.1	180	$1.2x10^{21}$	2.1	$1.5x10^{5}$	5
							4.1[1]	140	$1.2x10^{21}$	2.7	$1x10^{5}$	5
Cloud B +53 Pup OB2/ Rup 44	246°.	+0°.5	3°.	3°.5	+38	+60	5.2	290	$9.5x10^{20}$	1.0	$3.4x10^{5}$	5
							4.3[2]	240	$9.5x10^{20}$	1.3	$2.3x10^{5}$	5

[1] Photometric distance to NGC 2467 HII emission complex from Pişmiş and Moreno (1976)

[2] Photometric distance to Pup OB2 from Havlen (1976).

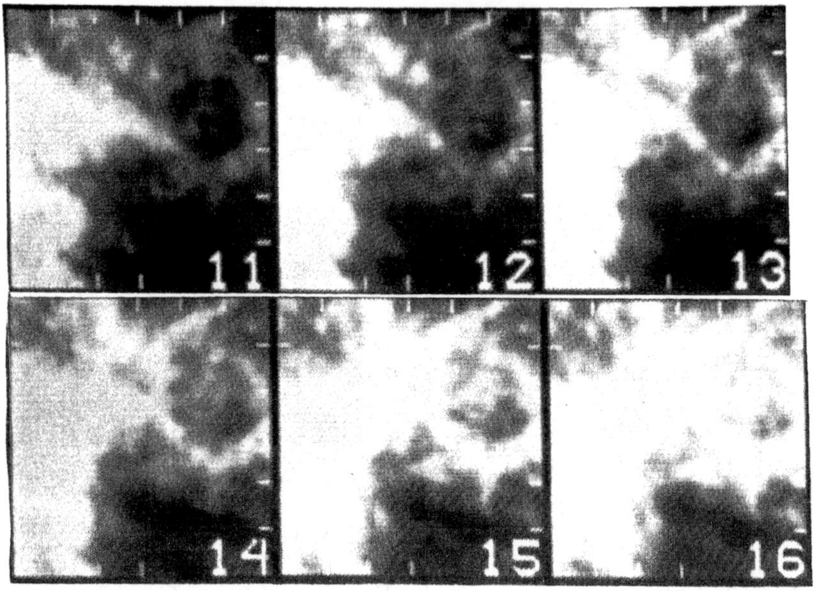

FIGURE 3.

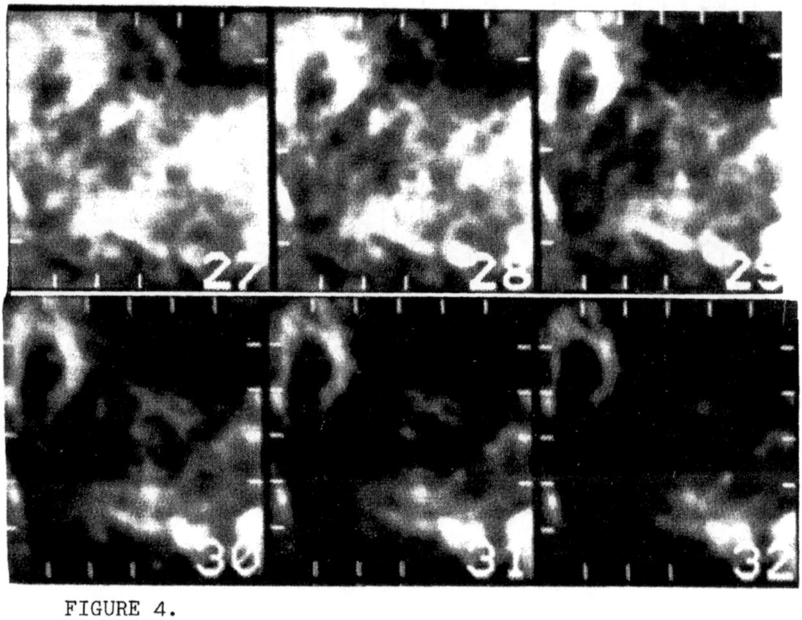

FIGURE 4.

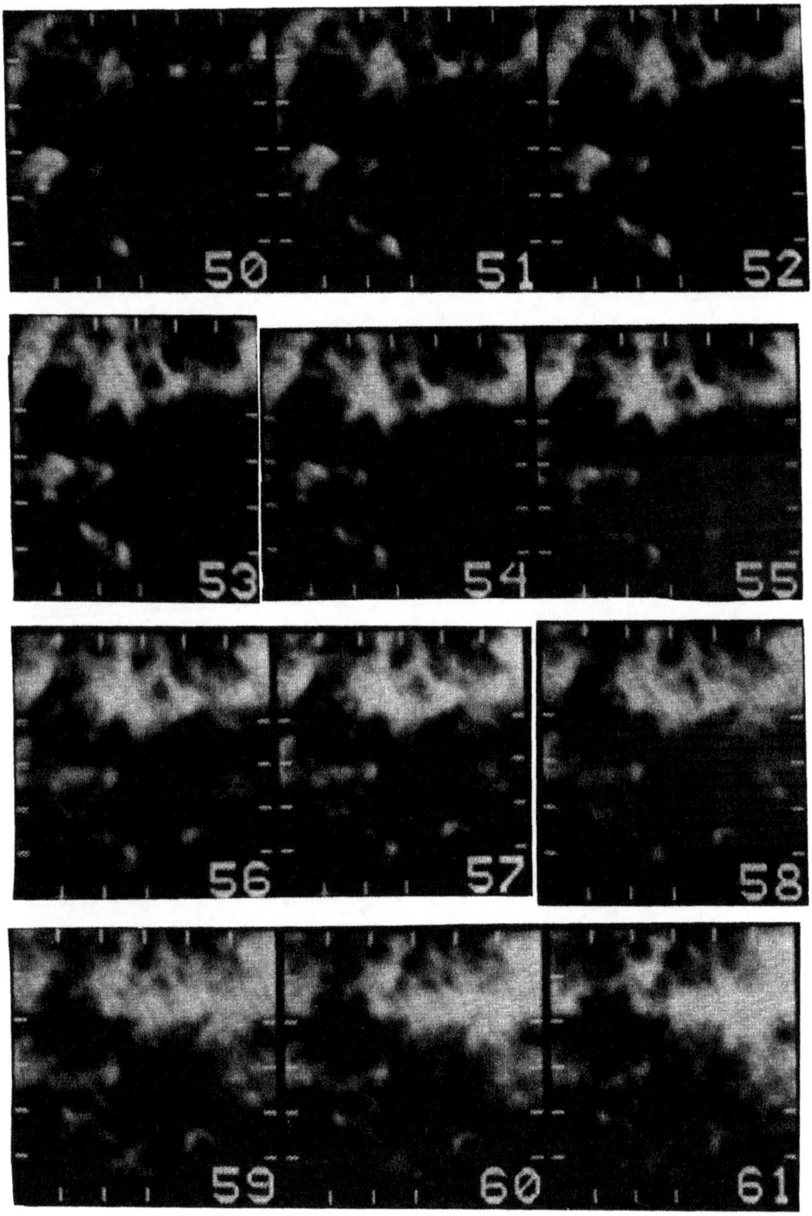

FIGURE 5.

NEUTRAL HYDROGEN TOWARDS TYCHO'S SUPERNOVA REMNANT

J.S. Albinson and S.F. Gull
Mullard Radio Astronomy Observatory
Cavendish Laboratory
Cambridge, U.K.

We have observed the radio remnant of Tycho's Supernova (AD 1572), 3C10, with the Cambridge Half-Mile-Telescope (HMT). Two complete syntheses, overlapping in velocity coverage, were made in September 1979. Details of the telescope, receivers and the survey are in Table 1. In order to derive accurate HI absorption measurements towards the source, it is important to include carefully the contribution from large scale emission. Data containing this large-scale structure, physically unobtainable with the HMT, were derived from the work of Weaver & Williams (1973) and Williams (1973), and added to the synthesis maps. Continuum emission was subtracted from these 'composite' maps to give the final channel maps (Plate 1). In Plate 2 we show a 'pie slice' representation of an RA-velocity plot through the centre of the field.

We derived absorption, spin temperature and column density profiles (Figs. 1a, b, c) for the HI along the line of sight to 3C10 in the usual way (Spitzer 1978). We took brightness temperatures off contour plots of our 'composite' and 'channel' maps. The derived parameters for the HI cloud seen in the first four channel maps, assuming that the cloud is local (within 500 pc), are: size $\leq$ 10 pc; column density $\gtrsim 6 \times 10^{20}$ cm^{-2}; spin temperature $\simeq$ 85K; number density $\gtrsim 20$ cm^{-3}; pressure, in units of $P/n \geq 1700$ cm^{-3} K. These are similar to values in the review by Salpeter (1979).

We estimate a value for the distance to 3C10 in the following way. From Fig. 1a we see that there is a peak of absorption at -48.8 km s^{-1} and that the negative velocity wing of this peak extends to -60.3 km s^{-1}. Fig. 1c, and the channel maps, show that the large-scale emission is becoming brighter over the same velocity range and beyond to -65 km s^{-1}. This suggests that 3C10 is embedded in the near edge of the gaseous part of the Perseus spiral arm. Stars form out of interstellar gas and it is reasonable to assume that the gas and young stars, (OB associations and open clusters, which are usually taken to be spiral tracers), are in close proximity (a few hundred parsecs). The distances to the OB associations (Humphreys 1978) and

R. S. Roger and P. E. Dewdney (eds.), Regions of Recent Star Formation, 193–199.
Copyright © 1982 by D. Reidel Publishing Company.

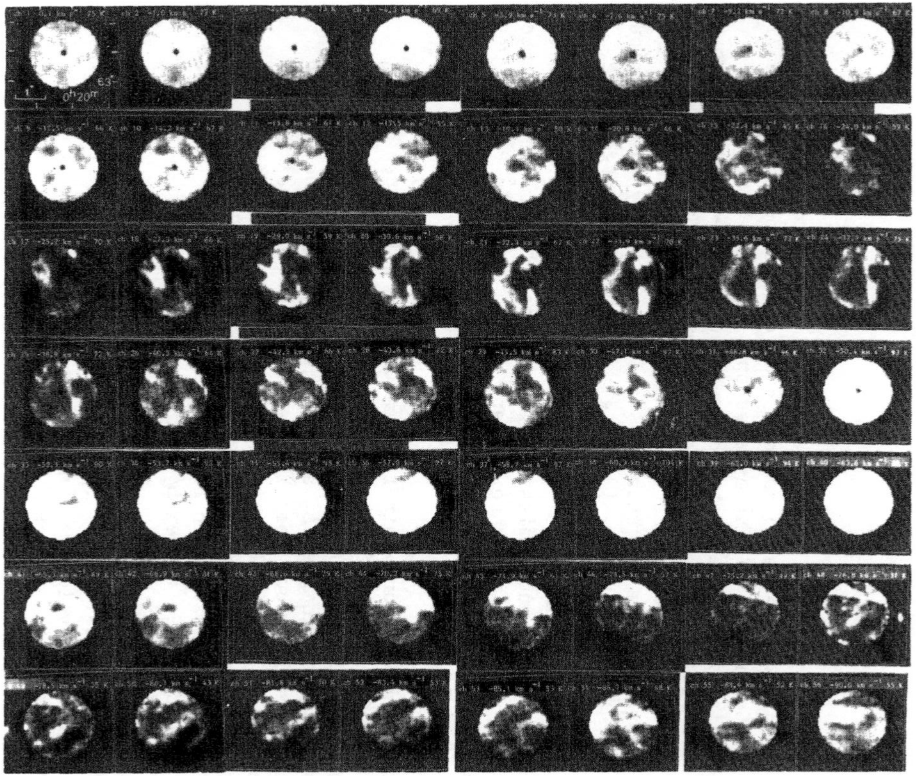

Plate 1. Photographic representations of the velocity channel maps. The radio images are positive (bright emission appears white) and cover a range in brightness of 5 to 100 K. The velocity of each channel (w.r.t. Local Std. of Rest) is marked above each channel. The field of view is sharply cut off at a radius of 67 arcmin. Notice the dark 'spot' in the centre of channels 1-11 and 30-32 caused by absorption in front of 3C10.

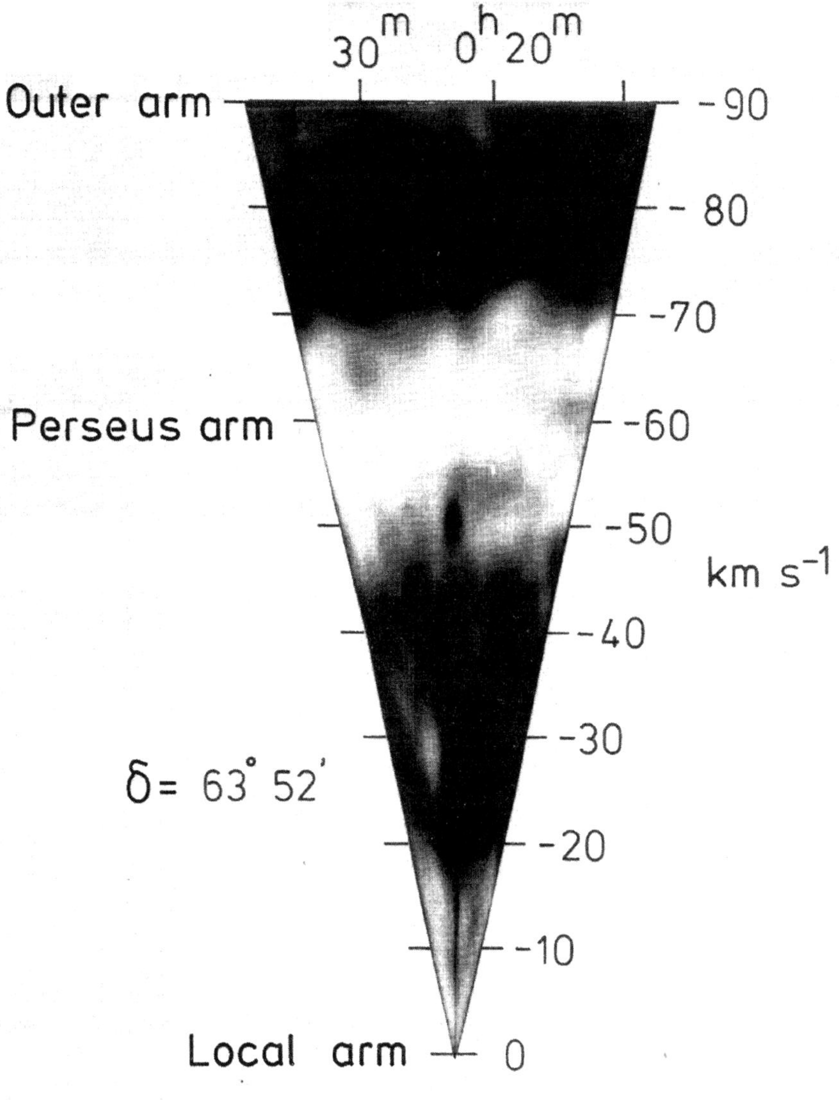

Plate 2. RA - velocity plot, showing HI emission from the Local, Perseus and Outer spiral arms. Clearly visible is a dark, vertical streak of absorption, which terminates suddenly at the beginning of the Perseus arm.

TABLE 1 Specification of the H.M.T. as used for observations of the 3C10 field

Primary Beam	94 arcmin HPBW (effective field of view = $2^\circ\!.25$)
Spatial resolution	7.1 x 7.9 arcmin (α x δ)
Receivers	Wide band (10 MHz) phase switched. Stokes Parameters observed I-Q, I+Q. Narrow band - Digital cross-correlator spectrometer. I+Q. 80 delay channels/spacing. 32 frequency channels for each synthesis. Bandwidth 0.25 MHz total.
Noise on synthesis channel maps	0.4 K
Velocity coverage (line)	+0.7 to -90.0 km s^{-1} w.r.t. LSR 56 informative channels
Velocity resolution	2 km s^{-1} (to half-power points)
Separation of velocity channels	1.65 km s^{-1}
Stokes parameters observed by spectrometer	I+Q
Position of field centre 1950.0	$\alpha = 0^h\ 22^m\ 30^s$ $\delta = 63^\circ\ 52'$

open clusters (Becker & Fenkart 1971) in this direction, which we believe to be associated with the Perseus arm, are in the range 2-3 kpc. We suggest, therefore, that 3C10 is at a distance of 2-2½ kpc. This value is less than has been assumed in the past, (Goss, Schwarz & Wesselius 1973; Williams 1973; Schwarz, Arnal & Goss 1980) but is in agreement with the estimate by Chevalier, Kirshner & Raymond (1980).

We must now recognise that the distance along the line of sight may be a many-valued function of velocity. The best velocity-distance curves in this class, to the best of our knowledge, are those due to Roberts (1972). We have made a composite, Fig. 2, of his Figures 3 and 4. The shape of our absorption profile can be explained if 3C10 lies behind a filament which has been freshly shocked. There would not be strong absorption due to gas which is in the main part of the arm at velocities between -40 and -45 km s^{-1}. No strong absorption is

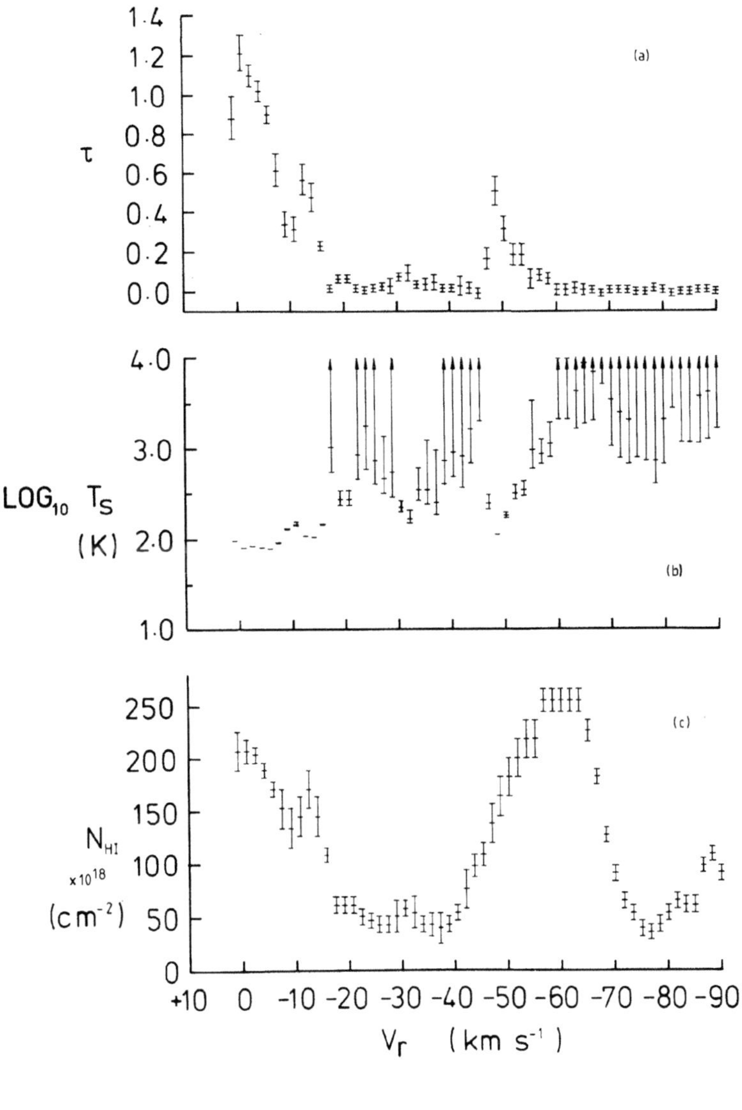

Figure 1a Plot of optical depth (τ) vs. velocity (km s^{-1}).

1b Plot of $\log_{10}(T_s)$ (spin temperature, K) vs velocity (km s^{-1}).

1c Plot of column density N_{HI} in units of 10^{18} cm^{-2} ($N_{HI} = T_b \times 2 \times 1.823 \times 10^{18}$ cm^{-2}) vs. velocity.

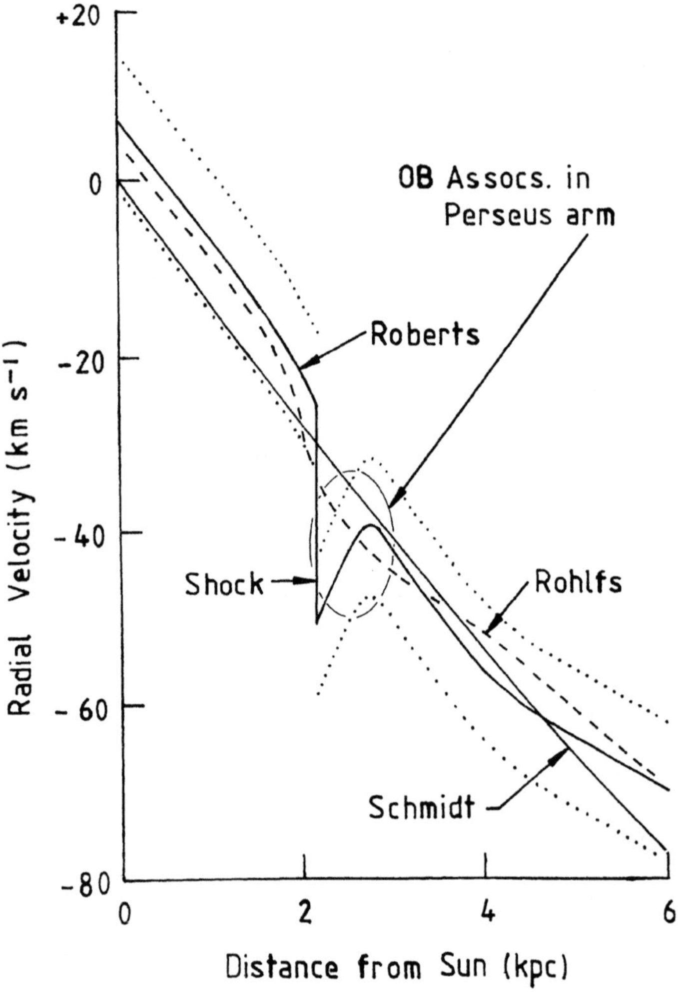

Figure 2. This is an interpolated velocity-distance curve for $\ell = 120^{\circ}$, derived from Roberts (1972) Figs 3 and 4. The curves due to Rohlfs (1974) and Schmidt (1965) are marked. The dotted lines mark a ± 8 km s^{-1} dispersion band about Roberts's curve.

observed. The weak absorption between -55 and -60 km s^{-1}, which would be strong if 3C10 is at $\sim$ 4 kpc, would be due either to HI in a turbulent state around the filament, or would be spurious due to small-scale variations in HI emission behind 3C10. Our estimate for the distance is, again, in the range 2-2½ kpc.

We note that the equilibrium Schmidt model (Schmidt 1965) for the galaxy, upon which Roberts' curve is a perturbation, may not be completely valid outside the Solar Circle. This will not greatly affect us because the difference between the latest rotation curve (Blitz, Fich & Stark 1980) and Schmidt's model is only a few km s^{-1} at a galacto-centric radius of 12 kpc.

REFERENCES

Becker, W. & Fenkart, R., 1971. Astr. Astrophys. Suppl. Ser., 4, 241.
Blitz, L., Fich, M. & Stark, A.A., 1980. IAU Symposium 87, 'Interstellar molecules', ed. B.H. Andrew, (Holland: D. Reidel) pp 213-220.
Chevalier, R.A., Kirshner, R.P. & Raymond, J.C., 1980. Astrophys. J., 235, 186.
Goss, W.M., Schwarz, U.J. & Wesselius, P.R., 1973. Astr. Astrophys., 28, 305.
Humphreys, R.M., 1978. Astrophys. J. Suppl. Ser., 38, 309.
Roberts, W.W., 1972. Astrophys. J., 173, 259.
Rohlfs, K., 1974. Astr. Astrophys., 35, 177.
Salpeter, E.E., 1979. IAU Symposium 84, 'The Large Scale Characteristics of the Galaxy', ed. W.B. Burton, (Holland: D. Reidel) pp 245-252.
Schmidt, M., 1965. 'Stars and Stellar Systems', V, (University of Chicago Press) eds. A. Blaaw and M. Schmidt, pp 513-530.
Schwarz, U.J., Arnal, E.M. & Goss, W.M., 1980. Mon. Not. R. astr. Soc., 192, 67p.
Spitzer, L., 1978. 'Physical Processes in the Interstellar Medium', (Wiley Interscience). New York.
Weaver, H. & Williams, D.R.W., 1973. Astr. Astrophys. Suppl. Ser., 8, 1.
Williams, D.R.W., 1973. Astr. Astrophys. Suppl. Ser. 8, 505.

THE VELOCITIES OF THE NEUTRAL AND IONIZED COMPONENTS OF HII REGIONS

Fich M., Treffers R.R., and Blitz L.
Radio Astronomy Laboratory
University of California, Berkeley.

ABSTRACT

We investigate statistically the relative velocities of the ionized gas in HII regions and the neutral gas in the molecular clouds associated with them for 151 optical HII regions in our Galaxy. The velocity of the molecular cloud is determined from CO observations at 2.6 mm and the ionized component from Hα spectra. The mean velocity difference between the molecular cloud and the ionized gas is 1.4 ± 0.4 km s^{-1} with a dispersion of 4.6 km s^{-1}. Both values are important for determining the mean flow velocity and they indicate that the bulk of the ionized gas in most HII regions flow away from the associated clouds at velocities of 5 - 10 km s^{-1}. They also indicate that a significant fraction of all HII regions will be visible optically even if they are on the far side of a molecular cloud.

1. INTRODUCTION

The interaction between the ionized gas of an HII region and the neutral gas of a surrounding cloud complex has been the subject of many studies in the past decade. Detailed studies of the motions within the Orion Nebula and other nearby HII regions have led to the formulation of the "blister model" of HII regions (Zuckerman 1973; Balick, Gammon and Hjellming 1974). In this model the ionized gas streams away from the neutral cloud with which it is associated at velocities of up to 40 km s^{-1} (ie. see Tenario-Tagle 1980). Since the optical depth of the molecular clouds is high, it might be expected that the HII regions on the far side of molecular clouds will be obscured and the radial velocity of the optical lines will be preferentially blueshifted relative to the velocities of their associated clouds by an amount of order the sound speed.

R. S. Roger and P. E. Dewdney (eds.), Regions of Recent Star Formation, 201–208.
Copyright © 1982 by D. Reidel Publishing Company.

We determine the velocity difference between the neutral and ionized gas statistically from data on 151 optical HII regions using CO observations for the velocities of the neutral clouds and Hα observations for the velocities of the ionized gas. We examine the effect of the different orientations of the HII regions and clouds with respect to the observer.

2. OBSERVATIONS

The CO velocities used are taken from the survey by Blitz, Fich and Stark (1981) and other work referred to in their catalogue. The maximum uncertainty in their velocities is typically 1 km s^{-1} and this value has been assumed for all of the data used here.

The Hα velocities come from two sources : the primary source is a survey by Treffers (1982) of the optical HII regions in the Galaxy. In addition, the survey by Georgelin (1975) is used to confirm velocities measured by Treffers and, in about 30 cases, to provide velocities where no other measurements are available. We estimate the uncertainty in the measurements of Hα velocities to be 3 km s^{-1} which is about one tenth of the line width. The Hα velocities are from observations taken of the brighter parts of the HII regions.

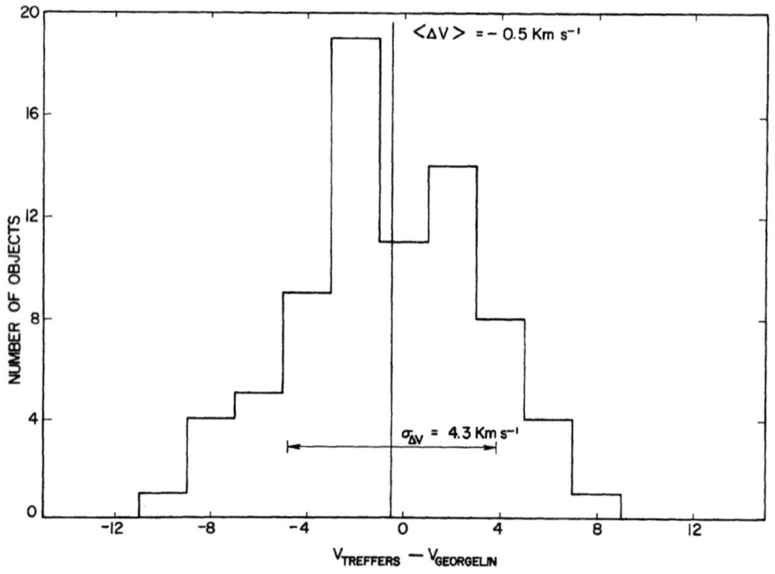

Figure 1 : Comparison of Treffers and Georgelin Hα Velocities

As a check, the Hα data of Treffers and Georgelin were compared (see Figure 1) and we find that the mean velocity difference is 0.5 ± 0.5 km s^{-1} with a dispersion of 4.3 km s^{-1}. This velocity difference indicates that there is no systematic difference between the velocities in the two data sets. Also, the observed dispersion is consistent with our estimate of a mean uncertainty of 3 km s^{-1} for an individual measurement in either data set.

3. RESULTS

Figure 2 shows the distribution of the difference between the CO and Hα velocities for the 151 objects. The mean velocity difference is $\langle V_{CO} - V_{H\alpha} \rangle$ (= $\langle \Delta V \rangle$) = 1.4 ± 0.4 km s^{-1} and the velocity dispersion ($\sigma_{\Delta V}$) is 4.6 km s^{-1}. While $\langle \Delta V \rangle$ is significantly smaller than the flow velocities expected in individual HII regions, it is different from zero by 3.5 σ. This indicates that there are systematic flows of the ionized gas relative to the the neutral component in HII regions. Israel (1979) has found a $\langle \Delta V \rangle$ of 3.4 ± 0.4 km s^{-1}, not in particularly good agreement with our result, but also significantly smaller than was previously expected.

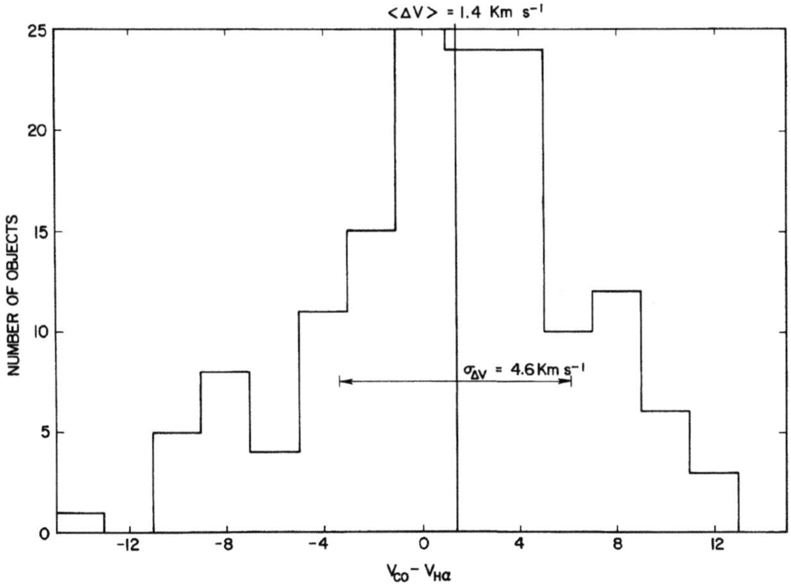

Figure 2 : Comparison of CO and Hα Velocities

The velocity dispersion has been corrected for the uncertainty in the measured CO and Hα velocities. In addition, we have considered the errors due to possible misidentification of objects, that is, some of the HII regions may not be associated with the cloud whose CO velocity was used in these calculations. For example, if there were 10 misidentified objects with velocity differences of 5 km s^{-1} and these were removed from the data set then the result would be to reduce $\langle \Delta V \rangle$ to 1.2 km s^{-1} and $\sigma_{\Delta V}$ to 4.5 km s^{-1}. It is unlikely that this many objects are misidentified and therefore we feel that misidentification of objects does not significantly affect our results.

The orientation of the cloud and the HII region relative to the observer is important in determining the velocity difference observed in an individual object. For example if the line joining the centers of the cloud and the HII region is oriented at 90° with respect to the line of sight and the ionized gas flows normally to the cloud surface, both components would have the same radial velocity (and the velocity difference would be zero).

One test of these models is to look at the velocity difference as a function of the relative separation in the sky between the HII region and the neutral cloud. We determine the ratio of the separation of the CO peak and the center of the HII region to the diameters of the HII region for all those objects which had been partially mapped in CO. The mean velocity difference determined for sets of objects grouped according to similar separation to diameter (S/D) ratios with is shown in Table 1. The dispersion has been corrected as discussed above.

Table 1

Velocity Difference and Separation of Components

n	S/D	$\langle V_{CO} - V_{H\alpha} \rangle$ ($\langle \Delta V \rangle$) km s^{-1}	$\sigma_{\Delta V}$ km s^{-1}
17	0.0–0.2	4.0 ± 1.2	3.5
12	0.2–0.4	0.5 ± 2.0	6.8
8	0.4–0.6	0.7 ± 1.9	4.5
7	0.6–0.8	1.6 ± 2.4	5.5

There are several points worth noting here. 1) In all cases the $\langle \Delta V \rangle$ is positive. 2) The only set of objects which have a $\langle \Delta V \rangle$ significantly different from zero is the one containing objects where the HII region is coincident in the sky with the the CO peak. 3) The dispersion ($\sigma_{\Delta V}$) is small in all cases.

To check the effect of extinction on these results we have compared the Hα velocities to the velocities determined using the H109α radio recombination line in the survey by Reifenstein, Wilson, Burke, Mezger, and Altenhoff (1970). Figure 2 shows the distribution of the velocity difference between the H109α and the Hα lines for the 21 objects which have been measured in both ways.

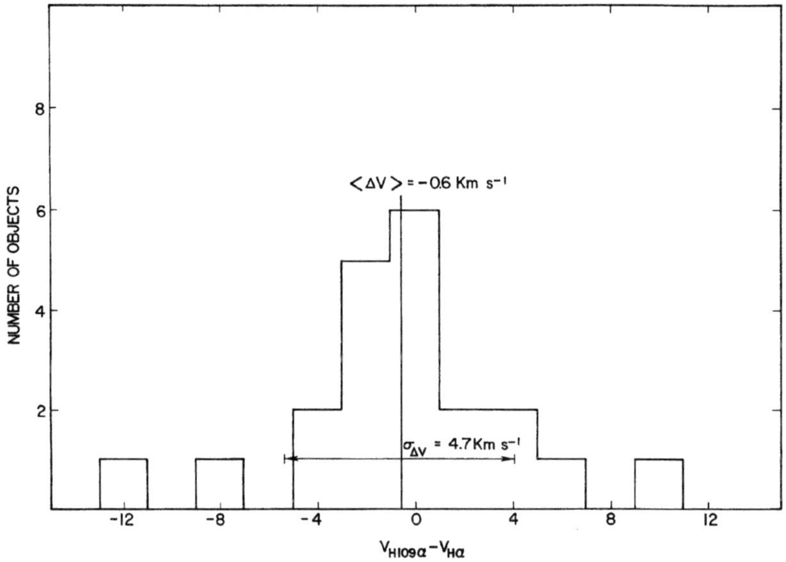

Figure 3 : Comparison of Hα and H109α Velocities

The mean velocity difference is 0.6 ± 1.0 km s^{-1} with a dispersion of 4.7 km s^{-1} which implies there is no systematic difference between the two sets of data. This is perhaps surprising considering that the radio line is observed with a much larger beam than the optical lines and thus samples a much larger volume of the HII region. The radio survey would sample gas hidden to the Hα surveys by the associated clouds. But from these results we can conclude that extinction is not important in determining the Hα velocities observed in the optical HII regions.

4. SIMPLE MODELS

In order to understand the results presented above we have looked at two basic models:

1] UNIDIRECTIONAL FLOW : In which the gas at the surface flows outward normal to the cloud .

2] OMNIDIRECTIONAL FLOW : In which the gas at the surface can flow outward through 2π steradians.

These two models represent the extremes in the flow patterns in which the gas may move away from the cloud. For both models we have considered two cases:

1] OBSCURED: The ionized gas cannot be seen optically if the HII region is on the far side of the cloud.

2] UNOBSCURED: The ionized gas can be seen optically no matter where it is.

Table 2 gives the $\langle \Delta V \rangle$ and the velocity dispersion expected for these models for an arbitrary ionized gas flow velocity V_F and for V_F = 10 km s^{-1}.

Table 2

Expected Velocities for Simple Models

	$\langle V_{CO} - V_{H\alpha} \rangle$		$\sigma_{\Delta V}$	
		V_F=10 km s^{-1}		V_F=10 km s^{-1}
UNIDIRECTIONAL FLOW				
Obscured	$V_F/2$	5.0	$V_F/\sqrt{12}$	2.9
Unobscured	0	0.0	$V_F/\sqrt{3}$	5.8
OMNIDIRECTIONAL FLOW				
Obscured	$V_F/4$	2.5	V_Fx.52	5.2
Unobscured	0	0.0	$V_F/\sqrt{3}$	5.8
OBSERVED		1.4±0.5		4.6

The observed mean velocity difference falls between the completely obscured and completely unobscured cases for both models. The velocity dispersion in the omnidirectional model is virtually independent of the obscuration and is somewhat higher than the observed dispersion for V_F=10 km s^{-1}. For both models the flow velocity cannot be much higher without conflicting with the observed dispersion. These models indicate that there are a significant number of objects where the HII region is not heavily obscured by its associated cloud (even if the orientation is such that this would be expected), and that the flow velocity for the bulk of the gas is ~ 8 km s^{-1}, regardless of the flow pattern.

5. CONCLUSIONS

Previous statistical studies of this problem (Israel 1979) have concentrated only on the velocity difference between the ionized and neutral components of the HII region. However we have shown that the velocity dispersion is just as important in gaining an understanding of the dynamics. Higher moments of the velocity distribution may also be important.

Our main result is that for a large sample of HII regions both the mean velocity difference (between the neutral and ionized gas) and the velocity dispersion are small. This may be explained in two possible, but not mutually exclusive, ways:

1) The extinction is not high. The HII region may have blown a hole in the cloud or clump in which it formed, creating holes or at least thin spots through which Hα is visible.

2) The HII region is larger than the cloud or clump in which it was formed. This would mean that the scale size of the HII regions would be an upper limit to the scale size of the associated clumps.

Although our models are simple they still require three parameters to describe them (the flow pattern, the flow velocity, and the amount of optical obscuration or the limits on the orientations visible optically). We have only produced two numbers: essentially the first and second moments of the velocity distribution. Therefore we can only set limits on the parameters in the models. The strongest limits we can place are on the flow velocity: Most of the ionized gas in the HII regions flows away from the neutral cloud with velocities in the range 5-10 km s^{-1}.

REFERENCES

Balick, B., Gammon, R.H. and Hjellming, R.M.: 1974, PASP, 86, 616
Blitz, L., Fich, M. and Stark A.A.: 1981, Ap. J. Suppl., in press
Georgelin, Y.M.: 1975, These, Univ. de Provence, Marseille.
Israel, F.: 1979, Astron. Astrophys., 70, 769
Reifenstein, E.C., Wilson, T.L., Burke, B.F., Mezger, P.G. and Altenhoff, W.J.: 1970, Astron. Astrophys. 4, 357
Tenorio-Tagle, G.: 1979, Astron. Astrophys., 71, 59
Treffers, R.R.: 1982, in preparation
Zuckerman, B.: 1973, Ap. J., 183, 863

DISCUSSION FOLLOWING PAPER BY FICH ET AL.

YORKE: A statistical study of this sort will strongly emphasize the high density regions. A lot of interesting information may be contained in the line profiles in low density, low emission regions. Would it be possible to study the associated low emission regions in a non-biased manner, say by always looking at a certain point located at a given projected distance from the ionizing source?

FICH: This will constitute a later part of the study. We plan to map individual HII regions and to look closely at the line profiles of the H_α.

SHULL: It may be useful to look for correlations of the CO and H_α emission velocities with interstellar absorption lines (either with IUE or optically.) High resolution IUE spectra would require stars 8th or 9th magnitude and brighter.

FICH: The present sample contains only Sharpless objects and about a quarter to a half have known exciting stars. We plan to add to the sample about sixty other optical HII regions, previously uncatalogued, some of which may have bright exciting stars.

A CO SURVEY OF 372 OPTICAL HII REGIONS

Leo Blitz
Michel Fich
Radio Astronomy Laboratory, University of California, Berkeley

Antony A. Stark
Bell Laboratories

It has been known for some time that the best place to find CO is toward HII regions. Nevertheless, there has been no unbiased survey of CO toward HII regions on even a limited basis. To ameliorate this situation, we have attempted to survey all of the optical HII regions accessible with northern hemisphere millimeter wave radiotelescopes. The initial motivation was to get as many points as possible in order to determine the Galactic rotation curve. We have subsequently found that the information contained in our survey has some inherent interest as has already been shown by Fich, Treffers and Blitz (1982). A catalogue of the results is on press in the Astrophysical Journal Supplement Series (Blitz, Fich and Stark, 1981).

The survey consists of observations of all but six of the 288 HII regions in the Sharpless (1959) catalogue in the J=1-0 transition of CO. Although the catalogue contains 313 entries, 19 objects have been found to be planetary nebulae or supernova remnants. One object which has been listed as a planetary (S128 also known as Abell 63 -Perek and Kohoutek 1967) is almost certainly an HII region on the basis of the strength and shape of its associated CO emission. Of the six objects we did not observe, five were so large and diffuse that even if CO were detected, there would be no way of associating it with the HII region. One object, S14, could not be found on the Palomar prints.

In addition, we searched the Palomar prints to see if there were additional uncatalogued HII regions. We found 67 candidate objects, only 9 of which had been previously catalogued. We catalogued nebulous objects near the galactic plane which were circular rather than filamentary and which were brighter on the red plate than on the blue. Of special interest was any red nebulous object which appeared to have associated dust obscuration. We observed 64 of these objects, the overwhelming majority of which appear to be HII regions based on the strength of the associated CO emission.

The CO observations were made with the 5m telescope at the

R. S. Roger and P. E. Dewdney (eds.), Regions of Recent Star Formation, 209–212.
Copyright © 1982 by D. Reidel Publishing Company.

Millimeter Wave Observatory at Fort Davis, Texas, and the 7m telescope at the Bell Laboratories in Holmdel, N.J. Unless the objects had been well mapped previously by other observers, we generally observed at least five points per HII region. For small objects (< 15 arcmin diameter) in the second and third quadrants, and at high latitude in the first quadrant, we performed a five point map centered on the HII region and at one radius north, south, east and west of the center. For large objects, we looked for places where there appears to be interaction with dust, such as in bright rims. If there was associated CO, we always found at least two positions where it was detected, but typically, associated CO was found at all five positions.

In order to determine whether the CO is associated with an HII region, we used all of the information available to us including line strengths, Hα velocities, longitude and latitude of the object, etc. Generally, CO antenna temperatures do not reach 10K except in regions of active star formation. CO lines with temperatures > 10K are evidence that an HII region is associated with a particular velocity component. Because CO is closely confined to the plane of the Galaxy (Cohen and Thaddeus 1977), detection of high latitude CO, especially in regions of dust obscuration is evidence for the association of CO with an optical nebula even if the observed line is weak. Other indicators of associations between CO and a particular HII region are a single velocity component along the line of sight which reaches a maximum near the HII region position, and the spatial coincidence of high velocity CO with a distant HII region. Mapping is the best way to assure the association of an HII region with a CO cloud and about 20% of the Sharpless objects have been at least partially mapped by us or by others. The results of our survey are summarized in Table 1.

TABLE 1

SURVEY OF CO TOWARD HII REGIONS

	Sharpless Objects	Newly Catalogued Objects
Detections	194 (68%)	48 (75%)
Cannot Associate	26 (9%)	3 (5%)
Questionable or Improbable Detection	30 (10%)	8 (12%)
No Detection	38 (13%)	5 (8%)
No Observation	5	
Planetary Nebulae	11	
Supernova Remnants	8	
Not Found	1	

Our results indicate that at least 70% of all HII regions have associated molecular clouds. The row "cannot associate" means that strong emission was found along the line of sight which could not be unamibiguously identified with a particular HII region, usually because several strong velocity components were found. This situation occurs most frequently in the first galactic quadrant. Most of these HII regions probably have associated molecular clouds so we conclude that ~ 80% of all HII regions have associated molecular clouds. "Questionable or improbable detection" means a weak CO line or lines was seen along the line of sight which is probably unrelated to the HII region.

Comparing our results with the continuum survey of Sharpless HII regions by Felli and Churchwell (1972), we find that both surveys failed to detect about 20% of the sample, and that the non-detections are well-correlated. 56% of the radio quiet HII regions (those with 21 cm continuum flux densities < 100 mJy) have no associated CO ("naked" HII regions). Felli and Perinotto (1974) argued that the radio quiet objects were likely to be excited by relatively late type (B1-B9) supergiants. These stars may have lived long enough to drift away from their placental molecular clouds and may now be ionizing random HI clouds which happen to be nearby.

Thirty four percent of the radio quiet HII regions have associated CO clouds. These may be very low emission measure HII regions or red reflection nebulae, but we have no independent data to confirm this hypothesis. Conversely, 13% of the radio loud HII regions are "naked". Some possible reasons are that i) our search technique does not find all of the molecular clouds associated with HII regions (more extensive mapping would test this), ii) these HII regions are excited by runaway O stars, or iii) some O stars form without the benefit of a placental molecular cloud.

The survey allows us to determine the distribution of optical HII regions in the Galaxy (Fich and Blitz 1981). When the positions of the HII regions are plotted in galactocentric polar coordinates, there is no obvious organization of HII regions into spiral arms. See, for example, figure 1 of Blitz, Fich and Stark (1980), which gives such a plot based on our preliminary data. However, a recent analysis of the distribution of atomic hydrogen in the outer galaxy (Blitz, Kulkarni and Heiles 1981) shows that the HI is organized into three spiral arms of nearly constant pitch angle. When the HII regions are superimposed on the plot showing these spiral arms, the HII regions correlate well with the HI arms.

By examining the clouds near the galactic center and anticenter, it is possible to determine the mean radial velocities of the HII regions in the π direction and their one dimensional velocity dispersion. The results toward the galactic center and anticenter are consistent and indicate that the clouds have an inward motion relative to the LSR of 4.5 ± 1.5 km s^{-1}. The uncertainty considers all sources of error. We find the velocity dispersion to be 6.6 ± 1.0 km s^{-1}, also considering all sources of error. We have checked this result for the effects of velocity streaming by considering only those objects with galactocentric distances of 11.4 ± 0.4 kpc and find the velocity dispersion unchanged. This result is also consistent with the value for the velocity dispersion in a direction perpendicular to the galactic plane suggested by Stark and Blitz (1978) based on the scale height of giant local molecular clouds, and by Stark (1979) based on observations of clouds in the 4-8 kpc molecular ring.

The survey provides an order of magnitude improvement in the measured radial velocities toward HII region/molecular cloud complexes. It therefore provides a good data base for studies of galactic kinematics and galactic structure. With the CO rotation curve, kinematic distances to HII regions with distances as large as 20 kpc from the galactic center are now available. Studies of the properties of HII regions as a function of distance from the galactic center can therefore be extended to very large distances.

REFERENCES

Blitz, L., Fich, M. and Stark, A.A., 1980, in "Interstellar Molecules", Proceedings of IAU Symposium #87, ed. B.H. Andrew, Reidel: Dordrecht p. 213.
Blitz, L., Fich, M. and Stark, A.A., 1981, Ap. J. (Suppl.), in press.
Blitz, L., Kulkarni, S. and Heiles, 1981, B.A.A.S., 13, 539.
Cohen, R.S. and Thaddeus, P., 1977, Ap. J. (Letters), 217, L155.
Felli, M. and Churchwell, E., 1972, Astron. and Ap. (Suppl.), 5, 369.
Felli, M. and Perinotto, M., 1974, Astrophys. and Space Sci., 26, 115.
Fich, M., Treffers, R.R. and Blitz, L., 1982, this volume, p. 201.
Perek, L. and Kohoutek, L., 1967, Catalogue of Galactic Planetary Nebulae, (Prague: Czech Institute of Science).
Sharpless, S., 1959, Ap. J. (Suppl.), 4, 257.
Stark, A.A., 1979, Ph.D. Dissertation, Princeton University.
Stark, A.A. and Blitz, L., 1978, Ap.J. (Letters), 225, L15.

DISCUSSION FOLLOWING PAPER BY BLITZ, FICH AND STARK

HARTEN: Have you considered the bias in the velocities resulting from selecting optically visible HII regions? One might expect a systematic shift, definitely in H_α but possibly in CO emission.

BLITZ: In a system of gas clouds where you have stars, ionized, atomic and molecular gas, the most massive component will be that associated with the CO emission. Thus the center-of-mass velocity will be the CO velocity. One expects a bias in H_α velocities, as described elsewhere by Fich, Treffers and Blitz (this volume).

J. DICKEL: Did you tie the distances to any reference other than the kinematic one, such as one deduced from reddening of the stars?

BLITZ: No. What we did was to find kinematic distances determined from a rotation curve based on a sub-set of the data (~90 objects) for which there are known optical distances.

HARTEN: In the region of the anticenter, there is a very large scatter between stellar distances and kinematic distances - the latter are almost useless.

HI AND CO OBSERVATIONS OF DISTANT HII REGIONS IN THE GALACTIC ANTICENTER

Edwin J. Grayzeck,
University of Nevada Las Vegas

Peter D. Jackson, James R. Sewall
University of Maryland

1. INTRODUCTION

Recent evidence has shown that the outer parts of the Galaxy have a significant fraction of young material including OB stars, HII regions, dust, and atomic and molecular gas in discrete clouds. Study of this material has shown that the rotation curve of the Galaxy is rather flat out to distances of 10 kpc from the Sun. Methods of studying the kinematics of these young objects have centered largely on optical photometry and spectroscopy, e.g. Moffat et al.(1979),and on studies in the associated CO emission, e.g. Blitz et al (1981) and Jackson and Sewall (1982). Once the existence of these young regions has been established and distances determined, it is possible to study their physical structure and dynamics. Viner and Jackson (1981) have done a radio continuum study of HII regions in this portion of the Galaxy. In this paper, we shall report on HI and CO distributions associated with some of these distant HII regions near the galactic anticenter.

2. OBSERVATIONS

The 21-cm HI observations were performed at the Arecibo observatory using the 305-m radio telescope and flat feed. This system was used in the frequency-switching mode (offset by 1 MHz) and split into two separate 504 channel receivers giving a 0.51 km/s velocity resolution. The two minute observations were taken in a grid which included the radio continuum and optical emission source with a spacing of 2 arcmin (one half the beamwidth). Declination strips were also observed in regions near the galactic equator that were also covered by the Maryland - Green Bank HI survey (MGB2C as designated by Harten et al.1975) so as to provide an external temperature calibration. The observations were gathered at low zenith angles to maximize consistency resulting in an internal uncertainty of 2%. From comparison to the MGB2C, the external errors are approximately 5%, based on a conversion factor of 1.50 ± 0.05. In the final presentation of profiles or contour maps, only brightness temperatures are used and the velocity resolution has been smoothed to 1.0 km/s.

The CO observations are described in more detail by Jackson and Sewall (1982), elsewhere in these symposium proceedings. The NRAO 11-m

R. S. Roger and P. E. Dewdney (eds.), Regions of Recent Star Formation, 213–220.
Copyright © 1982 by D. Reidel Publishing Company.

radio telescope was used. This system has a beam width of 65 arcsec at the 115 GHz frequency of the ^{12}CO line. The CO data has been smoothed to 0.51 km/s resolution and displayed as figures representing the grid positions observed (the normal grid spacing was 50 arcsec) , the distribution of $T_A{}^*$ (the peak intensity), the center or centroid line radial velocity, and the velocity halfwidth for each molecular cloud complex.

3. RESULTS

S241

The CO data for S241 is displayed in figures 1a, 1b, 1c, and 1d Towards the center of the region, there is a ridge of CO (as shown in figure 1a) that runs in an east-west direction, with a secondary peak centered at $6^h\ 00^m\ 36^s$, $30^o\ 10'$. The halfwidths of these features indicate a broad velocity spread and suggest that we are looking at the same cloud (see figure 1c). In figure 1d, the CO velocity centroid runs from -7 to -4 km/s across the region. The CO in general shows a good correlation with the spatial distribution of the HII region, especially with the northern rim of absorption that is readily seen in the PSS prints for this area.

The HI profiles for this region show an intermediate velocity feature at -35 km/s as well as multiple features associated with the central peak (-7.0, 0.5, 2.7, and 8 km/s). To best display the HI in this region, a position-position map of the moment integral $\int T_b\, dv$ over the velocity range of the feature (-7 to -4 km/s) was constructed as shown in figure 2. The rise that begins in the southern part of the HII region is further traced as a depression from $30^o\ 10'$ to $30^o\ 16'$ at an RA of $6^h\ 00^m\ 36^s$. The enhancement is approximately a 5% variation and as such is probably real.

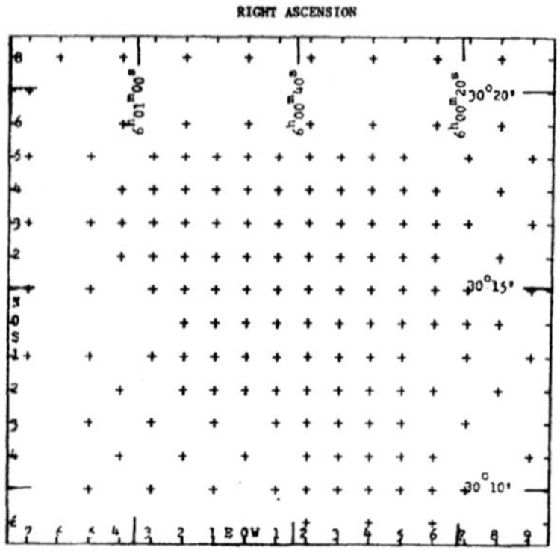

Figure 1a. Grid positions for CO observations of S241

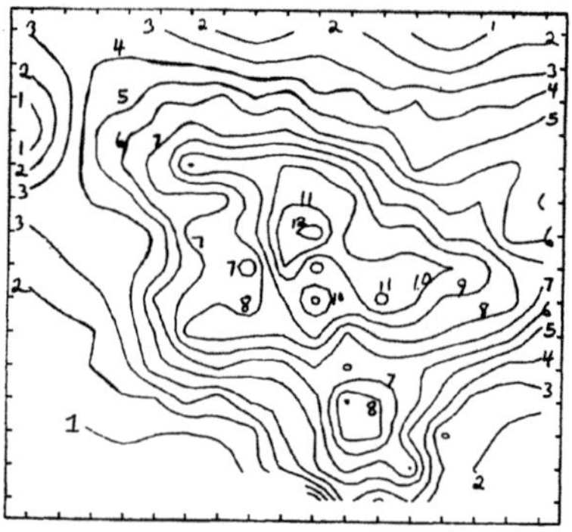

Figure 1b. Contour map for S241 in units of $T_A{}^*$.

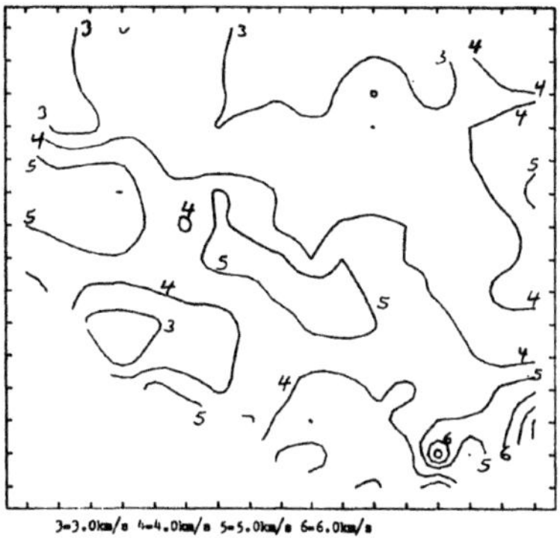

Figure 1c. Contour map of S241 in units of velocity halfwidth.

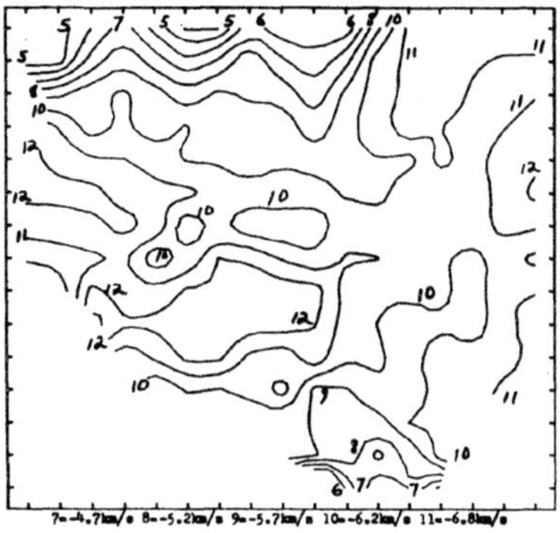

Figure 1d. Contour map of S241 for the centroid velocity.

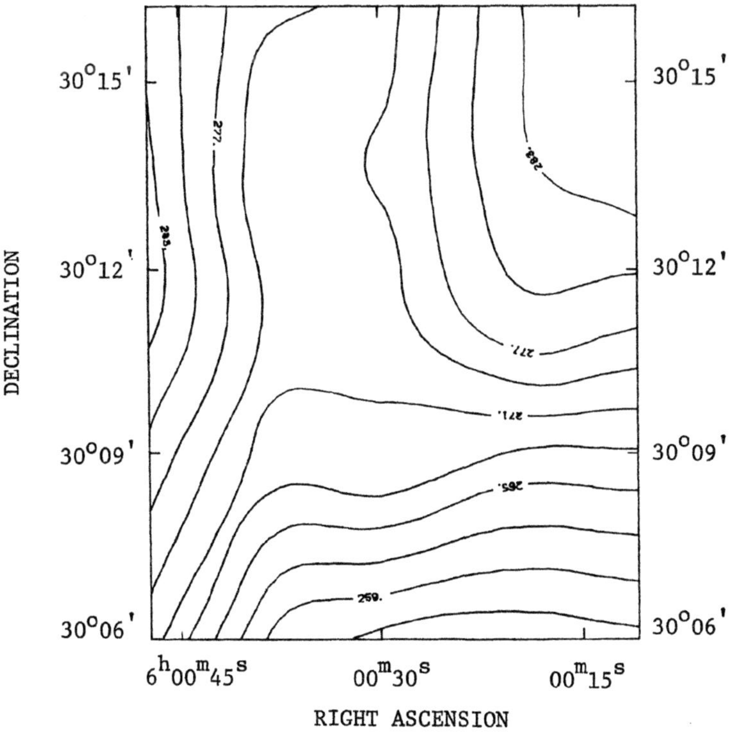

Figure 2. Moment integral for S241 over the velocity range -7 to -4 km/s.

S242

The CO data for this region is taken from Blitz et al. (1981) and represents only a few sample profiles in this direction, yielding a center velocity of 0.0 km/s and a velocity width of 2.5 km/s.

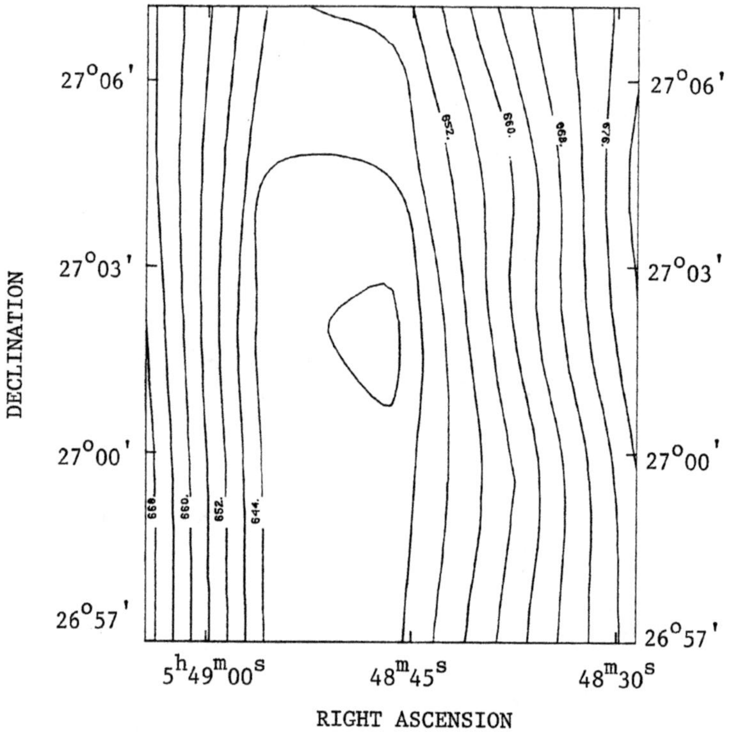

Figure 3. Moment integral for S242 over the velocity range -6 to 2 km/s.

The HI profiles in this region show strong features at -10, -1, and 5 km/s; none of these show any marked enhancement over the object. At the suspected CO velocity, there is one feature that appears to be correlated with the spatial position of the optical region (figure 3). It is interesting to note that this distribution does show a broad depression centered on both the CO position and the offset from the radio continuum peak.

S259

The CO emission from S259 extends over only a few arcmin. The maps of the CO parameters are given in Jackson and Sewall (1982). The map for T_A^* indicates that the CO is concentrated in a point-like source which is essentially unresolved at 65 arcsec resolution. Inspection of the PSS prints shows that there is a faint patch of nebulosity very near the CO position.

A few HI observations were taken across this region, but at the resolution of the 305-m telescope (3.95 arcmin), little variation could be associated with the HII region. The HI profile is shown in figure 4; clearly there is a feature that could be associated with the CO velocity for this region.

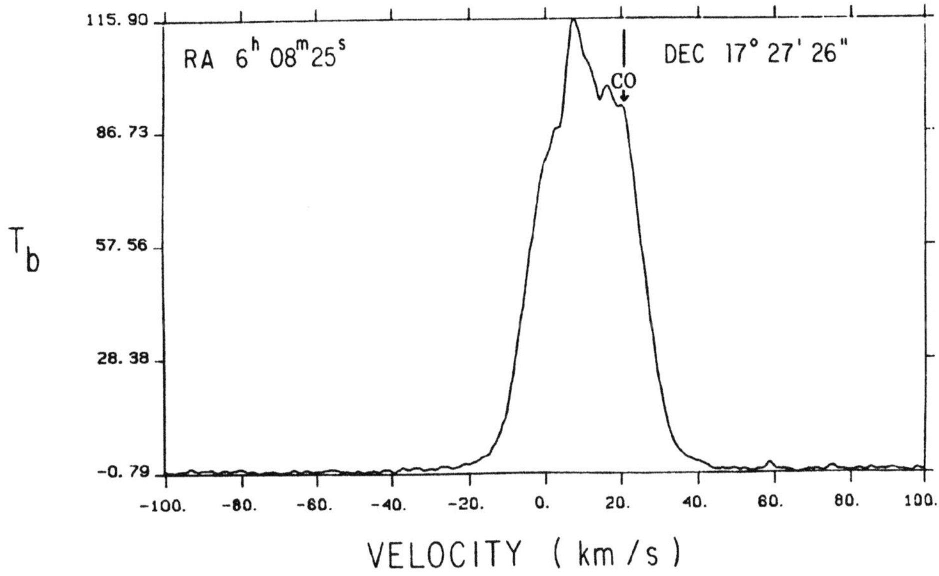

Figure 4. HI profile for the region S259 with CO velocity shown.

S271

For S271 (and nearby S272), the CO maps are shown in figures 5a, 5b, 5c, and 5d. A well defined CO ridge passes 1.5 arcmin south of S271 and 1.0 arcmin southeast of S272. The CO is located in a clear region on the PSS, although one suspects that there is considerable extinction in the area, as can be clearly seen obscuring part of S271 itself.

The HI profiles for this direction show not only features associated with the main peak (3.6, 8.1, 19.5, and 22.5 km/s) but also some intermediate velocity gas at 40 km/s. Constant velocity maps indicate that the 19.5 km/s feature splits in the direction of S271. A map of this feature, at 20 km/s, is shown in figure 6. There is a slight enhancement at $6^h\ 12^m\ 00^s$, $12^o\ 23'\ 00''$ for a one-sigma detection. In the southwest part of the map, a general depression is detectable.

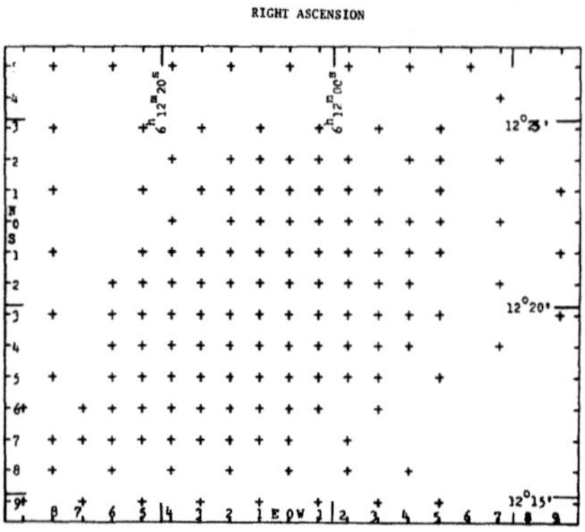

Figure 5a. Grid positions for CO observations of S271.

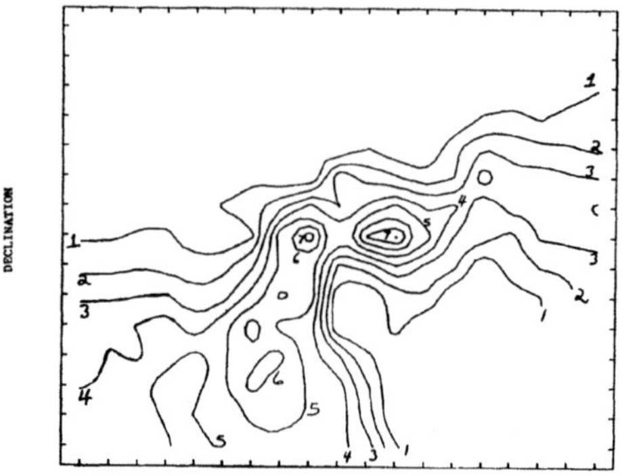

Figure 5b. Contour map of S271 in units of T_A*.

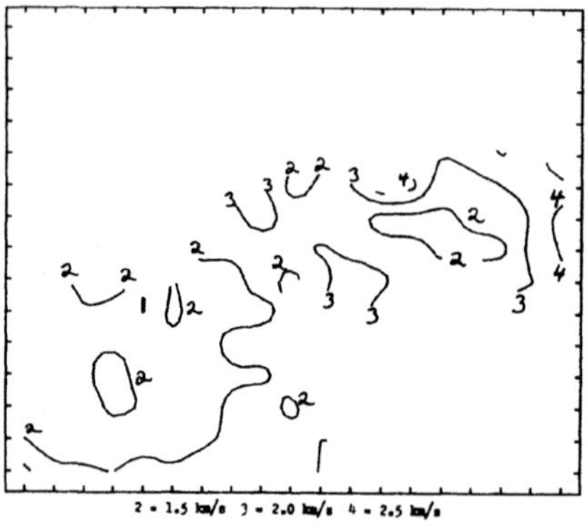

Figure 5c. Contour map of S271 in units of velocity halfwidth.

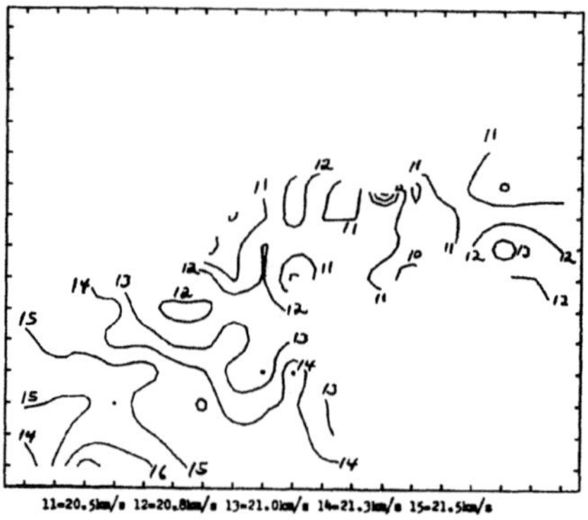

Figure 5d. Contour map of S271 for the centroid velocity.

4. DISCUSSION

In this study we have investigated four HII regions using both CO and HI information. The characteristics of this data have been compiled in Table 1 below.

Although these regions are small, they posess both dust and neutral gas as the remains of the star formation process. The detection of CO is often related to the age of the region, i.e. the young regions still have not completely thermalized their surroundings. This correlation was first pointed out by Bash et al. (1977) for a sample of 63 HII regions, located in the inner part of the Galaxy; it appears the same correlation holds for the anticenter region. The HI results are only

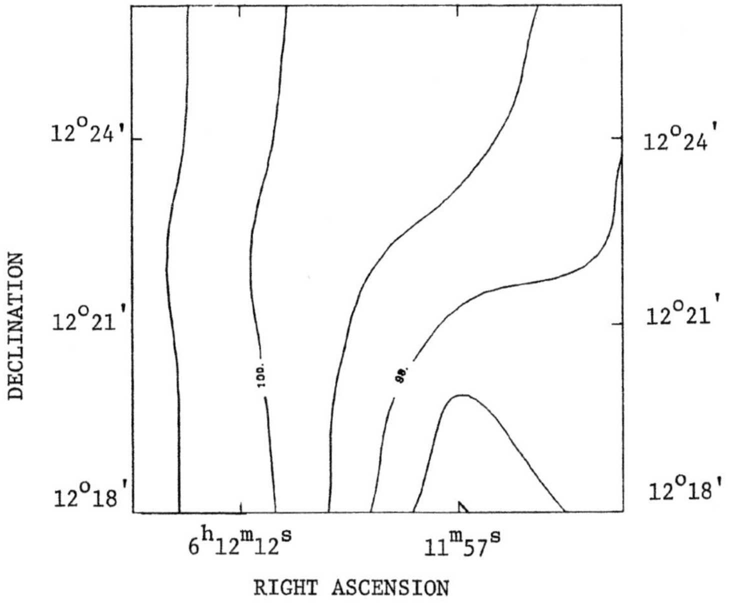

Figure 6. Moment integral for S271 over the velocity range 20 km/s.

Table 1

HII Region	Optical Size	d^1 kpc	Exciting[2] Star	Hα [2] velocity	CO velocity	HI velocity
S241	4' x 4'	4.7	O9V	-10 km/s	-6.5 km/s	-7 km/s
S242	4' x 5'	2.1	BOV		0.0	-2
S259	0'.7 x0'.6	8.3	B1V:		23.0	20.5
S271	0'.5 x2'.4	4.8	O9V	16	20.5	20:

Notes: [1] Moffat, FitzGerald and Jackson 1979
[2] Georgelin 1975

significant for S241 (although the evidence for S271 is suggestive) that the young HII regions do have enhancements of HI near their cores (see also Stacy and Jackson 1982). For these two regions, it is possible to determine mass estimates along the line of sight. In the case of S241, the average relation $N_{13CO} = 7.7 \times 10^{14}\ T_A^*$ (12CO) holds for the six positions for which 13CO was observed, and applying this mean relation to the value of T_A^*(12CO) found in the neighborhood of the HI rise in the southern portion of the S241 frame, yields $N(H_2) = 2.3 \times 10^{21}\ cm^{-2}$. For the HI in this area, assumed optically thin, the column density for the range -7 to -4 km/s is $N(H) = 4.2 \times 10^{20}$. See Jackson and Sewall (1982) for method of estimating $N(H_2)$.

These numbers indicate that a large fraction of the material is still in the form of molecules, reasonable considering the strong extinction in the northern part of the S241 region. For S271, we obtain $N(H_2)$ by direct observation of the ^{13}CO at the position of the HI variation: $N(H_2) = 4.8 \times 10^{20}$ cm^{-2}. The estimate for HI is also lower: $N(H) = 2.0 \times 10^{20}$ cm^{-2}, but indicates a more equal mix of neutral and molecular gas. The narrow line widths and high T_A*(12CO)/T_A*(13CO) ratios for S271 also indicate that the molecular cloud is becoming optically thin (Jackson and Sewall 1982).

The velocities of the CO and HI agree in general, although the Hα velocity is often more negative. If one takes the former velocity to be the undisturbed medium, then the ionized material has received an outward velocity of 3 to 6 km s^{-1} away from the parent cloud, agreeing with the 'blister' model of, e.g. Habing and Israel (1979). In the case of S241, this push could be the result of the exciting O9V star (LSV 30°31) although at the distance of this HII region, the Stromgren sphere would have to be large (10 pc). In some cases, the HII regions show dark lanes, and sharp boundaries with the surrounding medium. It is interesting to ask whether these isolated objects are still undergoing the phases of star formation, for which further observations at infrared and submillimeter wavelengths of the anticenter regions are needed.

ACKNOWLEDGEMENTS

The Arecibo Observatory is part of the National Astronomy and Ionosphere Center which is operated by Cornell University for the NSF. The National Radio Astronomy Observatory is operated by Associated Universities Inc. under contract to the National Science Foundation.

REFERENCES

Bash, F.N., Green, E. and Peters, W.L.: 1977, Astrophys. J. 217, 464.
Blitz, L., Fich, M. and Stark, A.A.: 1981, preprint.
Georgelin, Y.M.: 1975, Ph. D. thesis, U. Marseilles.
Habing, H.J. and Israel, F.P.: 1979, Ann. Rev. Astron. Astrophys. 17, 345.
Harten, R.H., Westerhout, G. and Kerr, F.J.: 1975, Astron. J. 80, 307.
Jackson, P.D. and Sewall, J.R.: 1982, this volume, p. 221.
Moffat, A.F.J., FitzGerald, M.P. and Jackson, P.D.: 1979, Astron. Astrophys. Suppl. 38, 197.
Stacy, J.G. and Jackson, P.D.: 1982, this volume, p. 185.
Viner, M.R. and Jackson, P.D.: 1980, in preparation.

DYNAMICS OF CO CLOUDS AROUND HII REGIONS IN THE OUTER GALAXY

Peter D. Jackson and James R. Sewall
Astronomy Program, University of Maryland

1. INTRODUCTION

Optical studies resulting in the determination of distances to a number of distant (up to 8 kpc from the Sun) HII regions in the outer Galaxy were made by Moffat et al. (1979). The present paper reports on a radio study, by means of the J = 1 → 0 transition of interstellar CO, of molecular clouds surrounding these HII regions, in order to compare sites of star formation between the outer Galaxy and the more heavily studied portions of our Galaxy near the Sun and interior to the Sun.

Molecular clouds were mapped around the seven HII regions shown in Table 1, which gives the galactic coordinates, distance from the Sun and galactocentric distance, R, for each HII region (based on $R_\odot$ = 10 kpc). Also given are the positions of the grid centers "E0N0" for the CO observing grids. These grid centers were chosen to be close to the positions of the radio thermal continuum for each region (see Viner et al. 1981).

2. OBSERVATIONS

All observations were made with the NRAO[1] 11-meter radio telescope, having a half-power beamwidth of 65 arcsec at the 115 GHz, J = 1 → 0 transition of 12CO.[2] The observations were performed over a mapping grid of spacing 50 arcsec in both the declination and right ascension directions. The spacing between grid points was increased near the edges of the clouds and the resulting contour maps (figures 1 to 5) should be treated with considerable caution in these areas.

---

[1] The National Radio Astronomy Observatory is operated by Associated Universities Inc. under contract from the National Science Foundation.

[2] We will use the notation 12CO for $^{12}C^{16}O$ and 13CO for $^{13}C^{16}O$.

R. S. Roger and P. E. Dewdney (eds.), Regions of Recent Star Formation, 221–230.
Copyright © 1982 by D. Reidel Publishing Company.

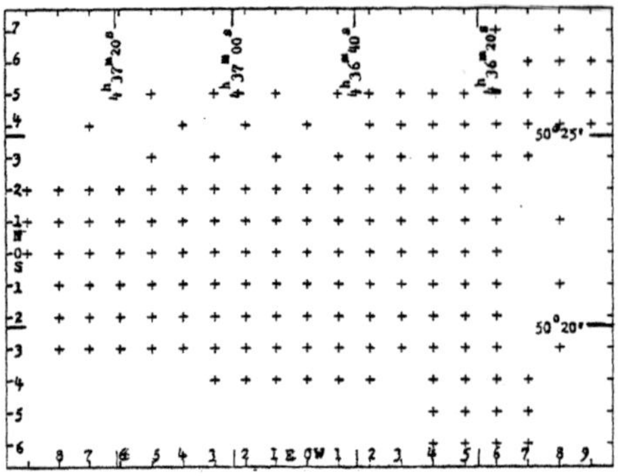

1a) Positions Observed

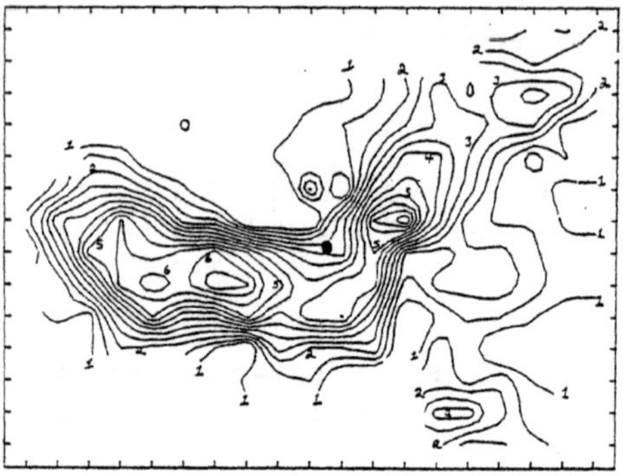

1b) T_A*

FIGURE 1 : S212

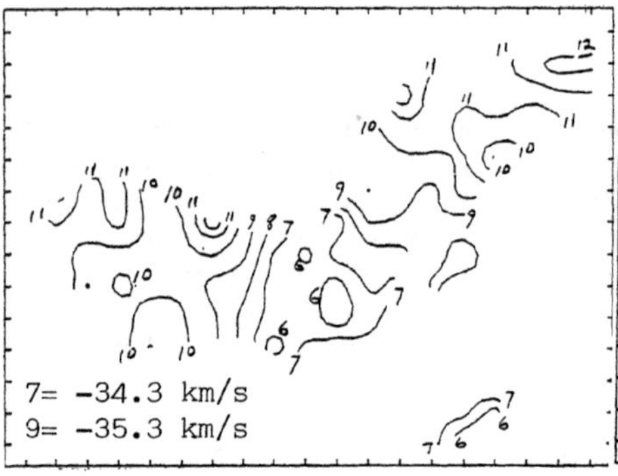

1c) Center Velocity

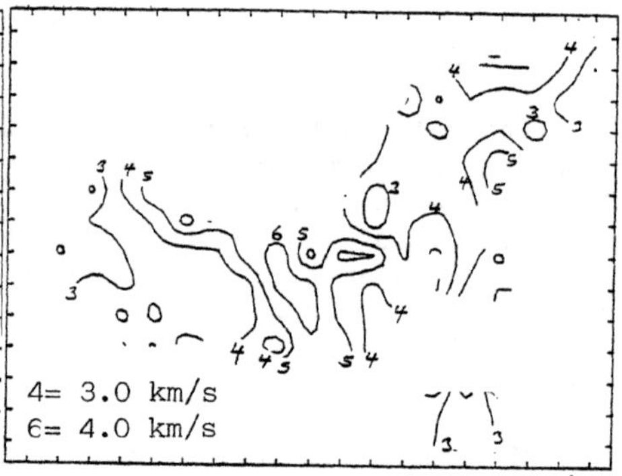

1d) Halfwidth

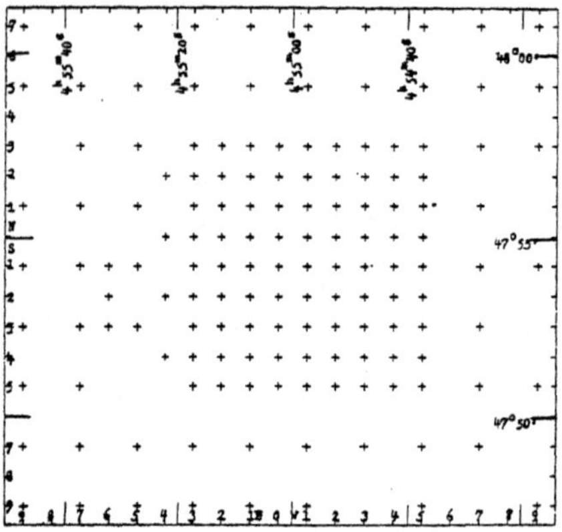

2a) Positions Observed

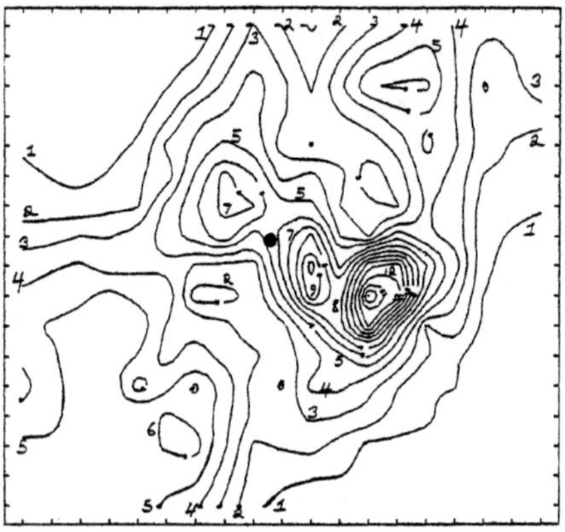

2b) T_A*

FIGURE 2 : S217

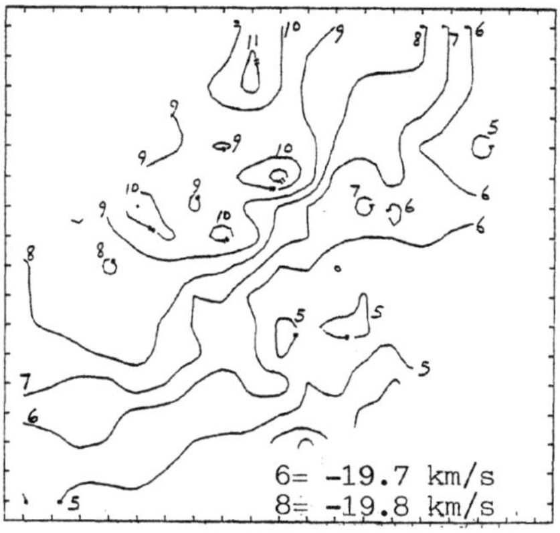

2c) Center Velocity

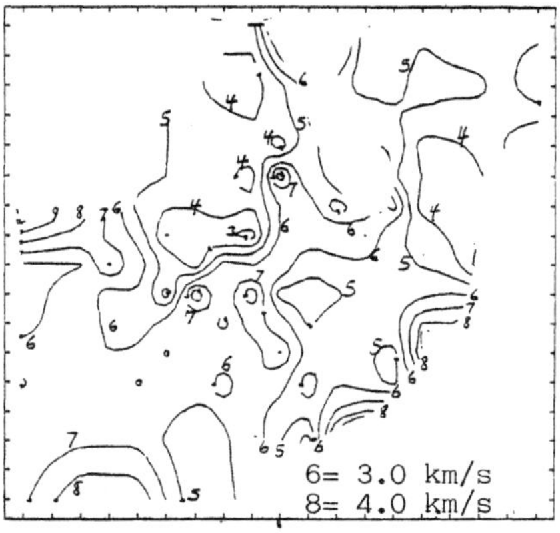

2d) Halfwidth

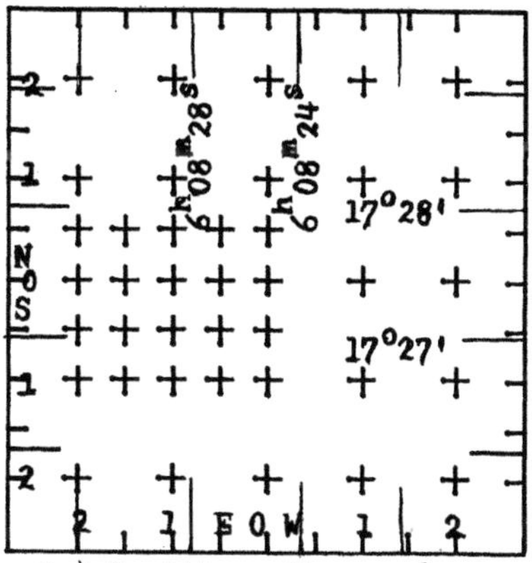

3a) Positions Observed

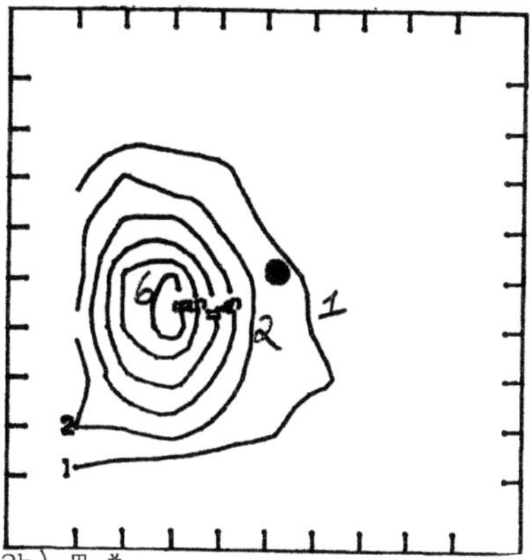

3b) T_A^*

FIGURE 3 : S259

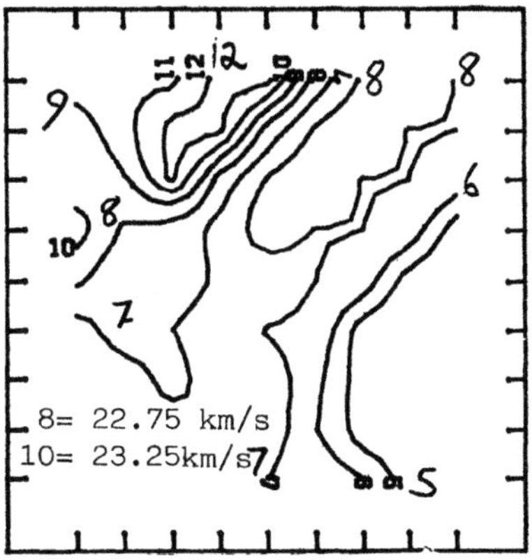

3c) Center Velocity

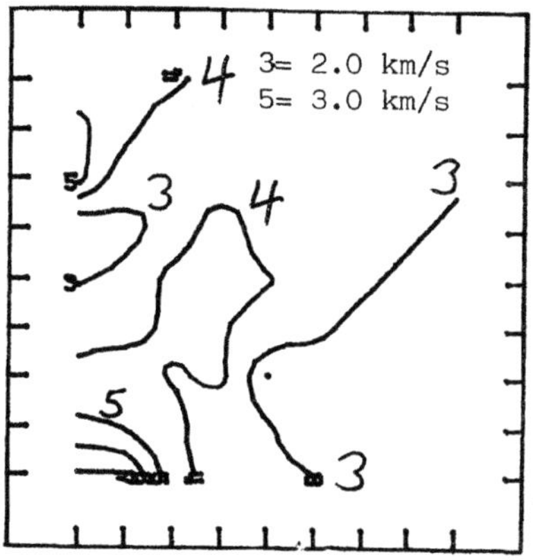

3d) Halfwidth

TABLE 1
HII Regions Observed

HII Region	ℓ	b	d^{a}_{pc}	R^{c}_{kpc}	Grid Center "EONO" α_{1950}	Grid Center "EONO" δ_{1950}
S212[b]	155.4	2.5	6.0	15.7	4^h 36^m 48^s	50 21 47
S217	159.2	3.3	5.2	15.0	4 55 02	47 55 02
S241	180.9	4.1	4.7	14.7	6 00 46	30 14 20
S259	192.9	0.0	8.3	18.3	6 08 25	17 27 26
S271	197.8	-2.3	4.8	14.6	6 12 05	12 22 16
S301	231.5	-4.4	5.8	14.3	7 07 45	-18 25 10
S305	233.7	-0.3	5.2	13.7	7 27 51	-18 26 38

a From Moffat et al. (1979) b NGC1624 c Assumes R_o = 10kpc

The 250 and 100 kHz/channel filter banks were used in series mode, one bank on each polarization mode. The profiles for the two banks were later averaged, after smoothing of the 100 KHz profiles, to give an effective velocity resolution of 0.58 km s^{-1}. All observations were conducted in absolute position-switching mode, with the reference position about 2° away in azimuth. An on-source integration time of 2 minutes per grid position was used, giving an rms noise of typically 0.4 K in the final averaged profiles. All observations were conducted in double side band mode; frequent checks were made to the positions of Ulich and Haas (1976) to confirm the intensity (T_A^*) calibration.

Observations of the 110 GHz transition of 13CO were later made towards about six to fourteen of the stronger 12CO profile positions for each molecular cloud. These 5-minute integrations had a typical rms noise of 0.15 K.

3. REDUCTION AND RESULTS

Gaussian fits were made to all detected line profiles. Figures 1 to 5 show, respectively, the maps of the 12CO Gaussian parameters for S212, S217, S259, S301 and S305. The maps for S241 and S271 are shown in Grayzeck et al. (1982). Each of figures 1 to 5 is divided into four portions a), b), c), d). These show (a) the actual grid of positions observed; (b) the peak intensity T_A^*; (c) the central LSR[3] velocity of the line; and (d) the half-power full velocity-width of

[3] Radial velocities are with respect to the Local Standard of Rest defined by a solar motion of 20 km s^{-1} towards α_{1900} = 18^h, δ_{1900} = +30°.

the profile. For (c) and (d), the contour labels are decoded on the figures. Maps of the total line emission (K km s^{-1}) are given in Viner et al. (1981). Table 2 gives the parameters for the positions observed in 13CO. For each position, physical parameters were calculated according to the following equations (see Sewall 1980 for derivation):

Excitation Temperatures:

$$T_x = 5.53 \left\{ \ln \left[1 + \frac{5.53}{T_A^*(12CO) + 0.83} \right] \right\}^{-1}$$

13CO optical depth:

$$\tau_{13} = -\ln \left[1 - \frac{T_A^*(12CO)}{T_A^*(13CO)} \right]$$

13CO column density:

$$N(13CO) = \frac{2.6\ \tau_{13}\ \Delta V\ (T_x + 0.9)}{1 - \exp(-5.29/T_x)} \times 10^{14}\ cm^{-2}$$

where ΔV is the 13CO line width in km s^{-1}.

The H_2 column density is

$$N(H_2) = 5 \times 10^5 N(13CO) \quad \text{(Dickman 1978)}.$$

The H_2 mass per grid point is

$$M(H_2) = d^2(kpc)\ N(H_2)/1.068 \times 10^{21}\ M_\odot$$

The total estimated H_2 mass for the cloud was estimated by multiplying the sum of all T_A^* values over the observed 12CO map (interpolating missing positions as necessary) by the mean ratio of $M(H_2)/T_A^*(12CO)$ for the positions listed in Table 2 for that HII region.

Table 3 shows for each HII region the number of grid points observed, the size of the CO cloud mapped, the peak T_A^* for the region and its LSR velocity, and the estimated H_2 mass. Also shown is the mean H_2 density (assuming for a cloud depth the geometric mean of the sizes in column 4), and the maximum H_2 density at the core of the cloud, assuming a depth of 2 pc.

4. DISCUSSION

Molecular clouds in the outer Galaxy share many properties in common with clouds closer to the galactic center. However, cloud masses appear to drop to below 10^4 $M_\odot$ near R = 15 kpc and below 10^3 $M_\odot$ near R = 18 kpc. Some or all of this apparent drop in mass may, however, simply reflect a changing CO to H_2 abundance ratio in the outermost parts of our Galaxy. The 13CO/12CO ratio does not vary

TABLE 2
13CO Data

HII Region	Grid Position	T_A^*	Half width km s^{-1}	Velocity km s^{-1}	τ_{13}	N(13CO) 10^{-15} cm^{-2}
S212	E4N0	0.88	1.35	-35.65	0.71	2.13
	E3N0	0.73	2.25	-35.68	0.13	1.79
	E5N0	0.88	1.42	-35.42	0.19	1.39
	E4S1	0.81	1.81	-35.46	0.15	1.61
	W3N1	1.61	1.97	-35.01	0.26	3.76
	W4N1	0.59	1.99	-35.13	0.18	1.32
	W3N2	0.99	2.15	-35.37	0.23	2.41
	W2N1	1.45	2.14	-35.05	0.27	3.63
	W3N0	0.50	1.36	-34.53	0.29	0.95
	E4N1	0.42	1.85	-36.36	0.22	1.02
S217	W4S3	0.73	2.33	-17.99	0.13	1.86
	W3S3	2.08	1.66	-17.88	0.19	4.64
	W2S3	1.07	1.90	-17.70	0.15	2.33
	W4S2	1.29	2.01	-18.19	0.11	3.42
	W3S2	2.74	2.06	-18.12	0.20	8.40
	W2S2	1.67	1.76	-17.82	0.25	3.52
	W4S1	2.42	1.86	-18.25	0.24	6.06
	W3S1	3.39	1.89	-18.23	0.38	8.99
	W2S1	1.11	1.89	-18.17	0.18	2.33
	E3N0	0.61	1.30	-20.86	0.12	0.85
S241	W1N0	2.16	3.70	-7.99	0.27	10.10
	W4N1	2.40	3.01	-6.51	0.27	9.45
	W1N3	2.04	2.80	-5.58	0.21	7.42
	W2N3	1.91	2.58	-5.91	0.20	6.33
	W3N3	1.85	2.92	-5.97	0.19	6.91
	W4N3	1.21	3.59	-6.06	0.17	5.04
S259	E1.5N0	0.79	1.19	+22.88	0.15	1.03
	E1N0	0.62	2.01	22.48	0.09	1.36
	E0.5N0	0.61	2.16	22.33	0.19	1.49
	E1.5S0.5	0.77	2.15	22.75	0.15	1.81
	E1S0.5	0.73	1.90	22.55	0.10	1.54
	E0.5S0.5	0.65	1.85	22.86	0.16	1.32
	E1.5S1	0.61	1.53	22.79	0.15	1.02
	E1S1	0.55	1.92	22.61	0.12	1.13
	E0.5S1	0.60	1.48	22.87	0.22	1.06

(table 2 con't)

HII Region	Grid Position	T_A^*	Half width km s^{-1}	Velocity km s^{-1}	τ_{13}	N(13CO) 10^{-15} cm^{-2}
S271	W2S2	1.19	1.63	21.07	0.16	2.25
	W3S2	0.92	1.63	20.93	0.18	1.67
	W1S2	0.46	1.91	21.13	0.06	0.96
	W2S1	0.38	1.39	20.67	0.07	0.56
	W2S3	0.78	1.51	21.11	0.22	1.35
	E1S3	1.09	1.40	20.96	0.22	1.73
	E2S4	0.78	1.25	21.17	0.14	1.07
	E3S5	0.94	1.23	21.08	0.16	1.29
S301	E4S1	0.36	2.53	52.78	0.15	1.09
	E4N0	1.08	2.11	53.10	0.20	2.57
	E4N1	2.06	2.71	52.82	0.27	6.97
	E4N2	2.63	2.83	53.09	0.40	9.68
	E4N3	2.18	2.00	54.06	0.44	5.52
	E3N0	0.84	2.40	53.95	0.15	2.22
	E3N1	0.96	3.08	52.73	0.16	3.30
	E3N2	1.78	3.25	52.91	0.25	7.02
	E5N1	2.44	2.14	53.10	0.34	6.69
	E7N1	3.28	1.94	52.73	0.48	8.75
	W2S7	2.80	1.45	51.86	0.39	5.36
	W3S8	2.94	1.60	51.93	0.44	6.28
	W3S7	2.45	1.26	51.93	0.37	3.95
	W4S7	1.65	1.72	52.52	0.25	3.39
S305	E4N1	2.38	1.48	44.93	0.35	4.49
	E4N2	2.40	1.85	44.74	0.31	5.70
	E4N3	1.98	1.62	44.36	0.28	3.95
	E3N1	3.36	2.37	44.29	0.36	11.20
	E3N2	2.79	1.70	44.34	0.37	6.27
	E3N3	2.14	1.52	44.07	0.26	4.12

outside the range seen for nearby HII regions, however. Comments on the results for individual regions follows:

S212: The CO velocity is clearly more positive near the HII region, showing the molecular cloud being pushed back by the HII region in front of it. This favourable geometry for optical study is confirmed by Moffat et al. (1979) who find a number of members of the OB cluster NGC1624 and a low, uniform reddening across the cluster.

S217: This is a large, complex molecular cloud with two strong but small CO emission centers near the HII region, and having a

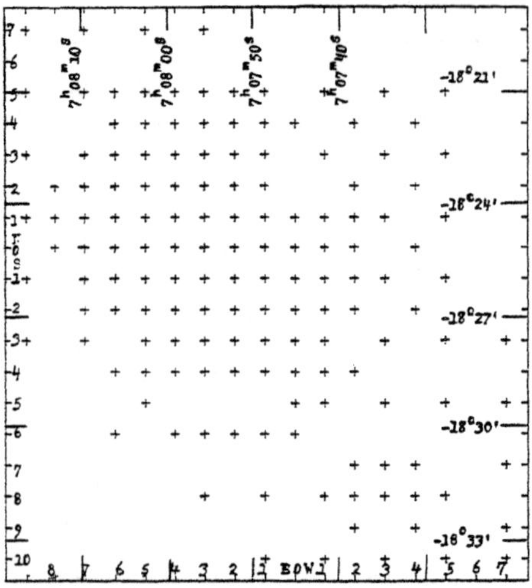

4a) Positions Observed

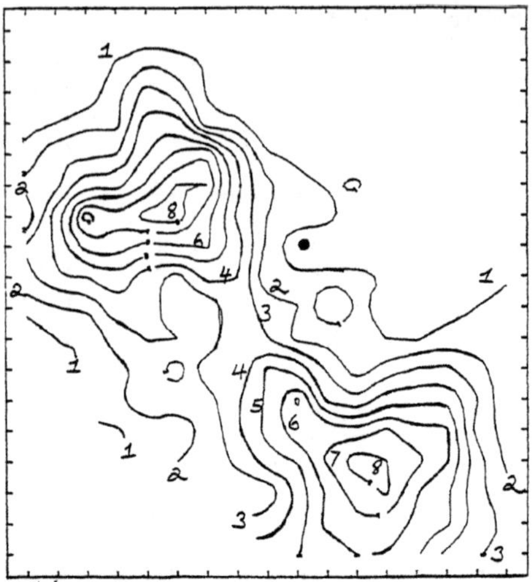

4b) T_A^*

FIGURE 4 : S301

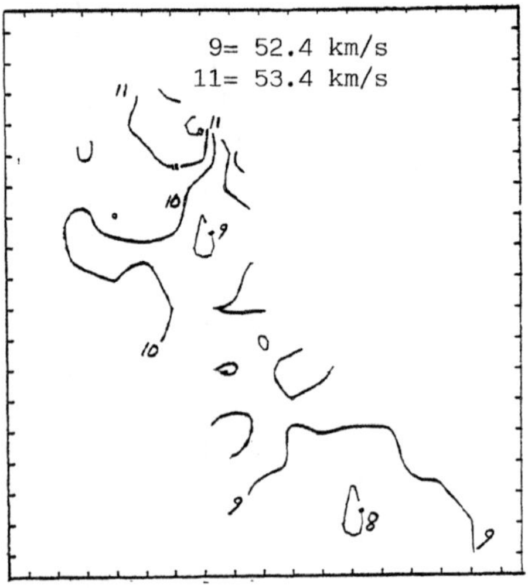

4c) Center Velocity

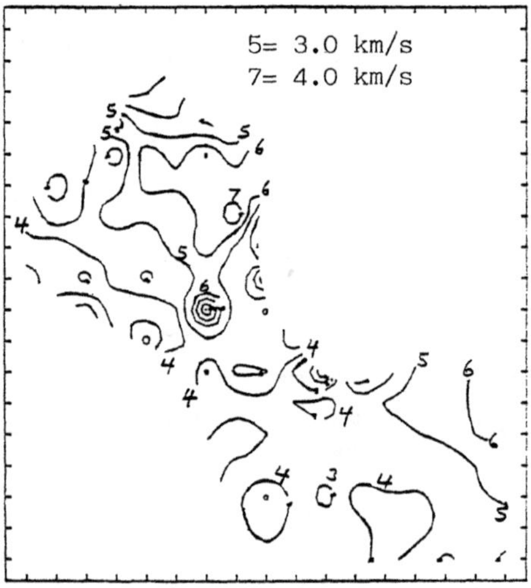

4d) Halfwidth

velocity near -18.7 km s^{-1}. The weaker NE component has a different velocity (near -20.8 km s^{-1}) and is probably well in front of the HII region, because of the more negative velocity, low line widths, and strong non-uniform reddening in this region.

S241: This is the most massive cloud of the group studied. The CO velocity is most negative near the HII region implying that the molecular cloud is in front of the HII region, confirmed by the irregular appearance (Moffat et al. 1979) and failure to find the

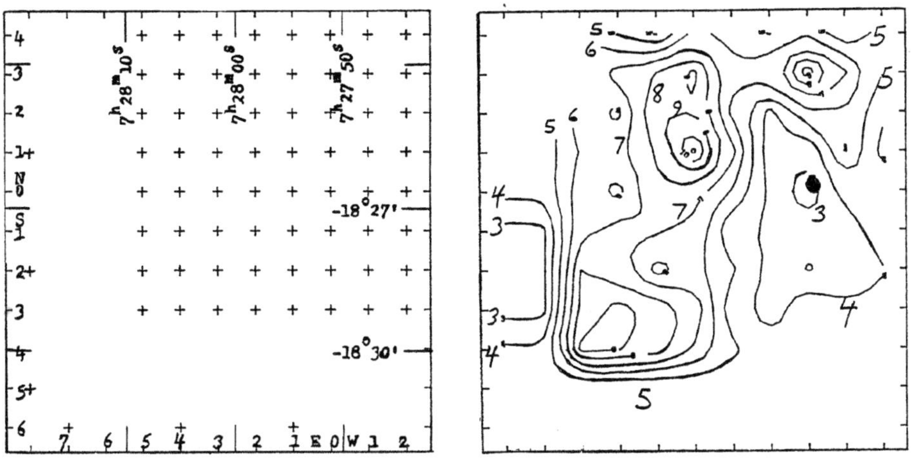

5a) Positions Observed

5b) T_A^*

FIGURE 5 : S305

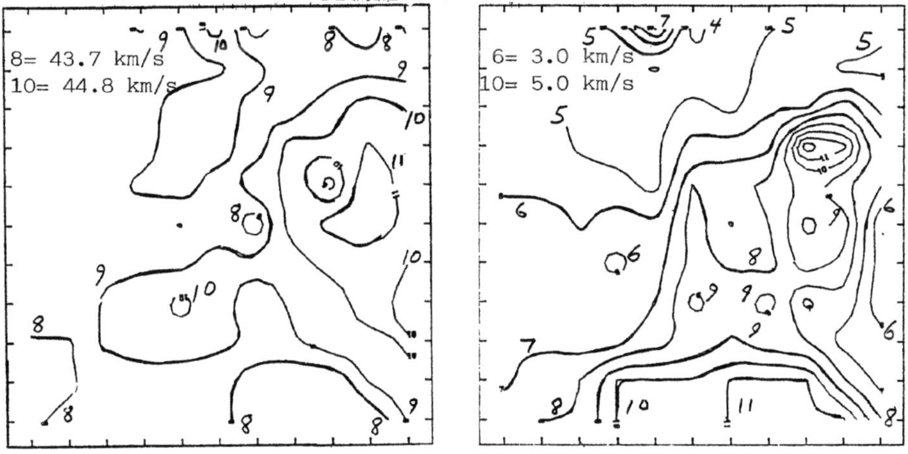

5c) Center Velocity

5d) Halfwidth

exciting star for the HII region. See Grayzeck et al. (1982) for details on HI associated with S241.

S259: This is the most distant, yet smallest of the clouds measured. It is essentially unresolved by the 11-m telescope.

S271: This is the most mature of our molecular clouds. Low linewidths and 13CO/12CO ratios indicate that column density is low, most of the molecules having been dissociated. See Grayzeck et al. (1982) details of HI associated with S271.

TABLE 3
CO Results

HII Region	No. of grid points observed 12CO	13CO	Size of CO cloud mapped (pc)	Peak T_A^*	12CO Velocity at Peak	Total H_2 Mass ($M_\odot$)	Mean H_2 density (cm^{-3})	Max H_2 density[1] (cm^{-3})
S212	153	10	28 x 7	7.0	-35.3	2854	17	300
S217	145	10	21 x 11	15.1	-18.2	4476	24	730
S241	147	6	19 x 12	10.9	-6.8	9535	55	820
S259	39	9	4 x 3	7.5	22.8	276	130	150
S271	145	8	22 x 5	8.1	20.7	1209	17	180
S301	147	14	24 x 10	8.7	51.9	7677	38	780
S305	70	6	11 x 9	11.2	44.5	5140	106	900

[1] Assuming 2 pc depth

S301: This unique dumb-bell shaped cloud is also probably mature. The clear separation of the clouds from the HII region and lack of dense concentrations imply an older, mostly dissociated region.

S305: This is the most poorly mapped of our clouds, and would perhaps be the most massive, if completely mapped. The CO velocities are more positive near the HII region, implying that most of the cloud is behind the HII, but large line widths and large stellar reddening for association members indicate that the cloud is very close to the HII region and may be partially enveloping it.

Acknowledgements:

The University of Maryland Computer Science Center provided computer time for the data reductions. Partial support for this project was provided by NSF grant AST 77-26898 to Prof. F.J. Kerr.

REFERENCES

Dickman, R.L. 1978, Ap. J. Suppl. 37, 407.
Grayzeck, E.J., Jackson, P.D., and Sewall, J. 1982, this volume, p. 213.
Moffat, A.F.J., FitzGerald, M.P. and Jackson, P.D., 1979 Astr. Ap Suppl. 38, 197.
Ulich, B.L. and Haas, R.W., 1976, Ap. J. Suppl. 30, 247.
Viner, M.R., Hughes, V.A., Jackson, P.D. and Sewall, J.R., 1981, in preparation.

CO J=2-1 OBSERVATIONS OF SOUTHERN GALACTIC PLANE HII REGIONS

Glenn J.White and J.P.Phillips.
Molecular Astronomy Group,Physics Department,
Queen Mary College,University of London,Mile End
Road,London EI 4NS,England.

Introduction

Much of our information on the structure and dynamics of star formation regions has come from observations of the CO J=I-0 molecular transition. Until recently these studies have been confined to those regions of the galactic plane lying north of declination -40 degrees,due to the lack of southern hemisphere millimetre wave telescopes. With the recent availability of large optical telescopes in the south,preliminary surveys,and some more extensive molecular mapping observations have been obtained. In this paper we have attempted to extend previous southern CO observations, and to search for objects exhibiting evidence for rotation, wide linewidths or self-absorption. The prescence of these effects in molecular spectra is extremely useful in modeling the thermal and dynamical structures of these star-formation regions. The present survey was biassed towards those objects showing evidence of recent star-forming activity, indicated by the presence of interstellar water vapour maser sources.

The Observations

The present data were obtained using the 2.5 metre du Pont telescope of the Las Campanas Observatory(altitude 2400 metres) at Las Campanas,Chile,during the period I8-23 February I98I. The QMC millimetre/sub-millimetre heterodyne spectral line receiver was positioned at the f/7.3 Cass focus,and matched to the telescope via polyethylene lenses, to give total system noise temperatures of 250-300 K, and a half power beamwidth of 2.3 arcminutes. Calibration of the beamshape and sensitivity was determined from observations of the moon,planets,ambient and liquid nitrogen loads. Regular monitoring of the atmospheric opacity was obtained using standard skydipping techniques,typical values of the atmospheric opacity at the zenith lying between 0.3-0.5 at

R. S. Roger and P. E. Dewdney (eds.), Regions of Recent Star Formation, 231–236.
Copyright © 1982 by D. Reidel Publishing Company.

230-244 GHz, and I.0-2.0 at 345 GHz. The atmospheric attenuation and sky fluctuations were considered to be too great for useful well calibrated data to be obtained at the latter frequency. In the 230-244 GHz range, we estimate the absolute value of T_A^* (the Rayleigh-Jeans equivalent temperature in excess of the cosmic background as measured by a lossless telescope above the terrestial atmosphere) to be within 20%. Regular checks of the absolute pointing of the telescope indicated an r.m.s. uncertainty of about 5 arcseconds. about 5 arcseconds.

The Survey

The present CO observations were biassed towards star formation areas selected as being associated with water vapour maser sources for which accurate positions are known (Batchelor et al I980,Scalise and Braz I980). Our objective was to select a subset of those sources exhibiting intense or highly variable maser emission for initial CO observations,and then to select a smaller group from those with interesting line profiles for further mapping. In table I the results of the survey are listed. We have compared the radial velocities with those obtained from other molecular species such as H_2CO, CS, HCO^+ and H_{109} (Wilson et al. 1970, Gardner and Whiteoak I978,Batchelor et al I98I).

TABLE I

SOURCE	RA (I950)	DEC (I950)	T_A^*(CO)	V (km s^{-1})	ΔV (km s^{-1})	$V_{HI09\alpha}$	V(CS)
ORION A	05 32 47	-05 24 20	86.0	9.0	6.5	6.I	9.7
COM GLOB	07 I4 30	-43 52 00	5.0	I.7	I.3	3.5	-
GLOB 2I0	08 24 I7	-50 42 08	I.2				
RCW 38	08 57 23	-47 19 24	II.9	SEE TEXT		3.0	-I.3
G269.I-I.I	09 0I 45	-48 I4 00	II.3	II.4	5.2	I0.I	-
G270.3+0.8	09 I4 58	-47 44 00	I5.7	9.4	3.9	3.7	-
G285.3-0.I	I0 29 37	-57 46 00	2I.5	3.0	5.2	2.3	-
CARINA	I0 4I 39	-59 I9 00	I8.3	-I8.0	22.0	-23.8	
RCW 57	II 09 42	-6I 0I 55	26.4	-24.0	I2.I	-25.8	-22.2
G309.8+I.8	I3 45 52	-60 07 I3	32.8	-38.0	8.I	-	-
G3I6.8-0.I	I4 4I 3I	-59 36 54	I7.5	-40.6	9.6	-37.I	-36.I
G326.6+0.6	I5 40 43	-53 56 29	32.9	-39.0	I0.0	-36.I	-
RCW 97	I5 49 I3	-54 28 I2	32.3	-47.0	6.3	-48.8	-46.8
G329.4-0.6	I5 59 4I	-53 0I 33	6.0				
G330.9-0.4	I6 06 30	-5I 58 I4	I8.9	-64.0	4.6	-62.6	-62.6
G33I.5-0.I	I6 08 2I	-5I 2I II	I8.8	-90.0	6.4	-89.3	-89.I
G333.3-0.4	I6 I7 47	-50 I9 06	32.4	-52.8	9.6	-	-52.I
G337.7-0.I	I6 34 47	-46 54 I3	2I.6	-46.0	4.5	-47.5	-

A selection of the individual spectra obtained are illustrated in figure I.

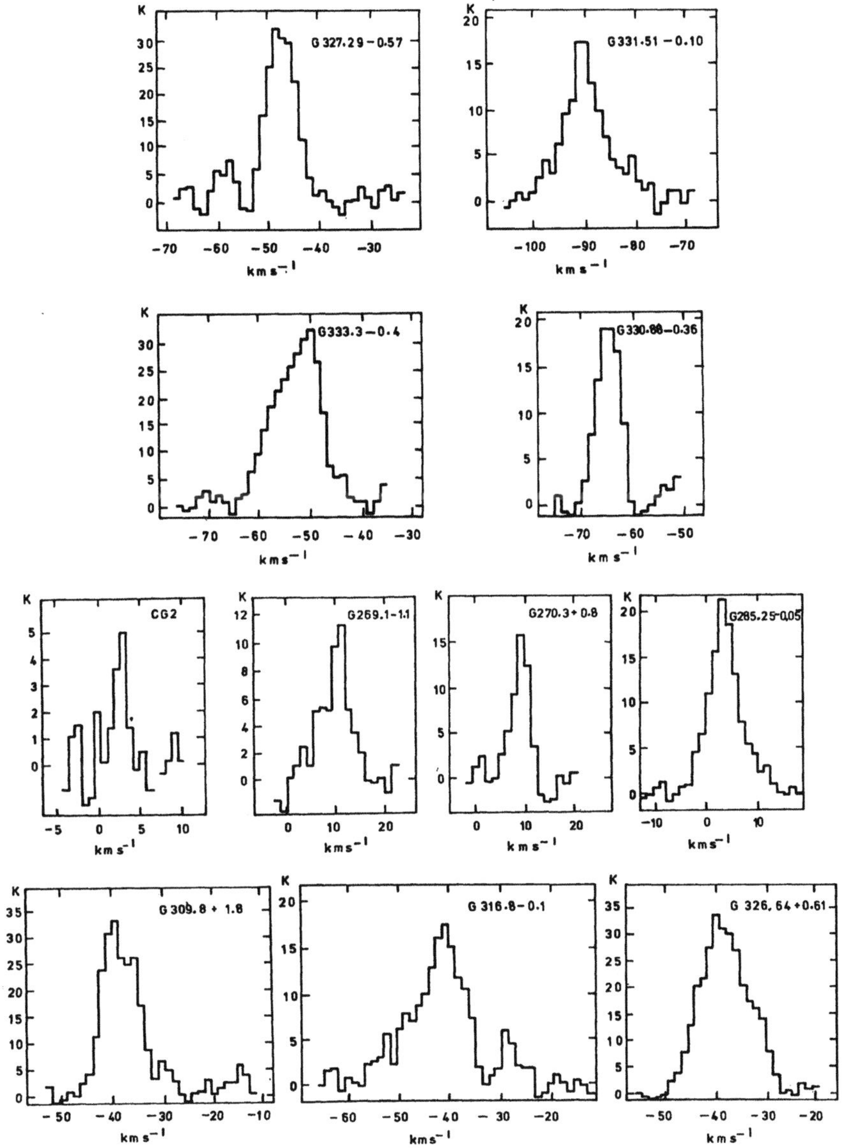

FIGURE I

CO J=2-I Spectra towards objects listed in table I

From the sources initially surveyed, we have chosen a small sample for further mapping observations;

RCW 38

This molecular cloud has been previously mapped in the CO J=I-0 transition by Gillespie,White and Watt (I979) and is seen to contain two compact hot spots. In figure 2 we show a spectrum obtained towards the IR complex associated with RCW 38E. The spectrum appears to show evidence for a deep self-absorbed feature at +3.0 km s^{-1}. In figure 3 we show a map of the central region indicating the source to have a compact central condensation of diameter 4 arc minutes.

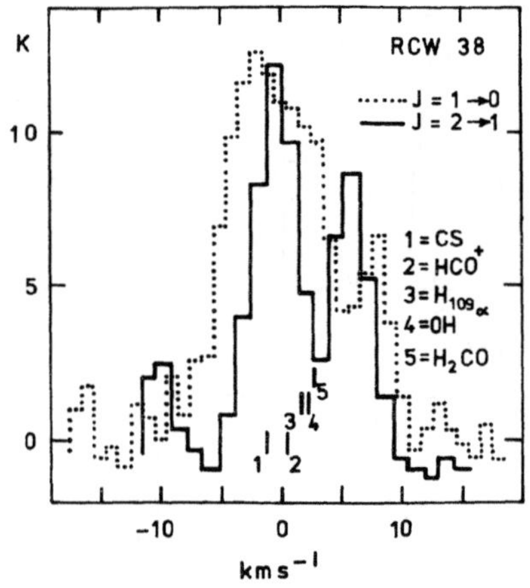

FIGURE 2 Spectrum towards the self reversed source RCW 38E

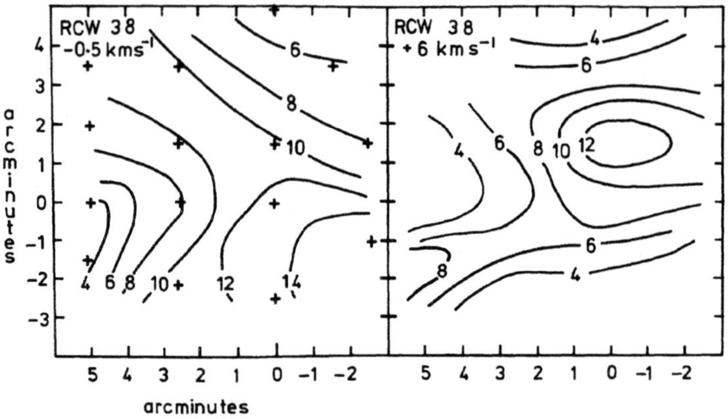

FIGURE 3 CO distribution around RCW 38E

CARINA NEBULA

The Carina Nebula is amongst the most prominent optical HII regions visible from the southern hemisphere. A preliminary CO J=I-0 map of the nebula has been obtained by White,Gillespie and Watt (in preparation). Spectra obtained in the CO J=2-1 transition are illustrated in figure 4.

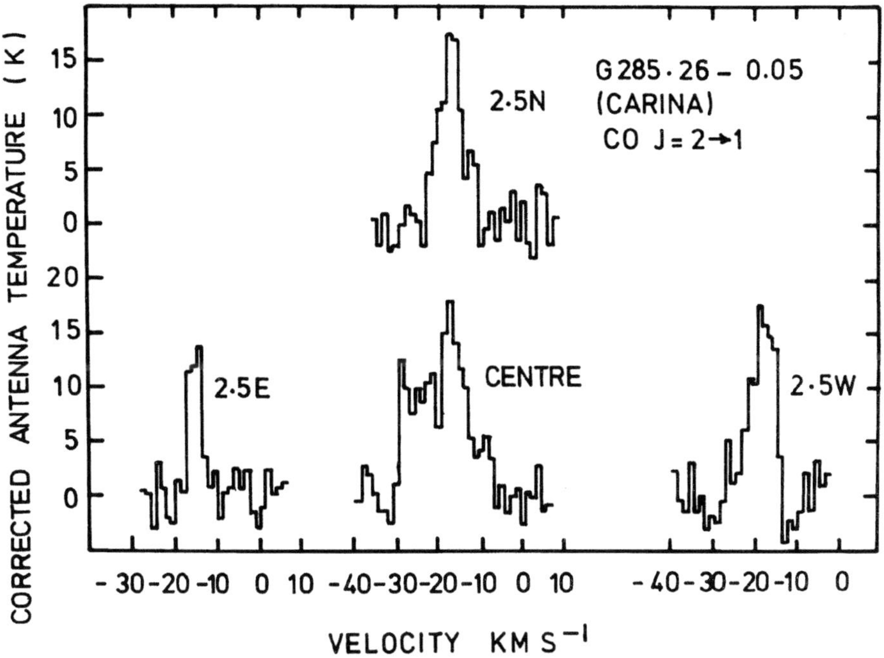

FIGURE 4 Spectra obtained towards the water vapour maser position in the Carina Nebula

The most striking feature of these spectra is the strong line broadening seen towards the central position relative to the profiles obtained one beam-width away. Similar behavior is also present in the CO J=I-0 profiles towards this position. It seems likely that an energetic dynamical process, such as mass outflow, may be responsible for the prescence of such complex spectra. Further high resolution observations of this molecular cloud are urgently needed.

In figure 5 we show the current sample of CO clouds

observed to date in this and previous CO surveys;

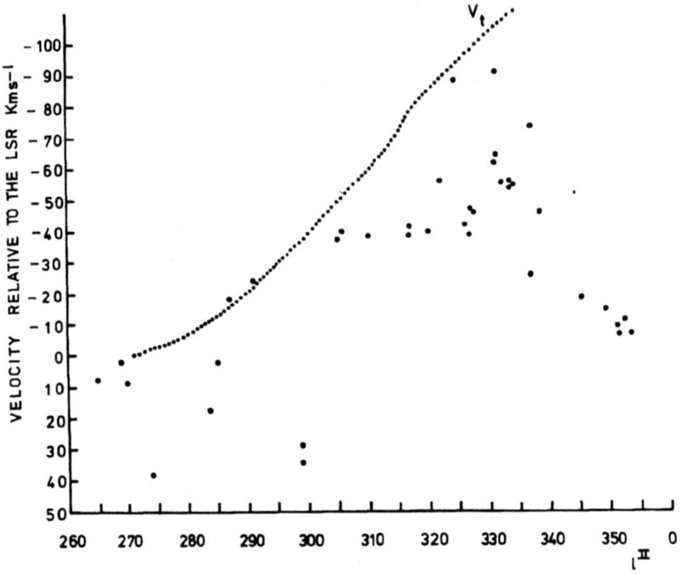

FIGURE 5 Structure of the southern galactic plane based on preliminary surveys of molecular clouds.

Acknowledgements

We extend our thanks to the Director and staff of the Las Campanas Observatory for observing time and technical support, the Science Research Council for financial support,and the authors of Gillespie,White and Watt (I979) for permission to include the CO J=I-O spectrum of RCW 38E.

References

Batchelor,R.A. et al.,I980. Preprint
Batchelor,R.A.,McCulloch,M.G. and Whiteoak,J.B.,I98I. MNRAS,I94,9II.
Gardner, F.F., and Whiteoak, J.B., 1978. MNRAS, 183, 711.
Gillespie,A.R.,White,G.J. and Watt,G.D.,I979.MNRAS,I86,383.
Scalise,E. and Braz,M.,I980. Astr.and Astrophys.,85,I49.
Wilson,T.L., Mezger,P.G., Gardner, F.F. and Milne,D.K., 1970. Astr.and Astrophys.,6,364.

OBSERVATIONS OF CO J = 3 → 2 EMISSION FROM MOLECULAR CLOUDS

Glenn J. White, J. P. Phillips and Graeme D. Watt
Molecular Astronomy Group, Physics Department,
Queen Mary College, Mile End Road, London E1 4NS, England.

1. INTRODUCTION

Observational studies of interstellar Carbon Monoxide have enabled astronomers to gather a wealth of information on the structure, energetics and kinematics of galactic molecular clouds. The majority of this information has been obtained from observations of the CO J = 1 → 0 rotational transition at 115.271 GHz (λ = 2.6 mm). Studies of higher rotational transitions have awaited the development of low-noise receiver systems for frequencies above 200 GHz. In this paper we report new data obtained using the 3.8 m United Kingdom Infrared Telescope (UKIRT) with the QMC and UKIRT submillimeter heterodyne receivers. These cryogenic heterodyne systems have system noise temperatures varying between 270 K at 220 GHz and 400 K at 370 GHz.

2. OBSERVATIONS

Spectra and maps have been obtained towards many galactic star formation regions. In Table 1 we compare some of the observed data in the J = 1 → 0, J = 2 → 1 and J = 3 → 2 CO transitions, where the data from lower transitions are collected from the literature using telescopes with similar beam sizes.

In all cases the kinetic temperature of the J = 3 → 2 line centre is similar or less than that measured in the lower transitions. Comparison of the peak line intensities suggests that it is reasonable to assume that the line centres are optically thick and thermalised in all three transitions.

In all cases except L 1551 the CO J = 3 → 2 half-power linewidth is broader than in the J = 1 → 0 or J = 2 → 1 transition.

In Figure 1 we show examples of spectra obtained towards NGC 1333 (IRS-1), NGC 2071, Mon R2 and AFGL 961. In all of these spectra, evidence can be seen for asymmetric line profiles, self-absorption or both.

R. S. Roger and P. E. Dewdney (eds.), Regions of Recent Star Formation, 237–244.
Copyright © 1982 by D. Reidel Publishing Company.

TABLE 1

Source	CO $J = 1 \rightarrow 0$			CO $J = 2 \rightarrow 1$			CO $J = 3 \rightarrow 2$		
	T_A^*(K)	ΔV(km s^{-1})	T_{ex}(K)	T_A^*(K)	ΔV(km s^{-1})	T_{ex}(K)	T_A^*(K)	ΔV(km s^{-1})	T_{ex}(K)
NGC 1333	19	4.7	26	-	-	-	11.2	10.4	18
L 1551/ HH-29	11	3.1	16	10	4.1	15	5.5	2.6	12
OMC-1	60	6.3	91	88	6.8	98	82	16.0	90
NGC 2023	40	2.9	51	34	4.0	40	31	-	39
NGC 2024	35	5.4	45	23	6.5	28	32	-	40
NGC 2068	40	2.2	51	-	-	-	42	-	50
MON R2	30	9.6	39	25	11.2	30	25	8.9	33
AFGL 961	8.3	5.8	15	-	-	-	8.3	10.8	15
DR 21	25	7.9	32	18	7.0	23	20	11.3	27

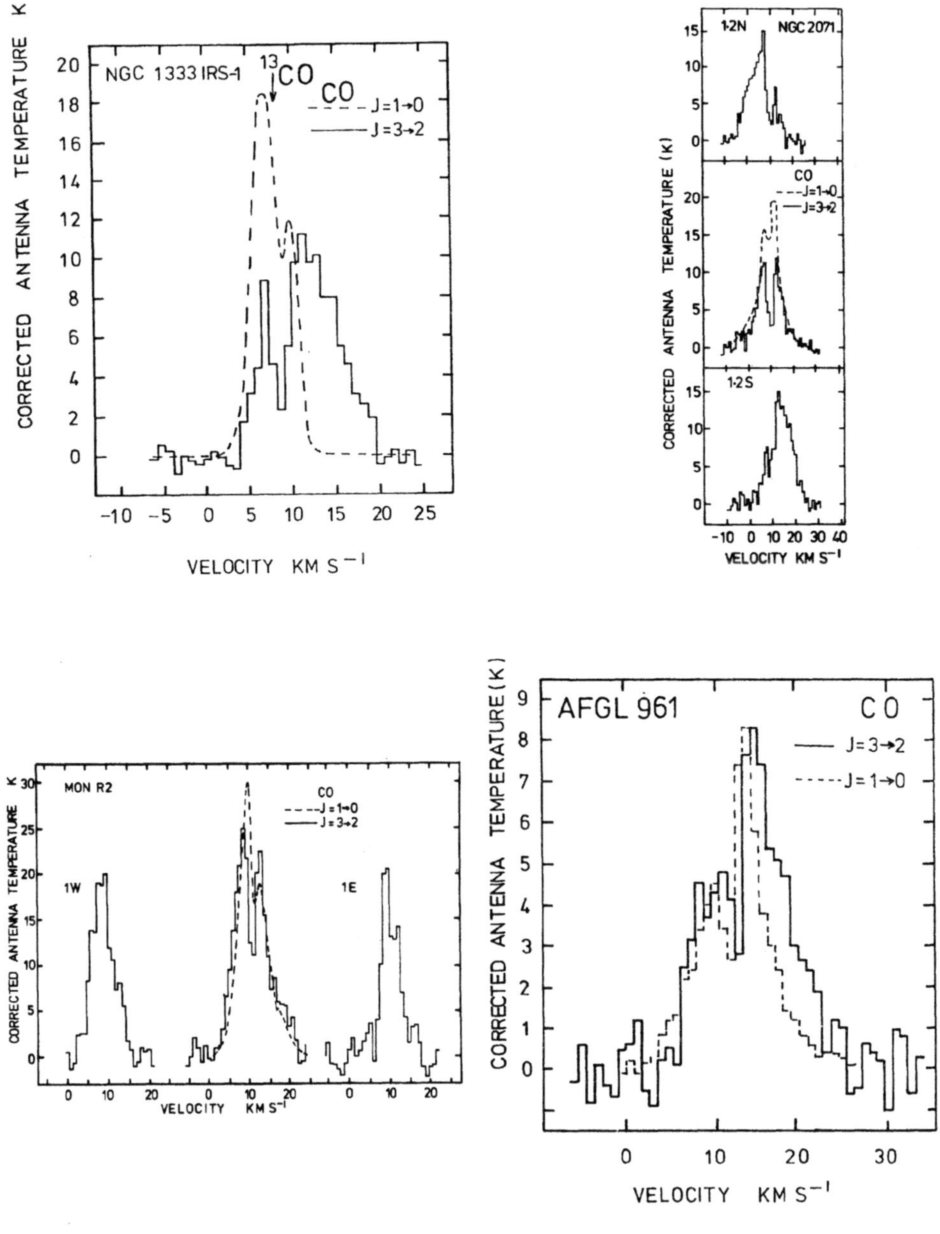

Figure 1. Spectra of the CO J = 3 → 2 transition towards objects showing self-absorption. The red-shifted excess observed towards NGC 1333 is due to the presence of hot, optically thin gas. Evidence for rotational motions are clearly seen towards NGC 2071.

A feature of several of the spectra is a tendency for the J = 3 → 2 line to be asymmetrical with an enhancement of the redshifted emission relative to the J = 1 → 0 or J = 2 → 1 profiles. In Table 2 are listed parameters for sources exhibiting self-absorption in their profiles, or asymmetries. In all cases a redshifted enhancement in the J = 3 → 2 line is observed, but only in half the cases is the blueshifted line enhanced.

TABLE 2

Source	T_A^* sabs (K)	T_K sabs (K)	Vel_{sabs} (Km s^{-1})	$V_{sabs} - V_{mol}$ (Km s^{-1})	Red wing enhanced	Blue wing enhanced	Sabs deeper	V sabs (Km s^{-1})	Size sabs (arc min)
NGC 1333	2.4	8.1	6.9	+1.2	Yes	No	Yes	1.9	2 x 3
NGC 2071	3.0	8.8	+10.0	+0.2	SEE TEXT		Yes	3.5	-
MON R2	11.3	18.4	+0.9	Yes	Yes	Yes	Yes	2.1	2 x 2
AFGL 961	2.7	8.5	+11.5	+0.4	Yes	No	Yes	1.5	-
DR 21	12.2	19.3	-8.0	-5.6	Yes	Yes	Yes	?	-

In Table 2 it is seen that in the cases of NGC 1333, NGC 2071 and AFGL 961, the value of T_A^* at the velocity of maximum self-absorption lies in the range 2 - 3 K, corresponding to a gas kinetic temperature T_{KIN} ~ 8-9 K, assuming this arises in an optically thick medium. For gas at these kinetic temperatures, the opacity in the J = 3 → 2 transition is several times lower than in the J = 1 → 0 and J = 2 → 1 transitions. These results have several consequences for the physical conditions in these clouds.

Unless the gas kinetic temperature appreciably decreases with decreasing source radius, or increasing depth of penetration into a foreground cloud, then the low velocity absorption region cannot fill the entire telescope beam (i.e., source dilution is $\eta < 1$). For dilution $\eta \ll 1$ (implying higher kinetic temperatures for the absorbing material, and optical depths $\tau(J = 3 \rightarrow 2) > \tau(J = 1 \rightarrow 0)$) and absorption occurring in an ensemble of foreground cloudlets, then each cloudlet would be required to possess a strong temperature gradient. This scenario appears improbable if these cloudlets represent structural subunits of typical interstellar clouds. Alternatively, it must be supposed that low velocity emission terms originate from a restricted area of the cloud close to the core, again leading to $\eta << 1$, but with the outer layers of the source blocking low velocity emission from the higher temperature inner regions. This would imply that the sources are strictly self-absorbing, and indeed there are now several self-absorbing cloud models which meet these criteria. For similar reasons systematic motions through the clouds must greatly exceed turbulence in determining line shapes.

Extensive mapping observations have been obtained towards many of the objects in this study. In Figure 2, typical maps with full spatial sampling towards NGC 2023 and NGC 2068 are shown. Both sources are spatially resolved and show a symmetrical distribution. For these two sources, the averaged variation of T_{ex}, the CO excitation temperature against projected radial distance, r, is illustrated in Figure 3.

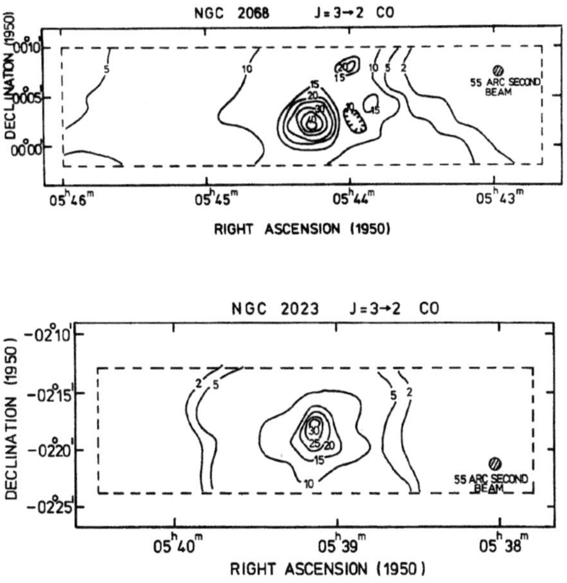

Figure 2. Maps towards NGC 2068 and NGC 2023 in the CO J = 3 → 2 transition.

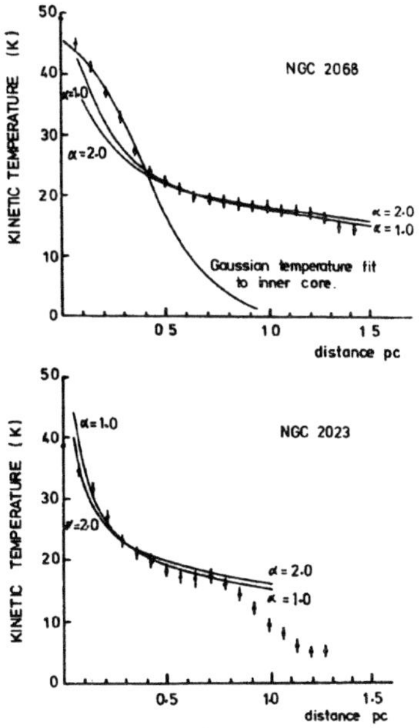

Figure 3. Comparison of T_{ex} calculated with observed data.

This value of temperature will also correspond closely to the grain temperature T_{gr}, providing beam dilution is small, and that the grains are in kinetic equilibrium with the surrounding gas. This latter condition implies (Scoville, Solomon and Penzias, 1975)

$$n_{H_2} > 10^3 \left\{ \frac{T_K^{3/2}}{(\Delta T/T_K)} \right\}^{\frac{1}{2}} \tag{1}$$

or for T_K = 20 K and $T/T_K \sim 0.1$, it indicates a value of H_2 number density $n_{H_2} \gtrsim 3 \times 10^4$ cm^{-3}. Here the parameter T_K represents the kinetic gas temperature, and ΔT is the difference in temperature between the grains and gas. For lower densities appropriate to the outer regions of the clouds, grains and gas will thermally decouple, leading to reduced values of T_{ex}; the possible cause of lower values for T_{ex} in NGC 2023 at radii > 1 pc.

In the case of a cloud with low grain absorption optical depth to the central irradiating source, and to the microwave background, the grain equilibrium temperature is determined by:

$$\frac{L_*}{4\pi r^2} \pi a^2 \bar{Q}_{abs1} + \sigma T_{BB}^4 4\pi a^2 \bar{Q}_{abs2} \sim 4\pi a^2 \bar{Q}_{em} \sigma T_{gr}^4 \tag{2}$$

where L_* is the central source luminosity
a is the grain size
$\bar{Q}_{abs\ 1,2}$ are the grain Planck mean absorption coefficients for the source and microwave background radiation
$\bar{Q}_{em}$ represents the grain Planck mean emission coefficient
T_{BB} is the microwave background temperature, 2.7 K.

The variation of Q_{abs} with wavelength is uncertain, but for $Q_{abs} \propto \lambda^{-\alpha}$, a value $1 < \alpha < 2$ seems to be implied by theory and far-infrared results (cf. Aannestad 1975; Day 1976; Herter et al. 1979; Erickson et al. 1977; Pipher et al. 1978). We adopt values of $\alpha = 1$ (Mezger et al. 1974) and $\alpha \simeq 1.9$ (Gilman 1974) as extreme cases to obtain

$$T_{gr} \simeq k \left[\frac{L_*}{10^{38}} \cdot \frac{\bar{Q}_{abs1}}{O_{\mu m}} \cdot \left\{ \frac{1pc}{r} \right\}^{\frac{1}{2}} \right]^{\frac{1}{4+\alpha}} + 2.7 \tag{3}$$

where K ≃ 60 for $\alpha = 1.9$ and 27.8 for $\alpha = 1$.

In Figure 3, we have compared the observed values of T_{ex} with values calculated from eqn. 2. The fits are generally good over large regions of the cloud for both values of α, although a value $\alpha \sim 1$ gives rather a better representation of the trend in T_{ex}. For the central irradiating sources we have used luminosities corresponding to the spectral class of the exciting stars as determined from recombination line or infrared studies, for NGC 2023 a B1.5 and for NGC 2068, a B1.5-B2.5 star.

SUMMARY

Extensive spectral observations in the CO J = 3 → 2 line have been obtained towards a sample of hot-centred molecular clouds which show evidence for self-absorption or high-velocity gas. The spectra bear general similarities to observations in the lower CO transitions, however there is a tendency for enhanced red-shifted high velocity gas to be present. The self-absorbed line centres are slightly red-shifted with respect to optically thin emission lines. Absorption velocity widths are observed to be between <1.5 to ~3.5 km s^{-1}. Observations of DR 21 indicate it to possibly show self-absorption, although this is not present in the J = 1 → 0 or J = 2 → 1 transitions. The CO J = 3 → 2 linewidths are systematically broader than those of lower CO transitions indicating the warm (~10 K) gas to have $\tau(J = 3 \rightarrow 2) > \tau(J = 2 \rightarrow 1)$ or $\tau(J = 1 \rightarrow 0)$. Peak kinetic temperatures at the line centres are generally similar to those estimated from lower transition, to within the absolute calibration uncertainties.

Mapping observations with the high angular resolution of this experiment have revealed accurate positions and intensity profiles for a number of sources. NGC 2068 and NGC 2023 are resolved, isolated and symmetrical clouds, whereas NGC 1333 and NGC 2024 are more complex areas containing multiple hotspots. Molecular cool-spots have been detected in the J = 3 → 2 transition confirming our earlier detections in the J = 1 → 0 transition, and may be related to turbulent cells in the outer layers of the cloud.

Analysis of the variation of gas kinetic temperature around NGC 2023 and NGC 2068 shows close agreement with modelling results of the grain kinetic temperature assuming single, central irradiating stars which are of similar spectral class to those indicated by radio and recombination line measurements. Evidence for possible thermal decoupling is seen in the outer layers of NGC 2023.

ACKNOWLEDGEMENTS

We thank the UKIRT team for assistance at the telescope and the many persons at QMC and Cambridge involved with the construction and operation of the heterodyne receiver systems.

REFERENCES

Aannestad, P.A.: 1975. Astrophys. J., 200, 30.
Day, K.L.: 1976. Astrophys. J., 210, 614.
Erickson, E.F., Strecker, D.W., Simpson, J.P., Goorvitch, D.,

Augason, G.C., Scargle, J.D., Caroff, L.J. and Witteborn, F.C.: 1977. Astrophys. J., 212, 696.
Gilman, R.C.: 1974. Astrophys. J. Suppl., 28, 397.
Herter, T., Duthie, J.G., Pipher, J.L. and Savedoff, M.: 1979. Astrophys. J., 234, 897.
Mezger, P., Smith, L. and Churchwell, E.: 1974. Astr. Astrophys., 32, 269.
Pipher, J.L., Duthie, J.G. and Savedoff, M.P.: 1978. Astrophys. J., 219, 494.
Scoville, N.Z., Solomon, P.M. and Penzias, A.A.: 1975. Astrophys. J., 201, 352.

PROPERTIES OF GIANT MOLECULAR CLOUDS IN THE GALACTIC MOLECULAR RING

W.L.H. Shuter and A. Szabo
Department of Physics
University of British Columbia
Vancouver, Canada.

We present results on the physical properties of giant molecular clouds based on a fully sampled J=1→0 CO survey at $b=0°$ from $\ell=29°-46°$ using the 4.6m millimetre wave telescope at UBC. The telescope beam-width was 2.6' and the velocity resolution (FWHM) was 2.6 km s^{-1}.

The definition of a cloud adopted here was that on a longitude-velocity diagram there should be a closed contour with $T_A^* = 4K$ and at least one inner contour at 5K or more. If there was only one maximum in T_A^* within the 4K contour the cloud was described as being isolated.

In our survey we obtained a catalogue of 71 giant molecular clouds of which 34 were isolated. Of the isolated clouds, 21 were sufficiently close to the tangent points that their distances can be regarded as being known unambiguously, and a full range of physical properties can be specified for them. These are summarized in Table 1, in which our parameters are compared with those obtained by Solomon, Sanders and Scoville (1979) and Liszt and Burton (1981). In deriving our molecular hydrogen density and column density mass we assumed (1) that ^{13}CO would 'mimic' ^{12}CO but with T_A^* values 5 times smaller, and (2) that the ratio of H_2 to ^{13}CO was 1.0×10^6 as given by Frerking, Langer and Wilson (1981). For the sake of uniformity in the comparisons we have expressed our result in the table in terms of a distance from the sun to the galactic centre of 10 kpc although we prefer the value of ∿ 8.5 kpc given by Graham (1979).

On comparing the results in Table 1 it appears that there is generally good agreement between us and Solomon et. al. Our effective radii which represent half-width to half-maximum in T_A^* are sufficiently uniform that we suggest it might well be possible when looking at longitudes near to the galactic centre to discriminate between clouds in the molecular ring on the near side of the centre and those on the far side on the basis of their angular diameters.

R. S. Roger and P. E. Dewdney (eds.), Regions of Recent Star Formation, 245–247.
Copyright © 1982 by D. Reidel Publishing Company.

TABLE 1
PROPERTIES OF GIANT MOLECULAR CLOUDS

	Shuter & Szabo	Solomon et. al.	Liszt & Burton
Effective radius, r_{eff}, pc.	13.3 ± 0.9	< 20	7.5 - 10
T_{ex}, K.	10	10	14
Molecular Hydrogen Density, n_{H_2}, cm^{-3}.	270	300	∿250
Virial Mass, M_V, solar masses.	7 x 10^5	5 x 10^5	≳ 10^4
Column Density Mass, M_{cd}, solar masses.	1.5 x 10^5	-	-
Internal Velocity Dispersion, σ_{vi}, km s^{-1}.	7.3	∿3-4	1.3
Radial Velocity Dispersion σ_{vr}, km s^{-1}.	6.5	-	3.5

Our masses inferred from estimates of column density are typically a factor of ∿5 lower than those derived by assuming the cloud is stable and applying the virial theorem. We suggest the primary cause of this discrepancy is our assumption of LTE, and that in fact the excitation temperature of ^{13}CO may be of ∿4K and not the values ∿10K assumed.

Our value of 7.3 km s^{-1} for the internal velocity dispersion of the clouds is alarmingly different from that used by Liszt and Burton and may be caused in part by our larger beam area (∿ 6 x that of the Kitt Peak Telescope) and our relatively poor velocity resolution (2.6 km s^{-1} FWHM). However it is expected that these effects would reconcile our results with those of Solomon et. al. but not with those of Liszt and Burton.

Finally, our value for one-dimensional velocity dispersion of 6.5 km s^{-1} is intermediate between the value of 9 km s^{-1} suggested by Stark and Blitz (1978) and 3.5 km s^{-1} used by Liszt and Burton in their modelling of longitude-velocity diagrams. Our value probably represents a slight overestimate, since it was obtained from deviations from a smooth tangent point line over the entire longitude range 29°-46°. Since the tangent point appears to show large scale wiggles over this range we estimate that if we had measured deviations from these our value may well have been reduced to ∿ 4.5-5 km s^{-1}.

In conclusion it appears that our results are in surprisingly good agreement with those of Solomon et. al. considering that the surveys upon which the results were based were done in different galactic longitude ranges using telescopes with different beam sizes and error beams and with different velocity resolution. There are still some important differences between our clouds and those used by Liszt and Burton in modelling longitude-velocity diagrams in CO, and possibly these could be resolved to some extent by assuming there are two major distributions of clouds in the inner galaxy. Giant clouds as described here largely confined to the distance range $0.5 \leq R/R_{\odot} \leq 0.7$ with typical masses $\sim 5 \times 10^5$ $M_{\odot}$, and less massive clouds $\sim 5 \times 10^4$ $M_{\odot}$, with a more uniform distribution.

This work has been supported by an N.S.E.R.C. grant. We intend to present a more comprehensive version later. We are grateful to Chris Chan for his work on the telescope and receiving system.

REFERENCES

Frerking, M.A., Langer, W.D. and Wilson, R.W.: 1981, Preprint.

Graham, J.A.: 1979, in "The Large-Scale Characteristics of the Galaxy", ed. W.B. Burton (Holland: D. Reidel), pp. 195-200.

Liszt, H.S. and Burton, W.B.: 1981, Astrophys. J. 243, 778.

Solomon, P.M., Sanders, D.B. and Scoville, N.Z.: 1979, in "The Large-Scale Characteristics of the Galaxy", ed. W.B. Burton (Holland: D. Reidel), pp. 35-52.

Stark, A.A. and Blitz, L.: 1978, Astrophys. J. Letters, 225, L15.

CHANGES OF THE STAR FORMATION RATE AND THE INITIAL MASS FUNCTION WITH GALACTIC RADIUS

J.L. Puget,
Institut d'Astrophysique, Paris
R. Gispert,
LPSP, Verrières
G. Serra,
CESR, Toulouse

Using the data of a survey of the northern part of the galactic plane at far infrared wavelengths (71-95 μm and 114-195 μm) made with a balloon born instrument (Gispert, Puget and Serra, 1981) we can compute the far infrared luminosity of the galactic plane as a function of galactic radius. The result is shown in Figure 1. The far infrared radiation is stellar light reemitted by dust, and is a good measure of the total energy output. The ratio of the far infrared luminosity to the amount of interstellar gas gives a measure of the star formation rate per unit mass of interstellar gas. Furthermore one can compute the "infrared excess" (ratio of the infrared luminosity to the energy in Lyman α photons in H II regions). The number of Lyman continuum photons is evaluated by taking the number of Lyman continuum photons from giant H II regions from Smith et al., 1978 and assuming that the fraction of Lyman continuum photons coming from giant H II regions is constant throughout the galaxy. The number of Lyman continuum photons evaluated this way is somewhat lower than the number obtained by Mezger (1978) using 1.4 GHz data. The infrared excess is found to decrease very significantly with galactic radius as shown in Table 1 which can be explained only by a change of the initial mass function of the stars formed. In the far infrared survey, 58 individual

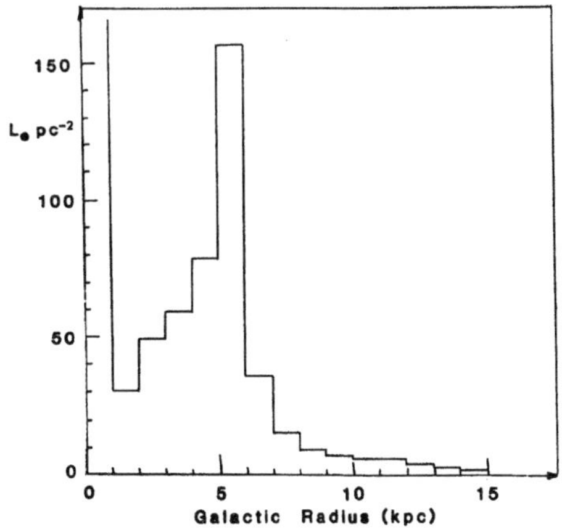

Figure 1: Far infrared luminosity as a function of galactic radius.

R. S. Roger and P. E. Dewdney (eds.), Regions of Recent Star Formation, 249–250.
Copyright © 1982 by D. Reidel Publishing Company.

Table 1. Star Formation Rate and Initial Mass Function

	L_{IR} $(L_{\odot} pc^{-2})$	M_{gas} $(M_{\odot} pc^{-2})$	Star formation rate $\propto \frac{L_{IR}}{M_{gas}}$ $(L_{\odot}/M_{\odot})$	Lyman cont. photons (pc^{-2})	Infrared excess $\frac{L_{IR}}{N_{Lyc} \times h\nu_{Ly\alpha}}$	Initial mass function
4-6 kpc ring (Scutum arm)	114	13.3	8.6	$1.2\ 10^{45}$	23.3	Luminosity dominated by intermediate mass stars.
7-8 kpc (Sagitarius arm)	34	10	3.4	$7.6\ 10^{44}$	6.8	
9-11 kpc	7	4.7	1.5	$3.2\ 10^{44}$	5.5	Luminosity dominated by ionizing stars.

extended sources can be isolated, many of them being associated with molecular cloud-H II region complexes. The properties of the complexes change depending of the distance of the complexes to the galactic center. A similar systematic change in the initial mass function is found. The complete results and the detailed discussion of the astrophysical implications is given by Gispert, Puget and Serra, 1981.

REFERENCES

Gispert, R., Puget, J.L., Serra, G.: 1981, Astron. Astrophys. in press

Mezger, P.G.: 1978, Astron. Astrophys. 70, 565

Smith, L.F., Biermann, P., Mezger, P.G.: 1978, Astron. Astrophys. 66, 65

INFRARED AND MASER SOURCES IN REGIONS OF STAR FORMATION

Reinhard Genzel
Department of Physics, University of California
Berkeley, California

Dennis Downes
Institut De Radio Astronomie Millimetrique,
Grenoble, France

I. INTRODUCTION

The emphasis of this review is on the mass loss phenomena associated with compact infrared and maser sources in regions of star formation. Most of the compact 2→20 μm sources associated with H_2O, and OH (and SiO) masers and ultra-compact H II regions in regions of formation of stars more massive than a few $M_\odot$ seem to have mass outflow at velocities between 10 and 250 km s^{-1}. We discuss the evidence for these motions in several sources (in particular, Orion-KL), and their dependence on luminosity and evolutionary stage. The mass loss phase in the early evolutionary stages of stars appears to cover the whole spectrum of stellar masses and to be of long duration. The impact on the surrounding interstellar clouds may be considerable. We also address the problem that if most of the infrared objects suspected to be "protostars" in the earlier literature are actually in a later, mass-loss state, then where are the true protostars which are still accreting mass?

The discovery of the high velocity molecular motions came with the first spectra of the H_2O maser source in W49 (1,2), and with the detection of broad wings in the profiles of millimeter molecular lines in Orion (3,4). An interpretation of the high velocity dispersions in the H_2O maser spectra in terms of mass outflow due to stellar winds and radiation pressure came as early as 1972 in the excellent and perceptive paper by Strelnitskii and Syunyaev (5). In this review, we summarize the progress since about 1976. (see also the discussion of theoretical concepts in 134, and a general discussion of masers in 39).

R. S. Roger and P. E. Dewdney (eds.), Regions of Recent Star Formation, 251–286.
Copyright © 1982 by D. Reidel Publishing Company.

2. The Orion-KL Region

2.1 The Infrared Cluster

The BN-KL cluster of infrared sources, 1' NW of the Trapezium stars and the centroid of the ionized nebula, has been investigated in detail between 1 and 1000 μm. A review of the observations up to 1977 has been given by (6). Far-infrared studies beyond 30 μm have established that the total luminosity of the cluster is 1.5 $10^5 L_\odot$ and that the dust temperature of the central 30" is 120 K, in agreement with the expected radiation heating from a central source with the observed luminosity (7). Observations at 1 mm indicate that the opacity of dust averaged over the central 1' corresponds to a column density of hydrogen molecules of $\sim 5\ 10^{23}$ cm^{-2}, and A_V=100→200 mag (8). Submillimeter (400 μm) data (9) show that the core of this dust cloud has a size (FWHM) of 30" x 45" (EW x NS) and is elongated NE-SW. The recent work at $\lambda \lesssim 30$ μm has concentrated on high resolution ($\lesssim$3") mapping and on infrared spectroscopy. Whereas a few years ago interest was focused only on the BN object and the cool, extended, KL nebula, newer observations indicate a more complex situation.

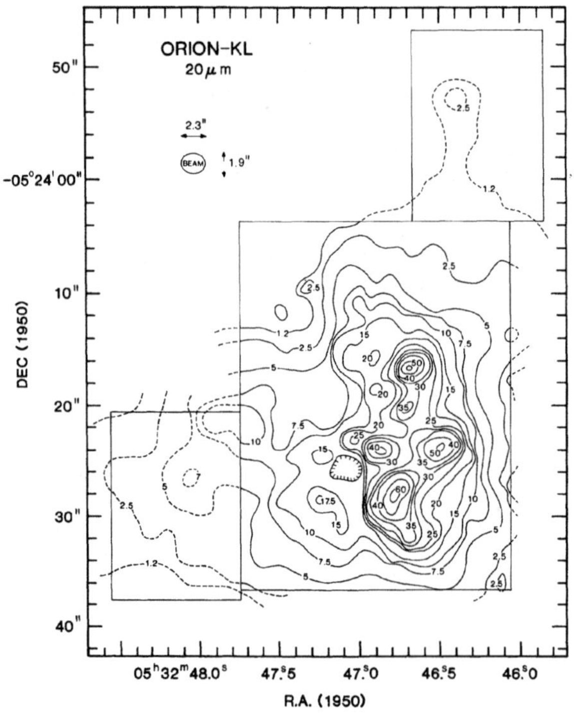

Fig. 1. Map of the Orion-KL region at 20 μm (from 10). The contour unit is 10 Jy per beam area. The most prominent sources at the center are (from W to E): IRc3, BN, IRc4, IRc5 and at the eastern edge of the central region (47.0^s,-23") IRc2.

Fig. 1 is a 20 μm map at 2" resolution, taken at the IRTF on Mauna Kea, Hawaii (10). A series of narrow band maps at 3" resolution has been made with the Wyoming infrared telescope across the silicate absorption feature (11). Spectrophotometry between 2 and 20 μm of individual sources has been reported by (10,12). The most important questions about these maps are i) do all (or which) of the near infrared condensations contain intrinsic sources of luminosity and what are the luminosities?; ii) are there any additional sources of high luminosity which are deeply buried in the cloud and have so far been missed in the near-infrared observations? Unfortunately, 90% of the luminosity is radiated between 20 and 100 μm and is contained in the extended dust component visible on the 20 μm map (Fig. 1). Hence, radiative transport and diffusion of radiation must play an important role, and an estimate of the luminosities of individual sources involves large (and therefore uncertain) correction factors.

a) The range $12 \leq \lambda \leq 20$ μm

The 20 μm map shows a large number of sources and peaks in addition to an extended background (30"). Except for BN (IRc1) and IRc2, the other sources (including the extended background) have similar 12.5 to 20 μm color temperatures (~100 to 125 K). The most prominent of these peaks, IRc3 and IRc4, are resolved by a 2" beam. Comparison of the 20 μm brightness with that estimated for a blackbody at the observed 12.5 to 20 μm color temperature indicates moderately high optical depths over the whole face of the 20 μm map ($\tau_{20\mu m} \sim 0.1$ to >1). Hence, emission peaks on the 20 μm map may represent small enhancements in dust temperature. The 12.5 to 20 μm color temperatures of IRc3, IRc4 etc. are consistent with radiative heating by a central source. If the dust absorbs radiation from a source of luminosity L with an efficiency $Q(\lambda) \propto \lambda^{-1}$, then the dust temperature at radius R from the source in a moderately optically thick dust cloud is given by (13)

$$T_d[K] = (165 \pm 20) \left[\frac{R}{4\ 10^{16}\ \mathrm{cm}}\right]^{-2/5} \left[\frac{L}{10^5 L_\odot}\right]^{+1/5} \qquad (1)$$

The scaling factor has been determined from the far-infrared color temperature (120 K) of the central 25" of the source (at the 480 pc distance of Orion-KL, 1"=7.2 10^{15} cm). While the color temperatures of most of the 20 μm peaks are in agreement with Eq.(1), the radial temperature gradient expected from Eq.(1) has not yet been observed. Alternatively, if IRc3 and IRc4 had internal heating sources of their own, their luminosities can be estimated from the observed sizes (=3"±1") and the color temperatures (~120±10 K), from the Stefan-Boltzmann law. This formula is applicable since the 20 μm optical depths of the sources are high. The opacities at λ<20 μm are even larger, and no radiation can escape without being absorbed. On the other hand, at λ>20 μm, where most of the radiation of IRc3 and IRc4 is emitted, the dust opacity decreases rapidly with increasing wavelength and any correction for extinction between the observer and the 110-150 K "photosphere" at R=1".5 is probably small. The luminosities of IRc3 and IRc4 can thus be estimated to be $5000^{+5000}_{-1000}\ L_\odot$, for color temperatures T_c between 110 and 150 K.

b) The range $\lambda \lesssim 12$ μm

Two unresolved (<1") objects dominate the maps below 12 μm. The brighter one, BN, has T_c (2 to 20 μm) ~550 K. Speckle interferometry has shown (14) that BN's dust shell has a size of $\Theta_s \lesssim 0.08''$, consistent with a black body at the observed brightness and color temperatures. The integrated infrared luminosity of BN is 1.5 $10^3 L_\odot$. IRc2, the other compact source, has a lower color temperature (~350 to 400 K). It is therefore much weaker than BN below 8 μm, but about half as bright as BN at 8 and 12 μm. Both of these objects must have intrinsic heating sources of luminosities $>10^3 L_\odot$. Since the optical depth of the nebula is large at the wavelengths where BN and IRc2 emit most of their energy, their true luminosities may be $\gg 10^3 L_\odot$.

IRc3 and IRc4 have color temperatures of ~400 K between 5 and 8 μm, obviously much higher than can be maintained by radiative heating from an external source (Eq.(1)). However, the integrated apparent luminosities between 5 and 8 μm in these sources are small (~20 $L_\odot$). The 5 to 8 μm emission may indicate the heavily reddened, warmer dust closer to the intrinsic heating sources within IRc3 and IRc4. Alternatively, the warm dust may have been heated by a nonradiative process. The mechanical luminosity in the mass loss discussed below, for example, is $\sim 10^{2.5 \pm 0.5} L_\odot$. If the source(s) of mass outflow were displaced from IRc3 and IRc4 by a few arc sec, 10 to 100 $L_\odot$ of this mechanical luminosity may be available for heating the dust in these sources (e.g. by shock excitation).

c) The silicate feature

The infrared spectra of the sources in BN-KL show an absorption feature between 8.5 and 12 μm which probably is caused by silicates. The depth of the feature varies within the nebula, and is deepest toward IRc2 and IRc4 (10,11,12). If the strength and shape of the silicate feature are known, the amount of dust and hence, the total extinction in front of the sources can be estimated, and de-reddened luminosities can then be derived. For BN, radiative transport calculations with differing assumptions (12,15,16,17) yield a de-reddened luminosity between 5 10^3 to 2 $10^4 L_\odot$. The situation in the other sources is more complicated, since the silicate features are very deep. No accurate model calculations have been performed, but the estimated luminosities are 5 10^3 to $\gg 10^4$ for IRc2 and $\lesssim 10^4 L_\odot$ for IRc4 (10,12). The broad shape and slight shift toward shorter wavelengths of the feature in IRc4 may indicate a range of dust temperatures, implying an internal energy source (12).

d) Other Evidence for Internal Heating Sources

IRc2: In the immediate vicinity of the source, there are v=1 and v=2 SiO masers, H_2O masers and OH masers (see discussion below and 10,18). This pattern and the double-peaked profiles of the SiO and H_2O masers are similar to those of the masers in envelopes of late-type giants and supergiants. One might conclude that IRc2 actually is an evolved giant or at least, has physical conditions similar to those in the envelopes of evolved giants (34). The models for pumping of SiO masers require strong near infrared radiation fields (T_K>2000 K, L>$10^4 L_\odot$) for radiative pumping (19), or a combination of high densities (>$10^9 cm^{-3}$) and temperatures (T_K>2000 K) for collisional pumping (20). In fact, the SiO maser luminosity of IRc2 and the estimated mass loss rate (21) are much greater than observed in the shells of late type giants (L≃$10^4 L_\odot$), and are similar to the values found for supergiants (L≥$10^5 L_\odot$).

IRc4: As discussed below, the brightest spot of the "hot core" line emission from NH_3 molecules at 1.2 cm is located close to IRc4. The brightness temperatures of the NH_3 lines are ∿200 K, and some of the excited non-metastable NH_3 inversion lines may indicate even higher temperatures (22). From Eq.(1), the NH_3 brightness temperatures probably imply that a source of high temperature and luminosity is close ($R \lesssim 3'' \pm^{3''}_{1''}$) to IRc4 (if the "hot core" gas is thermally excited).

e) Spectroscopy of BN

The recent dramatic progress in the 2 to 5 μm spectroscopy of BN, in particular due to work with the FTS on the Kitt Peak 4-m telescope, has been summarized in (23). Strong Brα and Brγ hydrogen recombination line emission indicates ionized gas around the object (24,25). The upper limit for radio free-free emission of 2 mJy shows that the ionized

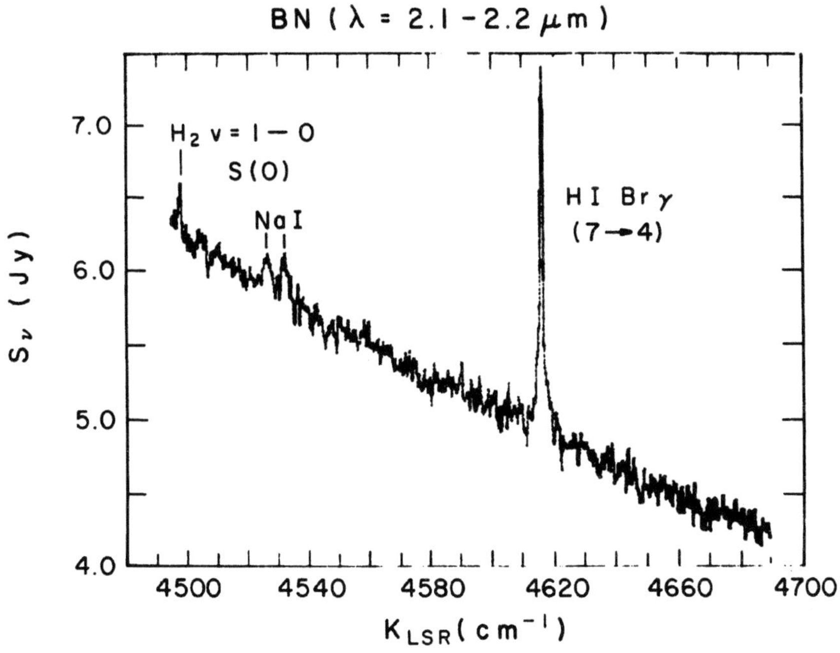

Fig. 2. 2 μm high resolution spectrum of BN, taken with the FTS on the 4-m Kitt Peak telescope (23).

gas is concentrated to radii R$\lesssim$4 10^{14} cm (26). In addition, a hot (T_K>3000 K), dense (n_{H_2}>10^{10} cm^{-3}) molecular core at a radius of 10^{13} cm is implied by the detection of overtone bandhead features of CO at 2.3 μm (27). The ionized zone around the star is probably at even smaller radii than this hot molecular core, and may be photospheric. The measured Brα and Brγ line fluxes (corrected for reddening) require an exciting star with a Lyman continuum flux of 7 10^{46} s^{-1} if the H II region is optically thin and photoionized. Hence, BN may be a B0 (ZAMS) star of luminosity $\sim 10^4$ $L_\odot$ (25). In principle, it could also be a star of later spectral type if the hydrogen lines are collisionally excited. In addition to a component of width 30 km s^{-1}, the Brγ line profile (Fig. 2) shows broad wings out to ±200 km s^{-1}. The NaI lines at 2.2 μm (Fig. 2) and the CO overtone lines at 2.4 μm also show this high velocity component. It is likely that these motions are caused by a stellar wind from the star within BN. From the emission measure, the limit to the size of the H II region and the expansion velocity, an estimate of the mass loss rate in the stellar wind from BN is $\dot{M}_{BN}\lesssim 2\ 10^{-6} M_\odot y^{-1}$ (23). Furthermore, the LSR velocity centroid of the hydrogen recombination lines and the CO lines is 20 km s^{-1}, that is, offset by 11 km s^{-1} from the LSR velocity of the molecular cloud. One possible explanation is (23) that this is actually the velocity of BN itself. If this velocity shift is not caused by orbital motion in a binary system, BN will move out of the central 1' of the cluster in only 10^4y! An alternative interpretation may be a shift of the 2 μm velocity centroids by scattering off an expanding dust shell (25). Finally, absorption in the v=0→1 band by CO at 4.6 μm has two components, at V_{LSR}=9 and -16 km s^{-1}. This gas probably belongs to the kinematic components in the molecular cloud at radii >2" from BN.

2.2. The Core of the Quiescent Molecular Cloud

High resolution data on the core of the Orion molecular cloud are scarce, since the brightness of the quiescent gas is low (making interferometric observations difficult) and many molecular lines are optically thick (decreasing the contrast). The most important results (28→32) can be summarized as follows. There are two dense clouds ($\sim 10^6$ cm^{-3}) NE and SW of the BN-KL cluster and overlapping at about the center of it (Fig. 3a). The ammonia "spike" feature shows an emission peak at the core of KL and elongated NE-SW (33). The two H_2CO clouds have sizes of 30" to 60" and masses of 30 to 80 $M_\odot$ (30). These estimates are in agreement with the total mass and column density derived from the dust emission (8). Since the opacity of these clouds is high (10 mag) even at 20 μm, this may explain the bay in the NE of the 20 μm map and the sharp gradient toward the SW. On a scale >30", the gas densities must fall off rapidly ($n \propto r^{-2}$ to r^{-3}) with distance from the infrared cluster (31).

2.3. The High Velocity Gas

In the past few years, several components of molecular gas with unexpectedly high velocity dispersion have been found in addition to the gas with low velocity dispersion. We now discuss these components and their characteristics (see also Table 1).

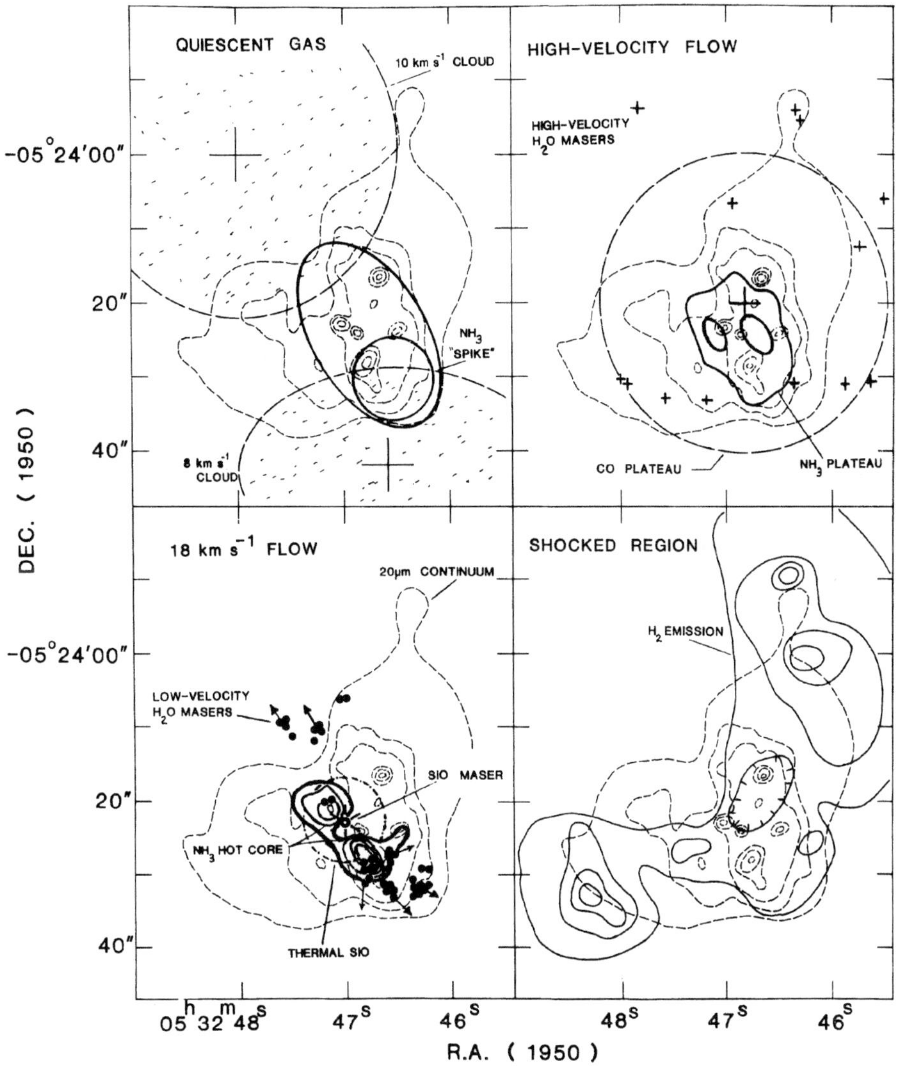

Fig. 3: The angular distribution of the different kinematic components in Orion-KL superposed on the 20 μm map (dashed). Top left(a): The quiescent gas: H_2CO(30), NH_3(33,119), HCN (32). Bottom left(b): The 18 km s^{-1} flow in H_2O masers (dark dots; 18), SiO masers and thermal SiO (dashed dotted; 21,37); the "hot core" NH_3 emission (heavy lines; 30,47). The OH masers (41) are located ±4" on either side of IRc2 (≙SiO maser). Top right(c): The high velocity ("plateau") gas in CO (dashed; 49,54), NH_3(heavy lines; 30) and H_2O masers (little crosses; 18). Only the HPW and centroid (cross) for CO, and the characteristic dimension of the two NH_3

knots and the extended NH_3 emission are indicated. Bottom right(d): The 2 μm S(1) line brightness distribution from (60). The J=21→20 CO emission (117) has a comparable size.

a) The "18 km s^{-1} Outflow" from IRc2

Fig. 4 shows the spectra of the 3 strong maser molecules toward Orion-KL: the v=1,2 J=1→0 SiO transition at 43 GHz, the $6_{16}\rightarrow 5_{23}$ line of H_2O at 22 GHz and the 1.7 GHz OH masers (21,34,35). The SiO maser spectrum is double peaked with a separation of about 24 km s^{-1} between the strongest maser spikes and a total velocity spread of 35±3 km s^{-1} (Full Width at Zero Power, FWZP). The velocity centroid is 5.5±1 km s^{-1}, that is, shifted by 3.5 km s^{-1} from the velocity of the molecular cloud. The H_2O and OH maser profiles also show the "shell features" which are reminiscent of the maser spectra from the expanding envelopes of evolved, late type giants and supergiants. Interferometric observations (36,37) have shown that these "shell type" maser features are associated with the envelope of IRc2, to within about ±0.″7. The SiO maser is a probe of the conditions very close (R∿ a few 10^{14} cm) to the infrared source. The "double shell" profile can be formed only if velocity gradients are small; it can be best interpreted as radiative transport in a shell expanding (or contracting) at constant velocity (21,36). Expansion seems more probable (see the arguments in (21)). Pure rotation is excluded since it would produce a triple-line profile (38). The data are, however, compatible with a rotating and expanding shell where $0 \lesssim v_{Rot}/v_{exp} \lesssim 1$. The physical conditions in the envelope of IRc2 at R∿2 10^{14} cm seem to be different from those in most evolved stars. In these stars--except for VY CMa--the SiO masers do not show the large velocity range (20 to 60 km s^{-1}) of the H_2O and OH masers further out in the envelope (39,40).

The H_2O "shell" features (shaded black in Fig. 4) probe radii of $\lesssim 10^{15}$ cm around IRc2 and have about the same velocities as the SiO masers. Finally, the OH masers at R∿1 to 3 10^{16} cm have strong maser features between v_{LSR}=3 and 10 km s^{-1} in addition to the blue and red-shifted "shell" features. The redshifted features are at larger velocities (17 to 24 km s^{-1}, rather than 13 to 21 km s^{-1}) and the LSR centroid of the spectrum is 7±1 km s^{-1}, shifted from the IRc2 centroid of 5.5 km s^{-1} toward the velocity of the molecular cloud velocity. The OH masers lie along position angle 68° E of N, within ±4" of IRc2, and the blue- and redshifted shell features are on opposite sides of the source (41). The emission of the "thermal" v=0 rotational lines of SiO (21,37,42) are also centered on IRc2 to within ±1" to ±2". The total velocity width and centroid of the v=0 SiO spectra are the same as those of the v=1 and 2 maser transitions. The source size is ∿10", with a brightness temperature of 150 to 200 K, estimated from the optically thick J=2→1, J=3→2 lines (see the profile in Fig. 5). The velocity range of the SiO maser, the OH maser and the v=0 SiO lines is the same, indicating a constant velocity of expansion (16 to 20 km s^{-1}) over almost 3 orders of magnitude in radius from the star.

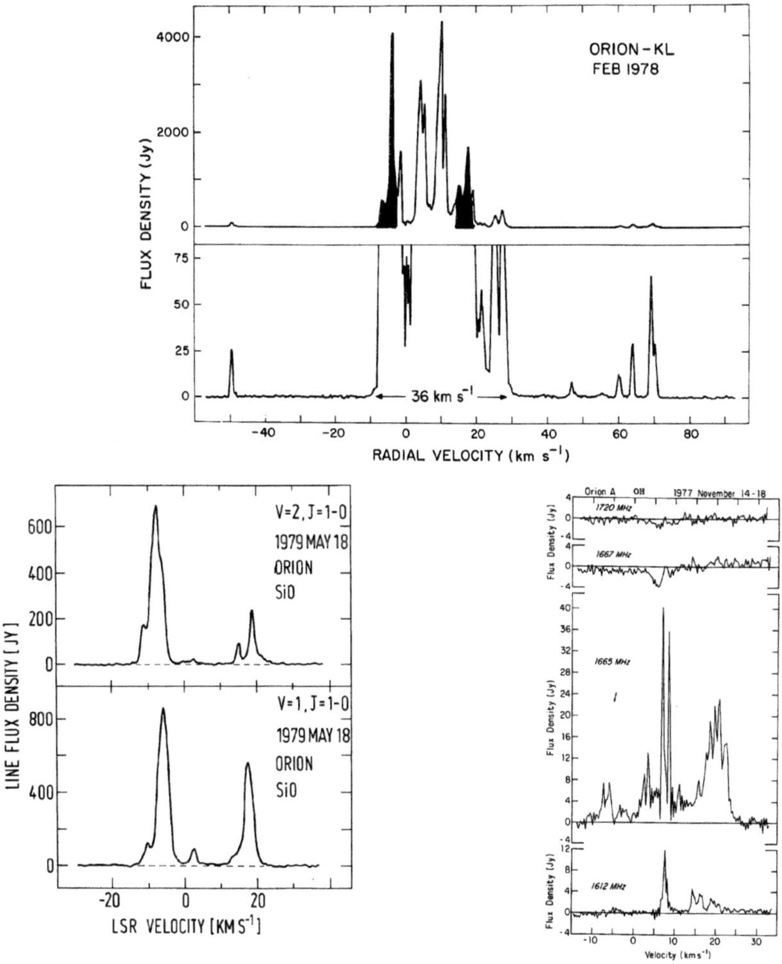

Fig. 4: Maser spectra toward Orion-KL. Top: 22 GHz H_2O spectrum (18), showing the strong low velocity masers and the high velocity flow. The H_2O "shell features" (shaded black) are coincident with IRc2 and have about 10 times larger size than the other H_2O features. Bottom left: SiO J=1→0 (43 GHz) masers toward IRc2 (21). The strongest peaks are separated by ∿24 km s^{-1}, but the total (FWZP) width of the (1612 and 1665 MHz) SiO emission is 35±3 km s^{-1}. Bottom right: OH masers associated with IRc2 (35). The total width is 36 km s^{-1}, but the centroid is shifted by ∿2 km s^{-1} to the red of the SiO profile (the H_2O spectrum is shifted by ∿4 km s^{-1}). The 1667 and 1720 MHz spectra (top) show absorption by thermal OH gas.

b) The Low Velocity H_2O Maser Outflow

The strong "low velocity" H_2O features ($-8 \lesssim v_{LSR} \lesssim 28$ km s^{-1}, 18, Fig. 4) are spread out over about 35" x 10" in p.a. 30° E of N around the center of the infrared cluster (Fig. 3b). The total velocity spread is thus the same within a few km s^{-1} as that of the "18 km s^{-1} flow" around IRc2. However, the velocity centroid is 10±2 km s^{-1}, and the profile as a whole appears shifted toward the red by ∿4 km s^{-1}. The H_2O features are concentrated in "centers of activity" ∿2" to 4" in size which have an average velocity spread of about half of the total spread. There is, in particular, a large concentration of maser features near IRc4/IRc5. The H_2O masers are dense clumps of gas ($n_{H_2} \gtrsim 10^8$ to 10^{10} cm^{-3}) with a size of ∿10^{13} cm. The detection of proper motions(18) has shown that these maser cloudlets are moving at velocities ≲20 km s^{-1}. The measurements indicate a large scale expansion for all masers from a common centroid somewhere between IRc2 and IRc4, and the data are fit best by a constant expansion velocity between radii of 2" to 25". The three-dimensional model suggests that the H_2O "centers of activity" may be due to spatial clumping on scale sizes of ∿4".

The radial velocity spread in these 4" clumps is large and about 40 to 90% of the expansion velocity, but the velocity centroids vary only by a few km s^{-1} from "center" to "center". This behavior can be explained by the geometry of a clumpy, expanding source. We think that the most likely interpretation of these measurements is that the low velocity H_2O flow is also associated with IRc2. The general shift to the red by ∿4 km s^{-1} from the flow directly around IRc2, and possibly the offset of the centroid of the proper motions from IRc2 may be caused by interaction with the molecular cloud. This point of view is actually supported by the OH maser observations. As noted above, the OH maser features are also slightly redshifted from the IRc2 center velocity (by ∿half the amount of H_2O), although they are still pretty clearly associated with IRc2. The clustering of H_2O spots near IRc4, and also toward the NE may be correlated with the distribution of the dense molecular gas shown in Fig. 3a. Hence, the excitation (and/or creation) of H_2O masers may be strongly coupled with the interaction of the flow with the molecular cloud.

c) The "Hot Core"

The 1.2 cm inversion lines of NH_3(in particular the $J \gtrsim 3$ transitions; Fig. 5) and other high excitation molecular lines (Table 1) show a feature at V_{LSR}∿5 km s^{-1} with FWHM ∿10 to 15 km s^{-1} corresponding to FWZP ∿25 to ≳30 km s^{-1}). An excitation analysis of several nonmetastable ($J \neq K$) NH_3 lines (22,43) indicates that the kinetic temperature of this component is high (≳200 K; hence the designation). From mapping of the (J,K)=(4,3) inversion transition with the VLA and the (7,6) line with the 100 m telescope an association of this component with IRc4 was suggested (44,45). The hot core component may represent collapsing or turbulent gas associated with a "protostar" within IRc4 (30,44,45). A recent high resolution VLA investigation of the (3,3) line (30,47) also

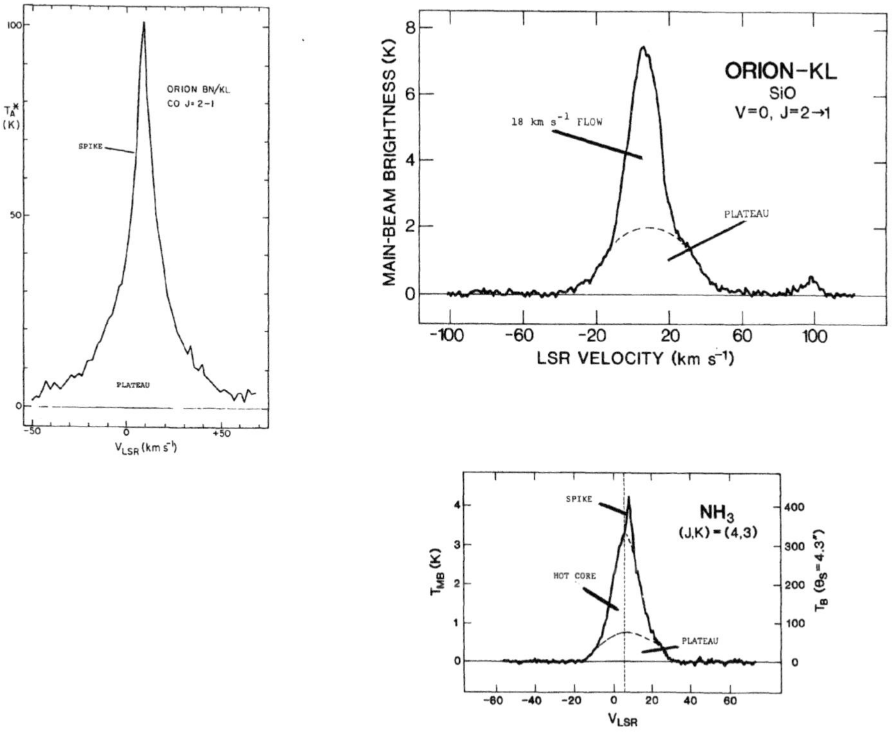

Fig. 5. "Thermal" molecular spectra toward Orion-KL. The CO (top left, 54), the SiO (top right, 42) and the NH_3 (45) lines show the different kinematic components: the quiescent gas ("spike"), the "18 km s^{-1}" and "hot core" components, and the high velocity "plateau". Note the varying widths of the "plateau" component in different molecules and the similarity between "hot core" and "18 km s^{-1}" components. The CO profile may also show this component as an asymmetry on the blue wing of the line.

shows the brightest spot of "hot core" emission (T_b=190 K) near IRc4. In addition there is, however, a second peak $\sim$3" to the NE of IRc2 (Fig. 3b), and an arc or ridge of weaker "hot core" emission is bending around the SE of IRc2 and connecting these two knots (47). Maps of the (2,2)→(1,1) SO emission with the Hat Creek interferometer (46,47) in about the same velocity interval also have emission extending along the NE-SW direction and centered between IRc2 and IRc4. The characteristic velocity centroid of 5 km s^{-1} (22), the width and also the position angle and elongation in NE-SW direction are reminiscent of the "18 km s^{-1} flow" associated with IRc2. Also note that the total size and brightness temperature of the NH_3 "hot core" feature in Fig. 3b are similar to the values for the v=0 SiO source (Table 1). Ammonia and the other, high

TABLE 1. PARAMETERS OF THE FLOWS IN ORION-KL.

a) The Low Velocity ("18 km s^{-1}") Flow and the "Hot Core"

MOLECULE	μ a) (d)	TRANSITION	λ_{vibr} b) (μm)	E_{diss} c) (eV)	v_{LSR} (km s^{-1})	Δv_{FWZP} d) (km s^{-1})	BRIGHTNESS R.A. 05^h32^m	CENTROID Dec. $-05°24'$	SOURCE SIZE (arc sec)	e) T (K)	f) n_{H_2} (cm^{-3})	g) TOTAL MASS ($M_\odot$)	REMARKS	REFERENCES
CO	0.11	J=1→0 115 GHz	4.6	11		~40						~$10^{0.5}$→10^{1} from compar. with high vel. flow	change in slope of high vel. wings at ±20 km s^{-1}	53
		J=2→1 230 GHz				≲50							change in slope & optical depth higher	54
NH_3	1.5	(J,K)=(1,1)to(9,9) Invers. Transit. 22→27 GHz	2.9, 3.0, 6.1, 10.5	4.3	5.5±1	~30	$46^s.9\pm0^s.1$	25"±2"	2"x12" posit. ang. 30° EN	150–250			brightness centroid of "hot core" of (3,3) map; brightest knot at $46^s.8$, 28"	22,44,30,43
SO	1.6	var. transit. betw. (J,N) levels 30–140 GHz	8.7	5.4	≲8	≳30	$46^s.0\pm0^s.2$	25"±4	~15"	~70				46,47,120,122
OH	1.7	$^2\Pi_{3/2}$ F=1→1 1.665 GHz	2.7	4.4	7±1	36±4	$47^s.0\pm0^s.07$	23"±1"	<2"x8" posit. ang. 60° EN				mas. emis. spread ±4" around central position	35,41
H_2O	1.9	$J_{K-,K+}=6_{16}\to5_{23}$ 22.2 GHz	2.7 6.3	5.1	5.5±1	≳26	$47^s.0\pm0^s.05$	23"±1"	~0".5				mas. emis., H_2O shell feat. rel. to 10 km s^{-1} maser. centroid of proper motions of strong low vel. features	18,35,103
					10±2	36±2	$46^s.85\pm0^s.07$	27"±2"	10"x35" posit. angle 30° EN					
SiO	3.1	v=0, J=1→0 43.4 GHz	8.1	8.3	5.5±2	35±4	$47^s.0\pm0^s.1$	23"±1".5	6"<θ_s<10"	100→150		<<$10^{-1.5}$ from SiO lines ([SiO]=[Si]) 10^{1} from compar. with high velocity flow	line optic. thin, mas. spike at −6km s^{-1}, measurement rel. to v=1 maser, optically thick	21,37,42
		v=0, J=2→1 86.8 GHz			6±1	36±3	$47^s.0\pm0^s.2$	23"±3"	12	150→200				
		v=0, J=3→2 130.3 GHz				≳32				~200				
		v=1,2, J=1→0 43 GHz			5.5±1	37±4			≲0".1				maser "shell feat." strongest maser spikes separated by 24±5 km s^{-1}	21,37
		v=1, J=2→1 86.2 GHz			5.5	36±5	$47^s.0\pm0^s.03$	23"±1"					maser	
HC_3N	3.7	J+1→J; J≥8 >80 GHz	3, 4.4, 4.8, 11.4, 15.1, 20		6±2	≥30			≲40	100→500	~10^{7}			123
CH_3 CH_2 CN	3.8 (1.4)	J+1→J, J≥9 88 to 115 GHz			4.5±1	≥25				50→200				124
CH_3 CN	3.9	J=12→11, 13→12 221, 239 GHz	3.4, 4.4, 6.7, 7.2, 10.9			≥15			<60	70→300			temperatures higher for higher levels	125

TABLE 1. Continued

b) The High Velocity Flow

MOLECULE	a) μ (d)	TRANSITION	b) λ_{vibr} (μm)	c) E_{diss} (eV)	v_{LSR} (km s-1)	d) Δv_{FWZP} (km s-1)	BRIGHTNESS R.A. 05^h32^m	CENTROID Dec. -05°24'	SOURCE SIZE (arc sec)	e) T (K)	f) n_{H_2} (cm-3)	g) TOTAL MASS ($M_\odot$)	REMARKS	REFERENCES
H_2	0	J=4→2, S(2) 12.3 μm	2.2	4.5	∿9	30→>160			∿80"	≥750		$10^{0.3}$	dramatic variat. of the profile from point to point, emis. sym. with respect to cloud vel.	56,69,57
		S(8) to S(15) 5 to 3.6 μm							≥40"	2000		$10^{-0.4}$	mass only of gas at 2000 K ($\tau_{2\mu m}$=4).	
		v=1→0, 2→1 ΔJ=2 (S-branch) ΔJ=0 (Q-branch) 2.0 to 2.4 μm			3→9	2x100			∿80"	2000		10^{-1} ($\tau_{2\mu m}$=4) 10^{-2} ($\tau_{2\mu m}$=1.2)	emis. mostly blueshifted, mass is gas at 2000 K.	59,60,61,66,67
CO	0.11	J+1→J J=0,1,2 115,230,345 GHz	4.6	11	9	65(3→2)→ 190(2→1)	$46^s.75±0^s.3$	19"±4"	36"(1→0) 40"(2→1) 30"(3→2)	∿100		$10^{0.5}$	[C]/[H_2]=0.1 assumed, velocity increases with radius.	48→54,71
		J=6→5 690 GHz				≥45				≥200				116
		J+1→J; J>20 77 to 124 μm							∿80"	750→2000	10^6		[C]/[H_2]∿0.5.	58,117
		v=0→1 R and P branch 4.6 μm				≥50				85±15		$10^{0.9}$	absorption in front of BN, mass derived using θ_s=40" and [C]/[H_2]=0.1.	25
NH_3	1.5	(J,K)=(1,1)to(9,9) Invers. Transit. 21 to 27 GHz	2.9, 3.0, 6.1, 10.5	4.3	5±2	85	$46^s.9±0^s.1$	25"±2"	10"	70			brightness centroid is mean of two brightest clumps, distribution is anticorrelated with IR.	30,44,118,119
SO	1.6	(J,N)=(2,2)→(1,1) 86 GHz	8.7	5.4	8±2	90	$45^s.9±0^s.2$	20"±3"						46,47,120
SO_2	1.6	$J_{K-,K+}$=$13_{1,13}$→$12_{0,12}$ 251 GHz	7.3, 8.7, 19.3	5.6	9	∿50			25"	70				54,21
H_2O	1.9	$J_{K-,K+}$=6_{16}→5_{23} 22 GHz	2.7, 6.3	5.1		180	$46^s.8±0.2$	26"±6"	60"	300→1000	>10^8		mas. emis. from dense clumps (10^{13}cm), proper motions indicate expans., vel. increase with (radius)$^{0.3}$.	18
CS	2.0	J=2→1, 98 GHz J=3→2, 147 GHz	7.8	7.4	∿9	≥50			<40"		10^6		plateau weak	31
HCN	3.0	J=1→0, 89 GHz	3.0, 4.9, 14.0	5.6	6→8	≥40	$46^s.9±0^s.2$	25"±4"	15"	150	4 10^6	$10^{0.5}$	temp. and dens. from model fitting HCN and HCO^+.	32
SiO	3.1	J=2→1, 86.8 GHz	8.1	8.3	9	115±5	$46^s.9±0^s.2$	21"±3"					assum. that the "18 km s^{-1}" SiO comp. is at $47^s.0$, 23", shift betw. red-and blueshifted wings.	42
HCO^+	3.3	J=1→0, 89 GHz	∿9		9.8	≥45			40"?				extent mostly N-S	32,46

Footnotes to Table 1:

a) electric dipole moment (1 D = 10^{-18} e.s.u.)

b) wavelengths of the v=0→1 vibrational bands in μm

c) dissociation energy of the molecules

d) full width at zero power of the line

e) rotational or kinetic temperature

f) determined from excitation analysis

g) derived from measured column density, source size and kinetic (rotational) temperature; assumed abundance is indicated; uncertain by ± one magnitude excitation molecules are very sensitive to collisional excitation, in dense gas and/or to infrared excitation--in particular at ∿10 μm, where the molecules have their vibrational bands (Table 1) and where the high opacity of the dust may enhance infrared excitation over a large region (46). Furthermore, the opacity in the observed NH_3 lines is considerable. Hence, an alternative explanation is that the "hot core" component is gas within the "18 km s^{-1} flow" or formerly quiescent gas which has then been pushed along with the flow (creating a swept up shell). If IRc4 were an intense source of infrared radiation and/or a peak in the density of gas, it's influence on the excitation and optical depth (and possibly the chemistry) of the "18 km s^{-1} flow" gas at some distance from the central source (probably IRc2) may be important. The brightness centroid of these molecular lines may then be shifted away from the center of expansion. The lack of NH_3 hot core emission E of IRc2 may then indicate that the density of the cloud toward BN is lower than near IRc4 (possibly due to "clearing" by BN).

d) The High Velocity "Plateau"

As can be seen from the spectra (Figs. 4 and 5) and Table 1, many molecular lines show smooth emission over a range between 50 to 190 km s^{-1} (the high velocity "plateau"). The high velocity "plateau" emission of most of the high excitation ($\mu \geq 1$ Debye) molecules is confined to about 10" to 15" near the center of the infrared cluster. Unfortunately, the "plateau" emission in the extreme wings is weak and the position centroids determined from the emission between -10 and +30 km s^{-1} may be contaminated by the "18 km s^{-1}" and "hot core" components. The positional centroid of the high resolution NH_3 (3,3) map (Fig. 3c) is defined best and is ∿1.5" West and 2" South of IRc2 (30). The NH_3 (3,3) plateau gas shows the same angular distribution for all velocities and an anticorrelation with the 20 μm infrared sources. The high velocity emission of SiO and SO, on the other hand, are centered a few arc sec NE of IRc2, toward BN (42,46,33). This latter position is close to the centroid of the CO "plateau" source, which is extended over ∿40 kms^{-1} (48→54). Absorption in front of BN from a component blueshifted by about 25 km s^{-1} has been seen in the observations of the 4.6 μm v=0→1 vibrational lines (25). The high velocity H_2O maser features also are spread over a large region (4 10^{17} cm ≃60", Fig. 3c). The proper motions detected for a few of these weak masers are again consistent with expansion, but the centroid may be anywhere between IRc4 and BN. As with the low velocity flow(s), the velocity centroid may be a "fingerprint" of the central source(s), and in fact, the situation is very similar to the low velocity gas (Table 1): The spatially confined plateau gas (NH_3 and possibly HCN), has a velocity of ∿5 km s^{-1}, and a spatial centroid within a few arc sec of IRc2, while the more extended plateau components have a velocity centroid, of ∿9 km s^{-1} and a spatial centroid slightly further north. Again, this may indicate two or more sources or else one dominant flow, interacting with the inhomogeneous quiescent gas further out.

It is clear that the high velocity "plateau"--as well as the

"18 km s^{-1} flow"--represents expanding motion. To keep the gas gravitationally bound at radii of 10" to 20" and velocities $\gtrsim$50 km s^{-1} would require a central mass $>10^4 M_\odot$, while estimates of the mass actually contained in the central 1′ are two orders of magnitude lower. The same holds for rotation (rotation is also inconsistent with the angular distributions of most of the molecular lines (e.g. the (3,3) NH_3)). Furthermore, for collapsing and rotational motions the velocities should decrease with increasing distance from the center (by about a factor of two between NH_3 and CO), while the contrary is observed. The high velocity expansion is clumpy, and the flow cannot be an expansion with constant velocity. The velocities of high velocity masers increase with distance from the center ($v_{ex} \propto R^{0.3}$). The shapes of the CO "plateau" and the angular size of the CO source as function of radial velocity also suggests that velocities must increase with distance from the center (51,54). An increase of velocities with radius could be the result of a large intrinsic velocity dispersion in the flow. The fastest "particles" would then automatically be furthest out. A large intrinsic velocity dispersion may be caused by the interaction of a stellar wind from the central star(s) with a surrounding cloud and/or by variability in the mass loss rate or expulsion velocity of the wind. Alternatively, the high velocity gas may still be accelerated at $R \sim 10^{17}$ cm. This idea is possibly supported by the fact that the CO "plateau" and the H_2O masers (at radii of $\gtrsim$20") have about a factor of 2 larger velocity range than the NH_3, SO and SiO "plateaus" (radii 5" to 10"; but note that lack of sensitivity may prevent us from detecting gas at $v \gtrsim 110$ km s^{-1}). The acceleration mechanism may then be a hydrodynamic effect of expansion of a channeled flow in a medium of steep density gradients (see section 7). The acceleration at radii of 10^{17} cm cannot be due to radiation pressure or stellar wind pressure in a spherically symmetric situation (in the optically thin limit). In this case, the velocities of accelerated clumps would rapidly reach a terminal value

$$v(R) = v_T \left(1 - \frac{R_o}{R}\right)^{\frac{1}{2}} , \qquad (2)$$

if the radii are much larger than the initial radius R_o (R_o may be about 10^{14} to 10^{15} cm, 5).

e) The Shocked Gas

The discovery of the infrared rotational-vibrational and quadrupole lines of H_2 (55,56,57) and the J$\gtrsim$6 rotational lines of CO (58,116) suggests the presence of very hot ($\gtrsim$200 to 2000 K) gas with an angular extent larger than the CO plateau source (116,58,117). For a detailed discussion, refer to (58,59,68). Fig. 3d shows the distribution of the 2 μm v=1→0 S(1) emission (60). The distribution of the H_2 emission on either side of the infrared continuum sources suggests that the hot gas may also be associated with the infrared cluster. High spectral resolution data (59,61) have proven that the "hot" gas has a velocity dispersion comparable with that of the "plateau" gas. An interpretation may be that the "hot" H_2 and CO consist of shock-excited "plateau" gas (in clumps) as well as shock excited gas within the quiescent molecular cloud (59,61). The details of the shock physics and in particular, the question if the shock is gas dynamic or has a strong magnetic precursor, are still under debate (62,63,64). A comparison between Figs. 3a and 3d suggests an anticorrelation between the location of the two dense molecular clouds (as estimated from the H_2CO and HCN measurements) and the hot H_2 (18, 30). Furthermore, there is a "hole" in the H_2 emission toward the center of the infrared cluster and there may also be a more detailed anticorrelation of the brightness of the 20 μm dust emission and the H_2 (11). The opacities of the dust in the molecular clouds (Fig. 3a) and in the KL nebula at 2 μm are probably large and hence, an outstanding question is the influence of this extinction on the observed brightness distribution of the H_2 quadrupole lines (note that $A_V \gtrsim 100$ mag for these molecular clouds, hence $A_{2\mu m} \sim 10$ mag.). An indication of the importance of extinction within the source comes from the shapes of the 2 and 12 μm profiles at the H_2 emission peaks (59,65). While the red-shifted wings are totally missing in the 2 μm v=1→0 S(1) line, the S(2) line at 12.4 μm is more symmetric, indicating $A_V \sim 10$ to 30 mag and $N_{H_2} \sim 10^{22} cm^{-2}$ at these positions. Additional extinction due to foreground material has been estimated between 5 and 30 mag. (25,61,66,67). Hence, one interpretation of the observed anticorrelation is that there is hot H_2 within the central ~10"→20" and toward the large quiescent clouds, but there is even more extinction there than toward the H_2 peaks. Alternatively, the anticorrelation may be due to an anisotropic flow of the "plateau" gas mostly toward NW and SE and there may be no additional H_2 close to the center of the cluster (possibly indicating the existence of a "cavity" of gas around the source(s) of outflow).

2.4 Discussion and Conclusions

1) The infrared observations show the presence of at least two, hot ($T \gtrsim 3000$ K) stars of luminosity 5 10^3 to $>10^4 L_\odot$, BN and IRc2. Circumstantial evidence suggests that IRc4, IRc3 and possibly some of the other cooler sources have intrinsic luminosities $\lesssim 10^4 L_\odot$. However, the available data are also marginally consistent with one object having most of the 1.5 $10^5 L_\odot$ of the cluster. This object may be IRc2, but could also be a heavily obscured source which so far has been missed in the 2→20 μm observations. There is also little hope of identifying such a hidden object at λ>20 μm since the diffusion and radiative transport effects of the radiation are severe, and the distribution tends to smoothen out.

2) There are at least two different outflows at the center of the Orion-KL cluster. The "18 km s^{-1} outflow" is almost certainly associated with IRc2 (within the error bars of ±1"). The velocity centroid of the "18 km s^{-1}flow" component is 5.5±1 km s^{-1} which may be the velocity of IRc2 itself. The H_2O low velocity maser features and the "hot core" component identified in the spectra of the NH_3 inversion lines may indicate two similar flows from different objects within the BN-KL cluster possibly IRc4). We favor, however, the idea that they basically represent the same phenomenon (outflow at low velocities from IRc2) and that

the observational differences (positional centroid, velocity centroid, total velocity range) are caused by the interaction of the flow with the material in the vicinity of the source of outflow.

3) The origin of the second, high velocity flow is at present uncertain. The brightness centroids of the maps of different molecules give different answers, ranging from IRc4 to BN. The data do show, however, that the high velocity gas can also be found close to IRc2, and the center of the infrared cluster, and probably represents--as the 18 km s^{-1} flow--a more or less continuous mass outflow. The low and high velocity flows may actually be close to each other at the center of the cluster.

4) There are some indications of anisotropy. The low velocity outflow is elongated in NE-SW direction, along the ridge of the quiescent molecular cloud. The high velocity gas, on the other hand, is distributed mainly in the perpendicular direction (NW-SE), possibly along the directions of steepest density gradients. If the low and high velocity flows came from the same object, then this possible anticorrelation may indicate anisotropic mass loss (for possible mechanisms, see section 7).

5) The masses contained within the flows or the implied mass loss rates are high and about the same for the two flows ($\sim$1 to 20 $M_\odot$ or 10^{-3} to 10^{-2} $M_\odot y^{-1}$, Table 1, 10, 18, 68→70). These large mass loss rates immediately exclude BN as a source of some important fraction of the high velocity flow since the current mass loss at the surface of the star is two to three orders of magnitude too low (68). The momentum supplied by the radiation of the infrared cluster also is probably not sufficient (by 1 to 2 orders of magnitude) to drive the flows by single scattering. A way out has been given in the model by (71) in which the photons are scattered and absorbed several times thereby enhancing the available momentum to $\tau L_*/c$. This may be sufficient to drive the flow for opacities $\tau_{IR}>10$. (for other explanations see 18, 72, section 7).

6) As mentioned before, the observed brightness distributions vary from molecule to molecule. This could be caused by the varying sensitivity of the molecular lines to infrared and collisional excitation and by the differences in opacity. In addition, chemistry and molecular stability may also play an important role (73, 136). A molecular collision at a velocity of >20 km s^{-1} corresponds to few eV which can cause the dissociation of all but the most stable molecules (Table 1). Furthermore, at higher temperatures (e.g. due to shocks), the chemical abundances of some molecules like OH, H_2O, SO and SiO may be enhanced.

3. Sources of Luminosity >5 $10^5 L_\odot$

3.1 W51

a) W51-IRS2 (W51-NORTH)

There are two bright H_2O maser sources associated with the well-studied W51 complex of infrared and radio continuum sources and giant molecular clouds at a distance of 7 kpc: W51-MAIN and W51-NORTH (74, 75). The distribution of the 8 and 20 μm radiation of dust and the radio free-free emission at 2 cm from the compact source W51-IRS2 are shown in Fig. 6 together with the H_2O maser spots. The region has a

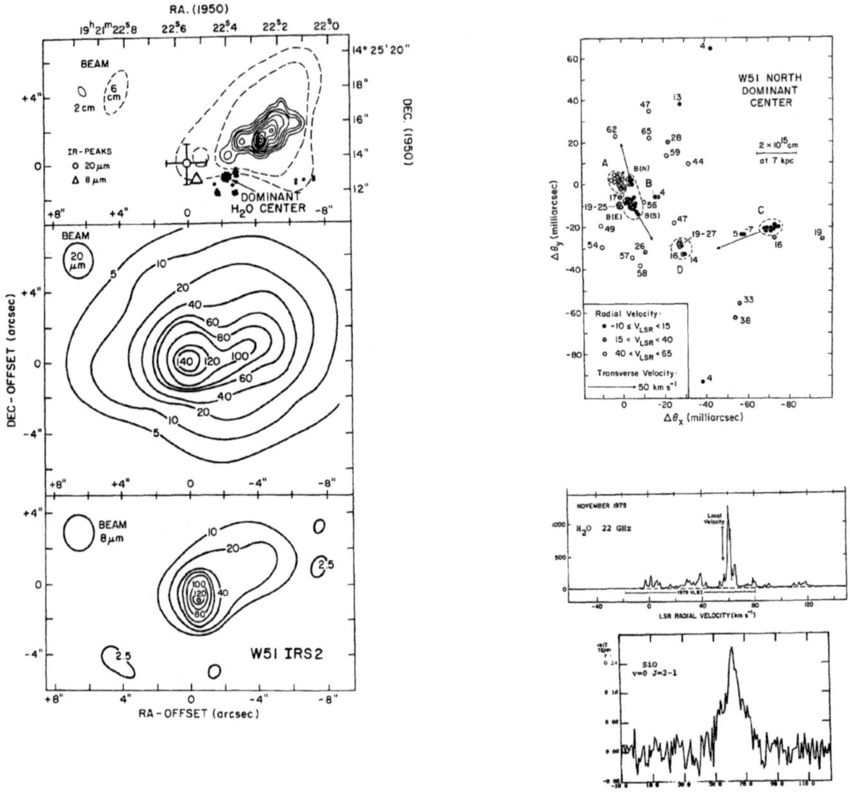

Fig. 6. Top left: 2 cm radio continuum emission, 6 cm radio emission (dashed) and H_2O maser spots (black dots) in W51-IRS2. The cross and triangle mark the peaks of the 20 and 8 μm dust emission. Middle and Bottom left: 20 and 8 μm maps (taken from 74). Top right: distribution of the H_2O maser spots and their proper motions within the "Dominant Center" (76). Middle right: H_2O maser spectrum; vertical scale is line flux density, Jy). Bottom right: v=0, J=2→1 spectrum of SiO taken with the Onsala 20 m telescope. Units are K of antenna temperature (multiply by 2.2 to get main beam brightness; from 42).

remarkable morphological similarity to the Orion-KL cluster: A bright infrared source is associated with the H_2O maser emission, and is close (4 10^{17} cm) to a compact H II region (74). As in the BN-KL cluster, the position and size of the infrared emission in W51-IRS2 change with wavelength, and the H_2O maser source (the "Dominant" Center in Fig. 6) is probably associated with the hotter infrared source(s), dominating between 2 and the 10 μm. The infrared source W51-IRS2 (H_2O) shows a deep silicate feature (74), while the adjacent infrared source associated with the H II region has a flat power-law spectrum. The H_2O maser spectrum, (Fig. 6) shows strong high velocity features over $\sim$140 km s^{-1}, and proper motions for a number of these maser features have been detected (Fig. 6, 76). Finally, the two other necessary "ingredients" have also recently been discovered: 2 μm v=1→0 S(1) emission from hot molecular hydrogen (77) and broad "thermal" molecular emission in the v=0, J=2→1 line of SiO (42, Fig. 6). The total width of the SiO profile is $\sim$30 km s^{-1}, but this is clearly limited by the available S/N ratio. The total infrared luminosity of W51-IRS2 (H_2O) is much larger than the luminosity of the Orion-KL source, and is between 5 10^5 to 2 $10^6 L_\odot$. The integrated strength of the SiO "plateau" component indicates that the mass and momentum contained in the high velocity gas in W51-IRS2 are also larger than in Orion by about one order of magnitude if one uses the same values for the size, kinetic temperature and SiO abundance as in the Orion-SiO source see discussion in (42)). The strength of the 2 μm H_2 quadrupole line, however, is much weaker than one would expect from the difference in distance between Orion and W51 (77). This may indicate more extinction, different geometry or true differences of the physical conditions in the two sources.

b) W51-MAIN

The second bright H_2O maser source, W51-MAIN, is $\sim$1' to the south-east of W51-IRS2, in a group of ultra-compact H II regions, OH maser sources and a few more H_2O maser clusters on the eastern edge of the extended H II region W51-IRS1. While the H_2O spectrum displays high velocity maser emission over about 150 km s^{-1}, similar to the range of the high velocity flow in Orion-KL, the VLBI observations indicate a more complex and more turbulent pattern for the high velocity motions in W51-MAIN than for Orion-KL (78).

i) The W51-MAIN source is very compact ($\lesssim$3 10^{16} cm), and the maser features are clustered on scale sizes of 10^{14} to 2 10^{15} cm.

ii) The radial velocity dispersion within these clusters is large (30 to 60 km s^{-1}) and is about the same as the total velocity dispersion in the source. Hence, the dynamical lifetime, $R/\Delta v$, of these clusters is short ($\sim$10y), and they probably have to be created in situ. The pattern of radial velocities within clusters is reminiscent of eddies on many scale sizes.

iii) This picture is supported by the proper motions. Except for a few "low velocity" maser features, there are no streaming motions and the data can be interpreted best by the break-up of turbulent cloudlets.

iv) The "low velocity" features ($|v-v_{cloud}|<20$ km s^{-1}) show proper motions comparable to those of the high velocity features and there may be no low velocity motions at all as in Orion.

v) Essentially all maser features associated with the maser source are redshifted with respect to the velocity of the molecular cloud. The v=0, J=2→1 "plateau" emission of SiO is also redshifted (42). Furthermore, there is no 2-20 μm infrared source at the position of W51-MAIN, to a limit of $\sim$1% of W51-IRS2 (which has the same maser luminosity). It is now clear from several independent indications that this negative

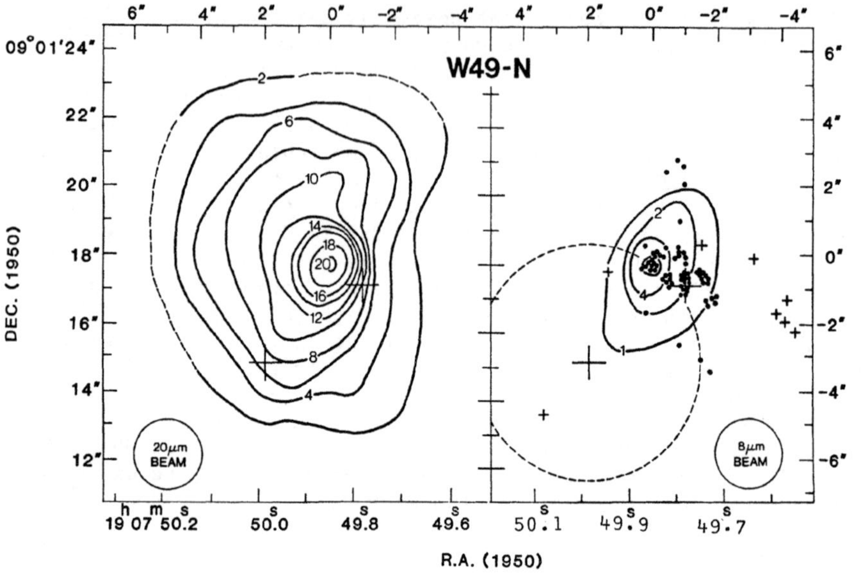

Fig. 7. Top left: 20μm map (83) of W49-N; right: 8 μm map and the distribution of H_2O maser features (82), OH masers (crosses, 41) and radio continuum emission (half-power circle, 126). The positional accuracy of the infrared peaks is ±1". Bottom right: Distribution of H_2O maser spots for three different LSR velocity ranges (82), indicating a separation of blue- and redshifted high velocity emission. Also note the "hole" in the distribution of maser features in the right part of the figure and the sharp fronts and boundaries of the maser distribution.

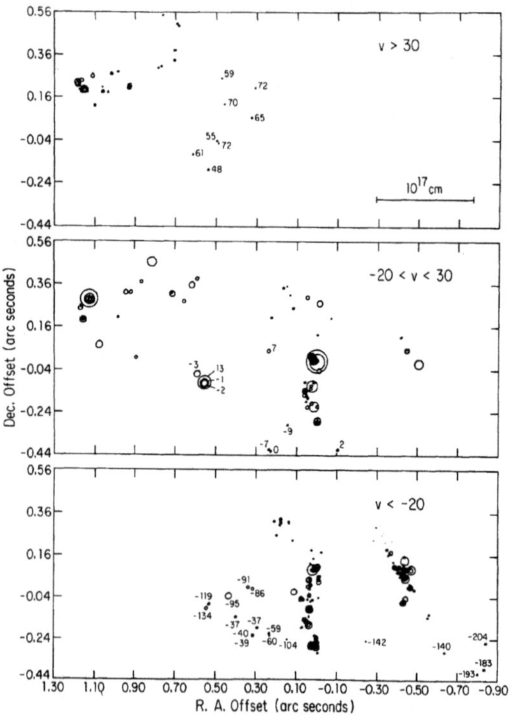

result is due to extinction by a compact molecular cloud associated with the maser source (42,74,79,80,81). The lack of 2-20 μm infrared radiation may be related to the preferential redshift of the high velocity emission. If there is a dense cloud ($n_{H_2}>10^6$ cm^{-3} $\ell \sim 10^{17}$ cm) in front of W51-MAIN, the extinction may be large even at 20 μm and the blueshifted high velocity gas may be stopped in the dense cloud. The interaction of a stellar wind with density inhomogeneities and cloud boundaries on the opposite side of the central source may also best explain the other characteristics of the H_2O maser data.

c) W49-N

W49-N is the brightest H_2O maser source in the Galaxy ($\sim$1 $L_\odot$ in the 22 GHz line if the emission is isotropic) and the total far-infrared luminosity is also large ($\gtrsim$5 10^6 $L_\odot$). The H_2O profile shows maser features over an extraordinary range ($\pm$250 km s^{-1}), but in general the characteristics of the maser features in W49 are similar to those in the W51 sources. The most recent VLBI map (with $\sim$400 maser spots!; Fig. 7, 82) shows clustering on scale sizes of 10^{14} to 10^{17} cm and there are sharp fronts and boundaries in the distribution of maser spots in several parts of the map. The local velocity dispersion of the maser features is comparable with the total velocity range. The most outstanding feature of the map shown in Fig. 7 is that the blue- and redshifted maser features appear to be separated toward opposite sides of the source, suggesting an anisotropic flow of the high velocity gas, or a inhomogeneous distribution of dense gas in the immediate vicinity of the source. "Thermal" high velocity emission has been found in the v=0, J=2→1 SiO line (42). The $\lambda<20$ μm emission from W49-N is weak. Recent IRTF maps (Fig. 7; 83) demonstrate that the 5 to 20 μm source is associated with the masers (to within ±1"). The weakness of the $\lambda<20$ μm emission is probably due to local extinction, since there is a deep silicate feature, since the dust emission from the H II region is much lower than expected (74,83) and since the source is extraordinarily bright at 30 μm.

d) W3-IRS5

So far we have discussed maser sources which show strong H_2O high velocity emission. The majority of the known H_2O maser sources, however, has only weak high velocity features or shows only "low velocity" masers ($\Delta v \lesssim 25$ km s^{-1}). An example for this latter class is W3-IRS5, a member of a cluster of luminous infrared sources at a distance of $\sim$2.5 kpc (L_{IRS5}=2 $10^5 L_\odot$, 84,85). The H_2O source has maser features over a velocity range of 24 km s^{-1} which are spread over 1" x 2".5 (RA x Dec) (38, 86). The infrared counterpart of the H_2O source has been shown to be double, with two components of nearly equal brightness, separated by 1.1" in position angle 37° E of N (87,88). There is also an ultracompact H II region (<1", 7mJy at 15 GHz) about 1" NE of the centroid of the H_2O masers (89). Observations of molecular lines, in the far-infrared and of the silicate feature indicate a peak of the dust and gas column density toward IRS5 (85). The weakness of the radio continuum emission, the lack of high velocity motions and evidence for shock excitation may then indicate that sources like W3-IRS5 are in an earlier evolutionary stage than Orion-KL, W51 or W49.

4. Sources of Luminosity $<10^4$ $L_\odot$

H_2O maser emission has also been detected toward regions of lower luminosity. Some of the recently studied sources are GL490 ($L\sim1.4$ 10^3 $L_\odot$, 90), OMC2 ($L\sim$ a few $10^2 L_\odot$, 91) and T-Tau ($L=20$ $L_\odot$, 92,93). T-Tau seems to be somewhat of an exception, since sensitive maser searches toward other T-Tau and "Orion Population" stars were unsuccessful (93). Although the masers in these less massive and lower density regions generally are much weaker, with a smaller number of maser features, nevertheless, high velocity masers have been often found and seem to be correlated with the presence of Herbig-Haro objects (94,95). It has, therefore, been suggested that Herbig-Haro objects and high velocity H_2O masers in these sources are manifestations of the interaction of stellar winds with the surrounding molecular clouds (94,133). The Herbig-Haro objects are much larger than the H_2O masers (by 2 to 3 orders of magnitude) and may be more evolved cloudlets which have moved further away from the central star(s) (94). Thermal high velocity emission in the CO line and H_2 quadrupole emission are seen toward several of the regions (see several other contributions in this volume).

Fig. 8 is a summary of observations of L1551 (HH29-IR). The CO data (96) delineate a double-lobed cloud of gas over ±0.5 pc, moving at $\sim$15 km s^{-1}. At the center of the CO cloud, a compact infrared source has been detected (IRS5, 97). Two of the Herbig-Haro objects within the south-western CO lobe show proper motions of $\sim$150 km s^{-1} (98). Their proper motion vectors intersect at the position of the infrared source. Radio continuum emission of a compact H II region elongated at an angle of $\sim60^\circ$ relative to the CO lobes has been found with the VLA (99). Since the total luminosity of L1551-IRS5 is only $\sim10^2 L_\odot$, the H II region is probably collisionally ionized. The CO lobes may, therefore, be a dense shell swept up by a stellar wind from IRS5. The mass loss rate is ~8 10^{-7} $M_\odot y^{-1}$ and the velocity of the wind may be $\gtrsim$200 km s^{-1}. The dynamical age of the CO shell, R/v_s, is about 3 x 10^4y, if the shell velocity v_s=15 km s^{-1} is constant. The Herbig-Haro objects are probably dense cloudlets accelerated by the wind within the less dense "bubble" inside of the CO shell. The double-lobed structure, the velocity gradient across the CO source, and the elongation of the H II region suggest that the stellar wind is ejected anisotropically, possibly in a bipolar flow.

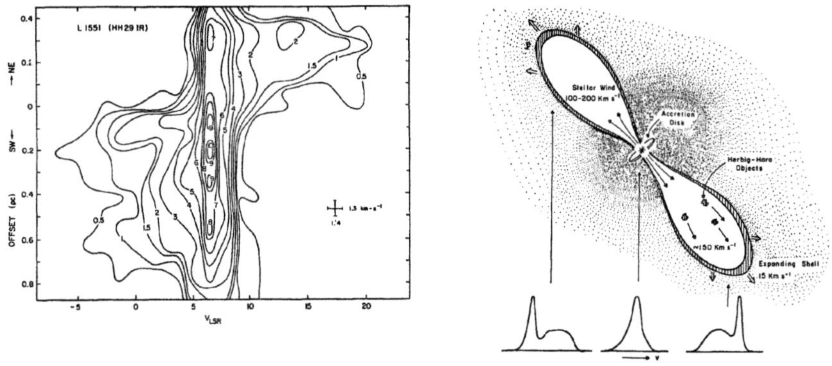

Fig. 8. The CO "lobes" in L1551 (96). Left: Velocity distribution along the lobes (NE-SW) indicating that the emission from the blue- and redshifted CO wings are separated by ∿1 pc. Right: Schematic diagram of the CO lobes and the Herbig-Haro objects (98) expanding from the infrared source IRS5 (97).

5) Later Stages, OH and H_2CO Masers

5.1 Cep A

The H_2O masers in Cep A are spread over ∿50 km s^{-1} and are clustered in two groups which coincide in position with the two brighest knots of radio continuum emission (size $\lesssim$1", 94,101). CO "plateau" emission with the same velocity width as the H_2O features has been detected from an extended, bipolar source (±0.5pc) surrounding the infrared and maser sources (102).H_2 quadrupole emission is similar in extent as the CO source and is strongest toward the blueshifted CO peak, underlining the importance of extinction (137). Cep A, and possibly NGC 7538-IRS1 and ON1 (193,104), may be examples of a later stage of a "mass-loss" source, where a small H II region already has formed, but H_2O maser emission is still present. An unusual fact in Cep A is (83,100) that the peak of the 20 μm infrared source is associated with neither H_2O masers nor the H II region. The infrared spectrum between 10 and 30 μm also has a remarkably low color temperature (83,100).

5.2 W3(OH)

Fig. 9 shows the distribution of 1665 MHz OH masers associated with the very compact H II region W3(OH) together with a high resolution map of the 2 cm radio continuum emission (105,106). The main result of the VLBI aperture synthesis map shown on the left is that the OH masers are

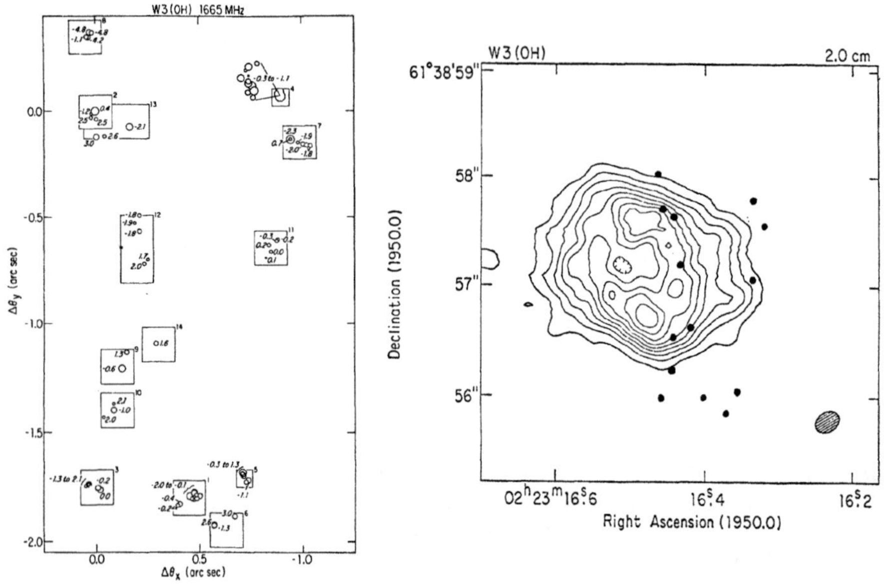

Fig. 9. Left: Spot map of an aperture synthesis of the 1665 MHz OH masers toward W3OH (105). Velocities are relative to -44.5 km s^{-1} (LSR). The masers have sizes of 10^{14} cm and cluster on scale sizes of 10^{15} cm which is probably the size of the OH cloudlets. Right: 2 cm VLA map of the radio emission of W3 OH (106). The source shows shell structure. The black dots are the OH clusters shown as rectangles in the left. The contours are in units of 880 K brightness temperature.

clustered on scale sizes of $\sim 10^{15}$ cm and are part of a dense, neutral cloud surrounding the H II region and moving toward the central star(s). The apparent velocity spread in the OH profiles of about 8 km s^{-1} probably is due to Zeeman splitting in a magnetic field of 2 to 10 mG (see also 132). The true kinematic velocities of the OH masers are then nearly constant across the face of the source. Since the velocity centroid of the OH masers is redshifted with respect to the H II region by 5.7±2 km s^{-1} and the H II region is optically thick at 1.665 GHz, the OH masers must be in front of the continuum source and moving toward it (105). These results are in contradiction to earlier shock-front models in which the OH masers were thought to be the expanding shocked shell just outside the ionized gas.

These data and high resolution mapping of other type I OH masers suggest that the OH features are always projected against an ultracompact, optically thick H II region (107,138). An exception is Orion-KL. Otherwise, one may almost predict the existence of a very compact (< a few arcsec) H II region from the presence of 1665 MHz OH maser emission. The appearance of the masers toward the surface of the H II region possibly is caused by amplification of the 10^4 K continuum background in a partially saturated maser (see the detailed model in 105). The kinematics in W3OH however, may not be common to all other sources. Aperture synthesis observations of the type I OH maser associated with an ultra-compact H II region in W75-N, for example, suggest that the OH cloud is rotating (108).

5.3 NGC 7538

Observations of the maser and infrared sources in NGC 7538 have been described in (109). Fig. 10 shows infrared and radio observations of the bright cluster in the northern complex. The infrared emission at 10 and 20 μm (75) is similar to the 6 cm radio continuum distribution (110), and probably comes from heated dust within the ionized gas. The most compact source (IRS1) is optically thick even at $\gtrsim$10 GHz (111) and is associated with OH and H_2O maser emission (103). Westerbork and VLA observations have shown that the H_2CO emission features discovered in NGC 7538 are also due to maser emission (112). The upper limit to the size and lower limit to the brightness temperature of the formaldehyde clouds are 0$''$.15 and 5 10^5 K and the H_2CO molecules may amplify the background of the optically thick H II region. The second H_2O and OH maser source, NGC 7538-S, about 90" south of IRS1, is much more deeply buried in the molecular cloud. Despite a total far-infrared luminosity of $\sim$2 $10^4 L_\odot$ ($\sim$10 times weaker than the northern cluster) the 20 μm flux density of NGC 7538-S is only $\sim$2 Jy, two orders of magnitude weaker than IRS1 (83). A very weak compact H II region has been found at 5 GHz with the VLA (4 mJy) coincident with the infrared and maser sources (110). The H_2O maser spectrum shows features over 15 km s^{-1} and a strong high velocity feature (Δv = -90 km s^{-1}) which is displaced by about 3" from the positional centroid of the "low velocity" H_2O masers.

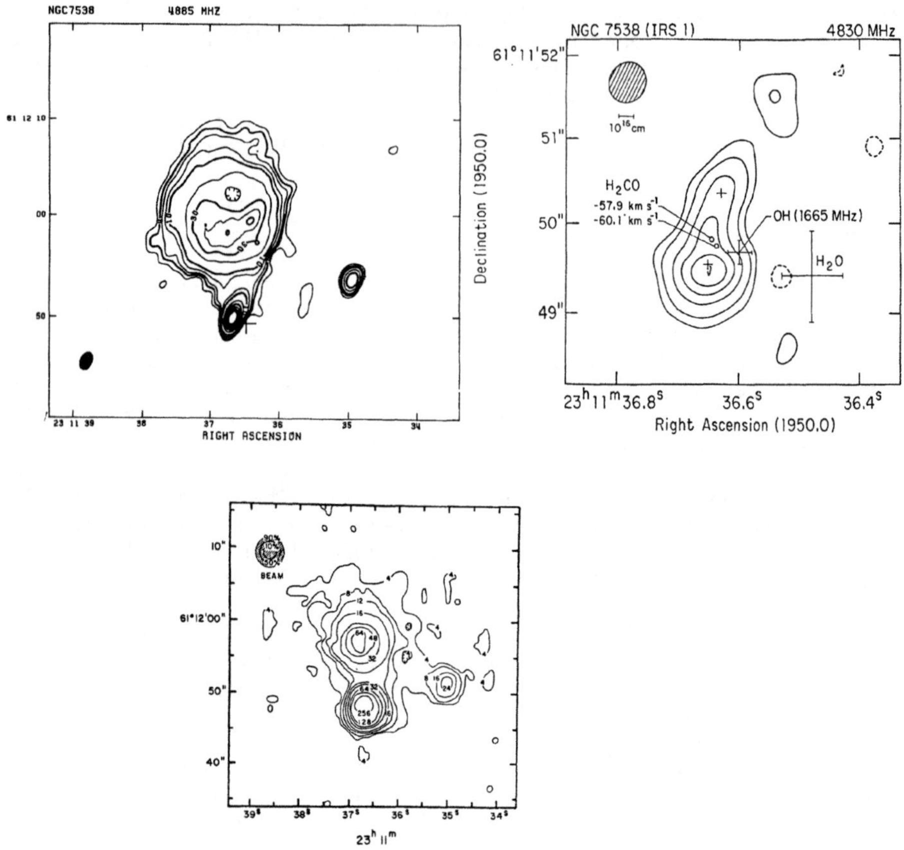

Fig. 10. Top left: 6 cm radio continuum emission from NGC 7538 110). The beam size is 1.7"x1".1 in p.a. 31°WN (shown in lower left corner). The contour units are 3, 4, 5, 7, 10, 20, 30, 40, 50, 60 mJy per beam. The cross to the W of IRS1 marks the position of the H_2O maser. Top right: Distribution of H_2O, OH and H_2CO masers (103,112) and the compact source IRS1. The contour units are -2.5, 2.5, 5, 10, 15, 20, 25 mJy/beam. Bottom: 10 μm map of NGC 7535 (3.5" beam, 75).

6. Influence of the Mass-Loss Phase on Molecular Clouds and the Interstellar Medium

6.1 Mass Loss - A Common Phenomenon

The examples given in the preceeding sections show that high mass loss at velocities of ~10 km s^{-1} appears to be a common phenomenon of newly-formed stars with masses greater than a few solar masses (O, B and A (?)). H_2O maser radiation, in particular, seems to be a good and easily detectable indicator of high velocity motions. Fig. 11

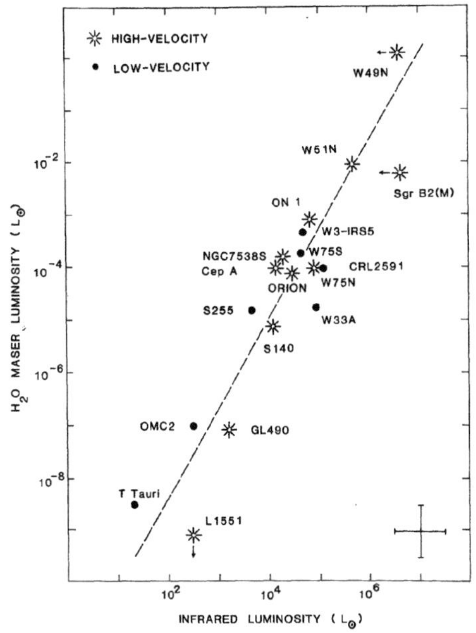

Fig. 11. Correlation of H_2O maser luminosity (integrated over all features and isotropic emission assumed) with the total infrared luminosity (83). The cross (bottom right) indicates the uncertainties. Asterisks denote that the source shows high velocity ($\Delta v > 20$ km s^{-1}) motions. The dashed line has a slope of 1.7.

shows that H_2O masers with high velocity features can be detected over more than four orders of magnitude of infrared luminosity. The diagram also indicates a good correlation of the integrated H_2O luminosity with the total infrared luminosity. The graph should, however, be treated with some caution since the H_2O masers may not radiate isotropically, the total infrared luminosity in a given region may differ from the luminosity of the infrared counterpart(s) of the maser (c.f. IRc2 vs. Orion-KL), and there probably are statistical selection effects. The maser luminosities seem to increase with about the square of the infrared luminosities. Since the mass outflow velocities are roughly constant ($\lesssim 100$ km s^{-1}) for almost all of the sources spanning 8 orders of H_2O luminosity, this trend suggests a strong increase of the mass loss rates with luminosity (if the masers are pumped collisionally using their own kinetic energy in the interaction with density inhomogeneities in the molecular cloud). An increase of the mass loss rates with luminosity is also suggested by the SiO and CO "plateau" data (42). An almost quadratic increase of mass loss rates with stellar luminosities has been found for the stellar winds of main-sequence O stars (113).

6.2 Impact on the Interstellar Medium

Table 2 is a summary of the characteristics of the outflow in Orion and presents a comparison of the mass outflows with other sources of energy provided to the interstellar clouds. The momentum contained in the Orion-outflows is large compared to the momentum in photons. The energy contained in the outflows, however, is small compared to the energy in radiation. Hence, while direct radiation pressure (in the optically thin limit) probably cannot explain the motions, a conversion of only a small part of the available luminosity into kinetic energy is sufficient. This is possible if the cloud is optically thick to the radiation (71) or if angular momentum and/or magnetic energy of the stars (in a binary system) can be converted into expanding motions (72; these energies may in turn be in some equilibrium with the radiative energy). Since the mechanical energy in the outflows is large compared to the "turbulent" energy in the quiescent gas ($\sim$3 km s^{-1}), the mass loss phenomena may be important for the "stirring" and stabilization of the dense molecular cloud cores. From the lower part of the Table 2, however, it is clear that the energy provided by these outflow phenomena to the interstellar medium as a whole is probably smaller than provided by H II regions, supernovae and O-star winds (c.f. 114,115) although the energy in these cases may only be transferred very inefficiently ($\sim$1%, 114,1).

TABLE 2: INFLUENCE OF THE MASS-LOSS PHASE IN YOUNG STARS ON MOLECULAR CLOUDS AND THE INTERSTELLAR MEDIUM

1) The Orion-Example

Stellar Parameters		Flow Parameters		Cloud Parameters	
Luminosity	$10^4 \to 10^5 L_\odot$	Mass in Low Velocity Flow ($v \lesssim 20$ km s^{-1})	3→30 $M_\odot$	Mass contained in Molecular Cloud Core (1')	200 $M_\odot$
Mass Loss Rate	$10^{-4} \to 10^{-2} M_\odot y^{-1}$				
Time Scale Mass Loss	$10^3 \to \gtrsim 10^4$ y	......High Velocity Flow ($v \sim 20$ to 100 km s^{-1})	1→10 $M_\odot$	Average Temperature	70 K
Energy Radiated in this time	$10^{49} \to 10^{50}$ erg	Momentum in Flows	10^{41} g s^{-1} cm	Thermal Energy	10^{45} erg
Momentum in Photons	3 $10^{38} \to$ 3 10^{39} g cm s^{-1}	Mechanical Energy	2 10^{47} erg	Gravitational Energy	10^{46} erg
		Energy radiated in H_2/CO Shocks	10^{47} erg	Turbulent Energy (Δv=3 km s^{-1})	10^{46} erg

2) Influence on Interstellar Medium

a) Mass Outflows from Young Stars	
i) energy density dumped by newly-formed stars $\gtrsim 5 M_\odot$ if each provides 2 10^{47}erg; star formation rate $\gtrsim 5 M_\odot$: 2 10^{-12} stars y^{-1} pc^{-3} (Miller and Scalo)	5 10^{-28}erg $s^{-1}cm^{-3}$
ii) energy density dumped if all formed stars loose 20% of their mass at velocities $\sim$100 km s^{-1}; total star for. rate $10^{-10} M_\odot y^{-1} pc^{-3}$ (Miller and Scalo)	2 10^{-27}erg $s^{-1}cm^{-3}$
b) H II Regions	
energy density provided by the Lyc-radiation of O-stars (Spitzer)	2 10^{-25}erg $s^{-1}cm^{-3}$
c) Supernovae	
energy density due to SN explosions (each 4 10^{50}erg and 1 per 25y; Spitzer)	2 10^{-25}erg $s^{-1}cm^{-3}$
d) Stellar Winds	
energy density from winds of stars > Chandrasekhar mass, if 10→50% of the mass is ejected (10^{-8} to $10^{-5} M_\odot y^{-1}$ and 1000 km s^{-1}, $\gtrsim 10^7$y)	5 10^{-26}erg $s^{-1}cm^{-3}$
e) Cloud-Cloud Collisions (Spitzer)	2 10^{-27}erg $s^{-1}cm^{-3}$

7. Summary: The Current Picture

How have the recent results on the mass loss phenomena modified our ideas on the evolution of protostellar and young stellar objects?

1) The accretion phase: The detection of true, massive protostars in the pre-nuclear burning and the pure accretion phases has not been possible up to now. Judging from the high obscurations found for some of the sources discussed in this review, the true protostars may be hidden behind A_V>100 mag, and hence, have escaped detection at $\lambda \lesssim 30$ μm. At far-infrared wavelengths, on the other hand, the combined effects of low angular resolution and diffusion of the radiation probably has inhibited the detection. It is possible that some "low velocity" H_2O masers are associated with the end of this phase of stellar evolution.

2) The mass-loss phase: After thermonuclear reactions have started, the envelopes of massive protostars have dimensions, temperatures and luminosities similar to those of late-type giants and supergiants. Instabilities of these envelopes due to the increased luminosity, the rotation and the magnetic field of the star (10,71,72) or instabilities due to the presence of a companion star may drive the mass loss at the base of the dense envelope. The wind has low velocities (10 to $\lesssim$300 km s^{-1}) and high mass loss rates. The high densities present make the cooling very efficient and may also inhibit the formation of a sizeable H II region. One of the most interesting facts about these mass flows is the apparent anisotropy of the flow (possibly including the low velocity--high velocity flow structure in Orion). This could be caused by an anisotropic density distribution (e.g. a disk) in the immediate vicinity ($\sim 10^{16}$ cm) of the source. Along the largest density gradients, the outflow would be accelerated in a way similar to supersonic nozzles or the Champagne model of expanding H II regions (127). In the limit of an isothermal source, the expansion velocity v_f at radius r_f and density ρ_f (r_f) will depend on the initial velocity v_i, and the density at the base of the flow ρ_i (r_i) in a logarithmic form,

$$v_f = v_i \left(a \ln (\rho_i (r_i) / \rho_f (r_f))\right)^{\frac{1}{2}} , \qquad (3)$$

where a is a constant of order 1. Hence, for a steep density gradient ($\rho_i/\rho_f \sim 10^4$) and an initial velocity of 20 km s^{-1}, the acceleration is considerable and v_f is $\sim$60 km s^{-1}. The interaction of this high velocity outflow with the dense surrounding cloud or disk, in turn, may feed material back to the base of the outflow (128). The interaction of the flow with density inhomogeneities further out into the molecular cloud (> a few 10^{16} cm) may produce high excitation molecular lines (e.g. NH_3) and create the bright H_2O maser spots, the "shocked" gas and Herbig-Haro objects. The high brightness of these objects at $\lambda \lesssim 20$ μm may be caused by the radiative transport through the dense expanding shell and the inhomogeneities in the surrounding cloud. In a few cases, the flow close to the star may be smooth, so that "shell type" SiO, H_2O and OH can be observed.

3) The H II-Region phase: The mass loss phase lasts for $\gtrsim 10^4$y (see also the statistics of H_2O maser sources in 129,130). When the strength of the wind dies down, a collisionally or photoionized H II region, first close to the star, may form. This phase can probably be observed in regions like BN, where broad-winged Brα and Brγ lines have been detected, but no radio H II region (131). As the rate of mass loss diminishes further, a radio H II region ($d \gtrsim 10^{15}$ cm) develops. OH masers located in the surrounding molecular cloud appear and are amplifying the 18 cm background, while H_2O masers are still found coincident in position with the H II region. Even later, the H II region expands to $\gtrsim 10^{17}$ cm and the H_2O maser phenomenon--having lasted for $\sim 5\ 10^{4\pm0.5}$y--disappears. Thereafter, the H II region becomes optically thin and expands into the region where the OH maser cloudlets are located and the OH masers also cease to exist. During this phase, the character of the wind may approach that observed toward main sequence O-stars.

Acknowledgment: The authors thank S.C. Beck, E.E. Becklin, J. Bieging, G. Grasdalen, S. Hansen, P. Ho, R. Martin, J. Moran, T. Pauls, T. Phillips, R. Plambeck, M. Reid, C.H. Townes, R.C. Walker, T.L. Wilson, C.G. Wynn-Williams and B. Zuckerman for helpful discussions and permission to use data prior to publication. R. Genzel is supported by the Miller Society for Basic Research in Science. D.D. acknowledges partial support from NATO research grant No. 1909. We wish to thank B. Allen and J. Bloomquist for their effort in preparing the manuscript.

REFERENCES

1. Knowles, S., Mayer, C., Cheung, A., Rank, D., Townes, C. 1969, Science, 163, 1055.
2. Meeks, M., Carter, J., Barrett, A., Schwartz, P., Waters, J., Brown, W., 1969, Science, 165, 180.
3. Thaddeus, P., Kutner, M., Penzias, A., Wilson, R., Jefferts, K., 1972, Ap.J. 176, L73.
4. Dickinson, D. 1972, Ap.J. 175, L43.
5. Strelnitskii, V., Syunyaev, R. 1972, Astron. Zh. 49, 704; Soviet. Astron. 16, 579 (1973).
6. Werner, M., Becklin, E., Neugebauer, G. 1977, Science, 197, 723.
7. Werner, M., 1977 IAU Symp. No. 75,"Star Formation",eds. de Jong and Maeder, Reidel.
8. Westbrook, W., Werner, M., Elias, J., Gezari, D., Hauser, M., Lo, K., Neugebauer, G. 1976, Ap.J. 209, 94.
9. Keene, J., Hildebrand, R., Whitcomb, S. 1981, in IAU Symp. 96*.
10. Downes, D., Genzel, R., Becklin, E., Wynn-Williams, C. 1981, Ap.J. 244, 869.
11. Grasdalen, G., Gehrz, R., Hackwell, J. 1981, in IAU Symp. No. 96*, p. 179.
12. Aitken, D., Roche, P., Spenser, P., Jones, B. 1981, MNRAS 195, 921.
13. Scoville, N. and Kwan, J. 1976, Ap.J. 206, 718.
14. Sibille, F., Lena, P. 1981, in IAU Symp. No. 96*, p
15. Bedijn, P., Habing, H., de Jong, T. 1978, Astr.Ap. 69, 73.
16. Finn, G., Simon, T. 1977, Ap.J. 212, 472.
17. Kwan, J., Scoville, N. 1976, Ap.J. 209, 102.
18. Genzel, R., Reid, M., Moran, J., Downes, D. 1981, Ap.J. 244, 884.
19. Kwan, J., Scoville, N. 1974, Ap.J. 194, L97.
20. Elitzur, M. 1980, Ap.J. 240, 553.
21. Genzel, R., Downes, D., Schwartz, P., Spencer, J., Pankonin, V., Baars. J. 1980, Ap.J. 239, 519.
22. Morris, M., Palmer, P., Zuckerman, B. 1980, Ap.J. 237, 1.
23. Scoville, N. 1981, in IAU Symp. No. 96*, p. 187.
24. Grasdalen, G. 1976, Ap.J. 205, L83.
25. Hall, D., Kleinmann, S., Ridgway, S., Gillett, F. 1978, Ap.J. 223, L47.
26. Simon, M., Righini-Cohen, G., Felli, M., Fischer, J., preprint.
27. Scoville, N., Hall, D., Kleinmann, S., Ridgway, S. 1979, Ap. J. 232, L121.
28. Ho, P., Barrett, A. 1978, Ap.J. 224, L23.
29. Evans, N., Plambeck, R., Davis, J. 1979, Ap.J. 227, L25.
30. Bastien, P., Bieging, J., Henkel, C., Martin, R., Pauls, T., Walmsley, C., Wilson, T., Ziurys, L. 1981, Astr.Ap., in press.
31. Goldsmith, P., Langer, W., Schloerb, F., Scoville, N. 1980, Ap.J. 240, 524.
32. Rydbeck, O., Hjalmarson, Å., Rydbeck, G., Elldér, J., Olofsson, H., Sume, A. 1981, Ap.J., in press.
33. Bieging, J., priv. comm.
34. Snyder, L., Buhl, D. 1974, Ap.J. 189, L31.
35. Hansen, S. 1980, Ph.D. Thesis, Univ. of Mass.

36. Genzel, R., Moran, J., Lane, A., Predmore, C., Ho, P., Hansen, S., Reid, M. 1979, Ap.J. 231, L73.
37. Baud, B., Bieging, J., Plambeck, R., Thornton, D., Welch, W., Wright, M. 1980, IAU Symp. No. 87,"Interstellar Molecules",ed. B. Andrew, Reidel, Dordrecht, Holland, p. 545.
38. Genzel, R., Downes, D., Moran, J., Johnston, K., Spencer, J., Walker, R., Haschick, A., Matveyenko, L., Kogan, L., Kostenko, V., Rönnäng, B., Rydbeck, O., Moiseev, I. 1978, Astr.Ap. 66, 13.
39. Reid, M., Moran, J. 1981, An.Rev.Astr.Ap., in press.
40. Spencer, J., Winnberg, A., Olnon, F., Schwartz, P., Matthews, H., Downes, D. 1981, Astr.J., in press.
41. Hansen, S., Johnston, K. 1980, BAAS 12, 824.
42. Downes, D., Genzel, R., Hjalmarson, Å., Nyman, L.Å., Olofsson, H., Rönnäng, B. 1981, Ap.J., submitted.
43. Ziurys, L. et al., in prep.
44. Matsakis, D., Hjalmarson, Å., Palmer, P., Cheung, A., Townes, C. 1980, BAAS 12, 824.
45. Zuckerman, B., Morris, M., Palmer, P. 1981, Ap.J., in press.
46. Welch, W., Wright, M., Plambeck, R., Bieging, J., Baud, B. 1981, Ap.J., 1981, 245, L87.
47. Bieging,J.,priv.comm.(Bieging,J.,Martin,R.,Pauls,T.,Wilson,T. in prep)
48. Phillips, T., Huggins, P., Neugebauer, G., Werner, M. 1977, Ap.J., 217, L161.
49. Solomon, P., Huguenin, G., Scoville, N. 1981, Ap.J. 245, L19.
50. Zuckerman, B., Kuiper, T., Rodriguez-Kuiper, E. 1976, Ap.J. 209, L137.
51. Kwan, J., Scoville, N. 1976, Ap.J. 210, L39.
52. Wannier, P., Phillips, T. 1977, Ap.J. 215, 796.
53. Kuiper, T., Rodriguez-Kuiper, E., Zuckerman, B. 1978, Ap.J. 219, 129.
54. Knapp, G., Phillips, T., Huggins, P., Redman, R. 1981, Ap.J., in press.
55. Gautier, T., Fink, U., Treffers, R., Larson, H. 1976, Ap.J. 207, L129.
56. Beck, S., Lacy, J., Geballe, T. 1979, Ap.J. 234, L213.
57. Knacke, R., Young, E., preprint.
58. Watson, D., Storey, J., Townes, C., Haller, E., Hansen, W. 1980, Ap.J. 239, L129.
59. Nadeau, D., Geballe, T., Neugebauer, G. 1981, Ap.J., in press.
60. Beckwith, S., Persson, E., Neugebauer, G., Becklin, E. 1978, Ap.J. 223, 464.
61. Scoville, N., Kleinmann, S., Hall, D., Ridgway, S. 1981, preprint.
62. McKee, C., Hollenbach, D. 1980, An.Rev.Astr.Ap. 18, 219.
63. Draine, B. 1980, Ap.J. 241, 1021.
64. Hollenbach, D., McKee, C. 1979, Ap.J.Suppl. 41, 555.
65. Beck, S. 1981, Ph.D. Thesis, Univ. of California, Berkeley.
66. Beckwith, S., Persson, S., Neugebauer, G. 1979, Ap.J. 227, 436.
67. Simon, M., Righini-Cohen, G., Joyce, R., Simon, T. 1979, Ap.J. 230, L175.
68. Scoville, N. 1981, in IAU Symp. No. 96*, p. 187.
69. Beckwith, S. 1981, in IAU Symp. No. 96*, p. 167.
70. Zuckerman, B. 1981, in IAU Symp. No. 96*, p. 275.
71. Phillips, J., Beckman, J. 1980, MNRAS, 193. 245.

72. Hartmann, L., McGregor, K., 1981 in prep.
73. Lada, C., Oppenheimer, M., Hartquist, T. 1978, Ap.J. 226, L153.
74. Genzel, R., Becklin, E., Wynn-Williams, C., Moran, J., Reid, M., Jaffe, D., Downes, D. 1981, Ap.J., in press.
75. Hackwell, J., Grasdalen, G., Gehrz, R. 1981, in prep.
76. Schneps, M., Lane, A., Downes, D., Moran, J., Genzel, R., Reid, M. 1981, Ap.J., in press.
77. Beckwith, S., Zuckerman, B. 1981, preprint.
78. Genzel, R., Downes, D., Schneps, M., Reid, M., Moran, J., Kogan, L., Kostenko, V., Matveyenko, L., Rönnäng, B. 1981, Ap.J., in press.
79. Phillips, T., Knapp, G., Huggins, P., Werner, M., Wannier, P., Neugebauer, G., Ennis, D., preprint.
80. Matsakis, D., Bologna, J., Schwartz, P., Cheung, A., Townes, C. 1980, Ap.J. 241. 655.
81. Becklin, E., Telesco, C., Hildebrand, R. 1981, in prep.
82. Walker, R., Matsakis, D., Garcia-Barreto, J. 1981, Ap.J., submitted.
83. Becklin, E., Downes, D., Genzel, R., Wynn-Williams, C. 1981, in prep.
84. Wynn-Williams, C., Becklin, E., Neugebauer, G. 1972. MNRAS, 160, 1.
85. Werner, M., Becklin, E., Gatley, I., Neugebauer, G., Sellgren, K., Thronson, H., Harper, D., Loewenstein, R., Moseley, S. 1981, Ap.J., in press.
86. Walker, R. 1981, A.J., in press.
87. Howell, R., McCarthy, D., Low, F. 1981, in IAU Symp. No. 96*.
88. Becklin, E., Neugebauer, G. 1981, in prep.
89. Colley, D. 1980, MNRAS 193, 495.
90. Lada, C., Harvey, P. 1981, Ap.J. 245, 58.
91. Gatley, I., Becklin, E., Matthews, K., Neugebauer, G., Penston, M., Scoville, N. 1974, Ap.J. 191, L121.
92. Knapp, G., Morris, M. 1976, Ap.J. 206, 713.
93. Thum, C., Bertout, C., Downes, D. 1981, Astr.Ap. 94, 80.
94. Rodriguez, L., Moran, J., Ho, P., Gottlieb, E. 1980, Ap.J. 235, 845.
95. Haschick, A., Moran, J., Rodriguez, L., Burke, B., Greenfield, P., Garcia-Barreto, J. 1980, Ap.J. 237, 26.
96. Snell, R., Loren, R., Plambeck, R. 1980, Ap.J. 239, L17.
97. Beichman, C., Harris, S. 1981, Ap.J. 245,589.
98. Cudworth, K., Herbig, G. 1979, A.J. 84, 548.
99. Cohen, M., Bieging, J., Schwartz, P. 1981, Ap.J., subm.
100. Beichman, C., Becklin, E., Wynn-Williams, C. 1979, Ap.J. 232, L47.
101. Lada, C., Blitz, L., Reid, M., Moran, J. 1981, Ap.J., in press.
102. Rodriguez, L., Ho, P., Moran, J. 1980, Ap.J. 240, L149.
103. Forster, J., Welch, W., Wright, M., Baudry, A. 1978, Ap.J. 221, 137.
104. Downes, D., Genzel, R., Moran, J., Johnston, K., Matveyenko, L., Kogan, L., Kostenko, V., Rönnäng, B. 1979, Astr.Ap. 79, 233.
105. Reid, M., Haschick, A., Burke, B., Moran, J., Johnston, K., Swenson, G. 1980, Ap.J. 239, 89.
106. Dreher, J., Welch, W. 1981, Ap.J. 245, 857.
107. Habing, H., Goss, W., Matthews, H., Winnberg, A. 1974, Astr.Ap. 35, 1.
108. Haschick,A.,Reid,M.,Burke,B., Moran,J.,Miller,G. 1981, Ap.J. 244,76.
109. Werner, M., Becklin, E., Gatley, I., Matthews, K., Neugebauer, G., Wynn-Williams, G. 1979, MNRAS 188, 463.

110. this paper, and Moran et al. in prep.
111. Harris, S., Scott, P. 1976, MNRAS 175, 371.
112. Rots, A., Dickel, H., Forster, J., Goss, W. 1981, Ap.J. 245, L15.
113. Abbott, D., Bieging, J., Churchwell., E., Cassinelli, J. 1981, Ap.J., in press.
114. Spitzer, L. 1978,"Physical Processes in the Interstellar Medium", Wiley.
115. Miller, G., Scalo, J. 1979, Ap.J.Suppl. 41, 513.
116. Goldsmith, P., Erickson, N., Fetterman, H., Clifton, B., Peck, D., Tannenwald, P., Koepf, G., Buhl, D., McAvoy, N. 1981, Ap.J. 243, L79.
117. Storey, J., Watson, D., Townes, C., Haller, E., Hansen, W. 1981, Ap.J. 247, 136.
118. Barrett, A., Ho, P., Myers, P. 1977, Ap.J. 211, L39.
119. Wilson, T., Downes, D., Bieging, J. 1979, Astr.Ap. 71, 275.
120. Gottlieb, C., Gottlieb, E., Litvak, M., Ball, J., Penfield, H. 1978, Ap.J. 219, 77.
121. Pickett, H., Davis, J. 1979, Ap.J. 227, 446.
122. Clark, F., Biretta, J., Martin, H. 1979, Ap.J. 234, 922.
123. Loren, R., Erickson, N., Snell, R., Mundy, L., Davis, J. 1981, Ap.J. 244. L107.
124. Johnson, D., Lovas, F., Gottlieb, C., Gottlieb, E., Litvak, M., Guélin, M., Thaddeus, P. 1977, Ap.J. 218, 370.
125. Loren, R., Mundy, L., Erickson, N. 1981, Ap.J.Lett., subm.
126. Dieter, N., Welch, W., Wright, M. 1979, Ap.J. 230, 768.
127. Tenorio-Tagle, G. 1982, this volume, p. 1.
128. Elmegreen, B. and Morris, M. 1979, Ap.J. 229, 593.
129. Genzel, R., Downes, D. 1979, Astr.Ap. 72, 234.
130. Jaffe, D., Güsten, R., Downes, D. 1981, Ap.J., in press.
131. Thompson, R. 1981, in IAU Symp. No. 96*, p. 153.
132. Moran, J., Reid, M., Lada, C., Yen, J., Johnston, K., Spencer, J. 1978, Ap.J. 224, L67.
133. Norman, C., Silk, J. 1979, Ap.J. 228, 197.
134. Shull, J. M. 1982, this volume, p. 91.
135. Krolik, J. and Smith, H. 1981, preprint.
136. Watson, W. and Walmsley, C.M. 1982, this volume, p. 357.
137. Lane, A. and Bally, J. 1982, this volume, p. 301.
138. Turner, B.E. 1982, this volume, p. 425.

*IAU No. 96: "Infrared Astronomy",eds. C.G. Wynn Williams and D.F. Cruikshank, Reidel: Dordrecht.

DISCUSSION FOLLOWING REVIEW BY GENZEL AND DOWNES

SNYDER: Apart from Orion, how many other SiO maser sources are associated with HII region/molecular clouds?

GENZEL: None. However, it may be an exceptional case just as is the Orion OH maser source associated with IRc2.

SNYDER: What error do you attribute to the 36 km s^{-1} total width for the SiO masers?

GENZEL: 36 ± 3 km s^{-1}. The velocity ranges for the SiO maser, the OH maser, and the thermal SiO lines are very similar. Nevertheless, it is not the 36, itself, that is so important. It is the fact that there are two flows, a low velocity flow and a high velocity flow.

KWOK: There are two very unusual parameters in your outflow picture; the mass loss rate, which is high and the velocity, which is low. It is true, as you suggested, that such mass loss rates cannot be driven by radiation pressure? The observed mass loss rate may be strongly influenced by swept up matter.

GENZEL: What you are saying amounts to a statement about mv. On the other hand you could postulate an extremely high velocity (1000 km s^{-1}) inner shell which is not observable at the moment, and transfer its momentum outwards. If it were there, however, we should be able to observe very high ionization at high velocities such as the winds common to O stars.

KWOK: Your mass loss limit or radiation pressure limit is derived from a photon momentum to mass momentum equation yet the quantity being transferred is really energy.

GENZEL: If you want to do it by energy transfer, then you need a pressure-cooker situation in which bubbles or blisters form. But our measurements indicate a temperature of 100 - 200 K which would have an insignificant blistering effect.

UNDERHILL: If you had flows of the order of 1500 ± 500 km s^{-1}, they would be associated with electron temperatures of the order of, or greater than, 20000 K. Emission lines from plasmas at such temperatures occur chiefly shortward of 1 μm where you cannot observe because of the large extinction.

GENZEL: Observationally you are correct, but theoretically, if there were temperatures of > 10^5 at these mass loss rates and these densities, the recombination rates would be too high to support an HII region.

FORSTER: The H_2O maser phase must be a transient phenomenon. Do we have a good idea of the duration of H_2O masers?

GENZEL: Our statistics at the moment indicate from 10^4 to 10^5 years, but there are many selection effects (e.g. observations at different angles). Also, an all-sky complete survey of H_2O masers has never been attempted.

KROLIK: I would like to merge your phases 2 and 3. In many HII regions there is evidence from Brackett line and some free-free radio continuum observations that there is both a substantial outflow and a high degree of ionization. I also would like to correct a common misconception that all compact HII regions are photo-ionized when in fact the densities are so high that thermal processes are dominant.

GENZEL: Well surely the L1551 case cannot be photo-ionization. It must be collisional ionization.

DEWDNEY: With lifetimes of the order of 10^4 years for H_2O masers, what sort of spatial extent do you expect these outflow regions to have?

GENZEL: Possibly between 0.1 pc and 1 pc with 0.1 pc being typical. The lower density regions are a bit more extensive.

SOMEONE: But using these lifetimes (10^4 years) and velocities of 100 km s^{-1}, the outflow regions are much bigger than 0.1 pc.

GENZEL: But there is considerable deceleration.

J. DICKEL: So it is only in the very core that the velocity is high, and the gas slows down quickly when it travels very far into the molecular cloud?

GENZEL: Yes.

HIGH VELOCITY CO LINE WINGS AND THE DYNAMICS OF STAR FORMING MOLECULAR CLOUD CORES

John Bally
Bell Laboratories
Holmdel, NJ 07733

ABSTRACT

Preliminary results of a search for high velocity CO in relatively nearby (d<3kpc) star forming cloud cores are presented. Detailed CO, CS, and VLA observations of the molecular outflow region associated with NGC 2071 suggest that a disk constrained stellar wind is responsible for the bipolar high velocity CO flow.

I. THE HIGH VELOCITY WING SURVEY

Thirty molecular cloud cores containing luminous buried infrared sources have been searched for the presence of high velocity CO emission with the Bell Laboratories 7-meter off-axis Cassegrain antenna. When used in a double position switching observing mode, this system provides exceptionally flat baselines, well suited for searches of faint, large velocity range emission. Each source was observed until a sensitivity limit of 100 mK in 1/4 MHz filters was achieved. The statistics of CO line widths found in this survey are presented in Figure 1. A number of intriguing patterns emerge when the infrared and radio properties of these sources are analyzed:

1) Out of 7 mapped sources, 6 are bipolar with the redshifted emission centroid spatially displaced from the blueshifted centroid. The bipolar flows are Ceph A (Rodriguez et al. 1980), L1551 (Snell et al. 1980), GL 490 (Lada and Harvey 1981), HH 7-11 (Snell and Edwards 1982), GL 961 (Lada 1981), and NGC 2071 (Bally 1981).

2) The highest velocity outflows have the smallest spatial extent. Spatial dimensions of the flows range from 0.1 pc (Orion) to 1.5 pc (GL 961).

3) Most sources contain evidence of ionized gas associated with the central IR source as revealed by Bα or Bγ recombination lines of hydrogen or radio free-free emission.

R. S. Roger and P. E. Dewdney (eds.), Regions of Recent Star Formation, 287–293.
Copyright © 1982 by D. Reidel Publishing Company.

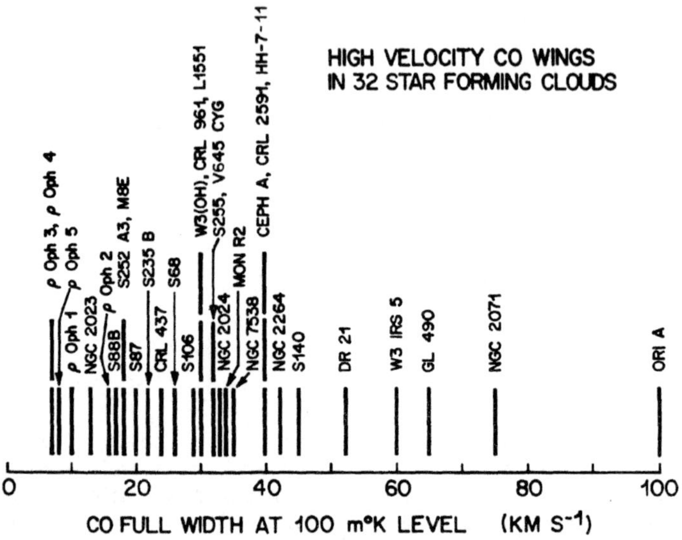

Figure 1. Statistics of CO Line Wing Full Widths

4) Most sources exhibit 2.12 μm H_2 S(1) line emission (Lane and Bally 1982) with stronger emission arising in the blueshifted region.

5) The mechanical luminosity of the CO flows correlates weakly with the total luminosity of the central sources.

6) The formation rate of high velocity wing sources is $R \gtrsim 0.1\ yr^{-1}$ in the galaxy. The kinetic energy and momentum that these sources provide is sufficient to power the supersonic turbulence in molecular clouds which can explain the observed line widths as well as prevent cloud collapse at the free-fall rate.

II. NGC 2071

An extended (3'x5') high velocity ($\Delta V \gtrsim 75$ km s^{-1}) outflow has been found to be associated with the infrared source in NGC 2071 (Bally 1981). This star forming region, located at a distance of 500 pc, is part of the Orion B molecular cloud. Figure 2 is a map of the redshifted and blueshifted gas generated from spectra obtained at 1 arcminute intervals in an 8'x8' grid centered on the IR source. The line wings at source center are detectable in ^{13}CO and in the J=2-1 transition of CS over a 25 km s^{-1} velocity range. Emission in this velocity range is probably formed in a clumpy medium exhibiting a small filling factor $f \sim T_A^*(v)/Tex \sim 0.2$ to 0.001 with each clump having a CO optical depth $\gtrsim 1$. Since most of the mass, energy, and momentum are represented by the lower velocity gas, these

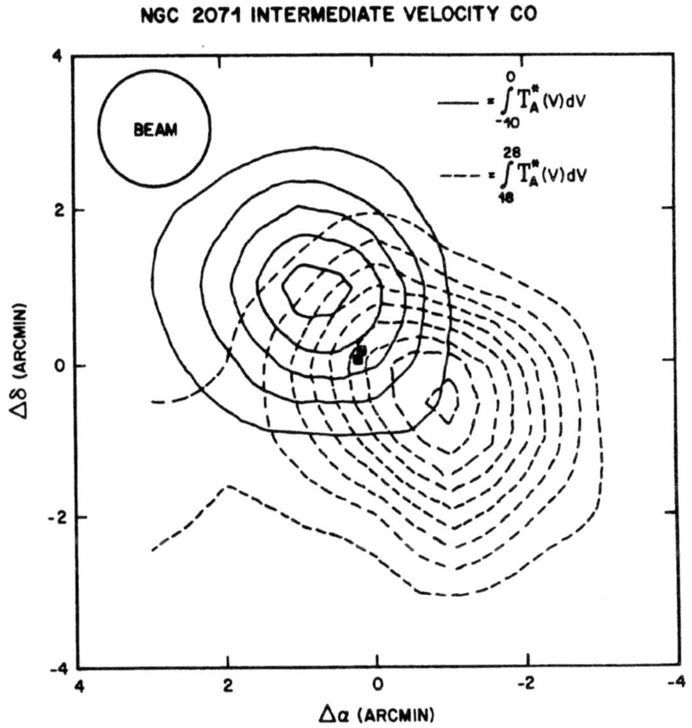

Figure 2. The blueshifted (solid) and redshifted (dashed) gas distribution in the NGC 2071 core. The two black boxes mark the positions of the pair of VLA sources shown in Figure 4. Coordinates are referred to $\alpha(1950)=05^h44^m30.1^s$ $\delta(1950)=00°20'40''$.

excitation conditions lead to a satisfactory estimate of the physical parameters of the entire flow. For T_{ex}=40 K the flow has a mass 22 $M_\odot$, a total momentum 4×10^{40} g cm s^{-1}, a kinetic energy 6×10^{46} ergs, and a dynamical lifetime $r/V_{max} \sim 10^4$ yrs. The kinematics is best explained as outflow since the cloud core mass is too small by two orders of magnitude to gravitationally bind the highest velocity gas observed.

The presence of an energetic bipolar outflow poses two questions: 1) What collimates the outflow into two oppositely directed streams? 2) What is the underlying engine? To approach the first question, the flow region was mapped in the 98 GHz J=2-1 transition of the CS molecule which probes the distribution and kinematics of dense ($n>10^4$ cm^{-3}) gas. The distribution of peak T_A^*(CS) (Figure 3), shows the presence of a ridge of emission orthogonal to the NE-SW

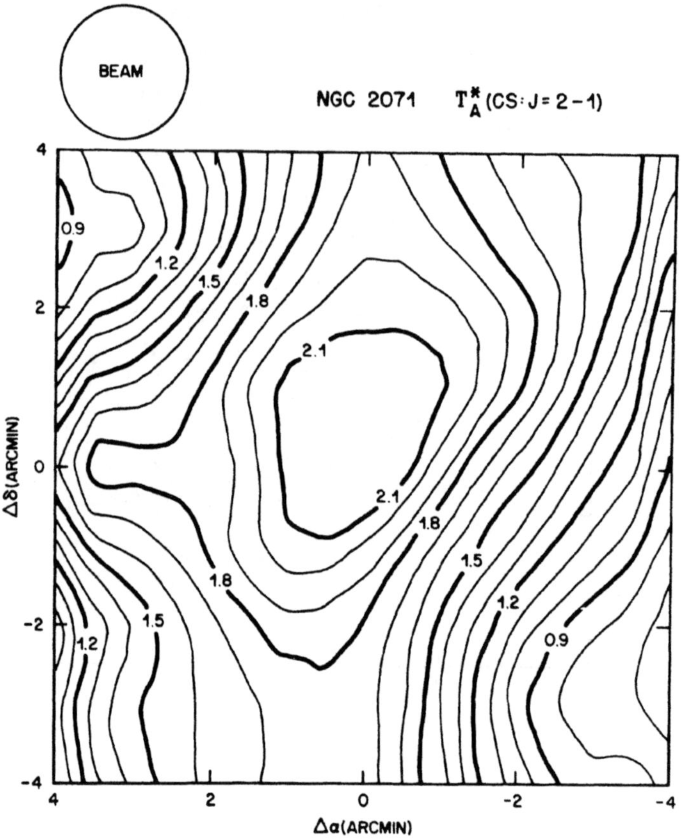

Figure 3. A CS J=2-1 map of the NGC 2071 star forming cloud core.

symmetry axis of the bipolar CO flow. Assuming a fractional abundance $X(CS)/(dv/dr)=10^{-10}$ (km s^{-1} pc^{-1})$^{-1}$ the mean density in the ridge falls in the range $3x10^4 cm^{-3} < n(H_2) < 2x10^5$ cm^{-3} (Linke and Goldsmith 1980). A total mass of $M_{ridge} \lesssim 400\ M_\odot$ is implied for this structure. A SE-NW velocity gradient is observed in the CS line core suggesting rotation around an axis closely aligned with the CO flow. These observations suggest the presence of a massive disk of dense gas which confines gas streaming from the IR source into lobes lying along the disk axis.

A pair of compact 5 GHz VLA sources were found by Bally and Predmore (1981) at the IR source position (see Figure 4). The two sources have flux densities S_{5GHz} = 8 mJy and S_{5GHz} = 3 mJy and

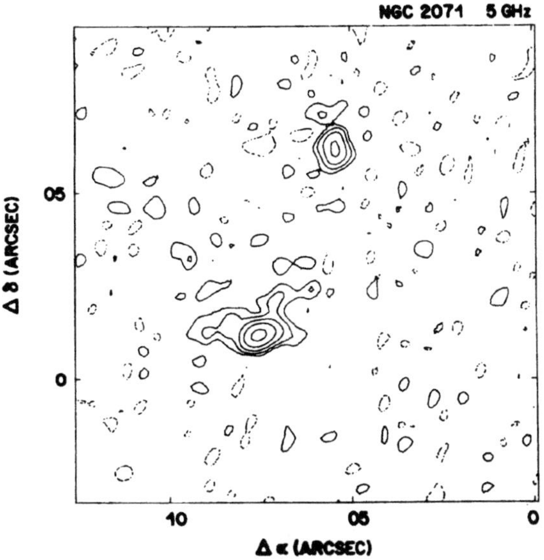

Figure 4. 5 GHz VLA map of NGC 2071.

are oriented in a direction closely aligned with the CS ridge. On the assumption that these sources are spherical, optically thin, photo-ionized compact HII regions a Lyman continuum flux of 1.3×10^{44} photons s^{-1} and 5×10^{43} photons s^{-1} is required to maintain photoionization equilibrium. This requires 2 ZAMS stars with a total luminosity $L_B \geq 4\times10^3$ $L_\odot$. For pre-main sequence stars the required luminosity is larger still. The total luminosity of NGC 2071 IR has been measured (Harvey et al. 1979) and found to be $L_B \sim 750$ $L_\odot$ only!

An alternative explanation for the VLA sources is that they are f-f emission from a high velocity, ionized, spherical stellar wind (Wright and Barlow 1975, Panagia and Felli 1975). For such a wind the continuum flux density is determined by

$$\dot{M}=4\times10^{-6}\left|\frac{S_{5GHz}}{mJy}\right|^{0.75}\left|\frac{d}{kpc}\right|^{1.5}\left|\frac{V_w}{10^3 km\ s^{-1}}\right| \quad (M_\odot\ yr^{-1})$$

where V_w is the wind velocity. Assuming that the observed CO flow is driven by the same stellar winds responsible for the VLA sources, a unique determination of the quantities $\dot{M}$ and V_{wind} can be obtained. From the CO observations, $\dot{M}=L_M/V_{wind}$ where $L_M \approx 3\times10^{-2}$ ($M_\odot$ yr^{-1} km s^{-1}) is the mechanical luminosity of the CO flow (=P/T). Combining the above two relationships results in $\dot{M}=1\times10^{-5} M_\odot\ yr^{-1}$ and $\dot{M}=5\times10^{-6} M_\odot\ yr^{-1}$ with $V_w=1500$ km s^{-1} for the winds associated with the two VLA sources.

(Assuming that the two winds have the same velocity). These high velocity stellar winds are blocked in the equatorial plane of the CS disk but can blow a pair of bubbles in the lower density molecular gas situated along the disk axis. The observed CO emission may arise at the turbulent and clumpy interface where the wind encounters the ambient cloud and forms an expanding shock wave.

REFERENCES

Bally, J. (1981) Ap J (submitted).
Bally, J. and Predmore, R. (1981) Ap J (in preparation).
Harvey, P. M., Campbell, M. F., Hoffman, W. F., Thronsen, H.A. and Gatley, I. (1979) Ap J 229, 990.
Snell, R. L. and Edwards, S. (1982) this volume, p. 173.
Lada, C. J. (1981) private communication.
Lada, C. J. and Harvey, P. M. (1981) Ap J 245, 58
Lane, A. P. and Bally, J. (1982) this volume, p. 301.
Linke, R. L. and Goldsmith, P. G. (1980) Ap J 235, 437.
Panagia, N. and Felli, M. (1975) A.A. 39, 1.
Rodriguez, L. F., Ho, P. T. P. and Moran, J. M. (1980) Ap J 240, L149.
Snell, R. L., Loren, R. B. and Plambeck, R. L. (1980) Ap J 239, L 17.
Wright, A. E. and Barlow, M. J. (1975) M.N.R.A.S. 170, 41.

DISCUSSION FOLLOWING PAPER BY BALLY

BLITZ: How do you know that what you are seeing is due to mass outflow and not to rotation?

BALLY: We calculate what we call the dynamic mass. Given the two velocity centroids (blue and red) we determine, for the highest velocities observed, the mass required, internal to that radius, to satisfy gravitational binding. For NGC 2071 you need $4 \times 10^4 M_\odot$ within 4×10^{17} cm for rotation. The mass calculated from ^{13}CO, $C^{18}O$ and CS column densities is only 400 $M_\odot$.

J. DICKEL: The plots for these sources all seem to show merely the flux density integrated in the red and in the blue. It should be possible to subtract the central emission from the spectra and fit a profile to the "wing" emission only, to show the gradient in mean velocity versus position. This might allow a better comparison of rotational and bipolar models.

BALLY: It might help but I suspect the discrepancy with a rotational model would still exist.

PHILLIPS: As you know, I've modelled this region in terms of a rotational model. However, for an outflow model with an accretion disk, the axis of the jets would probably be perpendicular to the rotational axis of the surrounding complex, if the accretion disk represents placental material. We find this to be the case for NGC 2071. In addition, such a phenomenon might be a useful diagnostic indicator for the presence of jets; there are for instance several puzzling cases where a rapidly rotating core has an axis perpendicular to the surrounding neutral envelope - perhaps such cores are really jet-outflow sources.

UNDERHILL: There is a well established relationship between terminal velocity and effective temperature (i.e. spectral type) for supergiants in the range A to O - the only stars for which you can measure terminal velocities from UV spectra. For A2 stars, luminosity 10^4 - 10^5 $L_{\odot}$, the velocity is ~200 km s^{-1}. An outflow velocity of 1500 km s^{-1} is always associated with luminosities near 10^5 $L_{\odot}$. The adopted luminosities and outflow velocities must be consistent.

BALLY: I suggest that, although one might expect some kind of relationship between wind velocity and luminosity, for these particular stellar winds it may not be the conventional relation for O and B type stars.

KINEMATICS OF MOLECULAR GAS IN ORION FROM OBSERVATIONS OF THE ^{13}CO J=2→1 LINE

Paul F. Goldsmith, Richard Arquilla,
F. Peter Schloerb, and N.Z. Scoville
Five College Radio Astronomy Observatory
Department of Physics and Astronomy
University of Massachusetts, Amherst

We have obtained spectra of the J=2→1 transition of ^{13}CO covering a 16 arcminute by 33 arcminute region centered on the KL object in the Orion molecular cloud. These spectra reveal a high degree of complexity in the emission at many positions, which suggests that the molecular gas has been significantly perturbed by the HII region and Trapezium stars located in front of it. The velocity features in the affected regions are generally quite narrow with $\delta v_{FWHM} \sim 1.5$ km s^{-1}. A number of well-defined but extended velocity features indicate that the gas motion in this region is inherently quite complex.

The data reported here were taken using the 4.9-m MWO antenna at Fort Davis, Texas. The receiver, built at FCRAO, is essentially a cooled version of the 200-350 GHz receiver described by Erickson (1981). The receiver temperature was typically 900K (SSB) throughout this period. The system temperature, including correction for emission and absorption by the atmosphere, was 1300 to 1500K. The beam pattern was somewhat elliptical, having FWHM size 1'.5 in RA by 1'.75 in Dec. The forward beam coupling efficiency to the moon was measured to be 0.85, while the main beam efficiency, assuming a gaussian beam defined by the half-power points given above, is 0.56. Thus, the relevant efficiency to use in correcting the ^{13}CO J=2→1 data is not very well defined, and should vary with position in the source. The emission region is quite asymmetric, having half-power dimensions of ~20' NS by ~4' EW. Even within this region there are appreciable velocity shifts, making the effective source dimensions significantly smaller. The spectra shown in Figure 1 give the antenna temperature above the background, corrected for atmospheric and antenna absorption. The corrections for coupling efficiency discussed above are not included. We estimate that for positions near the center of the cloud, an efficiency of 0.70-0.75 is appropirate, and hence the peak ^{13}CO J=2→1 intensity is 27K. The efficiency for positions at the outer portions of the map is ~0.65.

R. S. Roger and P. E. Dewdney (eds.), Regions of Recent Star Formation, 295–299.
Copyright © 1982 by D. Reidel Publishing Company.

Some of the spectra, taken from the central portion of the map, are shown in Figures 1a and 1b. The scale of antenna temperature has been adjusted to emphasize the complex nature of the profiles. What is particularly striking is that out of the entire map (including over 275 spectra) the (0,0) position is the only reasonably symmetric gaussian line! Some of the other features of the data are briefly noted below; some of these have been observed in studies of other transitions of other molecules (cf. Loren, 1979; Schloerb et al., 1982).

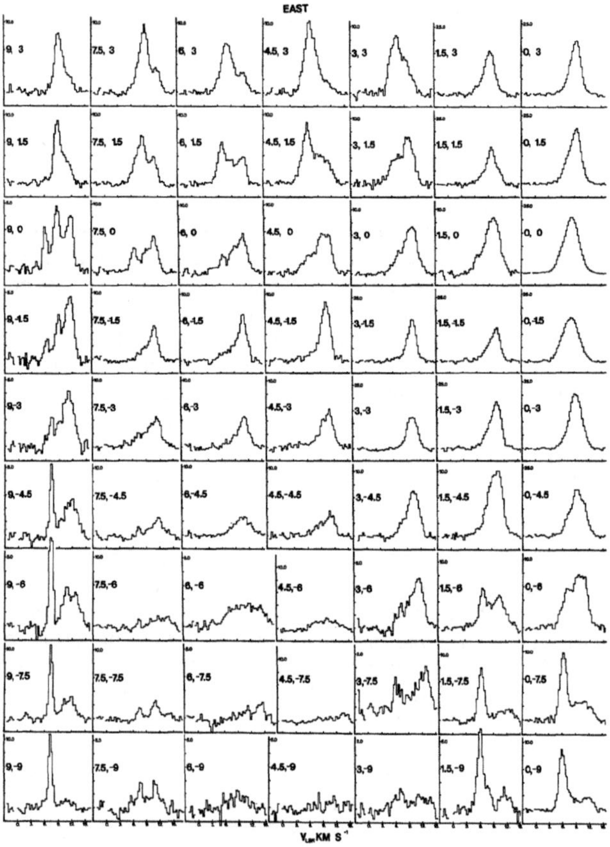

Figure 1a

Figures 1a, 1b. East-central and West-central portions, respectively of ^{13}CO J=2→1 map of the Orion molecular cloud. The offsets in right ascension and declination from the center of the map (KL) in minutes of arc are given in the upper left of each box. Note that the scale for the temperature of each spectrum varies throughout the map, and that the antenna temperatures are not corrected for coupling efficiency.

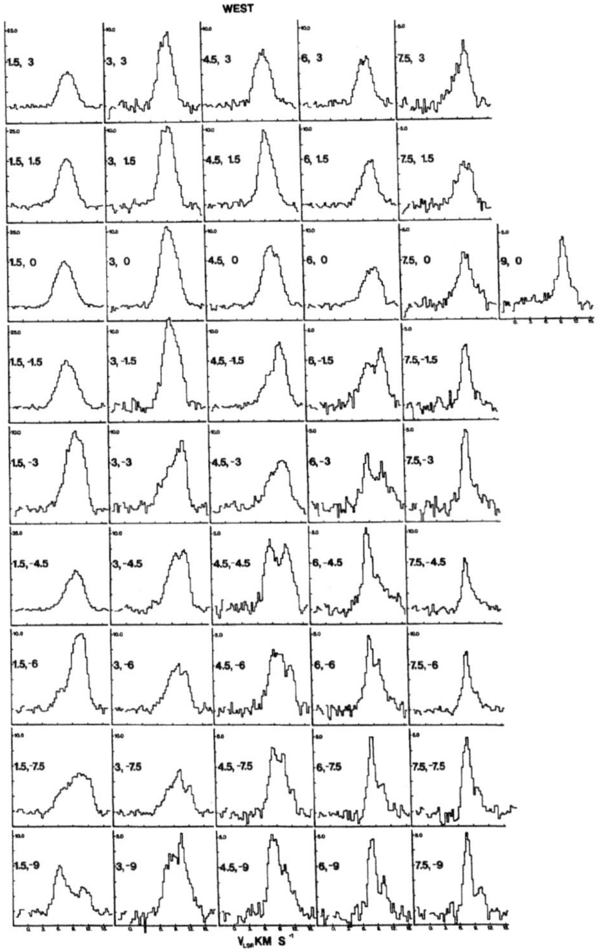

Figure 1b

Although most evident in the portions of the map to the North and South of the region covered by Figure 1, there is a general N-S rotation of the region. More precisely, the area with generally the highest velocities (~12 km s^{-1}) occurs in the NE corner of the complete map, and the lowest velocity region occurs in the SW. The velocity structure in the central region of the map appears to be dominated by perturbations due to the Trapezium stars and the HII region M42, located in front of the molecular cloud. In particular, this has resulted in red shifted (~12 km s^{-1}) material being present in an arcuate region extending from the NE clockwise to the W of the

Trapezium, having a radius of ~9' (1.3 pc). In several positions (particularly 9E, ON and 9E, 1.5S) triply-peaked (at ~6 km s^{-1}, 9 km s^{-1}, and 12 km s^{-1}) profiles are clearly seen, indicating that the molecular gas has been accelerated towards as well as away from us along these lines of sight. Strongly blueshifted (6 km s^{-1}) material is predominant around the OE, 7.5S position. Our data are generally consistent with the "eroding stellar bubble" model proposed by Balick, Gammon, and Hjellming (1974), Jaffe and Pankonin (1978), Pankonin, Walmsley, and Harwit (1979), and others. The linewidths of the individual velocity components in these regions are quite narrow, in some cases less than 1.5 km s^{-1} FWHM.

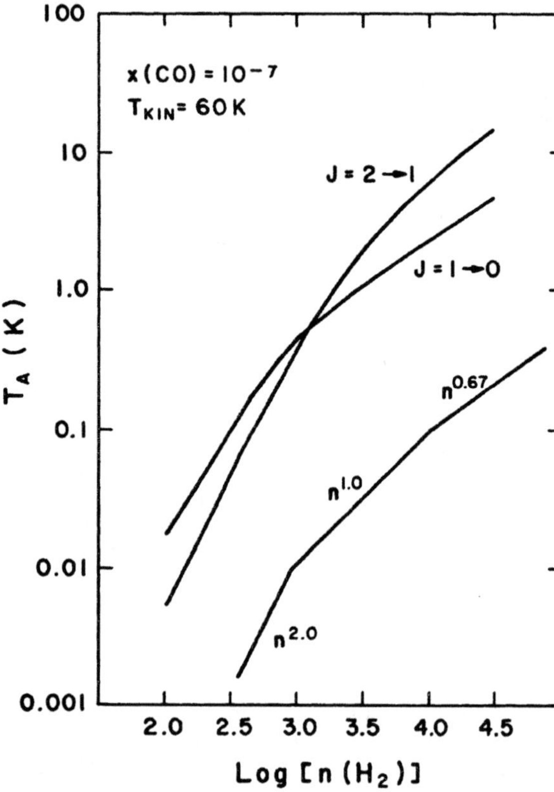

Figure 2. Antenna temperature for the two lowest transitions of CO as a function of hydrogen density, for a fixed fractional abundance typical of that found for ^{13}CO. The power-law dependence indicated in the lower right illustrates the greater sensitivity of the J=2→1 transition to changes in density.

The small spatial extent and sharp boundaries of many of the kinematical features seen is intriguing. Considering, for example, the change in the profile between the 9E, 6S and the 7.5E, 6S positions, for example, one is struck by the rapid change in the character of the emission. Reconciling this with the very narrow linewidths observed remains an important challenge.

It appears that the ^{13}CO $J=2\rightarrow1$ transition should be a relatively sensitive probe of changes in conditions. In Figure 2, we compare emission from the $J=2\rightarrow1$ CO line with that from $J=1\rightarrow0$, for a fixed fractional abundance of carbon monoxide characteristic of ^{13}CO. The dependence of T_A upon hydrogen density is noticeably steeper over the range $10^2 < n(H_2) < 10^4$ cm^{-3} for the higher transition, indicating that we will see enhanced variations from inhomogeneous regions. The ratio of the $J=2\rightarrow1$ to $J=1\rightarrow0$ intensity can, of course, be used to determine the hydrogen density. At the central position of the cloud, we determine $n(H_2) = 5000$ cm^{-3}, using the $J=1\rightarrow0$ data of Schloerb, Goldsmith, and Scoville (1981) with degraded spatial resolution. This is in good agreement with the result of 4000 cm^{-3} obtained by Plambeck and Williams (1979). While it is not clear to what region along the line of sight this density applies, analysis of density variations throughout the cloud should be a valuable tool in unravelling the structure of this complex region.

We wish to thank Neal Erickson for providing the receiver used in this work, the MWO for the use of their facilities, Dan Clemens for assistance in the observations, and Paul Duffey for help with the data reduction. Astronomy research at the Five College Radio Astronomy Observatory is supported by NSF grant AST 80-26702. This is contribution 486 of the Five College Astronomy Department.

REFERENCES

Balick, B., Gammon, R.H. and Hjellming, R.M. 1974, Publ. Astr. Soc. Pacific, 86, 616.

Erickson, N.R. 1981, IEEE Trans. Microwave Theory Tech., MTT-29, 557.

Jaffe, D.T. and Pankonin, V. 1978, Ap. J. 226, 869.

Loren, R.B. 1979, Ap. J. 234, L207.

Pankonin, V., Walmsley, C.M. and Harwit, M. 1979, Astr. Astrophysics 75, 34.

Plambeck, R.L. and Williams, D.R.W. 1979, Ap. J. 227, L43.

Schloerb, F.P., Goldsmith, P.F. and Scoville, N.Z. 1982, this volume, p. 439.

MOLECULAR HYDROGEN EMISSION FROM BROAD WING CLOUD CORES

Adair P. Lane
Department of Physics and Astronomy
University of Massachusetts, Amherst, Mass.

John Bally
Bell Telephone Laboratories
Holmdel, N.J.

ABSTRACT

We report observations of 2μ line emission from vibrationally excited H_2 in the vicinity of the high velocity molecular flows associated with Cepheus A, NGC 2071, and GL 961. The luminosity and spatial extent of the H_2 emission in these regions are compared with those of other known H_2 sources associated with high velocity molecular cloud cores.

INTRODUCTION

Broad CO line wings have now been observed in numerous nearby molecular cloud cores exhibiting recent or ongoing star formation (e.g. Scoville 1980, Rodriguez et al. 1980, Blitz and Thaddeus 1980, Lada and Harvey 1981, Bally 1981,1982). For most sources exhibiting line wings with Δv (full width at T_A = 0.1 K) > 30 km s^{-1}, the high velocity gas is best interpreted as an outflow of matter from the vicinity of a central infrared source, since the amount of material in the molecular cloud core appears to be too small to gravitationally bind these velocities. Shocks formed either within the flow region or at the interface between the outflow and the ambient molecular cloud may provide the energy (E/k > 6000 K) for the excitation of the 2μ lines of H_2. Detection of the shock-excited H_2 gas at the outer boundary of the high velocity flow can help to verify the outflow model for the gas dynamics and provide a tool with which to study the nature of energetic molecular flows in star forming regions.

We have searched for H_2 emission from broad wing molecular cloud cores with high sensitivity and low spectral resolution in order to facilitate detection of weak lines from the extended regions where high velocity CO emission is seen. In this paper, we present 2μ observations of four sources: Cepheus A, NGC 2071, GL 961, and GL 490. (A more detailed discussion of these observations is given in a paper to appear in the Astrophysical Journal.)

R. S. Roger and P. E. Dewdney (eds.), Regions of Recent Star Formation, 301–306.
Copyright © 1982 by D. Reidel Publishing Company.

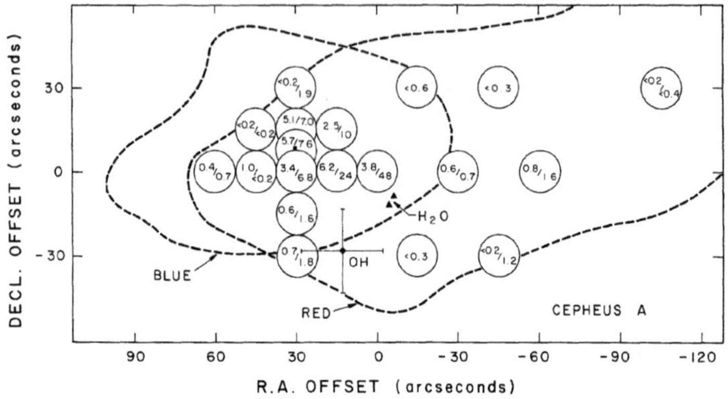

Fig. 1 - Map of the H_2 v=1-0 S(1) line emission toward Cepheus A. Flux values in the S(1) line and adjacent continuum are given at each position in units of 10^{-20} W cm^{-2}. Dashed contours outline regions of blue and red high velocity CO emission (Rodriguez et al. 1980). Coordinates are relative to the 2.2μ peak (α_{1950} = 22h54m19.9s, δ_{1950} = 61°45'56").

OBSERVATIONS AND RESULTS

Our observations were obtained with the Kitt Peak National Observatory 2.1m telescope in December 1980. Molecular hydrogen emission was searched for in the v=1-0 S(1) line at 2.12μ using the $\Delta\lambda/\lambda$=1.3% resolution circular variable filter (CVF) and a 15" diameter circular beam. At each position on the sky, the flux at three wavelengths was measured: first at the wavelength of the S(1) line and then longward and shortward by 0.03μ. In order to verify the detections, observations of the Q-branch emission near 2.4μ were also obtained using the CVF. The rms noise level typically obtained for each spectral point was 0.3 x 10^{-20} W cm^{-2}.

Detections of H_2 emission were made in three of the four sources we searched: Cepheus A, NGC 2071, and GL 961. The H_2 emission regions in these sources are extended on a scale of at least 0.2 to 0.3 pc, comparable to or somewhat greater than the size of the Orion H_2 source. Figures 1 - 3 present maps of the emission regions in Cep A, NGC 2071, and GL 961, with the flux in the H_2 S(1) line and in the adjacent continuum indicated at each observed position in units of 10^{-20} W cm^{-2}. Upper limits represent the rms noise level for the observation. The dashed contours on the maps outline the spatially separated emission regions in the blue and red high velocity wings of the ^{12}CO line. Fig. 4 shows CVF spectra from 2.1-2.5μ toward Cepheus A and NGC 2071 at positions marked in Figs. 1 and 2 with a solid square. The source GL 490 was also mapped at 23 positions over a region subtending several arcminutes and does not exhibit S(1) line emission above a flux level of 1 x 10^{-20} W cm^{-2}.

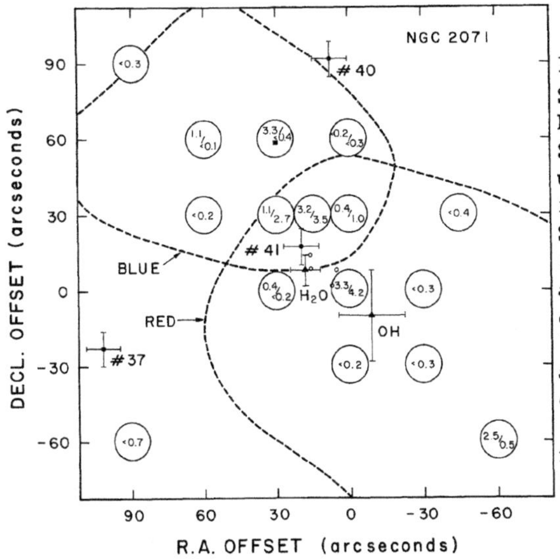

Fig. 2 - Map of the H_2 v=1-0 S(1) line emission toward NGC 2071. Flux values in the S(1) line and adjacent continuum are given at each position in units of 10^{-20} W cm^{-2}. Small open circles are 10μ sources detected by Persson et al. (1981). High velocity CO contours are from Bally (1981, 1982). Triangles indicate H_2O and OH maser positions. Black dots mark embedded 2μ sources (Strom et al. 1976). Coordinates are relative to α_{1950} = 5h44m30.1s, δ_{1950} = 0°20'40".

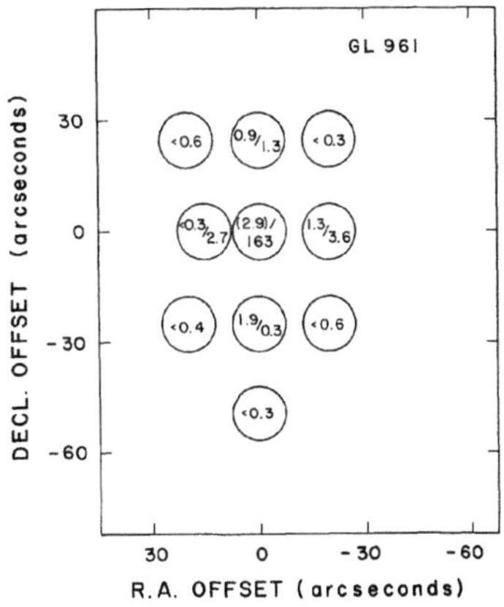

Fig. 3 - H_2 emission toward GL 961. Flux values in the v=1-0 S(1) line and adjacent continuum are given at each position in units of 10^{-20} W cm^{-2}. Coordinates are relative to the 2.2μ peak (α_{1950} = 6h31m58.9s, δ_{1950} = 4°15'07").

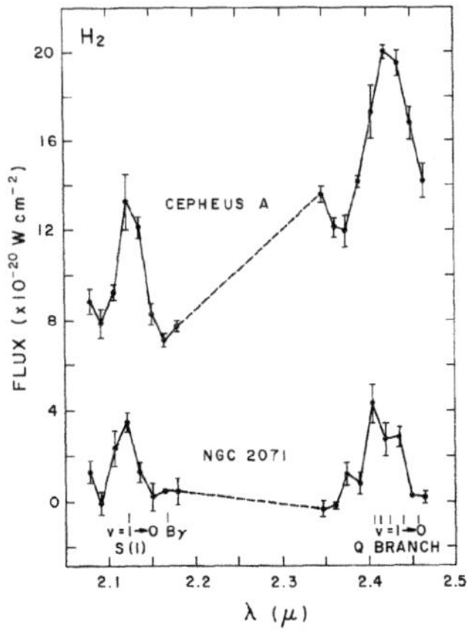

Fig. 4 - CVF spectra of H_2 emission in the v=1-0 S(1) line and (unresolved) Q-branch for Cepheus A and NGC 2071. Spectra were obtained at positions marked with a solid square in Figs. 1 and 2. Error bars are 1σ.

DISCUSSION

In Table 1, we list all currently known molecular cloud cores that exhibit both high velocity CO line wings and H_2 S(1) line emission. Also listed is our upper limit for the S(1) line emission for the broad wing source GL 490. It appears that conditions suitable for the excitation of the 2μ transitions of H_2 are commonly found in broad wing molecular cloud cores. Models of shock-heated molecular gas (Kwan 1977, Hollenbach and Shull 1977) indicate that shock velocities between 10 and 24 km s^{-1} and cloud densities $\geq 10^5$ cm^{-3} can account for the observed H_2 line intensities in Orion. In GL 490, either the excitation conditions are not satisfied or the extinction in front of the emission region is too large to allow detection of the S(1) line.

In most sources, the H_2 S(1) line is observable from an extended region coincident with the region where broad CO line wings are observed. In NGC 2071 and Cep A, where the high velocity CO emission is formed in a very extended region (3'x 5' and 1.5'x 5'), the S(1) source appears smaller by perhaps a factor of 2. In Orion, the reverse is true: the S(1) source is more extended than the CO plateau source and in fact peaks just outside its boundary (Beckwith et al. 1978, Solomon

TABLE 1.

Broad Wing Sources with Molecular Hydrogen Emission

Source	Distance (kpc)	$^{12}CO(J=1-0)$ full-width (km s^{-1})	CO ref.	Observed S(1) line luminosity[a] ($L_\odot$)	H_2 spatial extent (pc)	H_2 ref.
Orion	0.5	100	17,18	2.5	0.15	11,5
DR 21	3.0	50	3,7	2.2	≥1.0	8
NGC7538	3.5	35	3	1.2	≥1.0	9
NGC6334	1.7	55	7	0.3	≥0.5	7
Cep A	0.73	50	16	0.13	≥0.3	4,this paper
GL 961	1.6	40	6	0.09	≥0.3	4,this paper
NGC2071	0.5	70	1,2	0.05	≥0.2	4,this paper
S140	0.9	45	3	0.02	≥0.4	10
GL490	0.9	65	14	<0.01	-	4,this paper

Note to Table:
[a] Uncorrected for possible extinction.

et al. 1981). The spatial distribution of the high velocity CO gas in NGC 2071 and Cep A exhibits bipolar symmetry, with the blueshifted and redshifted emission centroids located on opposite sides of the central IR source. We find that in these sources, there is a tendency for the strongest H_2 emission to occur in regions where blueshifted CO wings are the most prominent component. Emission at the positions of redshifted CO wings is weaker or absent. This property of the H_2 emission is consistent with a model where the S(1) line is excited in shocks formed as a result of outflow from the central source. In such a model, H_2 line emission excited by shocks associated with blueshifted gas originates closer to the observer and suffers less extinction than emission associated with redshifted outflow. Although the Q-branch lines at 2.4μ were observed with the CVF (see Fig. 4), the blending of the lines and the effects of telluric absorption prevent determination of the extinction to the source from our data. The S(1) line luminosities listed in Table 1 are uncorrected for extinction and assume that the S(1) line emission region is of the size given in the following column (see Bally and Lane 1981). It is clear, however, that even for large extinctions ($A_V \sim 30$ magnitudes), the corrected S(1) luminosities for most sources are ~1 $L_\odot$, implying total H_2 luminosities roughly an order of magnitude larger.

The mechanical luminosity of the CO flow is sufficiently large in all of these sources to explain the observed H_2 emission by dissipation of energy in a shock. With typical parameters for the high velocity outflow ($M \sim 10\ M_\odot$, $v \sim 25$ km s^{-1}, $\tau \sim 10^4$ yrs), the mechanical luminosity available for excitation of H_2 emission is $L_M \sim MV^2/2\tau \sim 10$ to $100\ L_\odot$.

Comparison of the H_2 parameters of the sources in Table 1 with various other source parameters leads to a few interesting correlations. The total central source luminosity, as determined in the far infrared, is weakly correlated with the (unreddened) H_2 S(1) line luminosity. The sources DR21, Orion, and NGC 7538 have luminosities around $10^5\ L_\odot$ and S(1) line luminosities around 2 $L_\odot$ (30 $L_\odot$ for $A_V = 30$ mag). Both the S(1) luminosities and the total far infrared luminosities are one to two orders of magnitude lower in Cep A, NGC 2071, and GL 961. Orion also exhibits the largest CO velocity and is the most compact ($r \sim 0.1$ pc) and youngest ($\tau \sim 10^3$ yrs) source. In contrast, the flows associated with Cep A, NGC 2071, and GL 961 are more extended ($r \sim 0.5$ to 1 pc) and have lower CO velocities ($v \sim 30$ km s^{-1}) and longer dynamical ages ($\tau \sim 10^4$ yrs).

CONCLUSIONS

Emission in the 2μ lines of vibrationally excited H_2 is commonly found in association with broad wing CO sources in molecular cloud cores. In the sources NGC 2071 and Cepheus A, where the CO flows show bipolar symmetry, the H_2 emission is found to be preferentially associated with the blueshifted portion of the CO flow. The S(1) line luminosity is weakly correlated with total central source luminosity and anticorrelated with the source dynamical age. The mechanical luminosity of the molecular outflows is sufficient to power the observed H_2 line emission. Further observations of the H_2 lines at higher spectral and spatial resolution are required to better understand the foreground extinction to the emission regions and the relationship between the CO and H_2 morphology.

REFERENCES

Bally, J. 1981, submitted to Ap. J.
Bally, J. 1982, this volume, p. 287.
Bally, J. and Lada, C.J. 1981, in preparation.
Bally, J. and Lane, A.P. 1981, submitted to Ap. J.
Beckwith, S., Persson, S.E., Neugebauer, G. and Becklin, E.E. 1978 Ap. J., 223, 464.
Blitz, L. and Thaddeus, P. 1980, Ap. J., 241, 676.
Fischer, J. 1981, Ph.D. Thesis, State University of New York at Stony Brook.
Fischer, J., Righini-Cohen, G. and Simon, M. 1980, Ap.J. (Letters) 238, L155.
Fischer, J., Righini-Cohen, G., Simon, M., Joyce, R.R. and Simon, T. 1980, Ap. J. (Letters) 240, L95.
Gautier, T.N. III 1980, B.A.A.S. 12, 439.
Gautier, T.N. III, Fink, U., Treffers, R.P. and Larson, H.P. 1976, Ap. J. (Letters) 207, L129.
Hollenbach, D.J. and Shull, J.M. 1977, Ap. J. 216, 419.
Kwan, J. 1977, Ap. J. 216, 713.
Lada, C.J. and Harvey, P.M. 1981, Ap. J. 245, 58.
Persson, S.E., Geballe, T.R., Simon, T., Lonsdale, C.J. and Baas, F. 1981, preprint.
Rodriguez, L.F., Ho, P.T.P. and Moran, J.M. 1980, Ap.J. (Letters), 240, L149.
Scoville, N.Z. 1980, IAU Symp. #87, "Interstellar Molecules", ed. B. H. Andrew, D. Reidel Publ. Co., Holland, p. 33.
Solomon, P.M., Huguenin, G.R. and Scoville, N.Z. 1981, Ap. J. (Letters), 245, L19.
Strom, K.M., Strom, S.E. and Vrba, F.J. 1976, A.J., 81, 308.

ASYMMETRIC BROAD HCO^+ LINE WINGS IN CORES OF MOLECULAR CLOUDS

Aa. Sandqvist
Stockholm Observatory, Sweden
A. Wootten
Owens Valley Radio Observatory, Pasadena, California, USA
R.B. Loren
Millimeter Wave Observatory, The University of Texas at Austin, and McDonald Observatory, Fort Davis, Texas, USA
P. Friberg and Å. Hjalmarson
Onsala Space Observatory, Sweden

ABSTRACT

High resolution observations of HCO^+ in the cores of the W3, NGC 2071 and Cep MC-1 molecular clouds reveal asymmetric line wings extending over velocity ranges of 20, 35 and 55 km s^{-1}, respectively. Red and blue wings of the profiles are enhanced on opposite sides of infrared objects embedded in the cores of NGC 2071 and Cep MC-1. The lines attain their maximum breadths at the positions of these infrared objects. These facts imply that the infrared objects are the energy sources for the high velocity gas. Rotation cannot account for all observed features of the profiles. A model incorporating outflow centered on the infrared objects is suggested. W3, on the other hand, is more complex and evidence is presented which shows the presence of two molecular clouds in the core, one centered near IRS 5 and the other near IRS 4, the latter exhibiting a blue-shifted wing.

INTRODUCTION

Broad line wings in a double lobe pattern centered on an embedded star have recently been observed in CO profiles toward L1551 (Snell, Loren and Plambeck 1980), Cep MC-1 (Rodriguez, Ho and Moran 1980), GL 490 (Lada and Harvey 1981), NGC 2071 (White and Phillips 1981) and W3 (Brackmann and Scoville 1980). In two of these sources Rodriguez, Ho and Moran and Lada and Harvey proposed that the reason for the broadening was an energetic outflow from an embedded infrared source. In NGC 2071 and W3, White and Phillips and Brackmann and Scoville presented a model in which rotation about the central objects was the source of the line broadening.

HCO^+ also shows broad wings at the positions of infrared sources in W3, NGC 2071 and Cep MC-1 (Loren and Wootten 1980). In this paper we present detailed high resolution maps of the HCO^+ profiles in the cores of these three molecular clouds. A model

R. S. Roger and P. E. Dewdney (eds.), Regions of Recent Star Formation, 307–314.
Copyright © 1982 by D. Reidel Publishing Company.

incorporating outflow from the embedded objects in NGC 2071 and Cep MC-1 is outlined which can reproduce the general features of the observed profiles. In W3, the picture is more complex and strong evidence for the presence of two clouds in the core will be presented. One cloud is centered near IRS 5, the other near IRS 4 (IRS 4 and IRS 5 are compact infrared sources observed at 20 μm by Wynn-Williams, Becklin and Neugebauer 1972.)

OBSERVATIONS AND RESULTS

The observations of the 89.188523 GHz HCO^+ and the 86.754330 GHz $H^{13}CO^+$ lines were made with the Onsala (OSO) 20m and the NRAO 11m telescopes during March and June 1980 and May 1981. The resolution of the OSO telescope is 42", the main beam efficiency is 54% and the OSO data is presented here in terms of T_A^*. The resolution of the NRAO telescope is 76" while the forward beam efficiency is 72%. The NRAO data is presented here in terms of T_R^*. Line profiles were obtained with velocity resolutions between 0.2 and 3.4 km s^{-1}. The system temperature at OSO was about 350 K, at NRAO it was about 450 K. Some of the results are presented in figures 2, 4 and 6, which are velocity-right ascension HCO^+ temperature contour maps for W3, NGC 2071 and Cep MC-1. Figure 1(a) shows an HCO^+ profile near IRS 4 in W3. Self-absorption is evidently a prominent feature of the HCO^+ profile in all three sources. To determine the clouds' center of rest velocities, $H^{13}CO^+$ profiles were also obtained at several positions and are shown in figures 1(b), 1(c), 3 and 5 superimposed upon HCO^+ profiles.

DISCUSSION

The total velocity extent of HCO^+ emission in NGC 2071 is ~35 km s^{-1} and in Cep MC-1 is ~55 km s^{-1}. It is greatest near the position of the infrared sources. In W3 the total velocity extent is only ~20 km s^{-1} at any particular position and shows no tendency to peak at the IRS 5 position. In NGC 2071 and Cep MC-1 the region of broad wings is ~110" and 90", after deconvolution of the beam. At distances of 500 and 725 pc, this extent becomes .26 pc and .31 pc, respectively. Both red and blue wings are extended, and the ratio of integrated intensity in the wings changes by a factor of 15 in NGC 2071 from SW of the IR source, where the red wing dominates, to the NE, where the blue wing dominates. This change is a factor of 6 in Cep MC-1 from the SW, where the red wing dominates, to the NE, where the blue wing dominates. This behavior is not apparent in W3, perhaps as a consequence of its distance (.3 pc subtends ~30" at its ~2.4 kpc distance), or possibly as a consequence of its complex structure, for which we will suggest a new model. In both Cep MC-1 and NGC 2071, the deepest HCO^+ self-absorption lies at more positive velocities than the peak of $H^{13}CO^+$ emission, which is consistent with infall of the

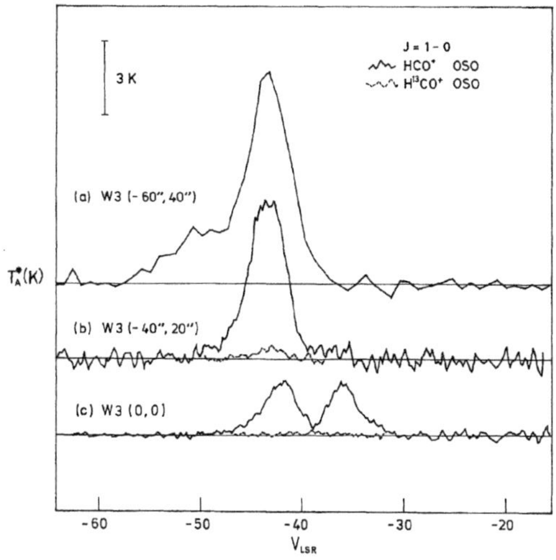

Fig. 1.
(a) HCO^+ emission near IRS 4 in W3, velocity resolution of 0.84 km s^{-1} .

(b) HCO^+ and $H^{13}CO^+$ emission near IRS 4 in W3, velocity resolution of 0.2 km s^{-1}.

(c) HCO^+ and $H^{13}CO^+$ emission near IRS 5 in W3, velocity resolution of 0.2 km s^{-1}.

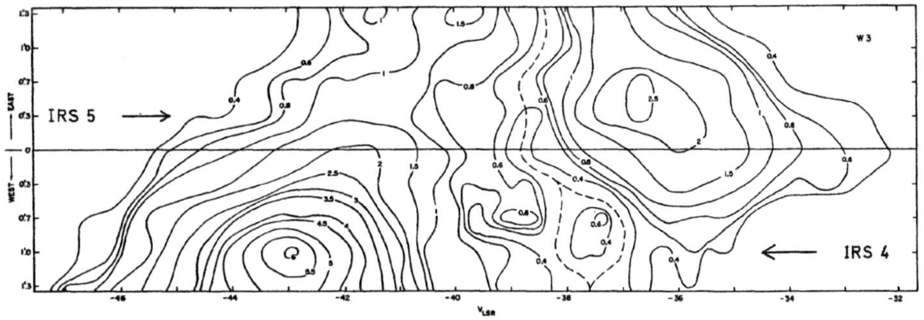

Fig. 2. Map of T_A^*(K) J = 1-0, HCO^+ emission in W3 ($02^h21^m51^s.0$, $+61^\circ 52'18''$) (1950.0), made at OSO with 20" spacing and 0.20 km s^{-1} velocity resolution. The right ascensions of the compact infrared sources IRS 4 and IRS 5 are indicated.

absorbing material. This is less apparent for the W3 profiles, which we attribute to its more complex structure.

W3 differs in many respects from NGC 2071 and Cep MC-1, and the high resolution (20" spacing) OSO observations suggest a new model for the molecular emission there. One difference, already mentioned, occurs in line width behavior. There appears to be no maximum near IRS 5, but near IRS 4 an extensive low velocity wing is present to ~ -60 km s^{-1}, blue-shifted from the general cloud emission at ~ -43 km s^{-1} (see figure 1(a)). Another difference appears in the $H^{13}CO^+$ data. Near IRS 4 this line reaches an intensity of ~ 0.4 (.06 rms per .2 km s^{-1} channel) K at a velocity of $V_{LSR} = -43$ km s^{-1} ; near IRS 5 48 hours of integration revealed a weak line of ~ 0.1 (.04 rms per .2 km s^{-1} channel) K at V_{LSR} = −39 km s^{-1} (see figure 1(c)). Observations of the J = 3-2 line of HCO^+ taken at MWO (in preparation) show a) lack of prominent self-absorption, b) a primary intensity maximum T_{peak} at V_{LSR} ~−43 km s^{-1} near IRS 4, c) a secondary maximum T_{peak} at V_{LSR} ~−39 km s^{-1} near IRS 5. Similar behavior occurs in the J = 2-1 line of ^{13}CO observed at OVRO with 26" resolution (Lichten, private communication). The behavior of T_{peak} (HCO^+, J = 1-0) is somewhat different. A primary maximum T_{peak} occurs at V = −43 km s^{-1} near IRS 4, but the secondary maximum T_{peak} occurs at V = −36 km s^{-1} near IRS 5 (see figures 1 and 2). These observations suggest that two molecular clouds are present in the core of W3 at velocities of V = −39 and V = −43 km s^{-1} associated with IRS 5 and IRS 4, respectively. The shift in the peak velocity of the HCO^+ (J = 1-0) emission with respect to −39 km s^{-1} is attributed to absorption by molecules within the cloud associated with IRS 5. The blue wing in HCO^+ profiles near IRS 4 has been noted above; the presence of a possible red wing analogous to those in NGC 2071 and Cep MC-1 is confused by emission from the IRS 5 cloud.

As mentioned earlier, the velocity structure of the HCO^+ (J = 1-0) profiles in NGC 2071 is centered on the infrared source which may provide the energy source for the mass motion. Observations of the optically thin ^{13}CO line by White and Phillips (1981) have shown the presence of a velocity gradient running from the NW to the SE with a magnitude of about 5 km s^{-1} over 9 pc, implying a rotation of the molecular envelope. The J = 3-2 line of HCO^+ has been mapped (in preparation) along an E-W strip through the core. The peak intensity shows a velocity gradient of 1 km s^{-1} over the 0.3 pc extent of this map with the same sense as the ^{13}CO gradient. The core then rotates in the same direction as the envelope. In this case, the line wings of the J=1-0 line cannot arise from rotation but are better explained as a flow of gas along the axis of rotation. Keplerian rotation and infall are ruled out since the velocity extent of the wings would require the presence of a central mass of greater than 24000 $M_\odot$ within a radius of 0.15 pc. We therefore conclude that the wings originate from outflow of matter from the core, along the axis of rotation.

Since HCO^+ requires n ~ 2000 cm^{-3} for excitation of the J = 1-0 line to 0.2 K at abundances $X(HCO^+)$~10^{-9} and T_{ex}~25 K, we can estimate a minimum mass in a spherical outflow emission region of M~1.4 $M_\odot$. The breadth of the line is ignored in this calculation, however, and a column density estimate indicates the mass in the region is ~ 10 $M_\odot$. The kinetic energy in the flow is then E~3.5 x 10^{46} ergs, and the momentum required is much greater than L/c if L~1000 $L_\odot$ (Sargent et al.1981). These parameters are very similar to those noted by Lada and Harvey (1981) for the high velocity CO flow in GL 490. The wings here arise in bipolar outflow along the rotational axis of the cloud, and not from rotation as suggested by White and Phillips (1981). The absorption in the HCO^+ profile, redshifted with respect to the $H^{13}CO^+$ peak, occurs as a result of material falling inward in the plane of rotation, while the

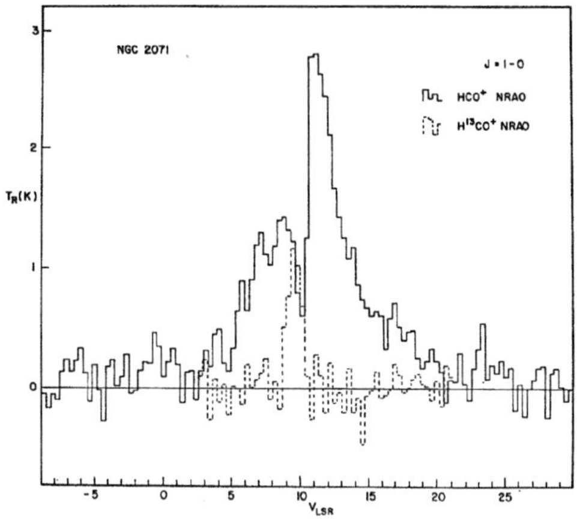

Fig. 3. HCO^+ and $H^{13}CO^+$ emission in NGC 2071, velocity resolution of 0.34 km s^{-1} at 89 GHz.

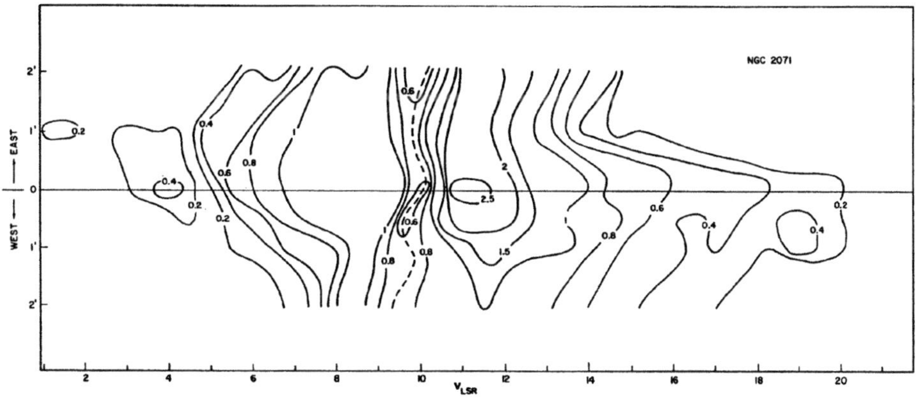

Fig. 4. Map of T_R^* (K) J = 1-0, HCO^+ emission in NGC 2071 ($05^h44^m30^s.0$, $+00^\circ20'17''$) (1950.0), made at NRAO with 60" spacing and 0.84 km s^{-1} velocity resolution.

more widespread absorption affecting the blue emission peak of the profile results from a dense foreground clump of cold gas asymmetrically located with respect to the IR source.

The geometry is surprisingly similar in Cep MC-1. The maximum HCO^+ linewidth occurs at the IR source, and, as Rodriguez, Ho and Moran (1980) have shown, the similar velocity extent of the CO emission requires the matter to be flowing outward. Our $H^{13}CO^+$ and J = 3-2 HCO^+ maps show only a small and marginally significant velocity shift redward toward the N and W, but this is consistent with a model such as that proposed for NGC 2071, with outflow along the rotational axis. From the minimum density required to excite the observed HCO^+ emission we estimate the mass in the outflow as at least 1.4 $M_\odot$. From column density measure-

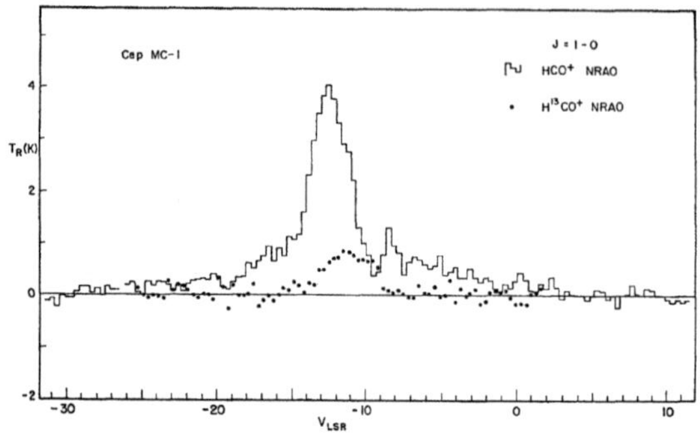

Fig. 5. HCO^+ and $H^{13}CO^+$ emission in Cep MC-1, velocity resolution of 0.34 km s^{-1} at 89 GHz.

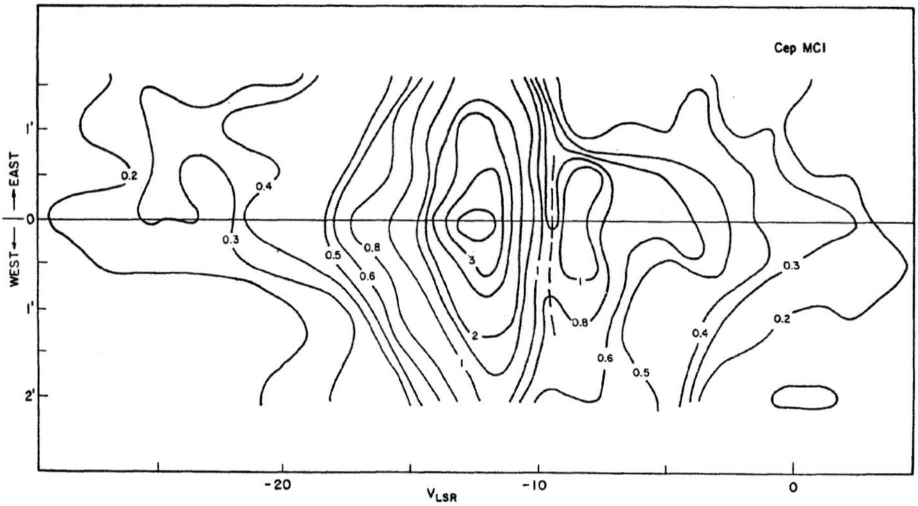

Fig. 6. Map of T_R^* (K) J = 1-0, HCO^+ emission in Cep MC-1 ($22^h54^m24^s.4$, $+61^\circ45'43''$) (1950.0), made at NRAO with 30" spacing and 0.84 km s^{-1} velocity resolution.

ments assuming $T_{ex}(HCO^+) = 25$ K and $X(HCO^+) = 10^{-9}$ we estimate a mass of ~20 $M_\odot$ in the outflowing material. The momentum required to drive this steady flow is much greater than L/c if L ~2.5 x 10^4 $L_\odot$ (Koppenaal et al. 1979). These parameters agree quite well with those derived by Rodriguez, Ho and Moran for the CO flow about Cep MC-1, as well as those for the CO flow in GL 490 observed by Lada and Harvey (1981) and the HCO^+ flow described in NGC 2071 in the previous paragraphs.

CONCLUSIONS

We have detected analogs in HCO^+ emission to the Cep MC-1 high velocity CO outflow reported by Rodriguez, Ho and Moran. A region in the core of the NGC 2071 cloud exhibits a similar outflow, though at somewhat lower velocities. We suggest a model for NGC 2071 in which the outflow is directed along the rotational axis of the cloud, and in which infall occurs in the rotational plane. Although rotation is less evident in the Cep MC-1 cloud, a similar model is suggestive also in that case.

New high resolution observations of W3 suggest that the observed HCO^+ profile (and by inference that of CO) is the result of superposition of emission from two dense molecular cloud components, one associated with IRS 4 at $V \sim -43$ km s^{-1} and one associated with IRS 5 at $V = -39$ km s^{-1}. Near IRS 4, the HCO^+ profiles in this component show a blue wing extending to $V \sim -60$ km s^{-1}.

ACKNOWLEDGEMENTS

The OSO is operated by Chalmers University of Technology with financial support from the Swedish Natural Sciences Research Council and the Swedish Board for Technical Development. The NRAO is operated by Associated Universities Inc., under contract with the U.S. National Science Foundation. This research has been supported by the Swedish Natural Science Research Council and the U.S. National Science Foundation (Grant AST 79-16815 to OVRO and grant AST 79-20966 to MWO).

REFERENCES

Brackmann, E., Scoville, N.Z.: 1980, Astrophys. J. 242, 112-120

Koppenaal, K., Sargent, A.I., Nordh, L., van Duinen, R.J., Aalders, J.W.G.: 1979, Astron. Astrophys. 75, L1-L3

Lada, C.J., Harvey, P.M.: 1981, Astrophys. J. 245, 58-65

Loren, R.B., Wootten, A.: 1980, Astrophys. J. 242, 568-575

Rodriguez, L.F., Ho, P.T.P., Moran, J.M.: 1980, Astrophys. J. Lett. 240, L149-L152

Sargent, A.I., van Duinen, R.J., Fridlund, C.V.M., Nordh, H.L., Aalders, J.W.G.: 1981, Astrophys. J. (in press)

Snell, R.L., Loren, R.B., Plambeck, R.L.: 1980, Astrophys. J. Lett. 239, L17-L22

White, G.J., Phillips, J.P.: 1981, Mon. Not. Roy. Astron. Soc. 194, 947-960

Wynn-Williams, C.G., Becklin, E.E., Neugebauer, G.: 1972, Mon. Not. Roy. Astron. Soc. 160, 1-14

DISCUSSION FOLLOWING PAPER BY SANDQVIST ET AL.

H. DICKEL: We have mapped H_2CO absorption against the continuum components of W3 with the Westerbork Synthesis Telescope. Two HII regions are near IRS 5 and two near IRS 4. We see the -39 to -40 km s^{-1} absorption against IRS 5 and both this and -43 km s^{-1} against IRS 4.

SOME OF THE PROBLEMS RAISED BY CO AND HCO^+ OBSERVATIONS IN THE RHO OPHIUCHI CLOUD

M. Pérault and E. Falgarone
Observatoire de Paris
92190 Meudon, France

The central parts of the Rho Ophiuchi dark cloud have been observed in the J=1-0 transition of HCO^+ and $H^{13}CO^+$ with the new cooled receiver which equips a 2.5m antenna at the Bordeaux Observatory. (For a detailed description of this instrument, refer to Baudry et al., 1981.)

The selected area, which extends over ~40'x40' (Fig. 1), is mainly characterized by the presence of:

i) a few compact HII regions and a diffuse background HII region which was only tentatively detected (Falgarone and Gilmore, 1981);

ii) a widespread emission of carbon and sulfur recombination lines (Falgarone et al., 1978);

iii) self-reversed CO profiles and, within localized areas, intense wings in the CO (J=2-1) profiles - at low velocity in the vicinity of the continuum source OPH 12 - at high velocity near OPH 10 (Fig. 1);

iv) several soft X-ray sources (Montmerle et al., 1981).

The purpose of these observations was threefold:

i) to find any correlation between the HCO^+ and the carbon recombination line emission;

ii) to compare the HCO^+ and the CO self-reversals;

iii) to display any peculiar HCO^+ chemistry in the vicinity of the X-ray sources and the compact HII regions.

A detailed discussion of the observations and the results will be the subject of a forthcoming paper. We limit ourselves here to a few conclusions.

I THE WEAKNESS OF THE HCO^+ LINES

The conspicuous weakness of the optically thick HCO^+ lines, compared to the CO and the $H^{13}CO^+$ lines (Fig. 2) seems to be a general rule.

i) In the sources where both the HCO^+ and CO lines have been observed with similar beamwidths (Loren and Wootten 1980) the ratios of

R. S. Roger and P. E. Dewdney (eds.), Regions of Recent Star Formation, 315–321.
Copyright © 1982 by D. Reidel Publishing Company.

the peak intensities lie between 4 and 12 with a sharp maximum near 10. This roughly constant ratio and the similarity in the shapes of the CO and HCO^+ lines strongly suggest that both lines, although emitted by quite different species, are formed in the same gas. The low abundance of HCO^+ almost exactly compensates the high value of the Einstein coefficients and the resultant optical depths for CO and HCO^+ are of the same order.

ii) In most sources where the HCO^+ and $H^{13}CO^+$ lines are available, the ratio $R=T_A^*(HCO^+)/T_A^*(H^{13}CO^+)$ is extremely low and depends on the spatial resolution of the observations. This appears clearly in maps of TMC1, TMC2 and L134N (see Guélin et al., 1981). If we refer to a sample of sources observed at Kitt Peak (HPBW~1.3') we find that R=19.5 for W3 (d=2.4 kpc), R=4.3 and 2.4 for Cep A (d=730 pc) and NGC 2071 (d=450 pc) and for ρ Oph, the closest source of this sample (d=160 pc), R~1 (Loren, private communication). These variations suggest that the HCO^+ and $H^{13}CO^+$ lines, in most cases, don't form in the same component. $H^{13}CO^+$ may arise in a clumped component and then suffer beam dilution for remote sources or large beamwidths.

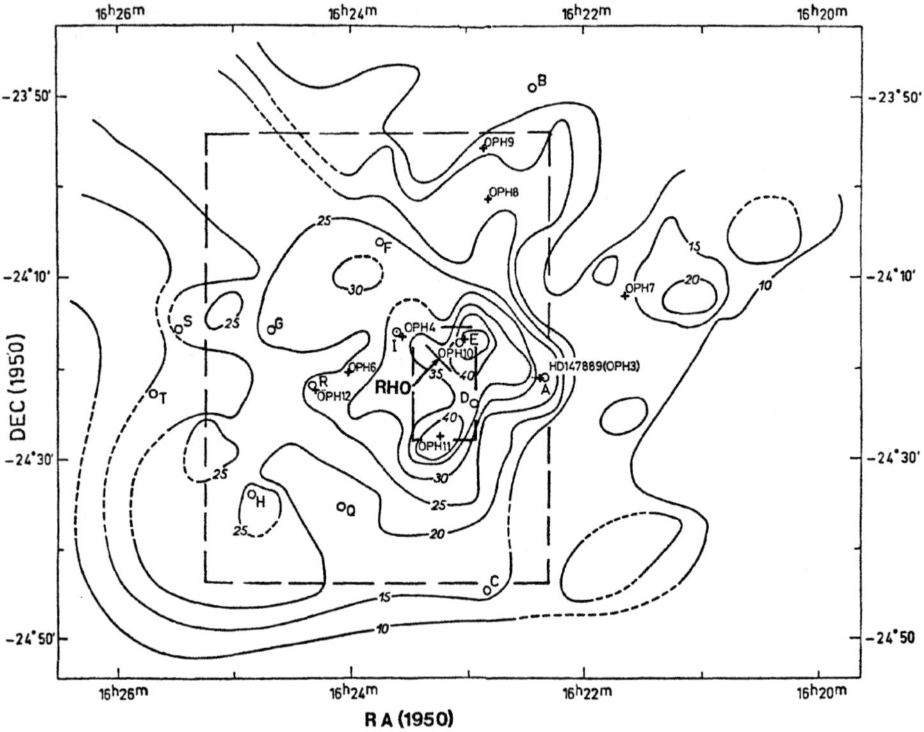

Figure 1. Contours of T_A^*(CO J=1-0). The larger frame indicates the area sampled in HCO^+, the smaller frame the extended HII region tentatively detected at 21 cm, the crosses (+) the compact HII regions, and the circles (o) the peaks of soft X-ray emission.

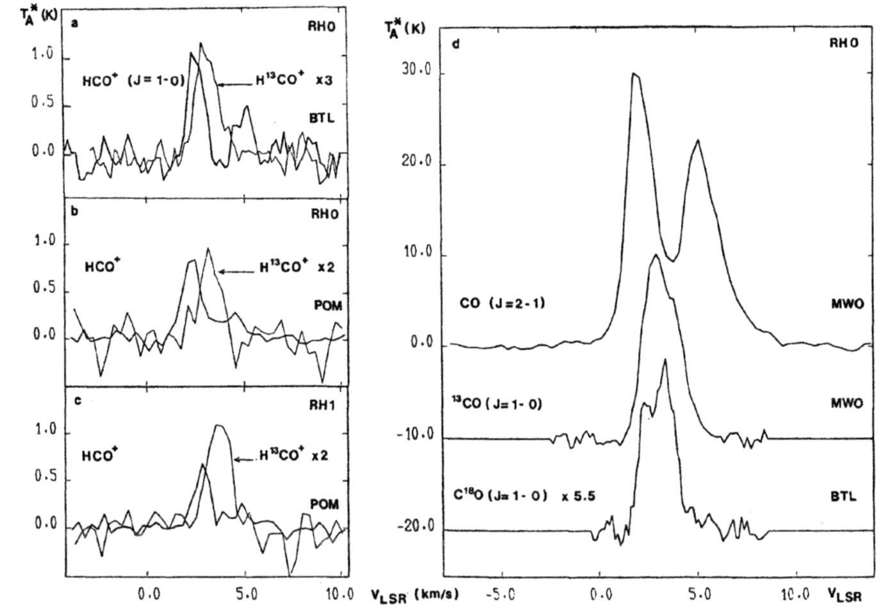

Figure 2. Lines obtained:

a) in the direction of RH 0 ($\alpha_{1950} = 16^h23^m15^s$, $\delta_{1950} = -24°19'$) with the Bell Telephone Laboratories antenna;

b) at the same point with the Bordeaux antenna;

c) in the direction of RH 1 ($\alpha_{1950} = 16^h24^m30^s$, $\delta_{1950} = -24°19'$) with the same antenna;

d) by smoothing to the angular resolution of the Bordeaux HCO^+ observations (5') the original maps obtained with the Millimeter Wave Observatory antenna (HPBW=1.4' and 2.7') and with the Bell Telephone Laboratories antenna (HPBW=1.7').

II ESTIMATION OF THE CLOUD PARAMETERS

The interpretation of the data was based on the spectra observed in the direction of RH 0 (Fig. 2) which are representative of the area where the compact HII regions are embedded. We adopted a simple treatment of line formation, with the following assumptions:

i) The adopted picture of the cloud is a superposition of homogeneous components, between which radiative coupling is neglected.

ii) For a given component the excitation of the molecules is spatially uniform and constant over the velocity profile: to determine this excitation we used classical "large (and constant) velocity gradient" calculations.

We briefly indicate below the line of argument followed.

1. The main component (the core) is responsible for the $H^{13}CO^+$

emission. We built the HCO^+ profile expected from the $H^{13}CO^+$ line, assuming $x(HCO^+)/x(H^{13}CO^+)=70$, $T_K > 10K$ and a gaussian velocity profile. The resultant line peaks at $T_A^* \simeq 10K$ and, even with a high H_2 density ($\simeq 2\ 10^5$ cm^{-3}), the expected HCO^+ brightness temperature at v=2.2 km s^{-1} is much higher than 1K (observed peak value).

2. This means that the core emission is absorbed at this velocity. There is a low excitation layer responsible for the deep reversals of the observed CO and HCO^+ lines (Fig. 2), but this layer cannot account for the HCO^+ absorption at v=2.2 km s^{-1}. If this were the case, the observed $T_A^*(CO) \simeq 30K$ at the same velocity, would imply $T_k(\text{core}) > 100K$ for reasonable values of $x(CO)/x(HCO^+)$ ($>10^4$). Then, the most straightforward way to reproduce the observations is to invoke an intermediate optically thick envelope, highly excited in CO, not in HCO^+.

3. Next we ask, "What is the contribution of this envelope to the observed HCO^+ emission?" Our answer is based on a comparison of the low velocity sides of the CO and HCO^+ line profiles.

In the extreme case of no contribution, three major constraints appear: i) a sharp cut-off towards the low values of the envelope velocity distribution, ii) a narrow range for the possible H_2 densities of the envelope ($<10^3$ cm^{-3}) and iii) not too high an HCO^+ core opacity.

In the opposite case, where the HCO^+ detected line forms in the totally opaque envelope, only the third constraint remains. A high core density is therefore required but it is consistent with the strong HCO^+ (J=3-2) observed line (Loren, private communication).

Table I displays a consistent set of parameters which allowed us to reproduce the observed lines. The adopted kinetic temperature for both components is T_k=40K.

The weakness of the HCO^+ line compared to $H^{13}CO^+$ might be due to a high saturation of the HCO^+ emission: i) in a cold component (T_k<5K), or ii) in a hot and clumped component ($f \simeq 0.1$). However, this interpretation conflicts with the intense HCO^+ (J=3-2) line detected 1.9'E, 3'N from RH 0 (Loren, private communication).

III MISCELLANEOUS REMARKS

1. One may be surprised by the values of $X(HCO^+)=x(HCO^+)/(dv/dr)$ which were estimated, in the direction of RH 0, for the dense core and the envelope (Table 1). Chemistry models (see, for example, Graedel, Langer and Frerking 1981) predict an increase of the HCO^+ abundance with the density, not the opposite.

The fact that $X(HCO^+)$ is lower in the core than in the envelope by a factor of 25, may simply reflect a higher velocity gradient in the core.

TABLE I

	MOLECULE	X	N/ΔV	TRANSITION	τ	T_B
CORE	HCO^+	$4\ 10^{-11}$	$2.4\ 10^{13}$	1–0	2	13
				3–2	6	6
$n_{H_2} = 2\ 10^5 cm^{-3}$	$H^{13}CO^+$	$6\ 10^{-13}$	$3.4\ 10^{11}$	1–0	0.05	0.4
$\frac{x(CO)}{x(HCO^+)}=1.5\ 10^5$	CO	$6\ 10^{-6}$	$3.6\ 10^{18}$	1–0	50	36
				2–1	180	34
$\Delta V=1.3\ km\ s^{-1}$	^{13}CO	$9\ 10^{-8}$	$5.1\ 10^{16}$	1–0	0.9	20
	$C^{18}O$	$1.2\ 10^{-8}$	$7.2\ 10^{15}$	1–0	0.1	4
INTERMEDIATE ENVELOPE	HCO^+	10^{-9}	$1.2\ 10^{13}$	1–0	8	0.8
				3–2	0.8	0.08
$n_{H_2}=4\ 10^3 cm^{-3}$	$H^{13}CO^+$	$1.4\ 10^{-11}$	$1.7\ 10^{11}$	1–0	0.15	0.01
$\frac{x(CO)}{x(HCO^+)}=2\ 10^4$	CO	$2\ 10^{-5}$	$2.4\ 10^{17}$	1–0	5	30
				2–1	16	30
$\Delta V=1.9\ km\ s^{-1}$	^{13}CO	$3\ 10^{-7}$	$3.4\ 10^{15}$	1–0	0.08	4
	$C^{18}O$	$4\ 10^{-8}$	$4.8\ 10^{14}$	1–0	0.005	0.6

$X=x/(dv/dr)$ is in $(km\ s^{-1}\ pc^{-1})^{-1}$, $N/\Delta V$ in $cm^{-2}(km\ s^{-1})^{-1}$ and T_B in K.

Since the half-power widths of the lines are comparable in both components (Table I) this would imply a clumpy structure in the core on a scale ≃ 0.02 pc which conflicts with our low-resolution observations.

The low value of $X(HCO^+)$ in the core may rather indicate a low HCO^+ abundance due to a high fractional ionization. We recall here that the central regions of this cloud emit intense carbon recombination lines. The parameters of the CII region near RH 0, inferred from the C158α and S158α lines (Falgarone et al. 1978) are $n_H > 3\ 10^4 cm^{-3}$ and $x_e \approx 10^{-4}$. Additional observations are required but we suggest that the $H^{13}CO^+$ lines arise in the partially ionized condensations (≈0.25 pc) which still surround the young compact HII regions embedded in the cloud and emit the recombination lines.

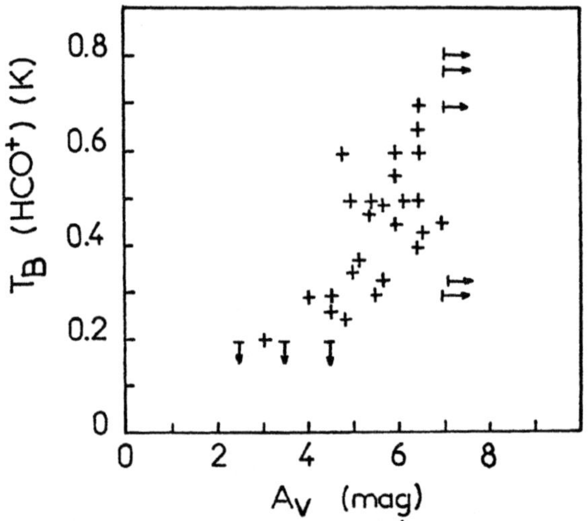

Figure 3. Plot of $T_B(HCO^+)$ vs. A_V.

2. There is some indication that the observed HCO^+ lines represent the emission of the envelope rather than what is left from the core emission. This is the possible correlation between $T_B(HCO^+)$ and A_V, the visual extinction derived from star counts (Fig. 3).

In the case where the HCO^+ lines form in a low density gas, T_B is proportional to the product of the density by the column density and the relation:

$$T_B(HCO^+)=50\ n_{H_2}\ X(HCO^+)$$

holds for $n_{H_2} X(HCO^+) < 4\ 10^{-5}$ and T_k=40K, even if $\tau(HCO^+)>1$.

Then, under the condition that the line of sight through the envelope is roughly the same for all the observed positions, $T_B(HCO^+)$ should be proportional to $A_V{}^2$:

$$T_B(HCO^+)=5\ 10^6\ X(HCO^+)\ L_{pc}{}^{-2}\ A_V{}^2$$

using $N_{Htot}=2\ 10^{21}\ A_V$ (Savage and Mathis, 1979). From our observations (Fig. 3) we find: $L \approx 0.6$ pc.

The same value for the line of sight through the envelope is derived from the data displayed in Table I. The corresponding abundance ratio for the envelope is $x(HCO^+)=3\ 10^{-9}$.

IV. CONCLUSION

We are left with the conclusion that, at least in the Rho Ophiuchi cloud, the density structure is such that the observed HCO^+ and $H^{13}CO^+$ lines cannot form in the same component. In addition, the high dipole moment molecule HCO^+ no longer appears, at low frequency, as a tracer of the high density regions in a cloud. The large optical depth of the J=1-0 transition implies that the associated emission is less weighted towards the high density components than is, for example, the J=1-0 line of ^{13}CO.

We are indebted to Dr. A.A. Penzias for kindly providing us the HCO^+ and $H^{13}CO^+$ lines displayed in Fig. 2.

REFERENCES

Baudry, A., Cernicharo, J., Pérault, M., de la Noë, J., Despois, D.: 1981, Astron. Astrophys. in press
Falgarone, E., Cesarsky, D.A., Encrenaz, P.J., Lucas, R.: 1978, Astron. Astrophys. 65, L13
Falgarone, E., Gilmore, W.: 1981, Astron. Astrophys. 95, 32
Graedel, T.E., Langer, W.D., Frerking, M.A.: 1982, to be published in Astrophys. J. Suppl.
Guélin, M., Langer, W.D., Wilson, R.W.: 1981, Astron. Astrophys. in press
Loren, R.B., Wootten, A.: 1980, Astrophys. J. 242, 568
Montmerle, T., Koch, L., Grindlay, J.: 1981, Proceedings 17th Int. Cosmic Ray Conference, Paris, Volume 1, pp 162-166
Savage, B.D. and Mathis, J.S.: 1979, Ann. Rev. Astron. Astrophys. 17, 73

CO J = 3 → 2 AND FAR INFRARED CONTINUUM OBSERVATIONS OF L1551, ORION KL AND IRC +10216.

J. P. Phillips, Glenn J. White, P. A. R. Ade and C. T. Cunningham
Physics Department, Queen Mary College, Mile End Road, London E1 4NS
E. I. Robson
Division of Physics and Astronomy, Preston Polytechnic, Corporation Street, Preston PR1 2TQ.
G. D. Watt
Department of Mathematics, UMIST, P.O.Box 88, Manchester M60 1QD.

The sources L1551, IRC +10216 and the Kleinmann-Low nebula possess microwave spectra indicative of mass-outflow at moderate to high velocities. In two of these cases the origin of outflow is probably to be located in newly formed or forming stars. For IRC +10216 however it seems clear that the progenitor of the outflowing envelope is a late-type variable star. We are therefore dealing in these three objects with a wide range of stellar evolutionary conditions, and this is probably reflected in a corresponding variety of mass-outflow mechanisms. In the following study, observations of these sources in the J = 3 → 2 transitions of CO are reported, together with photometry at wavelengths λ 377, 811 and 1136 μm. All results were acquired with the United Kingdom Infrared Telescope at Mauna Kea, Hawaii, with instrumental beam-sizes of 86 arcsecs for the photometry, and 60 arcsecs for the CO spectra.

In L1551 the CO J = 1 → 0 and J = 2 → 1 lines show two components; a spike feature of width ~1.5 km s^{-1}, and a weaker but much broader "wing" component. Snell et al. (1980) suggest the wings to originate in post-shock material, with the shock itself generated by a ~150 km s^{-1} stellar wind from a newly formed central star IRS 5 (Strom, Strom and Vrba, 1976). The fact that the wings are spatially confined to diametrically opposing lobes is attributed to the modulation of high velocity outflow by an accretion disk.

J = 3 → 2 CO spectra for three regions of this complex are shown in figure 1, together with lower frequency transitions taken with a comparable beam-size. A comparison of the earlier CO line measures with the data presented here confirms at least one of the regions to be optically thick. In the third zone centred on α(1950) = 04h 28m 31.6s; δ(1950) = 17° 59' 52" however the wing is optically thin in the

R. S. Roger and P. E. Dewdney (eds.), Regions of Recent Star Formation, 323–328.
Copyright © 1982 by D. Reidel Publishing Company.

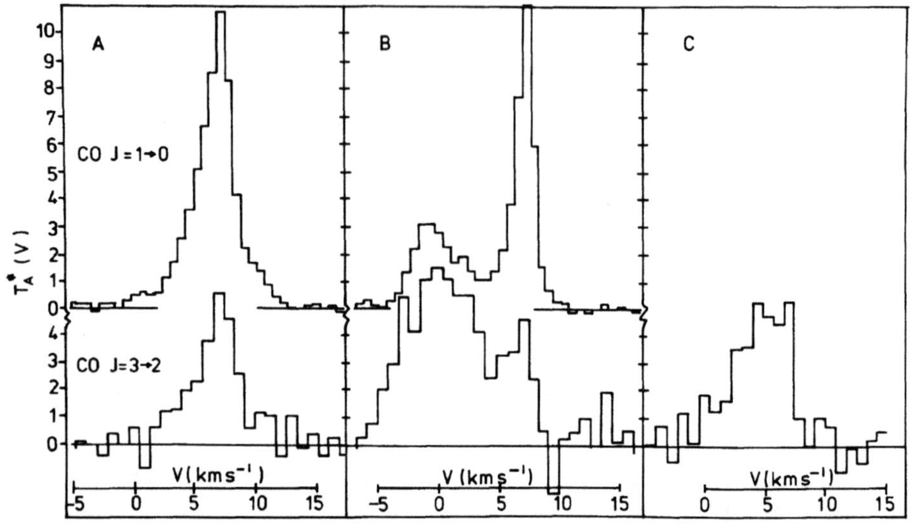

Figure 1. J = 3 → 2 CO spectra at three locations in L1551:
A. α(1950) = 04h 28m 40s, δ(1950) = 18° 01' 52";
B. α(1950) = 04h 28m 31.6s, δ(1950) = 17° 59' 52"
C. α(1950) = 04h 28m 06s, δ(1950) = 18° 01' 12"
For comparison are two CO J = 1 → 0 spectra (Snell et al., 1980) taken with comparable beam size.

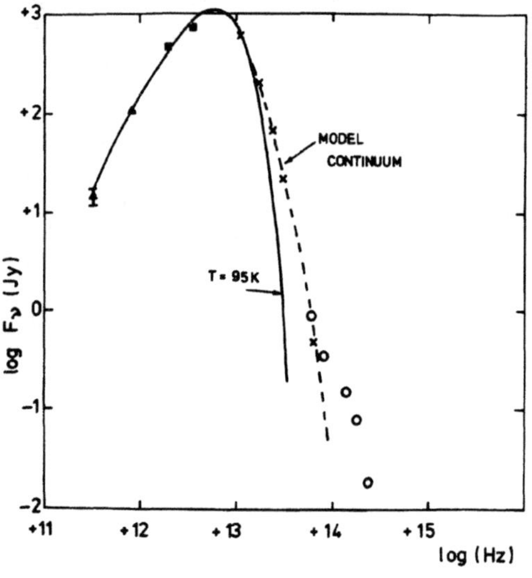

Figure 2. The infrared continuum of IRS 5 in L1551; ▲ - present results; ■ - Fridlund et al (1980); x - Beichmann and Harris (1981); o - Strom, Strom and Vrba (1976). Solid curve shows a T = 95K blackbody continuum; the dashed curve is a model continuum for a range of source temperatures.

$J = 1 \rightarrow 0$, but optically thick for $J = 2 \rightarrow 1$ and higher frequency transitions. A detailed analysis of this wing spectrum indicates that where gas kinetic temperature can be reliably evaluated it is broadly invariant, with a characteristic value $T_K \sim 25$ K. The present results further enable us to determine a typical beam dilution $W \sim 0.3$. Changes of optical depth are the prime cause of wing intensity variations, with a peak value of optical depth $\tau_{1\rightarrow 0} \sim 0.7$ in the $J = 1 \rightarrow 0$ transition occurring at $V_{LSR} \sim -1$ km s^{-1}. In combination with ^{13}CO $J = 1 \rightarrow 0$ results, this implies a relative abundance ratio $[^{13}C] / [^{12}C] \simeq 1/40$.

The narrow "spike" component in contrast shows a $J = 3 \rightarrow 2$ line strength which is much attenuated compared to lower frequencies. This can be shown to arise from photon-trapping effects, and a detailed LVG analysis indicates a velocity gradient $dV/dr \sim 56.3$ km s^{-1}, and molecular hydrogen density $n_{H_2} \sim 6.75 \times 10^3$ cm^{-3} in the line emitting zone, with uncertainties of approximately ± 0.1 dex in both parameters. The V_{LSR} of the spike feature does not however vary greatly with position in the cloud, suggesting a possible origin in pre-shock material. The large velocity gradient in the spike deduced from this analysis therefore comes as a surprise.

The central star, IRS 5, and presumed origin of this widescale kinematic structure also displays a large infrared excess, and it is tempting to identify this as originating in the accretion disk hypothesised by Snell et al. (1980). Our far-infrared results supplement photometry at shorter wavelengths, and indicate a substantially non-Planckian curve (figure 2). By taking a distribution of large, optically thick grains with a range of temperatures however, it is possible to simulate the observed continuum reasonably well. The model adopted assumes a grain temperature varying as $T_{gr} \propto r^{-\frac{1}{2}}$ with distance r from the central source, and a grain number density varying such that $N(r) \propto r^3$, where $N(r)$ is the number of grains within radius r. Whilst the model is therefore relatively simple, the accuracy of fit between observed and synthesised continua is gratifying.

Another compact region displaying high velocity microwave line features occurs near the Kleinmann-Low nebula. This too is a star-formation zone, with evidence for local maser activity and shocked H_2 emission, presumably arising at the interface between the high velocity outflow and OMC-1 molecular cloud. Our $J = 3 \rightarrow 2$ results clearly show a high velocity "Plateau" feature with space-variable asymmetry (figure 3). This is the first time that such characteristics have been attributed to this source, although similar features may be discerned in the high resolution $J = 2 \rightarrow 1$ data of Knapp et al. (1981). We estimate a FWHM of ~43 ± 5 arcsecs for the plateau zone, comparable with the $J = 2 \rightarrow 1$ estimate of ~40 arcsecs by Knapp et al., but significantly exceeding the previous $J = 3 \rightarrow 2$ estimates by Phillips et al. (1977) which corresponded to an harmonic mean size ~30 arcsecs.

From the present results (and those of Knapp et al.) it therefore seems that in the Kleinmann-Low source we are again seeing the

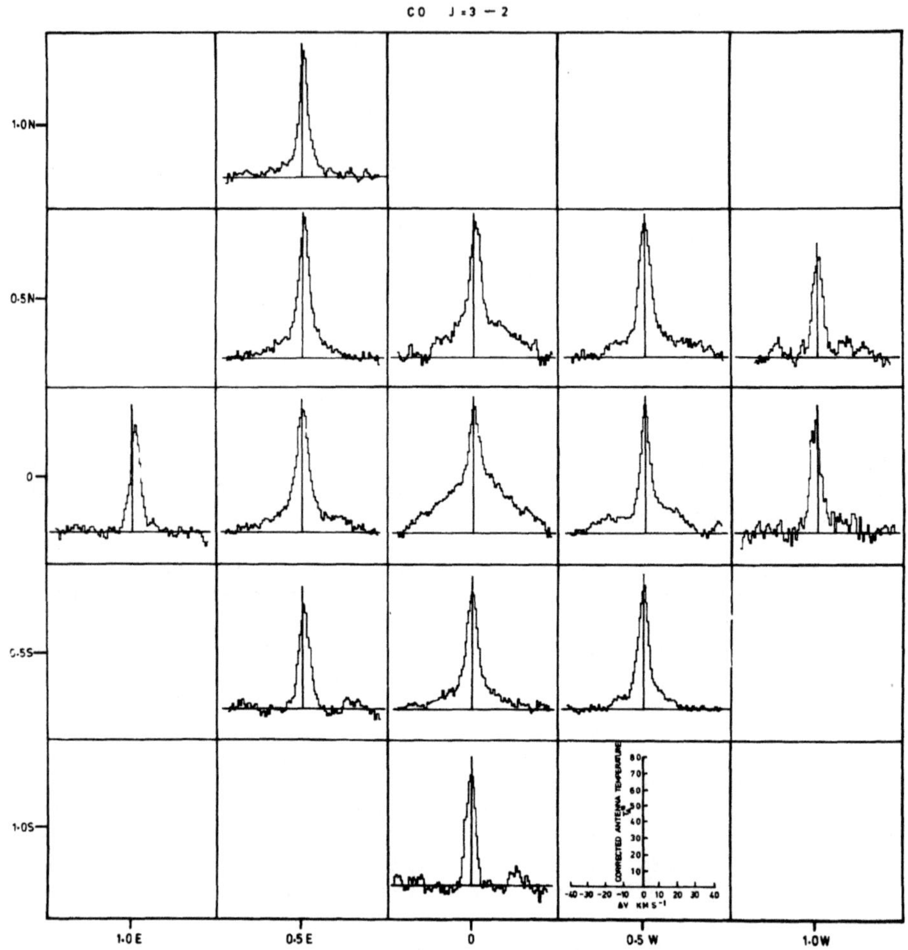

Figure 3. Grid sample of UKIRT J = 3 → 2 spectra around the nominal zero α(1950) = 5h 32m 47s, δ(1950) = -5° 24' 20". Each box is 0.5 arcminutes square.

consequences of asymmetric gas outflow, perhaps originating in a similar mechanism to that inferred for L1551.

Finally, we have observed a third source of vigorous mass-outflow, the carbon star IRC +10216. The distinctive parabolic microwave line shapes in the source indicate an envelope which is flowing outwards with more or less constant velocity. Recent J = 2 → 1 results (Wannier et al., 1979) however claim a much broader spatial distribution of line temperatures than can be accommodated by the model of Kwan and Hill

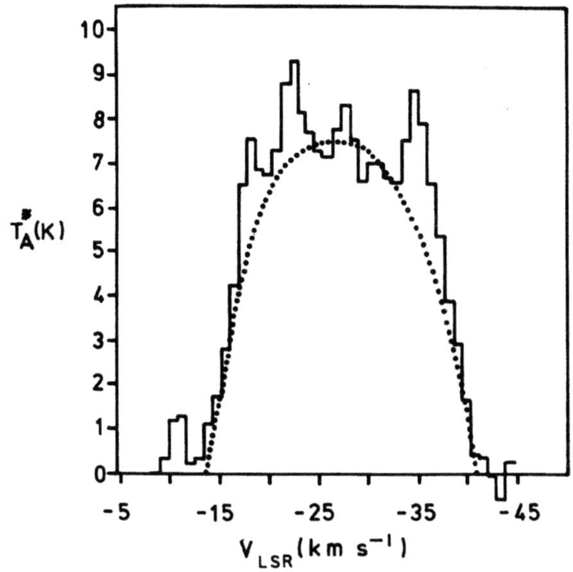

Figure 4. J = 3 → 2 CO spectrum of IRC +10216, with comparison model profile (see text).

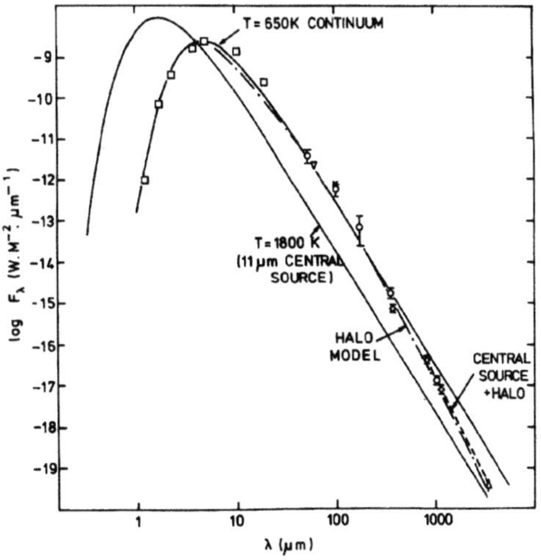

Figure 5. The infrared continuum of IRC +10216. ■ - Becklin et al. (1969) (as recalibrated by Toombs et al. (1972)); ∇ - Fazio et al. (1980); o - Campbell et al. (1976) (as recalibrated by Fazio et al.); Δ - Schwartz and Spencer (1977); ◇ - Present results; x - Shivanandan et al. (1977). The 1800 K continuum is based on the discussion of Sutton et al. (1979), and models the central star continuum.

(1977), possibly indicating a secularly decreasing mass loss rate. Our present $J = 3 \rightarrow 2$ results (figure 4) were therefore acquired with the aim of verifying the CO structure. If the lower rotational levels of CO are thermalised and the lines are optically thick, the line profiles for the present frequency and beamsize may be readily predicted from the $J = 2 \rightarrow 1$ results. The corresponding simulated curve is shown in figure 4, and the fit to the observed spectrum is seen to be excellent.

We have also acquired photometry at wavelengths λ 377 μm, 811 μm and 1136 μm. The continuum appears to fall as $F_\nu \propto \nu^{2.7}$, and can be modelled in terms of a distribution of optically thin grains with emissivity $\varepsilon \propto \lambda^{-1}$ (figure 5). The grain temperature is taken to vary as

$$T_{gr} = 230 \left(\frac{1 \text{ arcsec}}{r}\right)^{0.4} \text{ K}$$

and we have otherwise followed the prescription of Fazio et al. (1980). A comparison between the model and observed continua appears entirely satisfactory at longer wavelengths.

References

Becklin, E.E., Frogel, J.A., Hyland, A.R., Kristian, J. and Neugebauer, G.: 1969, Astrophys. J. 158, L133.

Beichman, C. and Harris, S.: 1981, Astrophys. J. (in press).

Campbell, M.F., Elias, J.H., Gezari, D.Y., Harvey, P.M., Hoffman, W.F., Hudson, H.S., Neugebauer, G., Soifer, B.T., Werner, M.W. and Westbrook, W.E.: 1976, Astrophys. J. 208, p. 396.

Fazio, G.G., McBreen, B., Stier, M.T. and Wright, E.L.: 1980, Astrophys. J. 237, L39.

Fridlund, C.V.M., Nordh, H.L., van Duinen, R.J., Aalders, J.W.G. and Sargent, A.I.: 1980, Astr. Astrophys. 91, L1.

Knapp, G.R., Phillips, T.G., Huggins, P.J. and Redman, R.O.: 1981, Preprint.

Kwan, J. and Hill, F.: 1977, Astrophys. J. 215, p. 781.

Phillips, T.G., Huggins, P.J., Neugebauer, G. and Werner, M.W.: 1977, Astrophys. J. 217, L161.

Schwartz, P.R. and Spencer, J.H.: 1977, Mon. Nat. R. astr. Soc, 180, p.297.

Shivanandan, K., McNutt, D.P., Daehler, M. and Moore, W.J.: 1977, Nature, 265, p. 513.

Snell, R.L., Loren, R.B. and Plambeck, R.L.: 1980, Astrophys. J., 239, L17.

Strom, K.M., Strom, S.E. and Vrba, F.J.: 1976, Astron. J. 81, p. 320.

Sutton, E.C., Betz, A.L., Storey, J.W.V. and Spears, D.L.: 1979, Astrophys. J. 230, L105.

Wannier, P.G., Leighton, R.B., Knapp, G.R., Redman, R.O., Phillips, T.G. and Huggins, P.J.: 1979, Astrophys. J. 230, p. 149.

CO IN THE HORSEHEAD NEBULA

Antony A. Stark and John Bally
Bell Laboratories
Holmdel, NJ 07733

ABSTRACT

Carbon monoxide observations of molecular gas at the ionization front associated with IC 434 show a corrugated structure with a periodicity of 1.4 pc along the entire 9 pc length of the front. The CO distribution may be explained by the Rayleigh-Taylor instability resulting from the rocket acceleration of the Orion B molecular cloud by Lyman continuum radiation from the Ori OB I association.

INTRODUCTION

The prominent ionization front between IC 434 and the Ori B molecular cloud is ideally suited for a study of the interaction of an old HII region with dense neutral material. The front is relatively nearby (500 pc) and appears to be propagating nearly orthogonal to the line of sight. The principle exciting stars of IC 434 are members of Ori OB Ib.

The CO map shown in Figure 1a was constructed from 600 spectra obtained with the Bell Laboratories 7-meter antenna. The map is fully sampled on a 2' grid and includes the Horsehead Nebula (B33, Barnard 1913, 1919), the reflection nebula NGC 2023, and the HII region NGC 2024. Figure 1b shows a photograph of the region covered by the CO observations. It is clear that the CO emission shows a close correspondence to the distribution of extinction in the photograph; the location of the molecular cloud edge and the ionization front coincide.

The edge of the molecular cloud appears wavy, with several protrusions, similar to the Horsehead nebula (the southernmost protrusion in the CO map) located along the ionization front. Some of these can be seen in photographs having the right exposure (Duncan 1921). A spatial-velocity contour map constructed from CO spectra lying along

R. S. Roger and P. E. Dewdney (eds.), Regions of Recent Star Formation, 329–334.
Copyright © 1982 by D. Reidel Publishing Company.

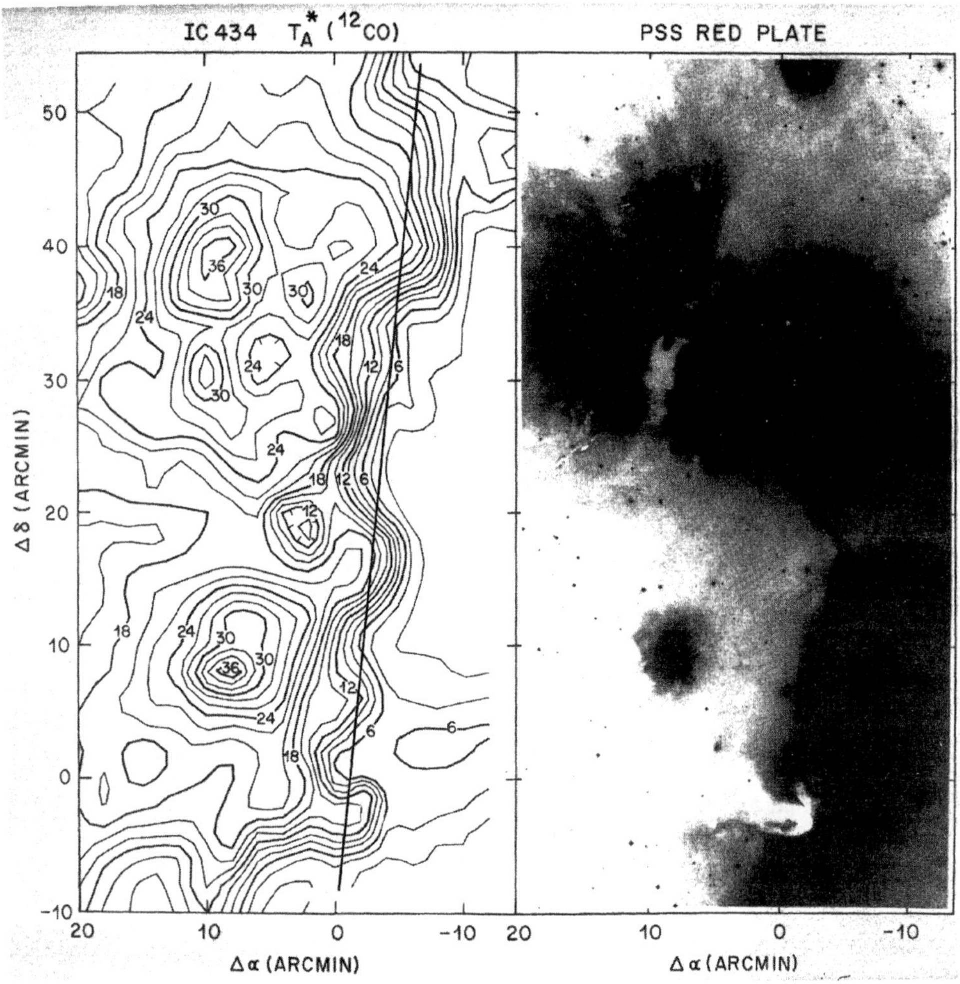

Figure 1. a) A ^{12}CO map of the ionization front associated with IC 434 showing peak antenna temperature T_A^*. Coordinates are referenced to $\alpha_{1950} = 5^h 38^m 36^s$ $\delta_{1950} = 2^\circ 26' 00''$. b) Photograph (PSS E-plate) of the same region.

the ionization front is shown in Figure 2. The corrugated edge of the molecular cloud shows up as a periodic structure in the spatial velocity diagram with a peak occuring roughly every 11 arc minutes (1.4 pc). Note in Figure 2 that there is little structure on length scales smaller than 11', even though the spatial resolution is 2'.

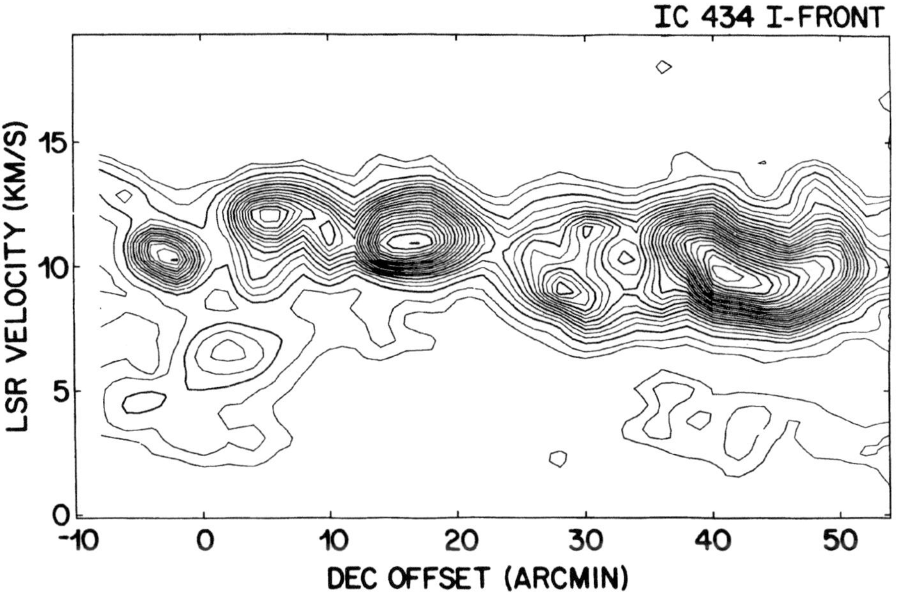

Figure 2. Spatial-velocity diagram constructed from ^{12}CO spectra obtained along the ionization front marked by the line in Figure 1.

DISCUSSION

The features in Figure 2 could be high density lumps which are being exposed as the material surrounding them is ionized. They might even be sufficiently dense to be gravitationally bound, in which case they could survive as cloudlets or globules: we know of no arguments to exclude this possibility. We will, however, pursue the hypothesis suggested by the periodicity of the corrugations, that the cloud edge is undergoing an instability. Possible mechanisms include the magneto-hydrodynamic instability suggested by Langer (1978) or the Rayleigh-Taylor instability (Spitzer 1978). Since very little is known about the strength and nature of magnetic-fields in molecular clouds, we will only consider the RT instability here.

The Rayleigh-Taylor instability operates whenever a dense fluid forms an overburden on a lower density medium. For a large density difference, the RT instability grows at a rate

$$t_{RT} \simeq \sqrt{\frac{\lambda}{a}}$$

where λ is the wavelength of a growing mode and a is the acceleration imparted to the dense layer from the direction of the low density medium. It is clear that the shortest wavelengths grow most rapidly. Short wavelengths, however, are stabilized by transport processes such as turbulence which moves gas a distance λ in a time

$$t_{STAB} = \lambda/\Delta V$$

where ΔV is the turbulent flow velocity in the gas. From the CO line widths, $\Delta V \sim 1$ km s^{-1}. The shortest wavelength instability which can grow (and it grows faster than any longer scale) is the one for which $t_{RT} = t_{STAB}$ or

$$\lambda = \frac{(\Delta V)^2}{a} \quad .$$

The evaluation of this expression requires an estimate of the acceleration acting upon the molecular cloud. The rocket effect caused by the ablation of gas from the cloud edge results in a force that pushes the cloud. In the approximation that the cloud acts rigidly, the acceleration is given by

$$a = \frac{n_e kT_e}{2m_H N(H_2)}$$

$$= 3x10^{-9} \left|\frac{L_c}{10^{49} \text{photons s}^{-1}}\right|^{1/2} \left|\frac{r_{II}}{5.4\text{pc}}\right|^{-3/2} \left|\frac{N(H_2)}{10^{22} \text{cm}^{-2}}\right|^{-1} \text{cm s}^{-2}$$

where n_e, the electron density is evaluated from the Strömgren condition at the I-front, L_c is an estimate for the Lyman continuum flux emitted by stars illuminating the I-front, and r_{II} is the projected distance to the dominant source of UV photons, σ Ori. $N(H_2)$ is the column density of the accelerating layer in a direction parallel to the acceleration vector. This acceleration holds if the shock wave driven by the I-front has completely penetrated the molecular cloud as is the case in the Pelican Nebula (Bally and Scoville 1980) and the cloud is supported by a non-thermal internal pressure such as turbulence or magnetic fields.

Observations indicate that clouds do not collapse in a free-fall time. To explain the star formation rate observed in the galaxy (Smith et al 1978) cloud collapse must be inhibited so that $t_{collapse} \sim 10^2 \ t_{f-f}$. An effective pressure other than thermal pressure must stiffen the clouds. The data do not show any evidence for a layer of enhanced density near the ionization front that would indicate the presence of a dense post shock layer predicted by the theory of isothermal ionization and shock-front systems (Spitzer 1978):

Observations of ^{13}CO in an E-W strip crossing B33 show that the column densities remain nearly constant in this direction. Supersonic transport mediated by either turbulence or magnetic fields may prevent the formation of a thin, quiescent, and compressed isothermal post-shock layer.

In order to determine the wavelength of the fastest growing Rayleigh-Taylor unstable mode, a reasonable value for the acceleration is $a=2x10^{-9}$ cm s^{-2} and a transport velocity $\Delta V=10^5$cm s^{-1} (=FWHM/2.71). Thus the dominant growing mode has a wavelength $\lambda=5x10^{18}$cm = 1.6 pc. This agrees well with the 11' scale-length of the observed corrugation seen in the IC 434 I-front.

How fast does the instability grow? Using the above parameters, one finds that $t_{R-T} \sim 1.5x10^6$ yrs. This is shorter than the OB association lifetime. The protrusions generated by the RT stability are not erroded by the difference in ionization rate between the crest and valley of the dominant mode provided the amplitude of that mode is not too great.

CONCLUSIONS

The above arguments suggest that the Rayleigh-Taylor instability might explain the Horsehead nebula and similar structures seen in the CO map of the IC 434 ionization front. In order for this mechanism to work, the observed CO line width must be interpreted as a mechanical transport process such as turbulence. Turbulence acts as an internal pressure in the molecular gas which prevents the compression of the post-shock layer by the large factors predicted by the theory of isothermal ionization-shock front systems and stiffens the entire molecular cloud so it can be accelerated by the rocket effect.

REFERENCES

Bally, J. and Scoville, N. Z. (1980) Ap. J. 239, 121.
Barnard, E. E. (1913) Ap. J. 38, 496.
Barnard, E. E. (1919) Ap. J. 49 1.
Blaauw, A. (1964) Ann. Rev. Astr. Ap. 2, 213.
Duncan, J. C. (1921) Ap. J. 53, 392.
Langer, W. D. (1978) Ap. J. 225, 95.
Smith, L. F., Biermann, P. and Mezger, P. G. (1978), A. A., 66, 65.
Spitzer, L. (1978) "Physical Processes in the Interstellar Medium." Wiley-Interscience, New York.

DISCUSSION FOLLOWING PAPER BY STARK AND BALLY

BEICHMAN: What mass do you calculate for these clumps - could there be a self-gravitation effect causing hot-spots in those regions?

STARK: They are of order 1 to 10 $M_\odot$ and the escape velocity on the surface is incompatible with their being gravitationally bound.

SHUTER: I suggest that for such clouds there may be acceleration towards the disc of the Galaxy (z-direction) with differentials, due to slightly different friction, that may be of a similar magnitude to those you calculate.

YORKE: The surface of the ionization front is two-dimensional and your analysis used only linear distances or projections of distance.

STARK: Yes, the geometry is particularly favorable. We are looking right along the edge of the ionization front.

YORKE: You don't know how deep it is, though?

STARK: No.

YORKE: So what effect would that have on the analysis?

STARK: Oh, square-root of three!

TENORIO-TAGLE: I would like to comment on the rocket-effect mechanism. A shock wave can propagate into a cloud just ahead of the ionization front and accelerate all swept-up cloud material. However, the rest of the cloud is unaware of this. The layer of shocked gas is very narrow ($\sim 10^{14}$ cm) and stays narrow although it grows in mass as time goes on.

STARK: After 10^7 years wouldn't you expect the shock to have advanced through the cloud at least once, and recoiled back?

TENORIO-TAGLE: Yes, but during that time it will have ionized the material between the fronts.

STARK: We don't really see any big build-up in density. We have a ^{13}CO map in a strip across the front which shows no evidence of a thin dense shell.

TENORIO-TAGLE: That means it must be faint, so we can't consider accelerations of very large portions of the cloud.

LARSON: Could it be that what one sees is just the result of an ionization front propagating into a "lumpy" cloud? It would propagate more slowly into the "lumps", leaving them behind to produce features like the "horsehead".

STARK: That's a good point. However, our impression is that where you can clearly see lumpiness in clouds, it doesn't have this appearance. This region appears very periodic. Furthermore, from physics, one would expect Rayleigh-Taylor instability here, and with the observed scale length.

SMALL SCALE CLUMPING IN THE ORION MOLECULAR CLOUD

P. Bastien[1], J. Bieging[2], C. Henkel[3], R.N. Martin[3], T. Pauls[4], C.M. Walmsley[3], T.L. Wilson[3], L.M. Ziurys[3]
[1]Astronomische Institute der Universität Bonn, D-5300 Bonn 1, F.R.G.
[2]Radio Astronomy Laboratory, University of California, Berkeley
[3]Max-Planck-Institut für Radioastronomie, D-5300 Bonn 1, F.R.G.
[4]Erstes Physikalisches Institut, Universität zu Köln, D-5000 Köln 41, F.R.G.

The results of new maps of the Orion Molecular Cloud in lines of NH_3 and H_2CO are presented. The angular resolution of these maps varies from 1' (for the $2_{11} - 2_{12}$ line of H_2CO at 14.5 GHz, using the 100-m telescope) to 2.2" x 1.6" (for the (3,3) line of NH_3, using the VLA). There are two quiescent molecular clouds NE and south of the BN/KL region. The contours of vibrationally excited H_2 (H_2^*) straddle the position of these clouds; we conclude that the physical confinement of the H_2^* between these clouds explains the observed distribution of the H_2^* contours. The quiescent clouds themselves have H_2 densities of $\geq 10^6$ cm^{-3} and masses close to the Jeans masses. From our VLA data, the kinetic temperature of the 5.5 km s^{-1} (so called "Hot Core") feature is $\lesssim$200 K, the size of the main peak is $\sim$3", and the maximum is centered on IRc4. We believe that this gas is heated by IRc4. An analysis of 100-m spectra for six metastable lines gives roughly consistent, but considerably more uncertain values of temperature and size. Our VLA data show that the Plateau feature ($\Delta V_{1/2} \geq 30$ km s^{-1}) has a main beam brightness temperature of 70 K, about twice the estimate from 100-m data. This discrepancy could be caused by a core-halo structure. The Plateau emission consists of two peaks; these are not centered on any single compact IR source. The NH_3 features are not obviously related to IRc2. In contrast to Downes et al. (1981), we argue that IRc2 does not have a primary role in the outflow of complex molecules such as NH_3 and SO in this region.

REFERENCES

Bastien, P., Bieging, J., Henkel, C., Martin, R.N., Pauls, T., Walmsley, C.M., Wilson, T.L. and Ziurys, L.M.: 1981, Astron. Astrophys. 98, L4
Downes, D., Genzel, R., Becklin, E.E. and Wynn-Williams, C.G.: 1981, Astrophys. J. 244, pp. 869-883

R. S. Roger and P. E. Dewdney (eds.), Regions of Recent Star Formation, 335–336.
Copyright © 1982 by D. Reidel Publishing Company.

DISCUSSION FOLLOWING PAPER BY BASTIEN ET AL.

SCHLOERB: Are you sure the two clouds are distinct clumps rather than parts of a ridge with a velocity gradient on it?

BASTIEN: Yes, with good velocity resolution we see two components, one at 8 km s^{-1} and one at 9.8 km s^{-1}.

FORSTER: What are the masses of the two cloudlets?

BASTIEN: A few solar masses each.

THE ON-1 CO CLOUD COMPLEX -- ONSET OF STAR FORMATION

F.P. Israel
Astronomy Division, Space Science Department of ESA,
ESTEC, Noordwijk, The Netherlands

H.A. Wootten
Owens Valley Radio Observatory, California Institute
of Technology, Pasadena, California, USA

ABSTRACT

Molecular line observations of the ON-1 region show the presence of a large CO cloud complex with overall dimensions of 25 x 60 pc at a distance of 1.4 kpc to the Sun. The complex consists of two major parts, CON-1 East and CON-1 West with sizes of respectively 20 x 40 pc and 15 x 35 pc. ON-1 itself is associated with a compact CO cloud of mean size 1.2 pc and a mass of about 750 $M_\odot$. The object coincides with a dense core, showing CO self-reversal and molecular densities in excess of 10^4 cm^{-3}. The apparently unique character of ON-1 -- being the only known site of star formation in the whole complex -- may provide an important clue to the nature of the process that triggers first generation star formation.

1. INTRODUCTION

The object ON-1 was one of the first main-line OH masers discovered in the Galaxy (Elldér et al., 1969). It is located at the edge of the heavily obscured Cygnus Rift region. At the position of the maser, an ultracompact HII region and a highly variable H_2O maser have also been found (Winnberg et al., 1973, Genzel and Downes, 1977). Therefore, ON-1 marks the site of very recent star formation. In the immediate vicinity of ON-1, no other signposts of star formation have been found. This is unusual, as star formation almost always occurs in clusters of activity centers (c.f. Habing and Israel, 1979).

In order to establish the nature of the neutral surroundings of ON-1, we have carried out a series of molecular line observations, using the Columbia 1.2 m telescope, the Texas 5 m MWO telescope, the OVRO and KPNO 10 m telescopes, and the NRAO 43 m telescope. The observations will be described elsewhere in detail (Israel and Wootten, 1981).

R. S. Roger and P. E. Dewdney (eds.), Regions of Recent Star Formation, 337–342.
Copyright © 1982 by D. Reidel Publishing Company.

2. MOLECULAR CLOUD COMPLEXES NEAR ON-1

The relatively large beamwidth of the Columbia telescope (HPBW = 8') enabled us to map an area of about four square degrees around ON-1 with one beamwidth spacing. ON-1 coincides with a local maximum in the CO distribution; the ON-1 CO cloud is part of a cloud complex with dimensions of 35' x 80' (marked CON-1 West in Figure 1), at velocity V_{LSR} = 11 km s^{-1}.

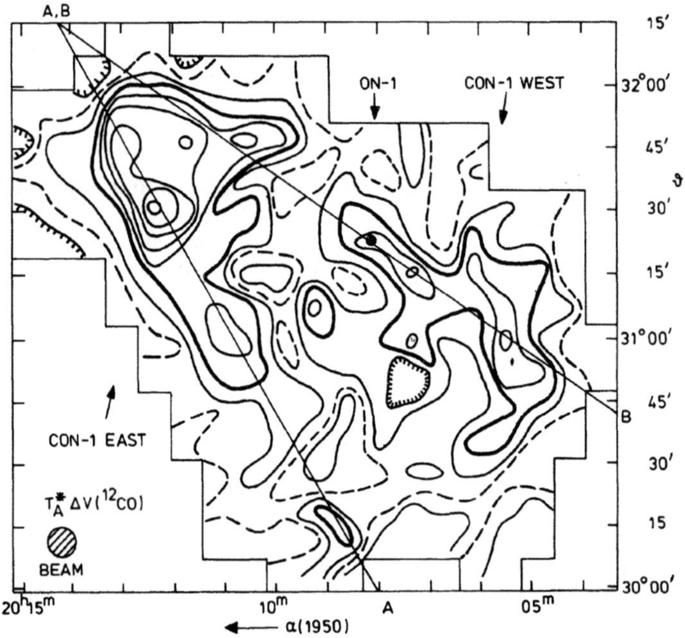

Figure 1. Integrated ^{12}CO map of the ON-1 molecular cloud complex, obtained with the Columbia telescope. The position of ON-1 is marked by a black dot. Contour values are in steps of 5.2 K km s^{-1}. For clarity, the 10.4 K km s^{-1} contour is dashed, and the 20.8 K km s^{-1} contour is drawn heavily.

Just east of this cloud complex there is a second one, with dimensions of 50' x 90', at about the same velocity (marked CON-1 East in Figure 1). These velocities correspond to distances of either 1.4 or 5.6 kpc.

However, the general outline of both cloud complexes (indicated by a heavily drawn contour in Figure 1) agrees closely with faintly visible obscuration on the red and infrared Palomar Sky Survey plates. From this we conclude that the near distance of 1.4 kpc is the appropriate one for the complex. This leads to linear cloud sizes of 15 x 35 pc and 20 x 40 pc respectively. If the two complexes are physically related, an overall size of 25 x 60 pc is indicated. Thus, we find that the ON-1 star formation site is part of a much more extended, giant molecular cloud complex.

3. THE ON-1 CO CLOUD

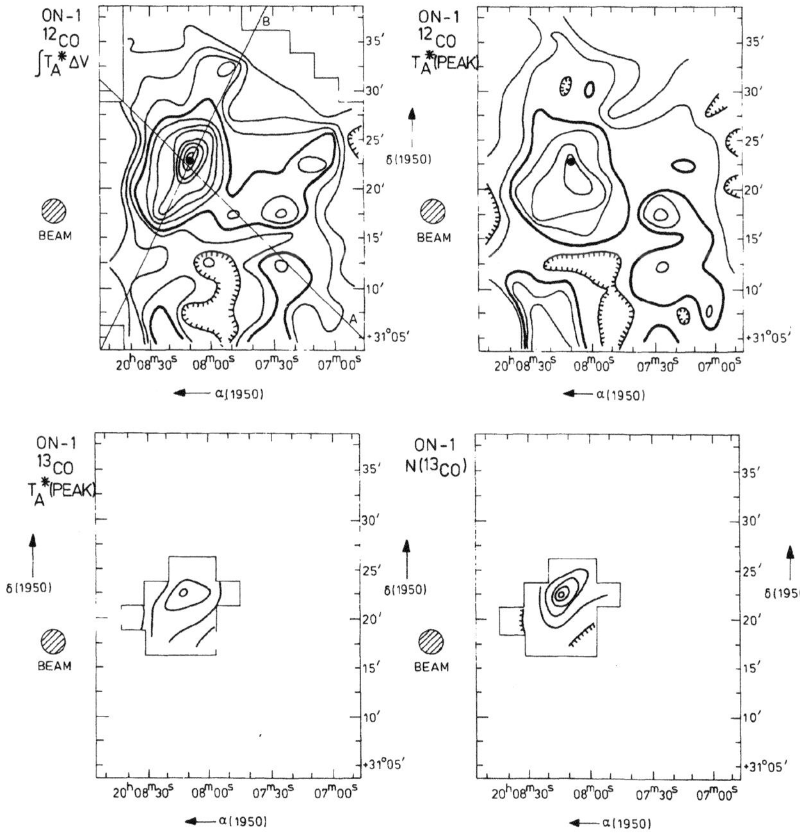

Figure 2. ^{12}CO and ^{13}CO maps of the ON-1 cloud obtained with the Texas MWO telescope. a. Integrated ^{12}CO map. The position of ON-1 is indicated by a black dot. Contour values are in steps of 3.25 K km s^{-1}; first contour is 6.5 K km s^{-1}. For clarity, the 16.25 K km s^{-1} contour is drawn heavily. b. Map of peak antenna temperature T_A^* (^{12}CO). Contour values are in steps of 1 K; first contour is 2K; 4 K contour is drawn heavily. c. Map of peak T_A^* (^{13}CO). Contour values are in steps of 1 K. d. Map of column density N(^{13}CO). Contour values are in steps of 3×10^{15} cm^{-2}.

The 3.5 times higher resolution of the Texas MWO telescope was used to map the immeditate surroundings (28' x 35', or 11 x 14 pc) of ON-1 in the J = 1-0 ^{12}CO and ^{13}CO lines in more detail (Figure 2).

ON-1 now coincides with the positions of peak $T_A^*(^{13}CO)$ and peak T_A^* (^{12}CO) V, but is displaced by about 2.5' from the position of peak T_A^* (^{12}CO). These positions represent the highest intensities in the map (T_A^* (^{12}CO) = 7.5 K -- see Figure 3). Observations at even higher resolu-

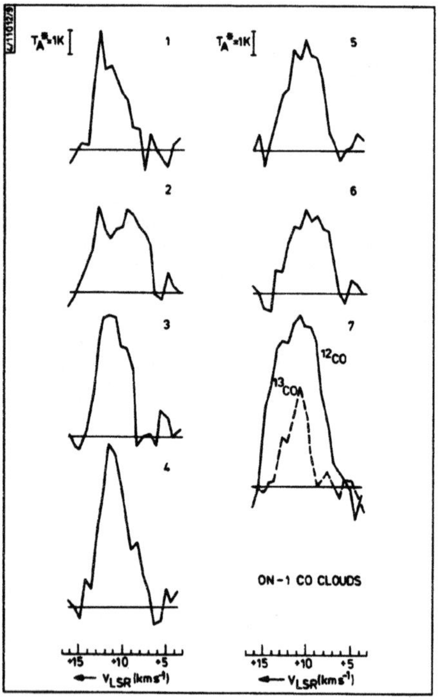

Figure 3. $^{12}CO(1-0)$ profiles at peak positions in Figure 2. Profile no. 7 corresponds to the ON-1 position. For ON-1, the ^{13}CO profile is also given on the same intensity scale.

tion with the OVRO and KPNO 10 m telescopes yield T_A^* CO(1-0) = 13 K (with 1' resolution) and T_A^* CO(2-1) = 11 K (with 0.5' resolution). The CO(1-0) linewidth increases very strongly at the position of ON-1; at the same time the peak intensity shifts to higher velocity and a broad low velocity wing develops. All these characteristics strongly suggest the presence of an appreciable amount of self-absorption in the ^{12}CO profiles at the position of ON-1, indicating a significant density enhancement on a scale of about one arcmin (0.4 pc). At the same position as the compact HII region and this molecular density enhancement, HCO^+ and H_2CO emission occurs on a scale of about 2' (0.8 pc). From the H_2CO observations we infer a molecular density of $n(H_2) \geq 10^4$ cm^{-3} with an abundance of $X(H_2CO) \cong 10^{-9}$. These values agree well with those of other molecular cloud cores (Wootten et al., 1980).

This dense core is embedded in a larger cloud, prominent in Figure 2, with ^{12}CO sizes of 2.1 x 0.7 pc. Under the usual LTE assumptions we derive a peak column density $N(^{13}CO) = 2 \times 10^{16}$ cm^{-2} at the position of ON-1; this would correspond to $N(H_2) = 10^{22}$ cm^{-2} (Dickman, 1978), so that a depth of 0.3 pc or less is indicated for the core region.

If we assume, on the other hand, that most of the molecular material is distributed along a line of sight equal to the mean diameter of the cloud (1.2 pc), we find a mean density for the cloud of $n(H_2) = 2500$ cm^{-3} and a mass $M(H_2) = 740\ M_{\odot}$. These values are likewise in the range expected for a compact, star forming molecular cloud.

4. THE NATURE OF ON-1

As noted in the introduction, ON-1 in unusual because it is the only isolated site of star formation known at present, i.e., it is not accompanied by other HII regions or signposts of star formation other than those associated with the source itself. Recent observations confirm this notion: very sensitive observations with the VLA still show a single, barely resolved radio continuum source (Turner and Baud, private communication), while infrared mapping at 20 microns reveals a double structure, coincident with the H_2O maser emission (Becklin, Downes, Genzel and Wynn-Williams, private communication), that nevertheless may be due to a single exciting source. Far-infrared observations show a single, unresolved source with $T_d \geq 30$ K (Sargent et al., 1981).

A WSRT search for further HII regions associated with the larger molecular cloud complex is in progress, but so far has failed to show any other HII region candidate. It thus appears that ON-1 is not just a site of recent star formation, but at the same time is the only site of star formation in the whole molecular cloud complex; its energy output corresponds to that of a single early B star.

The unique character of the ON-1 star formation site very strongly suggests that in this source we witness the very beginning of star formation in a molecular cloud complex. Several proposed scenarios, especially scenarios for secondary star formation are ruled out. These include triggering of star formation by cloud-cloud collisions, ionization shock fronts, and shock fronts due to expanding stellar wind or supernova bubbles.

Further observations, in particular at radio and far-infrared wavelengths are necessary to confirm the unique nature of ON-1. They are important, however, because they may provide the clue to a question that has not been answered yet: what process triggers the beginning of first generation star formation in galactic molecular clouds ?

REFERENCES

Dickman, R.L., 1978, Ap. J. Suppl. 37, 407.
Elldér, J., Rönnäng, B., Winnberg, A., 1969, Nature 222, 67.
Genzel, R., Downes, D., 1977, Astr. Ap. Suppl. 30, 145.
Sargent, A.I., van Duinen, R.J., Firdlund, C.M.V., Nordh, H.L., Aalders, J.W.G., 1981, preprint.
Winnberg, A., Habing,H.J., Goss, W.M., 1973, Nature, Phys. Sc. 243, 78.
Wootten, H.A., Snell, R., Evans, N.J., 1980, Ap. J. 240, 532.

CO OBSERVATIONS OF THE MOLECULAR CLOUD ENCOMPASSING SHARPLESS 222

R.A. Christie, W.H. McCutcheon, and C.P. Chan
Physics Department,
University of British Columbia,
Vancouver, B.C., Canada

ABSTRACT

^{12}CO observations around Sharpless 222 (Sharpless 1959) reveal a very wide region of CO emission at a temperature of about 10 K. There is a ridge of enhanced emission along the southern portion of the molecular cloud. HI emission associated with S222 lies above this ridge.

INTRODUCTION

Lk Hα101 is a B0.5 star (Allen 1973, Brown et al. 1976, Altenhoff et al. 1976, and Harris 1976) at least 800 persecs distant (Herbig 1971). There is a compact HII region less than one arcsecond across (Altenhoff et al. 1976, Harris 1976) and a more extended region of weaker emission out to 35 arcseconds (Altenhoff et al. 1976). The large nebulosity, easily seen on photographs, is a reflection nebula.

Dewdney and Roger (1981) mapped a large area around S222 in the 21 cm line of HI. We present ^{12}CO observations of the same region with similar spatial and velocity resolutions.

EQUIPMENT AND OBSERVING PROCEDURE

Observations were carried out between April and November 1980 using the 4.57 meter millimetre wave telescope located on the University of British Columbia campus (Shuter and McCutcheon 1974, Mahoney 1976). At the ^{12}CO J=1→0 transition frequency, the beam width is 0.044 degrees and the pointing repeatability of the telescope is ±0.015 degrees. The receiver uses a single ended mixer cooled to 20 K and the system noise temperature averaged 1200 K (single sideband).

Spectra were obtained using a 64 channel filter spectrometer with a velocity resolution of 0.65 km s^{-1} at the ^{12}CO J=1→0 transition.

R. S. Roger and P. E. Dewdney (eds.), Regions of Recent Star Formation, 343–348.
Copyright © 1982 by D. Reidel Publishing Company.

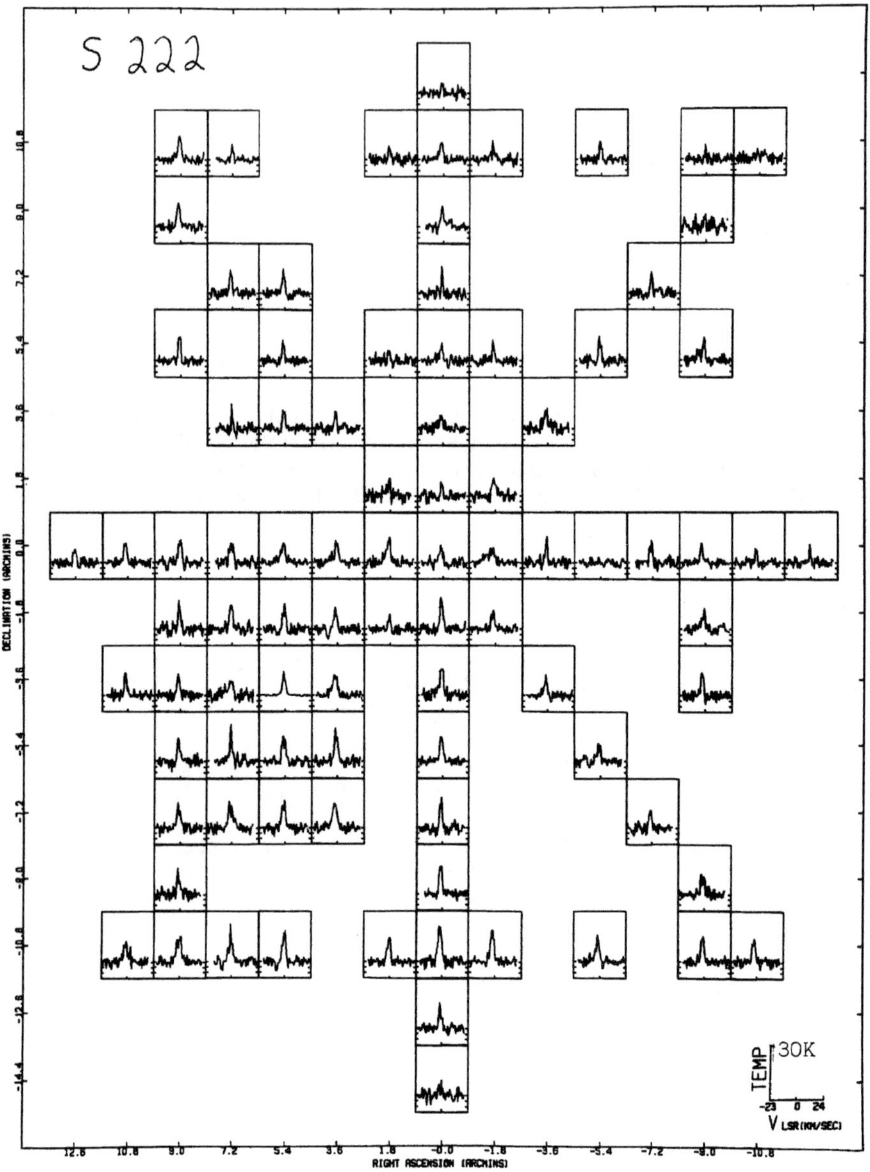

Figure 1 - ^{12}CO Profiles Obtained for Sharpless 222
(0.0,0.0) is $\alpha(1950) = 04^{h}26^{m}34^{s}.0$ and $\delta(1950) = 35°13'00''.0$

Eighty-nine positions were observed on a grid of spacing 0.03 degrees (Fig.1). Spectra were obtained by alternating between ON source scans and OFF source or reference scans in a load-switched receiver mode. Each spectrum is the sum of four 320 second pairs. The reference position is located one degree lower than Lk Hα101 in declination.

Atmospheric opacity was measured from regular antenna tipping curves and OMC1 was monitored daily to define an absolute temperature scale.

Our nominal centre (0.0,0.0) position was chosen as the approximate centre of the HI map of Dewdney and Roger (1981). As a result, the illuminating star, Lk Hα101, is located at (5.4,-3.6). Here the numbers refer to the offsets (arcminutes) of the right ascension and declination from our centre position. A small region around Lk Hα101 was fully sampled, as well as cross and diagonal strips over a wide extent of the cloud to determine where the ^{12}CO emission decreases.

^{12}CO RESULTS

The central profile velocity is -1.25 km s^{-1}, in good agreement with the results of Wilson et al. (1973) and Knapp et al. (1976). HI data of Dewdney and Roger indicate the presence of an absorption component whose velocity coincides very well with an observed dip in our ^{12}CO radiation temperature contours integrated over all velocities. The ^{12}CO results show emission at an average temperature of 10 K over a wide region. Only the north and west boundaries have been determined. Emission probably extends as far as the visual extinction (1 mag.) which covers a region almost one degree across and several degrees north to south.

The one distinguishing feature found from our observations was a ridge of enhanced emission ($T_A^* \sim 15$ K-20 K) at a declination offset of -10.8 arc min. (Fig. 1) relative to the map centre. The area around the ridge is poorly sampled and consequently the enhancement is only faintly visible in the contours in Fig. 2. However, the profiles at this declination, observed over several days, were consistently stronger than those at nearby positions.

This ridge of enhanced emission lies just outside the contours of HI emission of Dewdney and Roger (1981). These results indicate that there has been an increase in the temperature at the southern boundary of the dissociation wave responsibe for the HI. ^{13}CO observations are required to investigate this region further.

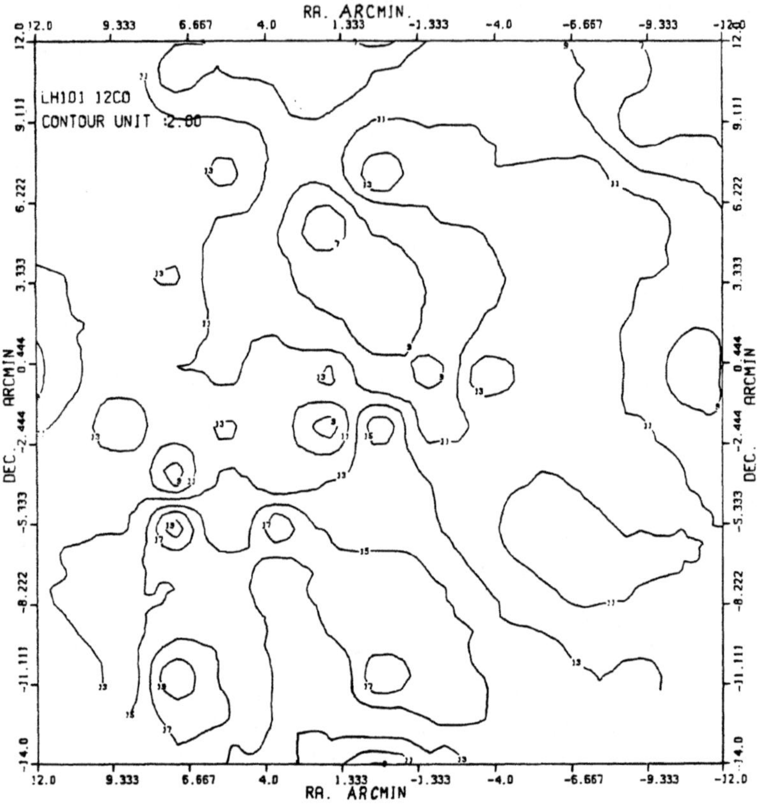

Figure 2 – ^{12}CO Radiation Temperature, T_A^* (^{12}CO), Contours for S 222. (0.0,0.0) is $\alpha(1950) = 04^h26^m34^s.0$ and $\delta(1950) = 35°13'00''.0$

Figure 3 gives the averaged visual extinction determined by us from star counts of the red and blue Palomar prints. The visual obscuration over much of the region is two magnitudes with small 'pockets' of stronger extinction. The maximum visual extinction measured is at least five magnitudes. Using the ratio of Bohlin et al. (1978) one would infer a mean total hydrogen column density of both atoms and molecules of 4×10^{20} atoms cm^{-2} with 'pockets' up to 10×10^{20} atoms cm^{-2}.

The extinction along the ridge at -10.8 arc min varies from one magnitude to five magnitudes indicating that the temperature enhancement occurs even in regions of minimum density.

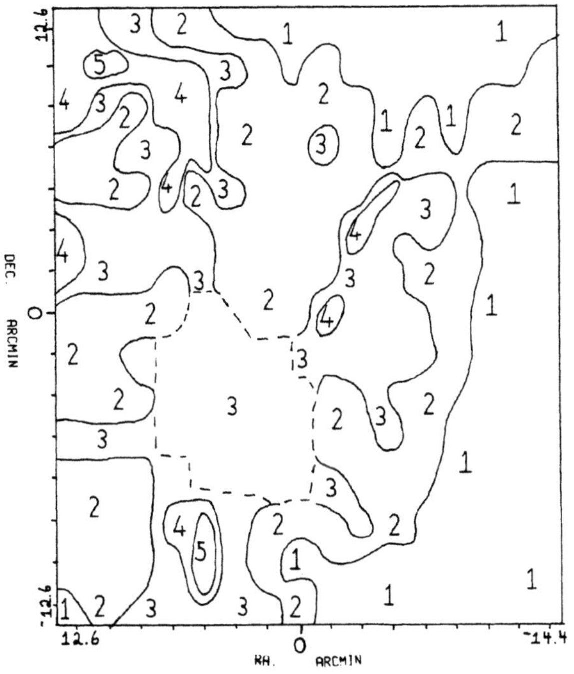

Figure 3 - Visual Extinction Contours of the Cloud Surrounding S 222. (0.0, 0.0) is $\alpha(1950) = 04^h26^m34^s.0$ and $\delta(1950) = 35°13'00''.0$. The numbers refer to the averaged visual extinction. For example 4 means the extinction is between 3 and 4 magnitudes.

SUMMARY

Our ^{12}CO results show a wide region of ^{12}CO emission ($T_A^* \sim 10$ K), but there is a ridge of enhanced emission ($T_A^* \sim 15$ K to 20 K) at a declination offset of -10.8 arc min. This ridge occurs at the southern boundary of the HI region (Dewdney and Roger 1981) and indicates that there has been a temperature enhancement at the interface of the HI and molecular regions. ^{13}CO observations are required to gain a further understanding of this region.

A visual extinction map indicates that the temperature enhancement in the ridge occurs even in regions of minimum density.

ACKNOWLEDGEMENTS

We thank Drs. P. Dewdney and R. Roger for their Sharpless 222 HI results in advance of publication. W.H.M. gratefully acknowledges grants in aid of research from the Natural Sciences and Engineering Research Council of Canada and from the University of British Columbia.

REFERENCES

Allen, D.A. 1973, Monthly Notices Royal Astron. Soc. 161, pp. 1P-2P.
Altenhoff, W.J., Braes, L.L.E., Olnon, F.M. and Wendker, H.J. 1976, Astron. Astrophys. 46, pp. 11-17.
Bohlin, R.C., Savage, B.D. and Drake, J.F. 1978, Ap. J. 224, pp.132-142.
Brown, R.L., Broderick, J.J. and Knapp, G.R. 1976, Monthly Notices Royal Astron. Soc. 175, pp. 87P-92P.
Dewdney, P.E. and Roger, R.S. 1981, Ap. J. (in press).
Harris, S. 1976, Monthly Notices Royal Astron. Soc. 174, pp. 601-607.
Herbig, G.H. 1971, Ap. J. 169, pp. 537-541.
Knapp, G.R., Kuiper, T.B.H., Knapp, S.L. and Brown, R.L. 1976, Ap. J. 206, pp. 443-451.
Mahoney, M.J. 1976, Ph.D. Dissertation, University of British Columbia.
Sharpless, S.L. 1959, Ap. J. Suppl. 4, pp. 257-279.
Shuter, W.L.H. and McCutcheon, W.H. 1974, Jour. Royal Astron. Soc. Canada 68, pp. 301-306.
Wilson, W.J., Schwartz, P.R. and Epstein, E.E. 1973, Ap. J. 183, pp. 871-881.

DISCUSSION FOLLOWING PAPER BY CHRISTIE ET AL.

BALLY: We also have observed the Lk Hα101 region at Bell Laboratories and have found very complex ^{12}CO spectra with multiple peaks. Yet the ^{13}CO lines are simple, single peak profiles. Evidently there is much ^{12}CO self-reversal over a large portion of the molecular cloud.

BEICHMAN: What do the Bell Laboratory observations show right at the position of Lk Hα101? Is there any evidence for asymmetries in the line?

BALLY: We do not detect any high velocity ^{12}CO line wings at the position of Lk Hα101 to a level of $T_A^* < 100$ mK. But there is a localized, compact ^{12}CO hotspot there.

FORMATION OF A B0.5 STAR DUE TO THE INTERACTION OF A SHOCK WAVE WITH A MOLECULAR CLOUD IN IC1805

V.A. Hughes
Sterrewacht, Postbus 9513, 2300 RA Leiden, The Netherlands.
Astronomy Group, Queen's University, Kingston, Ontario K7L 3N6

ABSTRACT. An isolated B0.5 star has been observed, using radio techniques, in the optically obscured region ahead of a shock front which is propagating into a molecular cloud. The shock is driven chiefly by stellar winds from an OB-association. The method for detecting the star is described as well as the parameters for the region; some implications to star formation are given.

I. INTRODUCTION

Various theories have been proposed to describe the way in which a cloud of interstellar gas can be induced to collapse and form stars. In the original mechanism by Jeans, irregularities form in which the self-gravitational force is greater than the internal pressure. More recently, additional factors have been recognized such as rotation, the presence of a magnetic field, the mixture of dust and gas, and possible dissociation of molecules, but some initial mechanisn is usually considered necessary in order to start the collapse. One, analyzed by Woodward (1976), involves the cloud encountering a shock wave such as that associated with the galactic spiral density wave; this has the effect of distorting the cloud and producing localized condensations on the outside, though the centre of the cloud can remain essentially undisturbed. Others involve cloud-cloud collisions (Loren, 1976), pressure waves produced by supernova remnants (Wootten, 1977; Herbst and Assousa, 1977) or HII regions (Elmegreen and Lada, 1977), or the formation in ionization fronts (Chevalier and Theys, 1975; Giuliani, 1980) or interstellar shocks (Welter, this meeting).

There is some difficulty in testing the theoretical predictions since when O-type stars form at the outside of clouds they produce a sufficiently large HII region that the cloud is dispersed rapidly, while later type stars are only seen optically when the original conditions have changed significantly. Recently, Hughes and Vallée (1978) made some progress in identifying young stars in HII regions. They showed that, in a number of cases, if the radio emission

R. S. Roger and P. E. Dewdney (eds.), Regions of Recent Star Formation, 349–355.
Copyright © 1982 by D. Reidel Publishing Company.

from the associated HII region is used to determine the excitation parameter of the exciting star, and through this its mass and luminosity by using the stellar models of Panagia (1977), then the luminosity so obtained was very close to the total infrared luminosity as measured by Fazio et al (1975). Thus, assuming that the stellar models are correct, and that all the radiation from the star is absorbed in the surrounding dust, it is possible to infer its spectral type, mass and luminosity even when it is in a region of very heavy optical obscuration.

One HII region at the edge of IC1805 was of particular interest, namely that designated IRS5 by Fazio et al. (1975) or as infrared source AFGL 333 by Price and Walker (1976). It is referred to here as G134.2+0.8. Its presence shows that an isolated B0.5 star has formed in the dense region ahead of a shock front, set up chiefly by stellar winds from the OB association within IC1805 (Vallée, Hughes and Viner, 1979), moving into the molecular cloud observed by Lada et al (1978). The OB association appears to be about 10^6 yrs old, and to have dispersed the surrounding gas. Further infrared measurements at λ50μm by Thronson, Harvey and Gatley (1979) essentially confirmed the presence of the star, but they pointed out the anomalous fact that though the star is expected to form in a region where the density is $\geq 10^7$ cm^{-3}, the electron density of the HII region is $\leq 10^3$ cm^{-3}.

Since, as mentioned previously, details of the region where stars form are of considerable interest, the region has been re-observed with the Westerbork Synthesis Radio Telescope (WSRT) at λ6cm with an angular resolution of about 9". This paper describes the results.

TABLE 1

EQUIPMENT PARAMETERS

Observing wavelength	6cm	
Single dish beam pattern HPW	11′	
No. of spacings	18	
Minimum baseline	36m	600λ
Maximum baseline	1260m	21000λ
Incremental spacing	72 m	
Highest resolution (RA x dec)	7"9 x 9"0	
Radius of first grating response	2'9 x 3'3 (RA x Dec)	
Observing time	12^h	
RMS noise at field center	0.1 mJy	
Field center, (1950.0)	$2^h24^m35^s$ $61^o15'00"$	

II. RESULTS

The equipment parameters used during the observations are given in Table 1; the reduction is described in more detail by Hughes (1981). The basic results are shown in the map of the area covering 9′ x 9′ in Figure 1, and the enlarged map covering 2'5 x 2'5 in Figure 2.

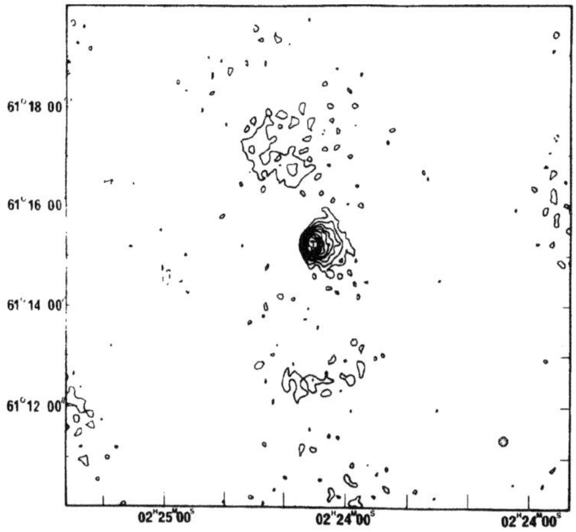

Figure 1.

A map of the area 9′ x 9′ at λ6cm around the HII region G134.2 + 0.8. Contour levels are at 1,2,3,4,5,6,7,8,9,10 mJy/beam area. The size of the antenna beam, 7"9 x 9"0, is shown by the shaded area in the bottom right hand corner.

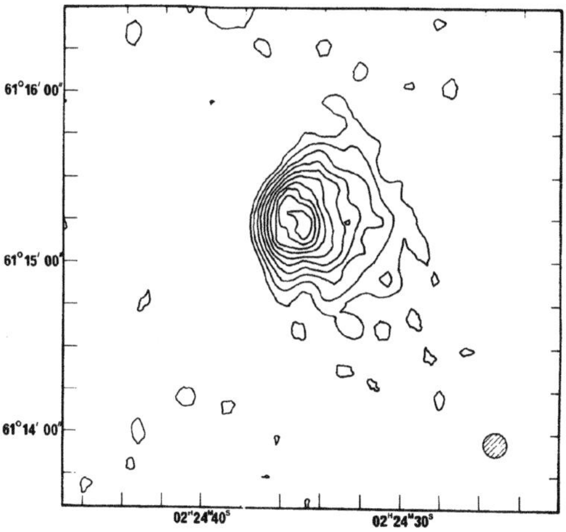

Figure 2.

Enlargement of the central 2'5 x 2'5 of the map in Figure 1. Contour values are as in Figure 1. The size of the antenna beam is shown by the shaded area in the bottom right hand corner.

Figure 1 shows the isolated HII region G134.2+0.8 together with some low level emission to the north and south of it. The large number of apparently small emission regions are believed to be due to noise superimposed on a ridge of emission running nearly north-south. The ridge does not appear over the whole of the map, since the minimum antenna spacing of 36m precludes regions of angular size greater than about 3′; the ridge is seen in more detail in the WSRT λ21cm results (Vallée and Hughes, 1978). From the combined WSRT results at λ6cm and λ21cm, and the λ4.7cm results from the Algonquin Radio Observatory (Vallée, Hughes and Viner, 1979), the ridge appears to be part of an ionized shell of gas of thickness 1 pc and electron density 50 cm^{-3}, on the inside of a denser, un-ionized shell; the ionization appears to be consistent with that which would be produced by UV-radiation from the OB association in IC1805.

The enlargement of the central region, as shown in Figure 2, is remarkabley lacking in structure except for the compression of the contours to the east, attributed to pressure of the stellar wind from the OB association which contains at least 9 O-stars. The integrated flux density at λ6cm is 154 mJy, which when compared with the value of 146±10 mJy obtained at λ21cm is consistent with a thermal source.

The flux density, S, can be used to determine the excitation parameter, since provided that the HII region is optically thin,

$$S = 3.88 \times 10^{6} \, [17.7 + \ln(T^{3/2}/\nu)] \, T^{-1/2} \, U^{3} \, D^{-2} \qquad (1)$$

where S is in mJy, T (K) is the temperature of the HII region, U (pc cm^{-2}) is the excitation parameter, and D (pc) is the distance. Assuming $T = 10^4$ K, D = 2.3 kpc, ν = 4.995 GHz,

$$S = 6.7 \times 10^{-2} \, U^{3} \qquad (2)$$

so that for S = (140-157) mJy, U = 13. The value of U is fitted to the stellar models by Panagia (1973) to determine the spectral type and total luminosity; the mass of the star is determined assuming that $\log(L/L_\odot) \simeq 4\log(M/M_\odot)$. Given the mass, the time taken to contract onto the main sequence is obtained from the work by Iben (1965). Parameters derived for the HII region and exciting star are shown in Table 2.

III. DISCUSSION AND CONCLUSIONS

The above analysis has determined the parameters of a newly formed star embedded in a region of high optical obscuration. The star has formed in or close to a thin shell of ionization produced by the Lyα emission from an OB association. This shell is on the inside of a larger diameter, denser region, which is expanding and powered by the stellar wind from the OB association, with a possible contribution from a supernova (Vallée, Hughes and Viner, 1979). Other evidence for the

TABLE 2

PARAMETERS OF HII REGION AND EXCITING STAR

HII REGION. G134.2 + 0.8	
Co-ordinates (1950.0)	RA = $02^h24^m35^s.26 \pm 0^s.15$
	Dec = $+61^o15'17''.1 \pm 0''.3$
Total Flux Density	173 mJy
Corrected for ridge	154 mJy
Angular size (RA x Dec)	25" x 29"
Linear Size	0.28 x 0.32 pc
Excitation Parameter	13 pc cm^{-2}
Electron Density	850 cm^{-3}
Infrared Luminosity*	1.4×10^4 $L_\odot$
EXCITING STAR	
Spectral Type	BO.5
Mass	13 M_o
Luminosity	3.0×10^4 $L_\odot$
Time to contract onto M.S.	5.5×10^4 yrs
IONIZED SHELL	
Thickness of shell	1 pc
Electron Density	50 cm^{-3}

* At $\lambda \geqslant 30\mu m$ (Thronson et al, 1979)

expansion is the apparent compression of the HII region to the east, and as pointed out by Vallée and Hughes (1978), by the additional presence of a band of optical obscuration to the west.

In a previous analysis, Hughes and Vallée (1978) showed that the luminosity of the star, as derived from the parameters of the HII region, approximated the $\lambda 100\mu m$ luminosity as derived by Fazio et al (1975). However, more recent observations by Thronson et al (1979) revised the value for luminosity to 1.4×10^4 $L_\odot$ at $\lambda 100\mu m$, compared with the value of 3.0×10^4 $L_\odot$ from the radio results. But they show also that the radio source is biassed towards the west of the infrared source, towards denser regions of the expanding shock wave. This immediately suggests that the difference in the two luminosities could be due to leakage of the infrared radiation towards the east, where the density of absorbing dust will be much smaller, and illustrates one possible limitation to the method for identification of stars.

An anomaly has been pointed out by Thronson et al (1979), namely, that the electron density of the HII region is remarkably small if it reflects the atomic density of the region from which the supposedly young star formed. However, since the excitation parameter is known, we can use the normal physical treatment for the expansion of an HII region (e.g., Spitzer, 1978), to estimate the original density if the age is known. Thus, if N is the present density, N_o the original density, U the excitation parameter and t_4 is the age in 10^4 yrs,

$$N \simeq 2.9\ N_o^{3/7}\ U^{6/7}\ t_4^{-6/7}\ \text{cm}^{-3}$$

$$\simeq 26\ N_o^{3/7}\ t_4^{-6/7}\ \text{cm}^{-3}$$

assuming that U $\simeq$ 13 pc cm^{-2}. Taking N = 850 cm^{-3}, if t_4 = 6, then $N_o = 1.2 \times 10^5\ \text{cm}^{-3}$, or if t_4 = 10, then $N_o = 3.4 \times 10^5\ \text{cm}^{-3}$. Since, subsequent to its formation, the HII region appears to have moved, relative to the shock front, into a much lower density region of $\simeq$50 cm^{-3}, where it is expected to have a more rapid expansion rate, the initial density would certainly have been much higher than the above. For example, if the initial density had been 10^7 cm^{-3}, the present density would have been expected to be about 6×10^3 cm^{-3}. Some evidence for this expanding gas may be seen around the HII region where the emission from the ridge is somewhat reduced.

The scenario for star formation in this particular case appears to be one where an expanding shock wave, produced by stellar winds from an OB association, moves into the denser region of a molecular cloud, leading to the collapse of a small region, ahead of the front, to form a star. The shock wave continues to move ahead and across the star, which is now situated close to the ionized shell behind the shock, produced by the Lyα radiation from the OB association. Thus it seems likely that the star formed at the edge of the molecular cloud where it is compressed to quite high densities by the shock wave, as shown by the fact that at present there is at least 12^m of extinction in the direction of the HII region (Vallée and Hughes 1978). It was not in an ionization front as in the mechanism by Giuliano (1980), or at a shock front with subsequent forward projection as in the mechanism by Chevalier and Theys (1975). Also, only one isolated B0.5V star formed, whereas a number of earlier type stars appear to form when shock fronts on a larger scale are involved, such as those associated with galactic spiral arms, or theoretically as in the mechanism by Elmegreen and Lada (1977). A number of later type stars could be forming, but there is no evidence for this.

There are two other reports on the observation of star formation in shock fronts. Loren and Wootten (1978) have observed a bright infrared source within a bright-rimmed dust cloud at the edge of the IC1848 HII region, which they attribute to a B0.5 or B1 star. More recently Felli, Johnston and Churchwell (1980) detected a radio point source in M17 which they associate with a B0 star which has been induced to form by a partially focused shock front which is propagating into a molecular cloud. The particular case mentioned in this paper gives more detail of the region of star formation. But in all the three observed cases, star formation is in small scale shock fronts, and it is stars of spectral type B0 to B1 that are formed. Some stars of later type may be forming, but if so they do not have sufficient Lyα output to produce a detectable HII region, and in any case they would not yet be expected to have come onto the Main Sequence. Of interest is the fact that in the Cep A condensation in a molecular cloud (Sargent, 1979), about 15 young stars have been found using the above technique, but all are of spectral type B3 (Hughes and Wouterloot, 1981). It would appear that stars of earlier spectral type form elsewhere, probably associated with larger scale shock fronts at the leading edges of spiral arms, where much more massive condensations can be produced, but where the dust content is much less.

The Westerbork Synthesis Radio Telescope is operated by the Netherlands Foundation for Radio Astronomy (SRZW), with the financial support of the Netherlands Organization for Pure Research (ZWO). Most of this work was carried out when on sabbatical leave at Leiden Observatory; some was under an Operating Grant from the Natural Sciences and Engineering Research Council of Canada.

REFERENCES

Chevalier, R.A. and Theys, J.C. 1975, Astrophys. J., 195, 53.
Elmegreen, B.G. and Lada, C.J. 1977, Astrophys. J., 214, 725.
Fazio, G.G., Kleinmann, D.E., Noyes, R.W., Wright, E.L., Zeilik, M. and Low, F.J. 1975, Astrophys. J., 199, L177.
Felli, M., Johnston, K.J. and Churchwell, E. 1980, Astrophys. J., 242, L157.
Giuliani, J.L. 1980, Astrophys. J., 242, 219.
Herbst, W. and Assousa, G.E. 1977, Astrophys. J., 217, 473.
Hughes, V.A. 1981, Astron. Astrophys. (Submitted)
Hughes, V.A. and Vallée, J.P. 1978, Astron. Astrophys., 69, 445.
Hughes, V.A. and Wouterloot, J.G.A. 1981, Astron. Astrophys. (Submitted)
Iben, I. 1965, Astrophys. J., 141,993.
Lada, C.J., Elmegreen, B.G., Cong, H-I. and Thaddeus, P. 1978, Astrophys. J., 226, L39.
Loren, R.B. 1976, Astrophys. J., 209, 466.
Loren, R.B. and Wootten, H.A. 1978, Astrophys. J., 225, L81.
Panagia, N. 1973, Astron. J., 78,929.
Price, S.D. and Walker, R.G. 1976, "AFGL Infrared Sky Survey" (Air Force Geophysics Lab., AFGL-TR-0208).
Sargent, A.I. 1979, Astrophys. J., 233, 163.
Spitzer, L. 1978, "Physical Processes in the Interstellar Medium"; John Wiley and Sons, New York.
Thronson, H.A., Harvey, P.M. and Gatley, I. 1979, Astrophys. J., 229, L133.
Vallée, J.P. and Hughes, V.A. 1978, Astrophys. J., 223, L97.
Vallée, J.P., Hughes, V.A. and Viner, M.R. 1979, Astron. Astrophys., 80, 186.
Woodward, P.R. 1976, Astrophys. J., 207, 484.
Wootten, H.A. 1977, Astrophys. J., 216, 440.

CHEMISTRY RELEVANT TO MOLECULAR CLOUDS NEAR HII REGIONS

W. D. Watson[1,2] and C. M. Walmsley[2,3]
[1]Astronomy Department, Univ. of Illinois at Urbana-Champaign
[2]Physics Department, Univ. of Illinois at Urbana-Champaign
[3]Max Planck Institut für Radioastronomie, Bonn

ABSTRACT

The successes and failures of models of interstellar chemistry are discussed with particular reference to molecular clouds, such as Orion, which are in the immediate neighborhood of HII regions. Some recent laboratory measurements at temperatures below 100 K are reviewed and their astrophysical consequences discussed. From a survey of abundance measurements towards Orion, it is concluded that composition differences between hot HII region clouds and cold dark clouds are not great. The influence upon the chemistry of shocks and mass outflow from young protostars is discussed. It is concluded that although some anomalies do exist, gas phase ion-molecule type schemes can account for most of the qualitative features of the observations.

I. INTRODUCTION

HII regions are beacons for star formation and we take the title of this review to imply a discussion of the chemistry of molecular clouds in "active" star forming regions; there are usually regions of ionized hydrogen nearby. Aspects of the chemistry in these regions will be contrasted with that expected for the (perhaps idealized) quiescent molecular clouds. The differences which we will focus on are: (a) Higher kinetic temperatures in the molecular gas under observation near active regions (~ 50-100 K versus ~ 10 K in dark molecular clouds). We are just beginning to appreciate the variation in rate coefficients for chemical reactions that can occur as the temperature is reduced from 300 K (at which essentially all the currently available data is taken) to the 50-100 K region, and on further to the 10 K neighborhood. (b) Non steady-state, non in situ effects on the molecular abundances near active regions, in particular shock waves and mass outflow from cool stars/proto-stars. For most of the past ten years during which the importance of gas-phase chemistry initiated by ionizations due to "high-energy" cosmic rays

R. S. Roger and P. E. Dewdney (eds.), Regions of Recent Star Formation, 357–377.
Copyright © 1982 by D. Reidel Publishing Company.

has been appreciated, studies of the chemistry of interstellar clouds have centered around the time-independent, in situ application of this "ion-molecule" reaction scheme. In the past few years, however, evidence for the occurrence of shocks in interstellar clouds has mounted (e.g., emission of vibrational radiation from H_2). Studies of the (molecular) mass outflow from cool stars (e.g., H_2O and SiO masers) and, in particular, of the BN-KL region in Orion have brought renewed attention to outflow as a source for interstellar molecules.

To appreciate the "differences", we will summarize in section II certain key aspects of the current ideas of the chemistry of the (supposed) quiescent regions. We then go on to discuss in section III certain recent results from laboratory studies of reaction rates with particular reference to their temperature dependence. In section IV we examine observed abundances in warm and cold clouds as well as make a comparison of the plateau and spike in Orion. In section V, we discuss three cases where it appears that we cannot fit the observations. In particular, we consider the consquences of new laboratory measurements of the H_3^+ deuteration reaction. In section VI, we review the results of shock chemistry and consider qualitatively what are likely to be the main processes governing the observed abundances towards Orion. Finally, in section VII, we summarize our conclusions.

II. ION MOLECULE CHEMISTRY

Ion-molecule reactions tend to have no activation energy barrier if the reaction is exothermic. Reactions with the H_2 molecule, the primary astronomical reactant, are also more often exothermic if the partner is positively ionized. Postive-ion molecule reactions usually occur at, or near, the classical collision rate relatively independent of collision energy. About ten years ago when only a few chemical reactions had been measured below 300 K and, altogether, only a limited number of the reactions of likely astrophysical importance had been measured at all, these virtues of positive ion molecule reactions were quite attractive. The key point for establishing that such reactions can be effective is the recognition that charge can be transferred efficiently from hydrogen and helium to carbon, nitrogen, oxygen, etc. The ionization rate for the latter can then increase, relative to the direct ionization rate, by a factor of $\sim 10^2$ to 10^4. One problem is that in even moderately dense clouds ($\sim 10^4$ cm^{-3}), the rate at which heavy elements hit dust grains exceeds the rate at which they can be ionized. Except for the formation of molecular hydrogen in which reactions on the surfaces of dust grains dominate under most astronomical conditions, there is no acceptable understanding of how molecules can be returned to the gas (Watson and Salpeter 1972). In fact, they may remain on the grains as suggested by certain observations (Wootten et al. 1978). The contribution of grains is usually ignored in recent studies.

The diffuse clouds (especially those toward Zeta Ophiuchi) have served as the laboratories for interstellar chemists because we can "see" through them and establish reasonably well the physical conditions in which to apply the chemical reactions. Except for those producing molecular hydrogen, the chemical reactions that produce the molecules in diffuse clouds reach steady state in a quite short time (~ 100-1000 yrs). Time independent, in situ formation and destruction is thus a reasonable assumption. Here, photoionization of elements with ionization potentials less than 13.6 eV determines the ionization (mainly C^+). For example, reactions in diffuse clouds that are important in the transfer of charge following ionization of H-atoms by cosmic rays are,

$$H^+ + D \rightleftarrows H + D^+ \tag{1}$$

and

$$H^+ + O \rightleftarrows H + O^+ \tag{2}$$

These lead to molecular HD,

$$D^+ + H_2 \rightarrow HD + H^+ \tag{3}$$

and OH and H_2O by a series of similar reactions with H_2 followed by dissociative electron recombination and photodissociation. Radiative association of carbon ions with H and especially H_2 plays a critical role in initiating the formation of carbon-bearing molecules. Once the hydrides are produced, chemical exchange reactions produce CO, CN, etc. These ideas have been incorporated into a refined chemical model for the gas cloud toward Zeta Ophiuchi (Black and Dalgarno 1977). The model reproduces the observed abundances quite well, though it has recently been argued that the actual physical conditions of the gas are rather different from those of the model (Crutcher and Watson 1981). A long standing problem is that the observed abundance of CH^+ is not reproduced by such chemical models for diffuse clouds.

When chemistry initiated by cosmic ray ionization is applied in molecular clouds which are thick enough that ultraviolet radiation can be ignored, the key initial ionizations are of H_2 and He by cosmic rays to yield H_2^+ and He^+. The H_2^+ reacts immediately,

$$H_2^+ + H_2 \rightarrow H_3^+ + H \tag{4}$$

Because of the low electron density in these clouds, the H_3^+ will frequently react to produce other molecular ions, especially,

$$H_3^+ + CO \rightarrow HCO^+ + H_2 \tag{5}$$

Thus an appreciable fraction of the cosmic ray ionizations of the gas produce HCO^+ so that it is a key indicator of the importance of ion-molecule reactions. The identification of a strong, widely occurring microwave line with HCO^+ (after its prediction from the above chemistry) as a result of the laboratory measurement of its frequency (Woods et al. 1975) has been the pivotal datum in establishing the relevance of a gas-phase chemistry oriented around positive ions for interstellar clouds. Subsequently, N_2H^+ was also detected.

The ionization of helium actually plays a more important role in the formation of most molecules because ionized helium reacts only very slowly with hydrogen. Instead, most of its ionization is transferred to carbon, nitrogen, etc.

Extensive networks of reactions have been solved by various investigators (Mitchell, Ginzburg & Kuntz 1978; Prasad and Huntress 1980a,b; Graedel et al. 1981) and show that most small molecules can be produced in abundances that are in reasonable agreement with observation, given the latitude provided by the various uncertainties.

In addition to generating the chemical species, ion-molecule reactions in steady state also lead to a dramatic isotope selection in favor of deuterium over hydrogen in molecules. Several reactions might be important in the selection, though

$$H_3^+ + HD \rightleftarrows H_2D^+ + H_2 + \Delta E \qquad (6)$$

seems likely to be of widest application. Its rate coefficient and exothermicity have recently been measured down to 80 K (Adams and Smith 1981). We discuss the question of deuterium enhancement in Section V.

Finally, we note that the above discussion has concerned itself purely with gas-phase chemistry. Yet, as mentioned earlier, at high densities the sticking rate for most molecules on dust grains exceeds their ion-molecule destruction rate. Also, grain surface recombination is required to understand the abundance of molecular hydrogen in diffuse clouds. It is therefore difficult to avoid the conclusion that grain surface reactions or at least molecule depletion onto grain surfaces must be important. Nevertheless, grains have been ignored for the sake of simplicity and this has perhaps been an error. The justification is that gas phase chemistry alone has been able to account reasonably well for observed abundances.

More extensive reviews of the basic steady-state, in situ interstellar chemistry are available (Watson 1976,1978).

III. TEMPERATURE DEPENDENT REACTIONS

In this section, we discuss some temperature dependent reactions which may be of importance for the chemistry. Molecular clouds near active regions seem to be warmer (50-100 K) than the more isolated dark clouds ($T \sim 10$ K). This difference can occasionally have a substantial effect on the relative importance of various contributing processes. We now discuss the laboratory data available for several types of reactions.

1. Dissociative electron recombination typically has a $T^{-1/2}$ dependence which will not cause dramatic effects, e.g., data by McGowan and co-workers (1979) down to energies corresponding to temperatures of ≃ 100 K. It has been argued that at lower temperatures, for example in the case of H_3^+ discussed by Carney and Porter (1977), that there should be significant deviations from the standard variation. There is some debate about the accuracy of sub-thermal measurements, especially for $CH^+ + e$.

2. Essentially no reliable rate coefficients for radiative association of atomic and molecular ions with neutrals are available. Since the investigation of Bates and Spitzer, the rate for $C^+ + H \rightarrow CH^+ + h\nu$ has been debated in connection with the abundance of CH^+. In recent years, $C^+ + H_2 \rightarrow CH_2^+ + h\nu$ has received more attention than any other radiative association. There is nevertheless about a three order-of-magnitude range of reasonable values for its rate coefficient (10^{-14}-10^{-17} cm^3s^{-1}). Refined laboratory studies are proceeding (Luine and Dunn 1981). We believe that it is fair to say that no radiative association has been unambiguously detected in the laboratory. Studies of 3-body associations have been used to infer information about radiative associations and these have been interpreted to indicate large rates which increase rapidly at low temperatures (e.g., $CH_3^+(H_2,h\nu)CH_5^+$, $\langle\sigma v\rangle \simeq 4 \times 10^{-13}$ cm^3s^{-1} @ 50 K and proportional to T^{-4}; Smith and Adams 1980). However, relating 3-body associations to radiative association is fraught with pitfalls (e.g., Herbst 1980a,b) and rate coefficients deduced in this manner should probably be considered as upper limits. On the other hand, some recent laboratory experiments have been interpreted as measuring two body associations directly. For $CH_3^+ + HCN \rightarrow CH_3HCN^+ + h\nu$, a rate coefficient 2×10^{-10} cm^3s^{-1} is found at 300 K (McEwan et al., 1980). It varies as $T^{-2\pm1}$ and is thus not inconsistent with the T^{-3} dependence predicted theoretically (Bates 1979a,b).

3. Chemical exchange reactions, i.e., $A + B \rightarrow C + D$ where A, B, C and D are atoms or molecules. These reactions are the backbone of the ion-molecule oriented, gas phase chemistry. However, until the past few years only a few such reactions had been measured below 300 K. There has always been the concern that small activation energy barriers are likely, even in strongly exothermic reactions, for the neutral-neutral case. In contrast, it has been expected that in reactions between positive-ions and neutrals, the rate coefficient would be constant below 300 K if the measured value were approximately the orbiting rate. Any rate coefficient that is significantly less than the kinetic collision rate has been assumed to have an energy

barrier that will make it negligible at $T \lesssim 100$ K. These probably remain good, general guidelines though as low temperature data accumulates for ion-molecule reactions it is clear that the deviations can be drastic.

One such case is that of $NH_3^+ + H_2 \rightarrow NH_4^+ + H$ which is the direct way to produce the widely observed and abundant NH_3 molecule in the ion-molecule scheme. Until recently, it was thought that this channel is ineffective because the measured rate coefficient at 300 K is about 10^{-13} $cm^3 s^{-1}$. By the usual reasoning, it would be expected to be completely unimportant at $T \lesssim 100$ K. New measurements (Luine and Dunn 1981) however indicate that the rate coefficient for this key reaction varies slightly but remains greater than 10^{-13} $cm^3 s^{-1}$ down to about 10 K. Below 10 K, it increases! The (seemingly) chemically straightforward reaction $N^+ + H_2 \rightarrow NH^+ + H$ also has a peculiar temperature dependence.

A key aspect of ion-molecule chemistry is that He^+ reacts extremely slowly with H_2 so that essentially all of the ionizations of helium are utilized in the chemistry of C, N, O etc. However, Johnsen et al. (1980) now find that the rate coefficient for $He^+ + H_2$ remains approximately 10^{-13} cm^3 s^{-1} down to 78 K and thus the reaction cannot be completely neglected.

The temperature variations in (1)-(3) have been studied in only a few cases so that it is premature to attempt to say how they will be reflected in differences in molecular abundances between warmer molecular clouds and cooler clouds. A situation where we seem to be on firmer ground is in the isotope exchange reactions -- primarily (^{13}C, ^{12}C) and (D,H). Here attention can be focussed on a single reaction and the chief temperature dependence is due to the energetics of the reaction.

In the case of $^{13}C/^{12}C$, the key reaction is $^{13}C^+ + {}^{12}CO \rightleftarrows {}^{12}C^+ + {}^{13}CO$ (Watson, Anicich and Huntress 1976). This has been measured at low temperatures by Smith and Adams (1980) and is exothermic by $\Delta E/k \cong 35$ K. It can thus cause substantial effects when $T < 20$ K, as in dark clouds, but should be relatively unimportant when $T \sim 50$ K. This seems consistent with observations (see e.g. Langer et al. (1980). The situation is different for deuterium enhancement which we discuss later (see Section V, but here, also, some temperature dependence does appear to be observed (Snell & Wootten (1979)).

IV. OBSERVED MOLECULAR ABUNDANCES

We now turn in more detail to the question of the observed molecular abundances. In particular, we consider the OMC1 region, especially the various velocity components in BN-KL. We consider that this may be a prototype for active regions of young, bright

Table 1: Molecular Abundances in Orion and L183

Molecule or ratio	Orion Ridge Abundances	Plateau Abundances	Ref.	L183 Abundance	Ref.
DCO^+/HCO^+	2(-3)	-	1	~.015	26
DCN/HCN	5(-3)	-	1		
NH_2D/NH_3	5(-2)	-	1		
HDO/H_2O	1(-2)-1(-3)	-	2		
Log N(CO)	19.1	~ 18.8	3,4		3
Log (Ratio to CO)					
H_2O	-1.0	(-0.5)	2,5	-	-
SO	-4.1	-2.8	6	-3.3	27
H_2S	-5.1	-3.8	6,7	-	-
SO_2	-4.3	-3.3	6,8	-	-
OCS	-4.5	-2.8	9	<-4.3	9
CS	-4.7	-4.6	10	(-3.6)	28
SiO	<-5.3	-4.0	11,12	-	-
CN	-4.3	<-4.1	3,13	<-4.0	3
HCN	-3.7	-2.3	14	<-4.8	29
HNC	-5.2	<-3.1	15,16,17	-4.1	30
NH_3	-2.6	3.0	18	-2.5	31
H_2CO	-3.0	<-4.3	19,20,17	-3.9	32
C_2H	-3.7	<-3.6	21	-4.5	33
HC_3N	-5.3	-5.2	22,23	-5.3	31
CH_3OH	-3.7	<-3.8	24	-	-
HCO^+	-4.0	-4.4	14	-4.0	30
N_2H^+	-5.8	<-5.8	25	-5.2	34
Temperature	~ 50K	50-200K		10K	
Density	$\sim 10^6 cm^{-3}$	$\sim 10^6 cm^{-3}$		$\sim 10^4 cm^{-3}$	

References To Table 1

1. Wannier (1980). 2. Waters et al. (1980). 3. Allen and Knapp (1978). 4. Phillips et al. (1977). 5. Phillips et al. (1980). 6. Gottlieb et al. (1978). 7. Thaddeus et al. (1972). 8. Snyder et al. (1975). 9. Goldsmith and Linke (1981). 10. Goldsmith et al. (1980). 11. Genzel et al. (1980). 12. Lovas et al. (1976). 13. Churchwell (1980). 14. Rydbeck et al. (1981). 15. Goldsmith et al. (1981). 16. Huggins et al. (1979). 17. Erickson et al. (1980). 18. Wilson et al. (1979). 19. Evans et al. (1979). 20. Wilson et al. (1980). 21. Tucker amd Kutner (1978). 22. Bujarrabal et al. (1981). 23. Loren et al. (1981). 24. Gottlieb et al. (1979). 25. Turner and Thaddeus (1977). 26. Guélin et al. (1981). 27. Rydbeck et al. (1980). 28. Linke and Goldsmith (1980). 29. Churchwell (priv. comm.). 30. Frerking et al. (1979). 31. Ungerechts et al. (1980). 32. Wootten et al. (1980b). 33. Wootten et al. (1980a). 34. Snyder et al. (1975).

stars and gas. Better information an OMC1 is presumably available because of its proximity.

In Table 1a, we have collected recent column density measurements for OMC1. For comparison, we also give similar results for the nearby cold (T ~ 10 K) dust cloud L183 (L134N) (Table 1b). We have where possible made the distinction between the broad lines centered at V_{LSR} ~ 9 km s^{-1} (plateau emission) and the narrow line centered at V_{LSR} ~ 8 km s^{-1} (spike emission). It seems to be the case that there are several sources for the "plateau" emission in Orion. Thus, both maser and non-maser SiO seems to come from a circumstellar shell surrounding the object IRc2. Ammonia (NH_3) and SO (Welch et al., 1981; Bastien et al. 1981) also show plateau emission but the origin of the emission seems to be offset from IRc2. HCO^+ plateau emission appears to be more extended and come from yet another position. Finally, VLA ammonia measurements show that the 5 km s^{-1} NH_3 line comes from circumstellar gas surrounding another embedded IR object, IRc4. Thus, in general, one concludes that there are several sources of plateau emission. Some of these are circumstellar but not necessarily all. To account for the circumstellar emission, quite high mass loss rates (see e.g., Downes et al. 1981) of 10^{-4} to 10^{-3} $M_\odot$ yr^{-1}) have been invoked for IRc2. Given that IRc4 seems to be the source of the 5 km s^{-1} emission, it is probable that several of the IR objects are losing mass rapidly. It is difficult to make statements about plateau abundances before interferometer observations are available that show which of the emissions come from shells and that give relative positions for the various broad line features. Table 1a is therefore intended merely as a rough guide to abundances in OMC1, at least as far as plateau emission is concerned. Experience and comparison with other estimates suggest that all values in Table 1 should be treated with skepticism (an uncertainty of 0.5 in the log). As far as the dark cloud (L183) results are concerned, we just note that foreground absorption by low excitation material (see e.g., Frerking et al. 1979) complicates abundance determination.

Comparison of the "ridge" abundances with those of L183 indicates that the abundances are generally similar; e.g., HCO^+/CO ~ 10^{-4}. The most striking difference occurs in the abundance of deuterated species. These are enhanced by factors of up to 10^4 in cold dark clouds such as L183 but by factors of ~ 100 in OMC1. Apart from this, there may be a difference in the ratio [HNC]/[HCN] in dark clouds and in HII region molecular clouds (Goldsmith et al. 1981). However, the main conclusion which we draw from Table 1 is that while composition differences may exist between hot "active" regions such as OMC1 and cold dark clouds, they are not large. This suggests that the basic processes determining the chemistry are the same in the two cases and are not very temperature sensitive. This is certainly consistent with the gas phase schemes which we discussed earlier but, of course, it is not unique to them. It is useful now to scrutinize the "plateau" abundances with the "ridge" abundances as benchmarks. There appear to be real differences.

1) Certain molecules seem to be about one order-of-magnitude more abundant in the plateau -- HCN, SiO and the sulfur molecules SO, H_2S OCS and SO_2 (but not CS).

2) Other molecules -- HCO^+, N_2H^+, C_2H, CN, CH_3OH, H_2CO and HCO are more abundant in the ridge (mainly ions and radicals).

3) HNC/HCN (plateau) is less than HNC/HCN (ridge). At present sensitivity limits, deuterated molecules seem to be absent in the plateau.

These differences appear to be real. However, they are not enormous and in some cases still within the "error bars". It seems to be an open question whether ridge material is gas which has been processed through a protostar envelope or whether it is "unsullied" molecular cloud gas. Conversely, plateau material may be outflowing gas or may have been swept up from the molecular cloud material by a stellar wind. Observationally, it is difficult to tell and the observed abundances of plateau and ridge at best suggest that there may be communication between the two.

V. SOME PROBLEMS WITH GAS PHASE CHEMISTRY

In this section, we will discuss three particular cases where low temperature, in situ gas phase chemistry has some difficulty in explaining observational data. These cases are the observed deuterium enhancement in various molecules, the observed HNC/HCN abundance ratio, and finally the problem of the oberved abundances of sulfur molecules. In none of these cases is it clear that gas phase chemistry "fails" but they are areas where the theory is in some trouble and hence they provide an indication that qualitatively different processes should be considered. We note also that the recent apparent detection of CO^+ (Erickson et al. 1981) cannot be explained on the basis of conventional ion-molecule theory.

Deuterium Enhancement

It has been generally assumed that reaction (6) is the fundamental cause of the deuterium enhancement observed in many species (see Table 1). Early estimates of the exothermicity (Watson 1976) suggested $\Delta E \approx 180$ K. However, as mentioned above, recent measurements by Adams and Smith (1981) have caused this number to be revised downwards to 90 K. This is probably still consistent with the deuterium enhancement observed in cold dark clouds such as L183 but it causes a problem when considering warmer clouds such as Orion. From ammonia measurements (see e.g. Bastien et al. 1981), the ridge temperatures are greater than 40 K and hence the maximum enhancement of

$H_2D^+(e^{\Delta E/kT})$ would seem to be only one order of magnitude. The observations, by contrast, seem to require enhancement by a factor of 100-1000 and hence there appears to be a contradiction with theory. However, the problem is complicated by the fact that (both in interstellar space and in the laboratory) the reactants in reaction (6) will in practice often be in excited rotational levels. Additionally, H_3^+ is a symmetric top molecule, whose "ground" (J = K = 0) rotational level is missing due to symmetry considerations (see e.g. Carney and Porter 1977), and which as a consequence has zero-point rotational energy which must be taken into consideration when estimating ΔE. All this would not matter if the laboratory experiment simulated the interstellar environment but it is not clear that it does. Also, theoretical estimates of ΔE, which take into account the H_3^+ zero point rotational energy, give considerably larger values for ΔE than found by Adams and Smith. Hence, we are not convinced that one can exclude reaction (6) as the basic deuteration mechanism in warm clouds. However, one should consider alternatives. The most obvious of these is the reaction:

$$CH_3^+ + HD \rightleftarrows CH_2D^+ + H_2 \tag{7}$$

Blint et al. (1976) estimate ΔE = 300 K for this reaction and so it potentially could explain the observed abundances of deuterated molecules. The difficulty here is to understand the degree of deuterium enhancement in molecules such as HDO and NH_2D which are not direct "descendants" of CH_2D^+. The enhancement in these molecules, as well as in DCO^+, will be diluted by reactions involving H_3^+. It would seem in fact that a necessary condition for the viability of a deuteration scheme based upon reaction (7) is that $[CH_3^+]/[H_3^+]$ should be large. This seems difficult to achieve in the conventional ion-molecule picture.

Another possibility is that the main DCO^+ formation route is:

$$D + HCO^+ \rightleftarrows DCO^+ + H \ (\Delta E \approx \text{large}) \tag{8}$$

We can make a rough estimate of how large an atomic hydrogen (and hence deuterium) concentration is required to explain the observed ratio of $[DCO^+]/[HCO+] \sim 2\ 10^{-3}$. We clearly require an HI concentration much larger than the steady state value appropriate when cosmic ray dissociation of H_2 is balanced by reformation on grains. However, since the time scale for this process is long, it is plausible that a non steady-state situation might occur. We estimate that $[H]/[H_2]$ must be at least 2×10^{-4} to explain the observed $[DCO^+]/[HCO^+]$ value. A consequence would be that the atomic hydrogen column density in the direction of OMC1 should be of order 10^{20} cm^{-2}.

One final possible explanation for the enhanced deuterium abundances found in Orion is that the deuterated molecules are to be found in a cold background region which is physically separated from the gas in the Kleinman-Low nebula where temperatures ~ 40-70 K are measured (Wilson et al 1979), Bastien et al. 1981). Over a more extended region, Ho et al. (1979) find that the ammonia rotation temperature is of order 25 K. This is probably a lower limit to the kinetic temperature. It seems possible that the deuterated molecules are to be found in this more extended colder region. Mapping, as well as analysis of higher rotational transitions of the deuterated species, may be able to test this hypothesis.

In summary, the small value of ΔE found by Adams and Smith (1981) for reaction (8), puts some pressure on models of gas phase chemistry. The question of possible internal excitation of reactants and products may need clarification both as far as the laboratory experiment is concerned and in the interstellar situation. On the astronomical side of the problem, we feel that the possibility that reaction (8) is of importance for deuterium enhancement is a strong argument, if one were needed, for high angular resolution 21-cm measurements of OMC1.

The HCN/HNC Ratio

One would hope that a reasonably uncomplicated prediction of the HCN/HNC ratio could be made theoretically. Simple considerations suggest that the ratio should be close to unity. Herbst (1978) has analysed the dissociative recombination of $HCNH^+$ using a statistical theory.

$$HCNH^+ + e \rightarrow HCN + H \qquad (9a)$$
$$\rightarrow HNC + H \qquad (9b)$$
$$\rightarrow CN + H_2 \qquad (9c)$$

Most gas phase chemistry models consider this reaction to be the main formation mechanism for HCN and HNC. Herbst finds that CN formation (branch c) should occur about fifty percent of the time and that the other fifty percent of the recombinations should produce about equal amounts of HCN and HNC. Destruction rates for HCN and HNC are liable to be very similar (reactions with C^+, He^+) and hence one concludes that the two species are liable to have almost identical abundances.

Recent observational evidence suggests that this is not the case. The situation has been summarized by Goldsmith et al. (1981) who find that in four "warm giant molecular clouds", [HNC]/[HCN] varies between 0.015 (Orion) and 0.4. These results are based upon a comparison of the rare isotopic species $HN^{13}C$ and $HN^{15}C$ with $H^{13}CN$ and $HC^{15}N$ and it is hoped therefore that the column density estimates are relatively independent of radiative transport effects. It is also suggested that in cold dark clouds such as L183, the situation is reversed and one has [HNC]/[HCN] > 1. This, if confirmed, would suggest that different formation channels predominate at low and high temperature or that more efficient isomerization occurs in the higher temperature regions.

Various reactions have been put forward which might account for these results. Goldsmith et al. (1981) suggest that the overabundance of HCN which they find might be understood if the reaction of CH_2 with N is fast at interstellar temperatures. Allen et al. (1980) (see also Brown 1977) have pointed out that the reaction of C^+ with NH_3 might lead preferentially to HNC via the isomer H_2NC^+. This form lies ~ 2 eV above the linear isomer $HNCH^+$ but it might be preferred since the most stable triplet configuration is H_2NC^+. Once produced, triplet H_2NC^+ might radiatively decay to singlet H_2NC^+ rather than isomerize to the linear form. Whether this happens in practice is problematic and depends upon the amount of vibrational excitation of the product ion. It is also not clear what fraction of the dissociations of H_2NC^+ would yield CN rather HNC. The extreme case would seem to be that HNC is favored by one order of magnitude over HCN as a consequence of the $C^+ + NH_3$ reaction.

It seems worth emphasizing however that, irrespective of the basic production mechanism, there will be a "reshuffling" of HCN and HNC due to reactions such as:

$$HCO^+ + HCN,HNC \rightarrow HNCH^+ + CO \tag{10}$$

$$H_3^+ + HCN,HNC \rightarrow HNCH^+ + H_2 \tag{11}$$

followed by dissociative recombination. It is important that, in this case, only $HNCH^+$ is energetically accessible. Also possible is direct conversion of HNC to HCN via proton exchange

$$HNC + H^+ \rightarrow HCN + H^+ \tag{12}$$

All of these processes compete with the destruction of HNC and HCN by ions (C^+,He^+) as well as, perhaps, by condensation onto grain surfaces. In a steady state situation, one can obtain an estimate of the importance of "reshuffling" by taking the results of Herbst (1978) cited above and assuming that the fifty percent of $HNCH^+$ recombinations leading to CN are destruction paths for HNC and HCN. Crude estimates that we have made on this basis suggest that one cannot obtain [HNC]/[HCN] greater than 3 or less than 1/3 if HCO^+ and H_3^+ are substantially more abundant than C^+ and He^+.

Finally, it may be significant that the largest overabundance of HCN found by Goldsmith et al. (1981) was in the Orion ridge source. Passage of gas through a shock might convert HNC into HCN although the details have not, to our knowledge, been worked out. Presumably outflow from a hot (2000 K) protostellar atmosphere would also produce preferentially HCN. It is well known that there is a considerable amount of shocked gas and protostar mass loss in the Orion region. It is tempting to speculate that the ridge gas may be being "contaminated" by the plateau. We will discuss this in more detail later but merely point out now that in the plateau (Table 1a), HNC is weak or absent.

Sulfur Gas Phase Chemistry

Sulfur chemistry under standard conditions has been discussed by Oppenheimer and Dalgarno (1974) and by Duley, Millar and Williams (1980). From these studies, it appears that there is a bottleneck in the production of sulfur bearing molecules. The problem is exemplified by considering the formation of H_2S in a gas phase scheme. This molecule is commonly detected in molecular clouds ($H_2S/H_2 \sim 10^{-9}$, Thaddeus et al. 1972) and it is sufficiently simple that one would expect its abundance should be a good test of theory. However, the following reactions are, for various reasons, "unsatisfactory"

$S^+ + H_2 \rightarrow SH^+ + H$ (13a) endothermic

$SH^+ + H_2 \rightarrow H_2S^+ + H$ (13b) endothermic

$S^+ + H_2 \rightarrow SH_2^+ + h\nu$ (13c) slow

$S + H_3^+ \rightarrow SH^+ + H_2$ (13d) does not produce $SH_2^+ + H$

$S^- + H_2 \rightarrow SH_2 + e$ (13e) very slow at best

Duley, Millar and Williams (1980) attempt to explain the discrepancy in terms of their model for reactions on grain surfaces. On the other hand, Hartquist et al. (1980) have pointed out that relatively high abundances of sulfur molecules can be produced in shock waves. In this case endothermic reactions such as:

$$S + H_2 \rightarrow SH + H \quad (14)$$

$$SH + H_2 \rightarrow H_2S + H \quad (15)$$

become important. It would be useful to have available observations of H_2S and other sulfur bearing molecules in cold, dark clouds where shocks and elevated temperatures do not seem to be prevalent. We note in passing that most sulfur molecules which might be expected to be abundant have in fact been observed and, in Orion, their combined abundance (Table 1) relative to CO is $\sim 2\ 10^{-4}$. This suggests that most sulfur is in atomic form.

VI. MAJOR NON-STEADY, NON in-situ EFFECTS; i.e., SHOCKS AND MASS OUTFLOW

If ultraviolet radiation can be ignored, the time to reach steady-state for the "ion-molecule" chemistry initiated by high-energy cosmic rays is about,

$$t_{im} \sim \left[\xi \cdot \frac{He}{CNO}\right]^{-1} \sim 10^6\text{–}10^7 \text{ yrs} \quad (16)$$

depending on the degree of depletion of CNO elements and on the cosmic ray ionization rate ξ per H-atom per s (10^{-17} adopted here). calculations, for example by Iglesias (1979), show this in detail.

A shock wave with a velocity of a few tens of km/s moving through a cloud with (pre-shock) density of 10^4 hydrogens cm^{-3} can thus leave behind it a column density ~ 10^{24} cm^{-2} or so in which the memory of the shock-produced abundances remains. Similarly mass outflow from a star at a rate ~ 10^{-5} $M_{\odot}$/yr produces a 10 $M_{\odot}$ region with a memory of the abundances in the stellar atmosphere. If the velocity of the flow is 10 km/s, the radius of this region can be a parsec or more.

If the cloud is thin enough that photoprocesses involving ultraviolet radiation are important, the time scale and hence the size of the region is reduced from the above estimates. For the interstellar averaged radiation field, a typical photodissociation time is 10^2-10^3 yrs. The above numerical values would then be reduced by a factor of 10^3 or more. Calculations by Scalo and Slavsky (1980) delineate this effect quantitatively for mass outflow from a cool star.

Both shocks and mass outflow have the effect of bringing into consideration endothermic reactions and producing molecular abundances that tend toward those that would be produced in a gas at thermodynamic equilibrium at an elevated temperature. The activity associated with HII regions which is likely to produce shocks and ejection of material by protostars associated with young stars in the neighborhood of HII regions may occur.

a) SHOCK CHEMISTRY

The chemical species produced in shocks have been calculated for various parameters by a number of investigators (e.g., Elitzur and Watson 1980; Iglesias and Silk 1978; Hollenbach and McKee 1979; Hartquist et al. 1980). The idea is that the high temperature behind the shock front drives the endothermic chemical reactions. Near thermodynamic-equilibrium ratios between reactants and products are obtained in most reactions. These can be altered as the gas cools (by radiation) back to molecular cloud temperatures and becomes more dense (see e.g., Spitzer 1978 for a discussion of the basic properties of shocks). After the post-shock gas has cooled, it will be proceeding toward the pre-shock region with a velocity equal to the shock velocity. The immediate post-shock, hot component will be moving in the same direction but with only about 3/4 of the shock velocity.

An effort has been directed toward establishing whether, in fact, the long sought cause for the high CH^+ abundance in diffuse clouds is shocks. Calculations (Elitzur and Watson 1980) show that column densities of CH^+ comparable to the large observed values will exist in the hot post-shock gas when a shock of 12 km/sec or so encounters the gas of a typical diffuse cloud. There should then be a tendency (depending upon the direction of the shock) to observe velocity shifts between the CH^+ line and those of the bulk of the cloud. There is evidence that this is the case (Chaffee 1975; Crutcher 1979; Frisch 1979; Federman 1980).

More extensive, in terms of molecular species, computations for the effects of shocks have been performed by Iglesias and Silk (1978) by Hartquist, Oppenheimer and Dalgarno (1980), and by Saito and Deguchi (1980). After the gas has cooled, Iglesias and Silk find that for a shock of 10 km/s, certain molecules are unperturbed (CO, NH_3, HCN and N_2), other molecules are greatly decreased in abundance (H_2CO, CN, HCO^+) and substantial amounts of other molecules are produced (H_2O, HCO and CH_4) relative to pre-shock, steady-state abundances. Shock chemistry has been extended by Hartquist et al. (1980) to the compounds of sulfur and silicon for which, as noted previously, the steady-state ion-molecule chemistry seems to be somewhat less rapid than for carbon and oxygen.

A second example of a gas in which the molecular abundances seem to have been altered by a shock is that of IC443 studied by DeNoyer and others (e.g., Dickinson et al. 1980; DeNoyer and Frerking 1981). Emission appropriate for a shock is also seen from H_2 (Treffers 1979). The apparent increase in the HCO^+ abundance through the shock, in contrast with expectation from calculations (Iglesias and Silk 1978), is a puzzle ($HCO^+/CO \sim 4\text{-}9 \times 10^{-4}$). Also OH/CO increases by by a factor ~ 100 through the shock. CS and HCN by contrast seem typical.

The shock in IC443 is also of interest because the apparent shock velocity is close to the critical value beyond which H_2 and most other molecules become dissociated. The problem has been studied in some detail by Dalgarno and Roberge (1979) and by Hollenbach and McKee (1980) (see also Hollenbach & McKee 1979). They conclude that the critical velocity for H_2 dissociation is ≈ 25 km s^{-1} for pre-shock densities greater than 10^4 cm^{-3}. However, for lower preshock densities, H_2 can survive shock velocities up to 50 km s^{-1} and this may be relevant for IC443. More recent calculations (Hollenbach, priv. comm.) include the influence of magnetic fields and of magnetic precursors to the shock (Draine 1980). The magnetic field causes the temperature structure of the shock to be modified. In a model with a preshock density of $5\ 10^5$ cm^{-3}, preshock field of 2 milligauss, and shock velocity of 55 km s^{-1}, the temperature behind the shock front is sufficiently depressed that molecular hydrogen can survive. This model also allows a reasonable fit to the relative intensities of vibrationally excited H_2 and rotationally excited CO lines seen towards Orion (Beckwith et al. 1978; Watson et al. 1980). The infrared H_2 lines show very large line widths (~ 100 km s^{-1}) and hence pose very directly the question of whether molecules can survive high velocity shocks. One possibility is that molecules may reform in the cool post-shock layer and this has been studied in the context of the galactic center gas by Saito and Deguchi (1980).

b) Orion Chemistry

Finally, we use Table 1 as a basis to make some comments about the chemical composition of the Orion molecular gas. We note the following:

1) The composition in Table 1 is in contrast to the study of Wootten et al. (1978) who find that the abundances of several common molecules (CO, HCO^+, HCN, HNC, H_2CO) scale inversely with density. For formaldehyde, Wootten et al. find

$$[H_2CO]/[H_2] = 3\ 10^{-4}\ (n(H_2))^{-1.26}$$

For a density of 10^6 cm^{-3} as might be appropriate in OMC1 (see e.g. Bastien et al. 1981), one finds on this basis $[H_2CO/H_2] \approx 10^{-11}$. This is clearly at variance with Table 1 and we tentatively conclude that the fraction of C, N, O elements already deposited out onto dust grain surfaces must be much smaller in OMC1 than in the majority of regions studied by Wootten et al. This might mean that the gas in Orion has been through some phase in which grain mantles have been evaporated (i.e. processed through protostar envelopes).

2) As discussed earlier, the observed high abundances of sulfur-bearing molecules are not reproduced in steady-state gas phase calculations. The contrast between plateau and ridge abundances of sulfur molecules together with the plateau linewidth suggests a high temperature origin for the plateau. Hartquist et al. (1980) obtain qualitative agreement with observation by postulating an 8 km s^{-1} shock. However, observed plateau line widths are considerably greater than this and it seems questionable whether the agreement would be maintained for faster shocks. There also appears to be a discrepancy between observed and predicted CS abundances.

3) The decrease in ion and H_2CO abundances in the plateau are also consistent with shock chemistry (Iglesias and Silk 1978). However HCN is predicted by Iglesias and Silk to be unaffected by a 10 km s^{-1} shock whereas it is observed to be more abundant in the plateau. Again, the theoretical predictions may be sensitive to shock velocity. It would be useful also to have available predictions for HNC as a function of shock velocity.

4) It appears that at least the SiO plateau consists of outflowing gas rather than shocked gas (Downes et al. 1981). It is not clear whether this is the exception or the rule. The region within the H_2 shock is ~ 10^{17} cm and may thus be reasonably filled by mass outflow according to our previous estimate. What can be said about the abundance spectrum in the outflowing gas? This is perhaps even more unclear than in the case of shocks because we neither know what type of stars we are dealing with nor do we know the nature of the outflow. Let's just look at what we have. The outflow calculations of Scalo

and Slavsky (1980) for a very low density/temperature initial condition (10^8 cm^{-3}, 1000 K) suggest that in addition to the usual high abundances of H_2O and SiO, SO and SO_2 also become abundant. We can also examine the thermodynamic equilibrium (giants and supergiants) by Tsuji (1964;1973), Dolan (1965) and Vardya (1966). Additional species -- H_2S -- are also seen to be abundant in these more extensive calculations. However, many others are not found to be abundant enough in the oxygen-rich atmosphere calculations, notably, CN, CS, OCS, NH_3 and HCN. Going to a carbon-rich atmosphere greatly increases CN, CS, HCN but seemingly not NH_3 and OCS. In a summary of data for IRC10216 -- a generally accepted case of the observed molecular gas being expelled from a cool, carbon-rich star -- McCabe, Connon-Smith and Clegg (1979) find that CN and NH_3 are two species that must be dramatically altered by the chemical reactions which occur in the outflow. They are increased by factors of 10^6 and 10^4, respectively, over the atmospheric abundances (see also McLaren and Betz 1980). Of course, the low HNC and absence of molecular ions would be equally indicative of a mass-outflow chemistry as it would be for shock chemistry.

VII. SUMMARY

The main conclusion of this review is that to a surprising extent, gas phase chemistry in situ has been successful. We have tried to outline some problem areas where the theory may be breaking down. There do appear to be real discrepancies between theory, as worked out to date, and observation but none where it is beyond doubt that an alternative to ion-molecule mechanisms is necessary. Certainly, one expects surface reactions, shocks, time dependent phenomena etc., to influence molecular abundances and, at some level, this might be the case. However, gas phase processes in steady-state have the advantage of allowing straightforward predictions and, in general, agreement has been reasonable. Moreover, the rough accord (Table 1) between "Orion abundances" and "dark cloud abundances" suggest that high temperature phenomena at least are not the major factor for chemistry. Hence, we conclude that for most purposes, HII region chemistry appears to be "normal chemistry". Despite the success of gas phase chemistry, we still acknowledge that the most abundant molecule in astronomy (the hydrogen molecule), but only this species, must be formed primarily on the surfaces of dust grains.

The authors' research is supported by the National Science Foundation.

References

Adams N., Smith D. (1981) preprint.
Allan M., Goddard, J. D., Schaeffer III, H.F. (1980) J. Chem. Phys. 73, 3255.
Allen M., Knapp G. R. (1978) Astrophys. J. 225, 843.

Bastien P., Bieging J., Henkel C., Martin R. N., Pauls T., Walmsley C.M., Wilson T. L., Ziurys L. M. (1981) Astron. Astrophys. 98, L4.
Bates D. R., Spitzer L. (1951) Astrophys. J. 113, 441.
Bates D. R. (1979a) J. Chem. Phys. 71, 2318.
Bates D. R. (1979b) J. Phys. B. 12, 4135.
Beckwith S., Persson S. E., Neugebauer G., Becklin E. (1978) Astrophys. J. 223, 464.
Black J. H., Dalgarno A. (1977) Ap. J. Suppl. 34, 405.
Blint R. J., Marshall R. F., Watson W. D. (1976) Astrophys. J. 206, 627.
Brown R D. (1977) Nature 270, 39.
Bujarrabal V., Guélin M., Morris M., Thaddeus P. (1981) Astron. Astrophys. (in press).
Carney G. D., Porter R. N. (1977) J. Chem. Phys. 66, 2756.
Chaffee F. H. (1975) Astrophys. J. 199, 379.
Churchwell E. (1980) Astrophys. J. 240, 811.
Crutcher R. M. (1979) Astrophys. J. 231, L151.
Crutcher R. M., Watson W. D. (1981) Astrophys. J. 244, 855.
Dalgarno A., Roberge W. (1979) Astrophys. J. 233, L25.
DeNoyer L. K., Frerking M. A. (1981) Astrophys. J. 246, L37.
Dickinson D. F., Rodriguez-Kuiper E. N., St. Clair Dinger A. Kuiper T. B. H. (1980) Astrophys. J. 237, L43.
Dolan J. F. (1965) Astrophys. J. 142, 1621.
Downes D., Genzel R., Becklin E. E., Wynn-Williams C. G. (1981) Astrophys. J. 244, 869.
Draine B. T. (1980) Astrophys. J. 241, 1021.
Duley W. W., Millar T. J., Williams D. A. (1980) Mon. Not. R. Astron. Soc.192, 945.
Elitzur M., Watson W. D. (1980) Astrophys. J. 236, 172.
Erickson N., Davis J. H., Evans III N. J., Loren R. B., Mundy L., Peters III W. L., Scholtes M., Vanden Bout P. A. (1980) in "Interstellar Molecules" (p. 25). IAU Symposium No. 87, ed. B. H. Andrew, D. Reidel Publ. Co., Dordrecht, Holland.
Erickson, N. R., Snell R. L., Loren R. B., Mundy L., Plambeck R. L. (1981) Astrophys. J. 245, L83.
Evans II N. J., Plambeck R. L., Davis J. H. (1979) Astrophys. J. 227, L25.
Federman S. R. (1980) Astrophys. J. 241 L109.
Frerking M. A., Langer W. D., Wilson R. W. (1979) Astrophys. J. 232, L65.
Frisch P. C. (1979) Astrophys. J. 227, 474.
Genzel R., Downes D., Schwartz P. R., Spencer J. H., Pankonin V., Baars, J. W. M. (1980) Astrophys. J. 239, 519.
Goldsmith P. F., Langer W. D., Schloerb F. P., Scoville N. Z. (1980) Astrophys. J. 240, 524.
Goldsmith P. F., Linke R. A. (1981) Astrophys. J. 245, 482.
Goldsmith P. F., Langer W. D., Elldér J., Irvine W., Kollberg E. (1981) Preprint.
Gottlieb C. A., Gottlieb E. W., Litvak M. M., Ball, J. A., Penfield H. (1978) Astrophysics J. 219, 77.

Gottlieb C. A., Ball J. A., Gottlieb E. W., Dickinson D. F. (1979) Astrophys. J. 227, 422.
Graedel T. E., Langer W. D., Frerking M. A. (1981) Preprint.
Guélin M., Langer W. D., Wilson R. W. (1981) Preprint.
Hartquist T. W., Oppenheimer M., Dalgarno A. (1980) Astrophys. J. 236, 182.
Herbst E. (1978) Astrophys. J. 222, 508.
Herbst E. (1980a) Astrophys. J. 237, 462.
Herbst E. (1980b) Astrophys. J. 241, 197.
Ho P. T. P., Barrett A. H., Myers P. C., Matsakis D. N., Cheung A. C., Chui M. F., Townes C. H., Yngvesson K. S. (1979) Astrophys. J. 234, 912.
Hollenbach D., McKee C. F. (1979) Astrophys. J. Suppl. 41, 555.
Hollenbach D., McKee C. F. (1980) Astrophys. J. 241, L47.
Huggins P. J., Phillips T. G., Neugebauer G., Werner M. W., Wannier P. G., Ennis D. (1979) Astrophys. J. 227, 441.
Iglesias E. R. (1977) Astrophys. J. 218, 697.
Iglesias E. R., Silk J. (1978) Astrophys. J. 226, 851.
Johnsen R., Chen A., Biondi M. A. (1980) J. Chem. Phys. 72, 3085.
Langer W. D., Goldsmith P. F., Carlson E. R., Wilson R. W. (1980) Astrophys. J. 235, L39.
Linke R. A., Goldsmith P. F. (1980) Astrophys. J. 235 437.
Loren R. B., Erickson N. R., Snell R. L., Mundy L., Davis J. H. (1981) Astrophys. J. 244, L107.
Lovas F. J., Johnson D. R., Buhl D., Snyder L. E. (1976) Astrophys. J. 209, 770.
Luine J. A., Dunn G. H. (1981) in "Electronic and Atomic Collisions" (P1035) (edited S. Datz).
McCabe E. M., Smith R. C., Clegg R. E. S. (1979) Nature 281, 263.
McEwan M. J., Anicich V. G., Huntress W. T., Kemper P. R., Bowers M. T. (1980) Chem. Phys. Letters 75, 278.
McGowan J. W., Mul D. M., D'Angelo V. S., Mitchell J. B. A., Defrance P., Froelich H. R. (1979) Phys. Rev. Letters 42, 373.
McLaren R. A., Betz A. L. (1980) Astrophys. J. 240, L159.
Mitchell G. F., Ginsburg J. L., Kuntz P. J. (1978) Astrophys. J. Suppl. 38, 39.
O'Donnell E. J., Watson W. D. (1974) Astrophys. J. 191, 89.
Oppenheimer M., Dalgarno A. (1974) Astrophys. J. 187, 231.
Phillips T. G., Huggins P. J., Neugebauer G., Werner M. W. (1977) Astrophys. J. 217, L161.
Phillips T. G., Kwan J., Huggins P. J. (1980) in "Interstellar Molecules" (p21): (IAU Symposium No. 87, ed. B. Andrew, publ. Reidel).
Prasad S. S., Huntress W. T. (1980a) Astrophys. J. Suppl. 43, 1.
Prasad S. S., Huntress W. T. (1980b) Astrophys. J. 239, 151.
Rydbeck O. E. H., Irvine W. M., Hjalmarson Å., Rydbeck G., Elldér J., Kollberg E. (1980) Astrophys J. 235, L171.
Rydbeck O. E. H., Hjalmarson Å., Rydbeck G., Elldér J., Olofsson H., Sume A. (1981) Astrophys. J. 243, L41.
Saito M., Deguchi S. (1980) Pub. Astron. Soc. Japan 32, 257.

Scalo, J. M., Slavsky D. B. (1980) Astrophys. J. 239, L73.
Smith D., Adams N. G. (1978) Astrophys. J. 220, L87.
Smith D., Adams N. G. (1980) Astrophys. J. 242, 424.
Snell R. L., Wootten, A. (1979) Astrophys. J. 228, 748.
Snyder L. E., Hollis J. M., Ulich B. L., Lovas F. J., Johnson D. R., Buhl D. (1975) Astrophys. J. 198, L81.
Spitzer L. (1978) "Physical Processes in the Interstellar Medium", Wiley-Interscience, New York.
Thaddeus P., Kutner M. L., Penzias A. A., Wilson R. W., Jefferts K. B. (1972) Astrophys. J. 176, L73.
Treffers R. R. (1979) Astrophys. J. 233, L17.
Tsuji T. (1964) Ann. Tokyo Astron. Obs. 2nd Ser. 9, 1.
Tsuji T. (1973) Astr. Astrophys. 23, 411.
Tucker K. D., Kutner M. L. (1978) Astrophys. J. 222, 859.
Turner B. E., Thaddeus P. (1977) Astrophys. J. 211, 755.
Ungerechts H., Walmsley C. M., Winnewisser G. (1980) Astron. Astrophys. 88, 259.
Vardya M. S. (1966) Mon. Not. Roy. Astron. Soc. 134, 347.
Wannier P. G. (1980) Ann. Revs. Astron. Astrophys. 18, 339.
Waters J. W., Gustincic J. J., Kakar R. K., Kiuper T. B. H., Roscoe H. K., Swanson P. N., Rodriguez-Kiuper E. N., Kerr A. R., Thaddeus P. (1980) Astrophys. J. 234, 57.
Watson D. M., Storey J. W. V., Townes C. H., Haller E. E., Hansen W. L. (1980) Astrophys. J. 239, L129.
Watson W. D. (1976) Revs. Mod. Phys. 48, 513.
Watson W. D. (1978) Ann. Rev. Astr. Astrophys. 16, 585.
Watson W. D., Anicich V., Huntress W. T. (1976) Astrophys. J. 205, L165.
Watson W. D., Salpeter E. (1972) Astrophys. J. 174, 321.
Welch W. J., Wright M. C. H., Plambeck R. L., Beiging J. H., Baud B. (1981) Astrophys. J. 245, L87.
Wilson T. L., Downes D., Beiging J. (1979) Astron. Astrophys. 71, 275.
Wilson T. L., Walmsley C. M., Henkel C., Pauls T., Mattes H. (1980) Astron. Astrophys. 91, 36.
Woods R. C., Dixon T. A., Saykally R. J., Szanto P. G. (1975) Phys. Rev. Lett. 35, 1269.
Wootten A., Evans N. J., Snell R., Vanden Bout P. (1978) Astrophys. J. 225, L143.
Wootten A., Bozyan E. P., Garrett D. B., Loren R. B., Snell R. L. (1980a) Astrophys. J. 239, 844.
Wootten A., Snell R., Evans II N. J. (1980b) Astrophys. J. 240, 532.

DISCUSSION FOLLOWING REVIEW BY WATSON AND WALMSLEY

H. DICKEL: Hartquist et al (1980) also predict that the SO_2 abundance will decrease behind a shock whereas in the KL "plateau" region of Orion SO_2 is enhanced. Also, Mitchell and Deveau (this volume) show abundances of sulfur molecules behind shocks different from those of Hartquist et al. because they have considered more species.

WATSON: Yes, Mitchell and Deveau allow small abundances of atomic species which apparently Iglesias and Silk (1978) did not. It especially makes a difference in the case of ammonia. The ammonia abundance was not enhanced in the previous shock calculations. Mitchell and Deveau introduce a small fraction of atomic nitrogen. Since whatever atomic nitrogen you start with in the shock almost always ends up as ammonia, the enhancement of the ammonia abundance depends upon the abundance of atomic nitrogen relative to the ammonia in the pre-shock gas.

PHILLIPS: We have recently observed $J=4 \rightarrow 3$ HCN lines which, when combined with $J=1 \rightarrow 0$ published results, indicate overall abundances in the range $X(HCN) \simeq 4 \times 10^{11}$ to 4×10^{12}. The regions investigated include DR21(OH), M17SW, NGC6334N and W51 - all compact, warm, dense dark clouds containing infrared sources and maser activity indicative of star formation. The abundances we obtain seem to be significantly lower than predicted by many dark cloud models.

~~~~~IVE PHOTODESTRUCTION OF CO ISOTOPIC SPECIES

William D. Langer
Bell Laboratories
Holmdel, NJ 07733
and
Princeton University
P. O. Box 451
Princeton, NJ 08544

John Bally
Bell Laboratories
Holmdel, NJ 07733

ABSTRACT

Observations of the relative abundances of the CO isotopes near the HII region S68 show enhanced $^{12}CO/C^{18}O$ and $^{13}CO/C^{18}O$ ratios. These large ratios can be explained by selective photodestruction of the isotopes of CO in which line absorption and self-shielding play a crucial role.

I. INTRODUCTION

Measurements of CO isotopes are used to determine many of the fundamental properties of molecular clouds (e.g. mass, density) and to study nucleosynthesis and chemical evolution of the Galaxy. Therefore it is important to establish the relationship between CO abundance and the physical conditions in interstellar clouds, such as density, radiation field, and extinction. To this end it is important to study in the laboratory the reactions which form and destroy CO and to make astronomical observations which test the models of the chemistry in interstellar clouds.

While the gas phase reactions which form and destroy CO, including chemical fractionation, are reasonably well established, the destruction of CO by UV radiation is not well understood. This situation is unfortunate because UV photodestruction is of paramount importance at the interface of a cloud with an HII region, near embedded sources, and in shocked gas. Molecules are photodissociated either by continuum or line absorption and the effect of the latter has, in general, been ignored in models of the CO abundance.

R. S. Roger and P. E. Dewdney (eds.), Regions of Recent Star Formation, 379–384.
Copyright © 1982 by D. Reidel Publishing Company.
~~~~~

In this paper we present observations of the relative abundances of CO isotopes which suggest that CO is destroyed by line absorption. The effect of self-shielding in the photodissociative lines makes the destruction rate of each species different; thus, in the presence of UV radiation, the relative abundances of all CO isotopes are enhanced by self-shielding. This effect explains puzzling features of several recent surveys of CO abundance versus extinction and CO isotopic ratios in interstellar clouds

II. OBSERVATIONS

The carbon monoxide isotopes, ^{12}CO, ^{13}CO, and $C^{18}O$, were mapped in the molecular cloud at the boundary of the HII region S68 using the Bell Laboratories 7m antenna. Values of $T_A^*(^{13}CO)/T_A^*(C^{18}O) \gtrsim 40$ were found at positions where the ^{12}CO data indicates a kinetic temperature $T_{kin} \gtrsim$ 25-28K along two parallel strips at the boundary of the cloud. At the same positions the ratio $T_A^*(^{12}CO)/T_A^*(C^{18}O)$ varies between 280 and 400 (see Bally and Langer 1982). A set of spectra at one representative position is shown in Figure 1.

These large ratios present a paradox in terms of the standard models of isotopic abundance and CO chemistry. If the ^{12}CO emission is optically thin then $T_{kin} \gg T_A^*(^{12}CO)$ and such high temperatures are incompatible with the explanation that the large ratio of $^{13}CO/C^{18}O$

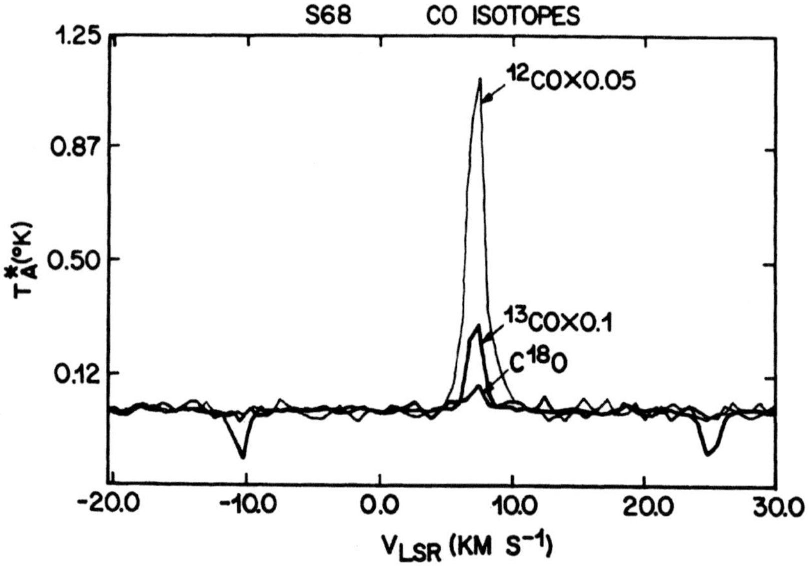

Figure 1. CO isotopes at $\alpha(1950)=18^h27^m57^s$, $\delta(1950)=01°14'39''$ showing $T_A^*(^{12}CO)$x0.05, $T_A^*(^{13}CO)$x0.1 and $T_A^*(C^{18}O)$.

is the result solely of chemical fractionation. On the other hand if the ^{12}CO emission is optically thick then the abundance ratio $^{12}CO/C^{18}O$ derived by correcting $T_A^*(^{12}CO)$ for saturation will be much larger than the terrestrial value of 500. Indeed, an estimate of the ^{12}CO opacity from that of ^{13}CO including chemical fractionation, yields abundance ratios $^{12}CO/C^{18}O$ > 700 to 1100 (see Bally and Langer (1982) for a more detailed discussion).

The large abundance ratios of CO isotopes determined for the molecular boundary layer of S68 suggests that some process other than chemical fractionation enhances CO abundances. We suggest that this mechanism is selective photodestruction of the carbon monoxide isotopes in which the CO photodissociation rate is different for each isotope due to the presence of: 1) photodestruction by line absorption; 2) self-shielding of photodissociating photons; and, 3) a wavelength shift for the photodissociating transitions due to the effect of isotopic mass difference.

III. THEORY AND MODEL

To explore quantitatively the effects of selective photodestruction on the abundances of CO isotopes we have made calculations of abundances as a function of extinction using a simple gas phase chemistry based on the discussion of Langer (1976). The major changes considered here are the inclusion of UV line absorption and self-shielding, for each of the CO isotopes. The most crucial, and least certain, aspect of the model calculations presented here is the photodissociation rate. The photodissociation of CO is reviewed below; for a more recent discussion of cloud chemistry see Prasad and Huntress (1980).

The absorption spectrum of CO has been studied in the laboratory and a large number of bands are observed for λ > 912 Å (see the review of Hudson 1971). Little is known about which bands lead to dissociation (the dissociation energy of CO is 11.1 eV corresponding to 1120 Å) and what dissociation strength, f, applies. Solomon and Klemperer (1972) suggest that two discrete transitions from the ground state lead to dissociation, namely; $X^1\Sigma^+(v=0)\rightarrow B^1\Sigma^+(v=1)$ and $C^1\Sigma^+(v=1)$. They estimate dissociation strengths for these transitions to be 2.7×10^{-3} and 1.4×10^{-3} respectively. In addition to these transitions at $\lambda \sim 1050$ Å there are a series of discrete features and an apparent continuum in the range 912 to 960 Å. The dissociation cross section in the continuum is estimated to be 10^{-17} cm^2. None of the discrete features are considered by Solomon and Klemperer to lead to dissociation. Hudson (1971) has noted, however, that the apparent continuum may be the result of the overlapping of the pressure broadened wings of the discrete lines. Thus the number of lines leading to photodissociation must be regarded as uncertain. Accordingly, we consider the total CO photodestruction rate $\Gamma(CO)$ to be composed of a continuum and n lines.

The attenuation of the UV photodissociating radiation is due to grains and, for the lines, self-shielding. In general $\Gamma(CO)$ will be different for the various isotopes of CO (except at the boundary of the cloud) if corresponding dissociating lines of the various CO isotopes do not overlap. For the $X^1\Sigma^+ \rightarrow C^1\Sigma^+$ transitions of ^{12}CO and ^{13}CO this appears to be true according to the measurements of Tilford and Vanderslice (1968).

To calculate the CO isotopic abundances as a function of extinction a one dimensional semi-infinite model of UV radiative transfer has been adopted. The photodissociation rate, based on the standard interstellar radiation field (Gerola and Glassgold 1978), has the following form,

$$\Gamma(CO) = (\Gamma_{cont} + \Sigma\Gamma_i e^{-\tau_i})e^{-\tau_g} ,$$

where τ_i and τ_g are the attenuating opacities in the line and the grains. The opacities of the lines discussed by Solomon and Klemperer are, $\tau \simeq 1.1\times10^{-14} N(CO)/\Delta v$, where $N(CO)$ is in cm^{-2} and Δv in km/s. Abundances are first determined at the edge and then stepwise into the cloud so that $N(CO)$ can be calculated self-consistently. Figure 2 shows plots of the isotope ratios for a cloud with density $n(H_2)=10^3$ cm^{-3}, temperature 20K, isotope abundances $^{12}C/^{13}C=89$ and $^{16}O/^{18}O=500$, for a line with velocity dispersion of 1 km/s. Large

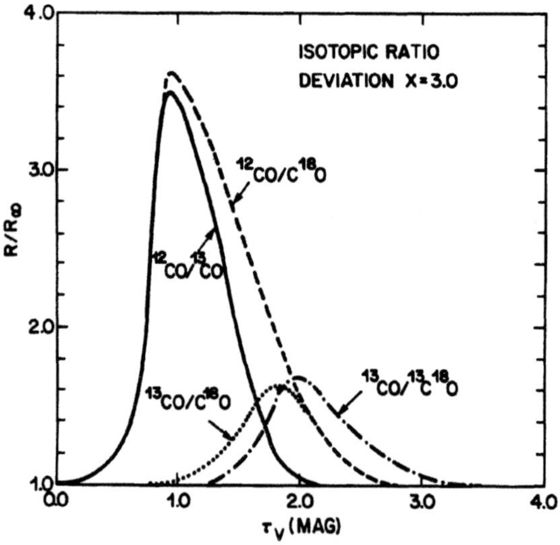

Figure 2. Isotopic ratio enhancements due to selective photodestruction relative to CO containing cosmic (unaltered) abundances of the various isotopes plotted as a function of visual extinction from the cloud edge.

$^{12}CO/^{13}CO$ and $^{12}CO/C^{18}O$ enhancement is evident at extinctions of $A_V \simeq 0.75$ to 2 magnitudes. In this case the maximum enhancement of the isotopic abundances is ∿3.5. The enhancement of $^{13}CO/C^{18}O$ does not become large until much further into the cloud (a larger column density of H_2 is required before $N(^{13}CO) \simeq 10^{14}$ cm^{-2}) and is at most 1.7.

DISCUSSION

Selective photodestruction of the isotopes of carbon monoxide along with chemical fractionation explains the relatively large ratio of $^{12}CO/C^{18}O$ and $^{13}CO/C^{18}O$ observed in the warm regions at the edge of the S68 molecular cloud. A large enhancement of isotope ratios is also evident in Taurus (Frerking, Langer and Wilson 1982) over the range A_V=1.2 to 2.2 mag. Self-shielding of CO results in a sharp increase in the abundance of any isotope whenever $N(^{x}C^{y}O) \approx 10^{14}$ cm^{-2} (for $\Delta v \simeq 1$ km/s), similar to the sudden onset in H_2 abundance seen in diffuse clouds. This jump can be seen in ^{12}CO for diffuse clouds (Federman et al. 1980) and in ^{13}CO and $C^{18}O$ in Taurus and ρ Ophiuchi (Frerking, Langer and Wilson 1982). The variations in the $^{12}C^{18}O/^{13}C^{18}O$ ratios observed in dark clouds (Wilson, Langer and Goldsmith 1981) and the isotope ratios deduced from line wings (Wannier et al. 1976; Penzias 1980) may be due, in part, to selective photodestruction.

The more abundant CO isotopic species are enhanced relative to the less abundant species if three conditions are met. 1) A significant portion of CO photodissociation occurs in discrete lines. 2) The isotope shift prevents the overlap of corresponding dissociative transitions in the various CO isotopic species. 3) The UV optical depth of the dissociative transitions is large enough so that the more abundant isotopic forms of CO self-shield before dust attenuates the radiation field. Selective photodestruction of CO isotopes can explain observations of anomolous isotopic ratios, as well as sharp transitions in N(CO) versus A_V seen in several sources. This result has several consequences which affect the use of CO for determining the properties of the interstellar gas; 1) <u>all</u> isotope ratios may be altered relative to cosmic abundances in the presence of UV; 2) the determination of mass from CO observations in regions of low A_V ($\lesssim 6$ mag) is sensitive (in a highly non-linear manner) to physical parameters such as density and velocity dispersion; and, 3) the determination of ^{12}CO opacity from ^{13}CO is complicated in the presence of UV since the ratio $^{12}CO/^{13}CO$ can be greater than the $^{12}C/^{13}C$ ratio. The largest uncertainty in our analysis is the lack of precise laboratory data on the photodissociation of CO by line absorption. Until better measurements are obtained, a precise quantitative determination of the enhancement of CO isotopic species can not be made.

REFERENCES

Bally, J. and Langer, W. D. 1982, Ap. J. in press.
Federman, S. R., Glassgold, A. E., Jenkins, E. B. and Shaya, E. J. 1980 , Ap. J. 242, 545.
Frerking, M. A., Langer, W. D. and Wilson, R. W. 1982 , Ap. J. in press.
Gerola, H. and Glassgold, A. E. 1978, Ap. J. Suppl. 37, 1.
Hudson, 1971 , Rev. Geo. and Space Physics 9, 2, 305.
Langer, W. D. 1976, Ap. J., 206, 699.
Penzias, A. A. 1980 , Science, 208, 663.
Prasad, S. S. and Huntress, W. T. Jr. 1980, Ap. J. Suppl. 43, 1.
Solomon, P. M. and Klemperer, W. 1972, Ap. J. 178, 389.
Tilford, S. G. and Vanderslice, J. T. 1968, J. Mol. Spec. 26, 419.
Wannier, P. G., Penzias, A. A., Linke, R. A. and Wilson, R. W. 1976 , Ap. J. 204, 26.
Wilson, R. W., Langer, W. D., Goldsmith, P. G. 1981 , Ap. J. (Letters), 243, L47.

DISCUSSION FOLLOWING PAPER BY LANGER AND BALLY

BLITZ: There is a possibility that your assumptions of cloud uniformity could affect the interpretation. If you are observing through several unrelated clumps whose velocities are such that you don't get shadowing of one cloud by another as you map across the source, then this could influence your interpretation of the isotope ratios.

LANGER: We have seen no evidence of that.

X-RAY IONIZATION AND THE CHEMISTRY OF THE ORION MOLECULAR CLOUD

Julian H. Krolik*
and
Timothy R. Kallman†
Center for Space Research, Massachusetts Institute of Technology

ABSTRACT

The collection of unusually strong stellar X-ray sources in the vicinity of the Orion molecular cloud together bathe the gas with such an intensity of X-rays that they, rather than cosmic rays, dominate the ionization and heating of the gas. We present estimates of the ionization rate and the elevation in temperature, and discuss the consequences for the gas chemistry. Strong small-scale inhomogeneities in molecular abundances and temperature are a distinguishing feature of ionization by stellar X-rays.

Recent Einstein observations of the Orion Nebula[1,2] and nearby molecular clouds[3] have discovered that the (presumably) young stars there are anomalously strong X-ray emitters: the brightest known X-ray emitting star anywhere in the Galaxy of each spectral class from A through M is in Orion (of course, even the brightest X-ray sources in Orion could not be seen if they were more than a few kpc away). These X-rays impinge on the neighboring molecular cloud in such intensity that they significantly affect its ionization equilibrium, thermal balance, and chemistry. Today, because our calculations of these effects are still in a preliminary stage, I will present only estimates of how drastically the X-rays affect the life of the molecular cloud, and some suggestions of how an X-ray-ionized molecular cloud differs qualitatively from one ionized by cosmic rays.

I'll begin by summarizing the X-ray observations. The total X-ray luminosity from the region observed by the Einstein Observatory is $\sim 10^{33}$ erg s^{-1}. There are two reasons why this number is probably substantially less than the true total X-ray luminosity: line-of-sight attenuation and the limited band of detector sensitivity. We know there is considerable reddening of the stars identified as X-ray sources; absorption of their X-rays follows as a corollary. The

*partially supported by NASA Grant NSG-7643
†partially supported by NASA Contract NAS8-30752

R. S. Roger and P. E. Dewdney (eds.), Regions of Recent Star Formation, 385–389.
Copyright © 1982 by D. Reidel Publishing Company.

typical unobscured X-ray star has a spectrum which is roughly exponential with a characteristic photon energy of ∿600eV[4]. The Einstein IPC band, which extends nominally from 0.5 to 4.5 keV, encompasses only a fraction of the emitted X-rays from such a spectrum. On the other hand, there may be some absorption between the stars and the molecular cloud. The balance is uncertain, but probably inclines toward a true X-ray luminosity a good deal greater than what we see.

The region (projected on the sky) in which these stars are found coincides fairly well with the peak of the molecular emission.[1,5] Both areas are 0.1pc in diameter. Of course, the local X-ray flux will vary considerably from point to point, depending on the distance to the nearest X-ray star.

Next, we estimate the ionization and heating rates due to the X-rays. Photons below 500eV are primarily absorbed by H_2, sometimes dissociating the molecule as well as ionizing it.[6] Above 500eV, K-shell ionization of oxygen is the dominant process. In the former case, the result is a single electron having slightly less energy than the X-ray photon; in the latter case, two electrons are released, one with the photon energy minus the K-shell ionization potential (533eV), and one Auger electron with the difference between the K and L-shell ionization potentials (∿500eV). The direct ionization rate per atom is:

$$\zeta_x \simeq \frac{6 \times 10^{-20}}{E\eta} \left[\frac{r}{0.1\mathrm{pc}}\right]^{-2} \left[\frac{kT_*}{500\mathrm{eV}}\right]^{-1} \left[\frac{\tau_*}{100}\right]^{-1} \mathrm{s}^{-1} \qquad (1)$$

where $E(\simeq 0.1 - 0.01?)$ is the fraction of the total X-ray flux perceptible to the Einstein IPC, $\eta(\sim 1?)$ is the ratio between the transmissivity of our line-of-sight to that of the line-of-sight to the molecular cloud, kT_* is the characteristic photon energy in the emitted spectrum, and τ_* is the optical depth at the point of observation for photons of energy kT_*. The actual scaling of ζ_x with optical depth is more complicated than shown in Eq. 1, but it will do for estimates. A column density of 10^{23} cm^{-2} (the usual estimate for the thickness of the Orion molecular cloud[7]) presents an optical depth of ∿100 to 500eV photons. The total ionization rate is, however, much larger than the direct photoionization rate, for the fast electrons produced by photoionization give up a large part of their energy to secondary ionization:

$$\zeta_T \simeq \frac{1 \times 10^{-17}}{E\eta} \left[\frac{\bar{E}_s}{30\mathrm{eV}}\right]^{-1} \left[\frac{r}{0.1\mathrm{pc}}\right]^{-2} \left[\frac{f_\zeta}{f_T}\right] \left[\frac{\tau_*}{100}\right]^{-2/3} \mathrm{s}^{-1} \qquad (2)$$

where $\overline{E}_s$ is the mean energy used to create each secondary electron (the ionization potential of H_2 is 15eV), and f_ζ/f_T ($\simeq$1/3-1/2, ref. 8) is the ratio of continuum to total oscillator strength of the molecular gas. Note that the total rate is always considerably greater than the direct rate, and also decreases more slowly with optical depth. The heating rate may be immediately recovered by multiplying ζ_T by $(\overline{E}_s - \overline{I}_p)f_T/f_\zeta$, where $\overline{I}_p$ is the mean ionization potential, and we assume that all radiative transitions excited by the fast electrons are thermalized.

Our standard of comparison is the rate of ionization attributed to cosmic rays, ususally taken to be $\sim 10^{-17}$ s^{-1}.[9] Even in the deepest part of the Orion cloud, ζ_T exceeds the cosmic ray ionization rate; in portions particularly near an X-ray star, ζ_T is several orders of magnitude greater.

We now examine what difference the X-rays make, beginning with the effects that may be regarded as if they were due to simple quantitative increases in the cosmic ray ionization rate. The electron density scales as $\zeta_T^{1/2}$ as long as a single class of species dominates the ionic population and the gas remains mostly neutral (as it always does in the molecular cloud), but the effective recombination coefficient rises sharply when there are so few neutral metallic atoms (e.g. Mg, Na) that molecular ions (e.g. H_3^+) which recombine dissociatively rather than radiatively provide the greatest part of the total recombination rate.[12] Such a situation arises when either the metals are all sequestered in grains or the ionization rate is so high that they are all ionized. In the course of the transition from metal to molecular ion recombination, the electron density is nearly insensitive to ζ_T because dissociative recombination coefficients are much larger than radiative ones. If the metals are present in cosmic abundance, the Orion molecular cloud probably lies in just that transitional state. Ions such as He^+, C^+, etc. which quickly neutralize by charge exchange increase in proportion to ζ_T.

Those molecular species created by ion-mediated reactions can be expected to increase in abundance in X-ray ionized clouds. For example, examination of the reaction networks given by Prasad and Huntress [10] shows that $[HCO^+]/[CO] \propto \zeta_T^{1/2}$, $[H_3O^+]/[H_2O] \propto \zeta_T^{1/2}$, $[CH^+] \propto \zeta_T$, and $[HCN^+] \propto \zeta_T$.

As we have pointed out, the heating rate increases as ζ_T/ζ_{CR}. The cooling function scales as T^3 (ref. 11), so we can expect temperature increases in strongly irradiated regions of factors of a few. Such temperature variations should be readily observable.

In fact, the temperature variations due to local X-ray heating point up what is probably the principal qualitative contrast between cosmic ray-ionized and X-ray ionized molecular clouds: the latter should exhibit a great deal of small-scale inhomogeneity in both temperature and abundances, while the former should be comparatively smooth. The range of a 1 GeV proton is $\sim 10^{27}$ cm^{-2}, while the mean absorption column density of a 500eV photon is $\sim 10^{22}$ cm^{-2}. We predict, then, that the scattering of discrete X-ray stars within and about the Orion molecular cloud should lead to sharp contrasts in molecular column densities and line temperatures on a scale of tens of arc seconds.

As a final note, I would like to point out another consequence of X-ray photoionization which in principle may play a role in defining the molecular abundances, and is unique to X-rays. When a medium atomic number atom (i.e., $3 \leq Z \leq 20$) is bound in a molecule and has been ionized by absorption of an X-ray, the chances are that the electron was removed from its K-shell. The excitation energy due to the inner-shell vacancy is then given to a valence-shell electron which is also expelled from the molecule (this process is the Auger effect). Few small molecules remain bound when twice-ionized, so the molecule splits apart, usually into two singly-charged fragments. The rate of destruction of molecules by this process is usually quite small compared to chemical reaction destruction rates, but in the case of those molecules which participate in nearly closed reaction loops, the small drainage to X-ray photoionization may be able to significantly tilt the balance, and may lead to effects uniquely attributable to X-ray ionization.

REFERENCES

[1] Ku, W. H. and Chanan, G. A., 1979, Ap. J. Lett. 234, L59.
[2] Chanan, G. A., 1981, in preparation.
[3] Pravdo, S. and Marshall, F., 1981, preprint.
[4] Harnden, F. R., Jr., Branduardi, G., Elvis, M., Gorenstein, P., Grindlay, J., Pye, J. P., Rosner, R., Topka, K. and Vaiana, G. S., 1979, Ap. J. Lett. 234, L51.
[5] Beckwith, S., Persson, S. E., Neugebauer, G. and Becklin, E. E., 1978, Ap. J. 223, 464.
[6] Backx, C., Wight, G. R. and Van der Wiel, M. J., 1976, J. Phys. B9, 315.
[7] Zuckerman, B., 1973, Ap. J. 183, 863.
[8] Douthat, D. A., 1979, J. Phys. B12, 663.
[9] Black, J. H. and Dalgarno, A., 1977, Ap. J. Suppl. 34, 405.
[10] Prasad, S. S. and Huntress, W. T., Jr., 1980, Ap. J. Suppl. 43, 1.
[11] Goldsmith, P. F. and Langer, W. D., 1978, Ap. J. 222, 881.
[12] Oppenheimer, M. and Dalgarno, A., 1974, Ap. J. 192, 29.

DISCUSSION FOLLOWING PAPER BY KROLIK AND KALLMAN

WOOTTEN: High velocity gas near the supernova remnant IC443 appears to have very high $x(HCO^+)/x(CO)$ abundance ratios (observed by Dickinson et al., Ap. J. Lett. 237, L43, 1980). The observers have attributed this to a shock chemistry. Might this be more clearly understood as the result of an X-ray enhancement of the ionization rate due to the supernova remnant interacting with the cloud?

KROLIK: It would depend on the strength of the X-rays and on how far the molecular region is from the supernova remnant.

J. DICKEL: The emission is fairly soft. The temperature is probably less than 10^6 K. Would that be strong enough?

KROLIK: The 10^6 K corresponds to 100 eV.

WALMSLEY: If the source of X-rays is exterior to a molecular cloud there is a possibility that you might detect hydrogen recombination emission corresponding to the X-ray flux. There would be an atomic region between the source of X-rays and the cloud.

TENORIO-TAGLE: We have thought about this problem and the atomic hydrogen should have a column density of about 10^{20} cm^{-2}.

KROLIK: With that column density only soft X-rays up to about 100 eV will be absorbed corresponding to a small fraction of the total energy.

SNELL: Could X-ray ionization explain the reported detection of CO^+ in Orion?

KROLIK: I am not very optimistic but I can give you nothing quantitative.

TURNER: This suggested identification of a single line at λ1.2 mm as CO^+ should be viewed with caution until other transitions of it are seen.

METHYL CYANIDE AS A PROBE OF THE TEMPERATURE AND DENSITY IN $SgrB_2$; QUASI-EQUILIBRIUM IN MOLECULAR ROTATIONAL LEVELS

Richard A. Linke
Bell Laboratories
Holmdel, New Jersey 07733

Sally E. Cummins, Sheldon Green and Patrick Thaddeus
Goddard Institute for Space Studies
New York, New York 10025

ABSTRACT

Observations of 14 rotational transitions of methyl cyanide have been used in conjunction with new estimates of the collisional excitation cross sections under H_2 impact to determine the kinetic temperature and density in the central 2 arcminutes of $SgrB_2$. An H_2 temperature near 90 K and an H_2 density range of $6 \rightarrow 16 \times 10^4$ cm^{-3} were obtained. This high value of kinetic temperature contrasts with much lower values ($10 \rightarrow 20$ K) implied by apparent rotational equilibrium in a number of molecular species; it is shown that this quasi-equilibrium in high J levels is an expected consequence of subthermal excitation.

INTRODUCTION

Observations of molecular rotational transitions have been used extensively as probes of the temperatures and densities within molecular clouds. In this paper we analyze observations of CH_3CN to obtain unambiguous and independent measurements of the kinetic temperature and hydrogen density in the galactic center source $SgrB_2$. Owing to the symmetry of the CH_3CN molecule allowed radiative transitions do not change the component of rotational angular momentum K along the molecular axis. Therefore the total population of a given K ladder is established only by collisions with the ambient H_2 gas, so that the relative populations of these ladders are <u>not</u> a function of the density but only of the kinetic temperature of the H_2, the dominant species in an interstellar molecular cloud. The population distribution within a K ladder is a function of both the kinetic temperature and the density of the H_2, but as the kinetic temperature is known, the density can also be determined uniquely from the excitation. Solomon et al. (1971) made the first observations of CH_3CN and derived a kinetic temperature of 150 K and a hydrogen density of 10^6 cm^3 for $SgrB_2$.

R. S. Roger and P. E. Dewdney (eds.), Regions of Recent Star Formation, 391–397.
Copyright © 1982 by D. Reidel Publishing Company.

Later observations (Solomon et al 1973) yielded a kinetic temperature of 100 K. The lack of well calibrated lines from different J levels put a large uncertainty on the derived density and the high kinetic temperature obtained is in apparent conflict with results from other molecules which seem to show rotation equilibrium temperatures of 10→20 K (c.f. Linke, Frerking and Thaddeus 1979; Frerking, Linke and Thaddeus 1979). In these cases rotational level pairs are seen to have nearly identical excitation temperatures - a condition normally expected only when all level populations are thermalized at the kinetic temperature - but this temperature is found to be quite low. In view of the better data now available as well as the fuller understanding of collisional rates (Green, 1979) it seems appropriate to re-examine the excitation of CH_3CN.

OBSERVATIONS

As part of an ongoing survey with the Bell Laboratories 7-meter telescope of the millimeter-wave spectrum of $SgrB_2$ at a spectral resolution of 1 MHz, observations were made of the J=4→3, 6→5, and 7→6 transitions of CH_3CN. Lines with K as large as 5 have been detected. The integration times for the three spectra were 30 min., 1.8 hrs., and 3.9 hrs., respectively. The approximate frequencies and corresponding half-power beamwidths are 73.6, 110.4, and 128.8 GHz and 2.9', 1.9' and 1.6'. The antenna beam efficiency was taken to be 0.95, 0.79, and 0.72 at these frequencies. The J=7→6 line is shown in Figure 1 uncorrected for beam efficiency.

COLLISION RATES

For the purpose of estimating the CH_3CN-H_2 collision rates a very useful simplification can be obtained by assuming that the rotational energy spacings are small compared with the collision energy. Then the entire matrix of collision rates can be expressed in terms of a few fundamental rates that are simply related to excitation out of the lowest J=K=0 level. For a symmetric top the rates are given by (Green 1979)

$$R(JK\rightarrow J'K') = (2J'+1) \sum_{L=|J-J'|}^{J+J'} \begin{pmatrix} J' & L & J \\ K' & K-K' & -K \end{pmatrix}^2 Q(L,K-K')$$

the large bracket being a 3-j symbol. This formula is often found to be a rather good approximation for downward transitions even when the rotational energy spacings are a significant fraction of the collisional

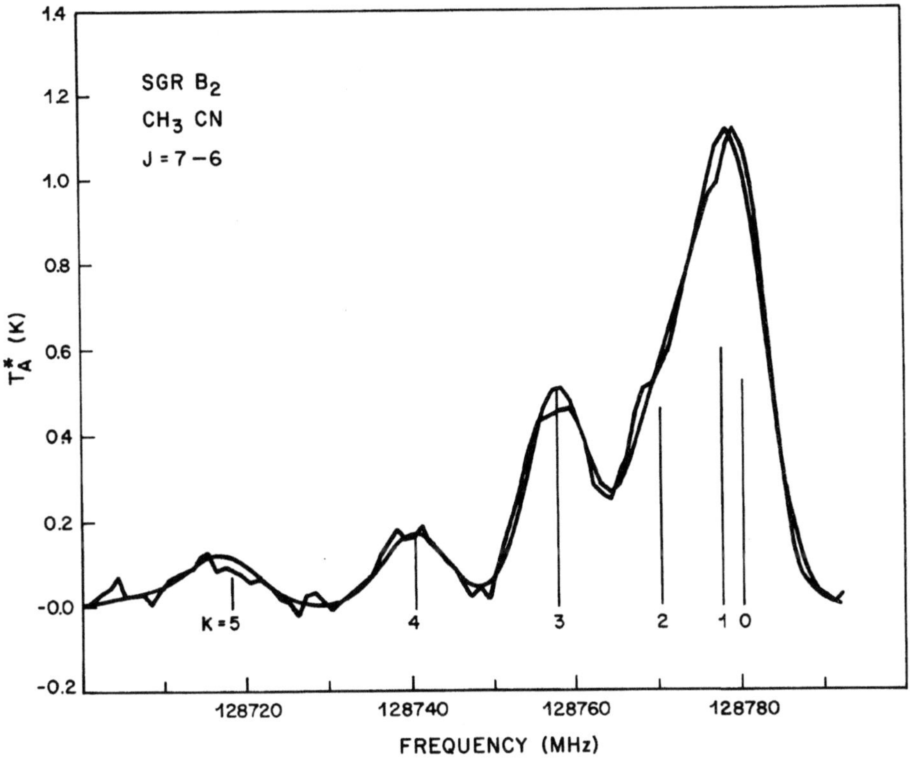

Figure 1 - The Methyl cyanide J=7-6 transition in SgrB$_2$ observed from the BTL 7 meter antenna. The strengths shown (bar lengths) at the frequencies of K components, as well as the smooth curve through the data, are the output of a statistical equilibrium excitation analysis for the best fit conditions listed under set 1 in Table 1. The K=5 component is blended with K=0 and 1 components of $CH_3{}^{13}CN$ (not indicated). The antenna temperature shown has not been corrected for the beam efficiency (η_B=.72).

energy, as at present, and the corresponding rates for upward transitions are readily obtained from detailed balance.

Two sets of values for the fundamental rates Q(L,K) were chosen in an effort to determine the sensitivity of our results to the collisional model. The values in the first model were adapted from calculations for OCS which is structurally similar to CH_3CN. This model (rate set 1) gives rates which decrease rather sharply with increasing ΔJ. The second model (set 2) was selected to give a higher probability for larger ΔJ changes. Because radiative rates are constrained to $\Delta J=1$, one can find (Goldsmith 1972) qualitatively different behavior in the statistical equilibrium calculation depending on whether collisions are dominated by $\Delta J=1$ transitions or by larger ΔJ transitions. Values for the Q(J,3) and Q(J,6) - which correspond to $\Delta K = 3$ and 6 respectively - were adapted from a consideration of calculations for NH_3. For both models the fundamental rates were scaled so that the total rate of collision into the J=K=0 level is the same. The value for this total rate was set by an estimate of the geometric cross section, and is likely to be correct to a factor of two. Data for pressure broadening of CH_3CN by H_2 could provide an accurate value for the total rate, but is not available.

ANALYSIS

We have analyzed the data in terms of a statistical equilibrium calculation for $J\leq16$ and $K\leq5$ using the two approximations to the collision rates discussed above. Seven parameters were adjusted to obtain a best fit to the data. These are: kinetic temperature T_K, hydrogen density $n(H_2)$, column densities of ortho and para forms of CH_3CN, N_{ortho}, N_{para}, the source linewidth ΔV, the source velocity with respect to the local standard of rest V_{LSR}, and the $^{12}C/^{13}C$ isotope ratio. This last parameter was included since transitions of $CH_3{}^{13}CN$ are blended with K=5 transitions of CH_3CN. The source was assumed to fill the antenna beams and the lines were assumed to be Gaussian in shape and optically thin (the results of the calculations were consistent with this last assumption). The results for the two sets of collision rates are given in Table 1. Figure 1 shows the data and the fit for the $J=7\rightarrow6$ transition using rate set 1. The fits were in general very good.

TABLE 1

	RATE SET 1	RATE SET 2
T_{kin}(K)	88.7	92.1
n_{H_2}(cm^{-3})	$1.6x10^5$	$6.2x10^4$
N_{ortho}(cm^{-2})	$6.3x10^{13}$	$6.2x10^{13}$
N_{para}(cm^{-2})	$7.9x10^{13}$	$7.7x10^{13}$
Δv(km/sec)	21.0	21.1
v_{LSR}(km/sec)	59.5	59.6
$^{12}C/^{13}C$	19.8	20.2

The line intensities predicted by the model are determined predominantly by three parameters: T_K, $n(H_2)$ and N. The remaining four parameters were included simply to improve the quality of the fits in frequency. In a separate calculation the parameters N_{ortho}/N_{para}, ΔV, V_{LSR}, and $^{12}C/^{13}C$ were fixed at values typical of a large sample of molecules in $SgrB_2$ (1, 22.7 kms^{-1}, 61.6 kms^{-1}, and 20) and the results for T_K, $n(H_2)$, and N were found to differ from those given in Table 1 by less than 15%.

The kinetic temperature of $SgrB_2$ required by the CH_3CN data is about 90 K consistent with the results of Solomon et al. (1973). It is therefore evident that rotation temperatures of 10-20 K do not represent true equilibrium. Figure 2 which gives the excitation temperature for a pair of rotational levels versus J_{upper} shows how this confusion can arise. The results are from the statistical equilibrium calculation using rate set 1 and show that above $J \simeq 7$ the excitation temperatures are nearly equal (or somewhat increasing with J) at a value which is much less than the kinetic temperature. Observational selection which occurs by virtue of the higher J transitions falling in a desirable portion of the frequency spectrum

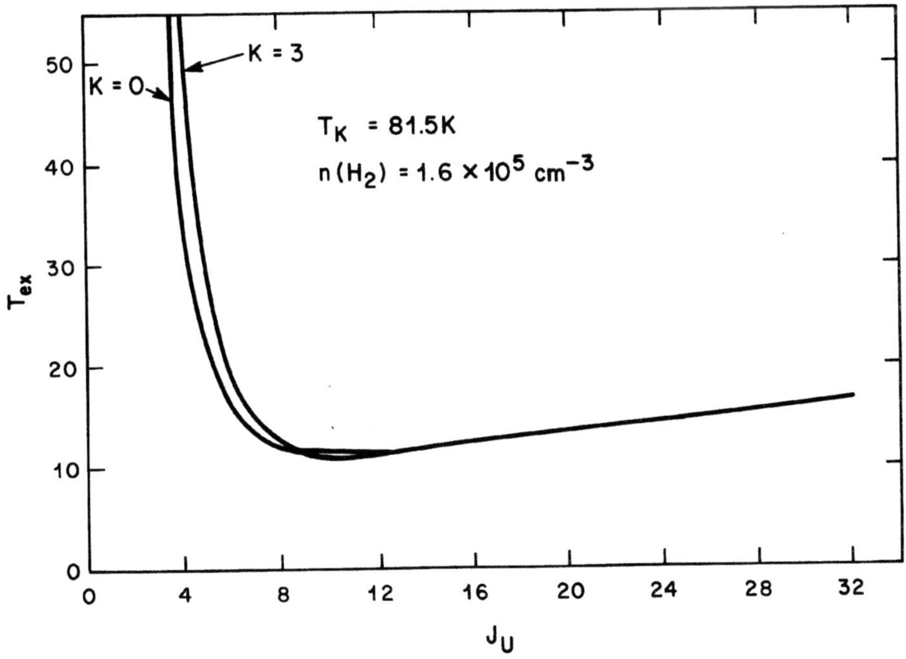

Figure 2 - Excitation temperature versus J_{upper} derived from a statistical equilibrium excitation including J levels up to J=40. The broad minimum and long tail can give the appearance of constant T_{ex} (i.e. thermalized levels) at a temperature much below the true kinetic temperature.

will lead to the erroneous conclusion that all T_{ex} are equal and therefore that rotational equilibrium is established. The somewhat surprising result that T_{ex} does not drop off with J is found to be qualitatively unchanged for our second set of rate constants.

Our derived density range of $6 \rightarrow 16 \times 10^4$ cm^{-3} is considerably lower than the 10^6 cm^{-3} value of Solomon et al. (1971) but close to a value of 3×10^4 cm^{-3} obtained by Linke and Goldsmith (1980) from observations of CS. Our derived density range implies a mass of $3 \rightarrow 8 \times 10^5 M_\odot$ for the central 2.0' source.

The ratio $^{12}C/^{13}C$=20 is in agreement with that found for HC_3N (Wannier and Linke 1978) and OCS (Goldsmith and Linke 1981) in $SgrB_2$. The linewidth ΔV = 21 km s^{-1} is consistent with most other molecules in $SgrB_2$. The V_{LSR} = 59.5 km s^{-1} is somewhat less than that for many molecules in this source, namely 62 km s^{-1}, although there are other molecules for which V_{LSR} = 60 km s^{-1}, such as HC_3N, H_2CS, SO, SO_2.

In conclusion we find that the rotational levels of CH_3CN and probably many other molecules in $SgrB_2$ are radiatively relaxed and that the kinetic temperature and $n(H_2)$ corresponding to the central source are ~90 K and ~10^5 cm^{-3}.

REFERENCES

Frerking, M.A., Linke, R.A. and Thaddeus, P.: 1979, Ap. J. (Letters) 234, L143.
Goldsmith, P.F.: 1972, Ap. J. 176, 597.
Goldsmith, P.F. and Linke, R.A.: 1981, Ap. J. 245, 482.
Green, S.: 1979, J. Chem. Phys. 70, 816.
Linke, R.A., Frerking, M.A. and Thaddeus P.: 1979, Ap. J. (Letters) 234, L139.
Linke, R.A. and Goldsmith, P.G.: 1980, Ap. J. 235, 437.
Solomon, P.M., Jefferts, K.B., Penzias, A.A., and Wilson, R.W.: 1971, Ap. J. (Letters) 168, L107.
Solomon, P.M., Penzias, A.A., Jefferts, K.B. and Wilson, R.W.: 1973, Ap. J. (Letters) 185, L63.
Wannier, P.G. and Linke, R.A.: 1978, Ap. J. 226, 817.

DISCUSSION FOLLOWING PAPER BY LINKE ET AL.

SNYDER: We have just finished a similar study of CH_3CCH in Sgr B2 (Hollis, Snyder, Blake and Suenran, Ap. J. in press) and we find a similar result - low rotational temperature and high kinetic temperature.

WALMSLEY: The ammonia observations of Sgr B2 at Bonn gave a temperature 50-100 K. The formaldehyde observations indicate a density $\sim 10^4$ cm^{-3}.

SHULL: Could you comment on the similarity between the kinetic temperature of ~90 K in this molecular cloud (Sgr B2) and that inferred for diffuse clouds from observations of the Copernicus satellite? Astrophysicists are hard-pressed to explain the source of heating for diffuse clouds and I should think these molecular cloud temperatures would be even harder to explain. Mechanisms such as ambipolar diffusion or photoelectric emission from grains may not suffice.

LINKE: I have been wondering whether there might be some effects of this 90 K that one might be able to observe - for instance, in the infrared. However, Sgr B2 is such a dense region, with so much cold dust surrounding it, that it might be difficult to find any consequences of the 90 K.

BEICHMAN: Are the lower J-levels useful to test predictions of your model?

LINKE: Yes, some lower frequency transitions should be checked - our results would imply strange things at the lower J-levels.

HIGH SPATIAL RESOLUTION OBSERVATIONS OF HCN IN S 235B

G. Sandell
Stockholm Observatory, Sweden and Observatory and Astrophysics Laboratory, University of Helsinki, Finland
B. Höglund
Onsala Space Observatory, Sweden
A.G. Kislyakov
Applied Physics Institute, Gorky, USSR

ABSTRACT

HCN has been mapped with 42" spacing (=HPBW) around the BN type IR source S 235B (=IRS 4). The HCN hyperfine structure is well resolved and departs from LTE. The ratio R_{12}, which in LTE lies between 0.6 and 1.0, is$\sim$0.28, i.e. far outside the LTE range. $H^{13}CN$ has also been detected in two positions. The HCN column density toward S 235B is $>8.0\ 10^{13}\ cm^{-2}$.

1. INTRODUCTION

S 235B is a BN type source located about 10' south of the bright H II region S 235 (M1-82). The area has been subject to numerous studies both in the radio and in the infrared (see e.g. Evans and Blair, 1981). There are two large molecular cloud components in the region: one at about -20 km s^{-1} centered somewhat east of S 235 and one at -17 km s^{-1}, which encompass a dense hot region of active star formation. The densest part of the -17 km s^{-1} component coincides with a region of self-reversed CO, two compact H II regions S 235A and B, as well as an H_2O maser tentatively assigned to S 235A. About 4' south of A is a third fainter H II region, S 235C (Israel and Felli, 1978).

S 235B is seen as a near and far infrared source which coincides with a faint nebulous object (Evans et al., 1981). Although it has no detectable radio continuum, Bγ recombination line emission (Tokunaga and Thompson, 1979) shows that it is surrounded by a compact H II region.

2. OBSERVATIONS

All observations have been made with the radome covered 20 m telescope at Onsala Space Observatory in March 1980 (HCN) and in

R. S. Roger and P. E. Dewdney (eds.), Regions of Recent Star Formation, 399–406.
Copyright © 1982 by D. Reidel Publishing Company.

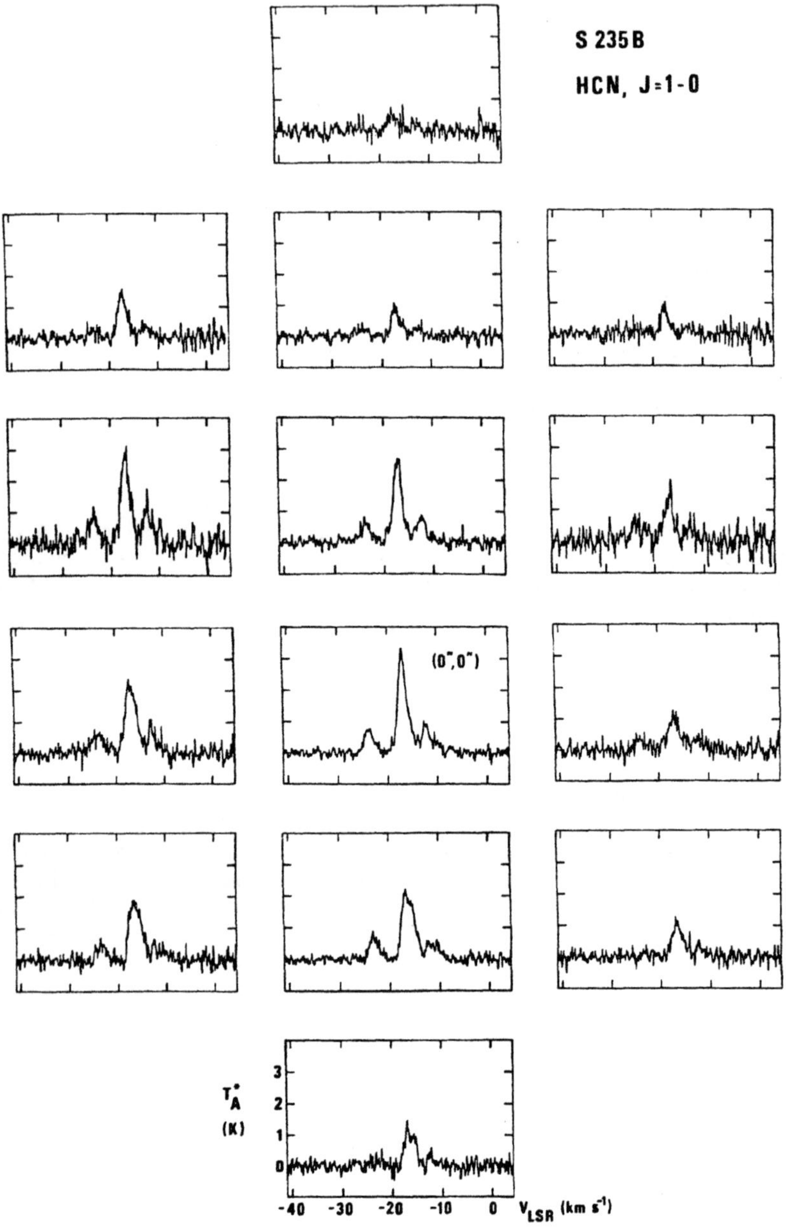

Figure 1. HCN J=1-0 spectra toward S 235B. The reference position (0",0") is: $\alpha(1950) = 05^h\ 37^m\ 30\overset{s}{.}4$, $\delta(1950) = 35^o\ 39'\ 57''$. The spacing between different map positions is 42".

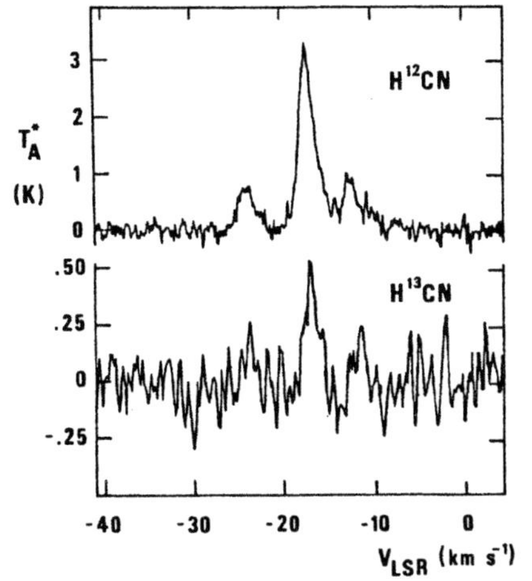

Figure 2. HCN and $H^{13}CN$ spectra toward S 235B, i.e. (0",0") position. Note the different temperature scale for $H^{13}CN$.

TABLE 1

Observational data toward (0",0") and (0",-42") obtained by least squares Gaussian fits. The spacing between the hyperfine components was kept constant and equal to the rest frequency values.

Molecule	Hyperfine component	T_A^* [K]	v_{LSR} [km s^{-1}]	Δv a) [km s^{-1}]
(0",0")	two velocity components			
HCN	F = 2-1	2.51	-17.2	1.7
		0.89	-16.1	3.4
	F = 0-1	0.57		1.7
		0.33		2.3
	F = 1-1	0.65		1.2
		0.41		3.1
	one velocity component			
	F = 2-1	3.00	-17.3	2.1
	F = 0-1	0.74		2.2
	F = 1-1	0.70		4.3
$H^{13}CN$	F = 2-1	0.47	-16.8	1.7
	F = 0-1	0.22		0.9
	F = 1-1	0.24		0.9
(0",-42")	two velocity components			
HCN	F = 2-1	1.53	-17.0	1.5
		1.55	-15.5	2.2
	F = 0-1	0.31		1.6
		0.44		2.4
	F = 1-1	0.55		1.3
		0.44		1.6
	one velocity component			
	F = 2-1	2.11	-16.3	2.8
	F = 0-1	0.68		2.3
	F = 1-1	0.42		3.4
$H^{13}CN$ b)	F = 2-1	0.36	-16.3	2.2
	F = 0-1	0.27		1.6
	F = 1-1	-		-

a) Width given as FWHM
b) Eye estimate only

Jan. 1981 ($H^{13}CN$). The rest frequencies were taken from Lovas et al. (1979). At 88 GHz the antenna HPBW is about 42" and the aperture efficiency 0.55. The front end was a cooled mixer receiver with a total system noise temperature of 300-400K (SSB) followed by a 512 channel autocorrelator (in Jan. 81 only 256 ch.) with 15 MHz bandwidth. The velocity resolution is therefore 0.12 km s^{-1} (35.2 kHz) for HCN and 0.24 km s^{-1} for $H^{13}CN$.

The observations are corrected for atmospheric attenuation and the intensity scale is given in main beam brightness temperature (T_A^*). The calibration was checked relative to Orion A, for which we assumed T_A^* = 30K (Rydbeck et al., 1981). The HCN map (Fig. 1) was made as several five point maps with adjacent ones overlapping in at least one position. By such a procedure all spectra were scaled relative to the adopted center position (0",0"): $\alpha(1950) = 05^h\ 37^m\ 30^s.4$, $\delta(1950) = 35^\circ\ 39'\ 57''$, i.e. the position of S 235B (Olofsson, 1980: private communication). The scaling corrections were usually small.

3. RESULTS

HCN is readily seen in all positions observed. A comparison with CO (Evans and Blair, 1981) shows that the -17 km s^{-1} cloud component extends much further than the HCN map in Fig. 1. The HCN map covers about half the area within the ^{12}CO 20K brightness temperature contour of Evans and Blair. From their analysis we infer a kinetic temperature of 30K and a total gas density of $\sim 4 \cdot 10^3$ cm^{-3} for the region mapped in HCN. The region with maximum CO antenna temperature (28K) lies just outside the southmost HCN profile (0",-84"). Evans and Blair have also made a partial mapping of HCN toward S 235A,B. Their data have a lower intensity than ours, probably because of filter and beam dilution. St. Clair Dinger et al. (1979) find a temperature of 2.7K toward the H_2O maser position, which is in good agreement with our observations.

3.1. Velocity structure

Fig. 2 gives the HCN and $H^{13}CN$ profiles toward the adopted cloud center position, S 235B. The HCN profile is clearly asymmetric and cannot be adequately fitted by a single Gaussian. The same tendency is also seen in the neighboring position (42",0"), while the profiles further south indicate that this asymmetry is due to an intrusion of another velocity component at a somewhat higher velocity (see Fig. 1, especially positions (0",-42") and (0",-84")).

The velocity of the northern part of the cloud is remarkably constant, the velocity centroid $\bar{v} = -17.2 \pm 0.2$ km s^{-1} and the linewidth $\overline{\Delta v} = 2.3 \pm 0.4$ km s^{-1}. If the spectra in the southern part of the cloud (except (-42",0") and (-42",42")) are fitted with two component Gaussians, the velocity centroid remains the same for the whole area. This procedure reveals another velocity component in the southern part of the cloud with $\bar{v} = -16.1 \pm 0.6$ km s^{-1} and

$\overline{\Delta v} = 2.5 \pm 0.7$ km s^{-1}.

Table 1 gives one and two component Gaussian fits for both (0",0") and (0",-42"). $H^{13}CN$ is very weak and therefore a two-component fit is not justified. The distribution of each velocity component is uncertain, because the two velocity components are blending with each other.

Our map does not extend far enough to show whether the -17.2 km s^{-1} feature disappears and whether the -16.1 km s^{-1} component gets stronger. One possibility is that the component might be associated with S 235C, or that it reflects some kind of shock interaction with the two compact H II regions A and B. In order to determine the significance of the -16 km s^{-1} component, the HCN map should be extended to cover at least the ^{12}CO temperature maximum.

3.2. Hyperfine line ratios

The departure from LTE between the hyperfine components is striking. The intensity ratio $R_{12} = T_A^*(F=1-1)/T_A^*(F=2-1)$ lies in the range 0.23 - 0.35 with a typical value of 0.28, which is much below the LTE value for the optically thin case, 0.6. R_{12} approaches 0.6 only in the position (0",126"), but there the line is very weak and the ratio unreliable (see Fig. 1). The line ratios in the southern part of the cloud are more uncertain, because of the two velocity components, but they are still far below the optically thin case.

The ratio $R_{02} = T_A^*(F=0-1)/T_A^*(F=2-1)$ ranges from 0.11 - 0.35 with a typical value of 0.25, i.e. close to the LTE ratio 0.2. We have only two positions with R_{02} below 0.2, which would indicate a real anomaly. Both these positions have large uncertainties and could easily be within the LTE range.

$H^{13}CN$ appears to depart from the optically thin case, but if we take into account the low S/N-ratio of the line, this apparent departure is most likely due to noise. Earlier $H^{13}CN$ observations of molecular clouds (Gottlieb et al., 1975) find $H^{13}CN$ hyperfine ratios equal to the optically thin values.

Our data therefore suggest strong departure from LTE in the ratio R_{12}, even larger than what has been previously observed (Gottlieb et al., 1975; Baudry et al., 1980). The ratio R_{02}, on the other hand, is only slightly above the optically thin value.

3.3 HCN column density

The column density determination of HCN is difficult because the line is not thermalized and yet optically thick. However, $H^{13}CN$ can be assumed to be optically thin and we can therefore use $H^{13}CN$ to solve for the column density. For an excitation temperature of 7K (only small differencies for T_{ex} = 5 - 10 K) and with the assumption that all levels have the same excitation temperature and are populated according to their statistical weights we get the total $H^{13}CN$ column density from

$$N_{H^{13}CN}\ [cm^{-2}] = 1.6\ 10^{12} \int T_A^* dv\ /\ F \qquad (1)$$

where F is the beam filling factor and the line integral is given in [K km s^{-1}]. With F = 1, which means that we assume that the cloud fills the beam uniformly, we get $N_{H^{13}CN} = 2.0\ 10^{12}\ cm^{-2}$ at the (0",0") position. This has to be taken as a lower limit, because we have a sharp density gradient toward south and west (Fig. 1), and furthermore the $H^{13}CN$ could be clumped, which also would affect the beam filling.

If we adopt a $^{12}C/^{13}C$ ratio of 40 (Wannier et al., 1976) this would give a total HCN column density of 8 10^{13} cm^{-2} and a total HCN optical depth of about 5-10. The $^{12}C/^{13}C$ ratio is a matter of dispute (see e.g. McCutcheon et al., 1980) and could easily equal the terrestrial value 89, in which case the column density could be twice as high.

Because the HCN optical depth in any case is moderate as judged from the similarity in HCN and $H^{13}CN$ linewidths, the line integrals reflect the density distribution of HCN. If we do not account for the two velocity components but only look at the total line integrals, the density maximum would appear to be at S 235B. However, because the −17.2 km s^{-1} component at B is already blended with the −16 km s^{-1} component, the maximum for the −17 km s^{-1} component is shifted to the compact H II region S 235A and the H_2O maser, or possibly somewhat east of these.

4. DISCUSSION AND SUMMARY

The effects of radiative trapping on HCN have been subject to several studies (Kwan and Scoville, 1975; Gottlieb et al., 1975; Guilloteau and Baudry, 1981): Guilloteau and Baudry show that Doppler width overlap between the hyperfine components to a large extent determine the anomalies in the hyperfine transitions. According to their analysis most of the hyperfine anomalies in the J=1-0 state can be explained by close line overlap of the J=2-1, F=2-1 and F=3-2, which will result in a weakening of the F=1-1 line and an enhancement of the F=2-1 line as soon as the J=2-1 transition becomes optically thick.

Unfortunately Guilloteau and Baudry present only a few numerical results. If we compare their model results for $n = 10^5\ cm^{-3}$ and T_k = 30K, which should be close to the conditions we have in our case, we notice, that in order to reproduce the hyperfine ratios, the column density, and hence the optical depth should be quite low ($\tau \sim$ 4-6). Our observed brightness temperatures are lower than what the model predicts, which easily could be explained by beamfilling effects.

Comparison with the models by Gottlieb et al. (1975), which also account for line overlap, although not as accurately as the model by Guilloteau and Baudry (1981), indicate that a lower gas density might result in better agreement between theory and observations. This again contradicts the density determination by Evans and Blair (1981), who use H_2CO observations and an LVG-model to derive a gas density of $n = (5 \pm 2)\ 10^5\ cm^{-2}$ slightly southeast of S 235B. It would therefore be instructive to obtain model results from the model by Guilloteau and Baudry for different densities and temperatures and see whether a homogeneous model can give a satisfactory fit.

To summarize: We have performed HCN J=1-0 observations with high spatial and velocity resolution of the molecular cloud associated with S 235B. The results indicate that the $-17\ km\ s^{-1}$ cloud component in the southern part of the cloud is affected by a velocity component at about $-16\ km\ s^{-1}$. The HCN line integral reaches its maximum at S 235B but because of blending with the $-16\ km\ s^{-1}$ component the gas density maximum appears to be closer to S 235A. The total HCN column density at S 235B, deduced from $H^{13}CN$ observations, is $>8.0\ 10^{13}\ cm^{-2}$.

Our observations indicate large anomalies in the hyperfine ratio R_{12}, even larger than previously observed. The model calculations by Guilloteau and Baudry (1981) show that these anomalies result from line overlap effects, when the J=2-1 transition becomes optically thick.

A more detailed paper, including also HCN observations of dark clouds is in preparation (Sandell et al., 1981). This work has been supported by the Swedish Natural Science Research Council. Onsala Space Observatory is operated by Chalmers University of Technology with financial Support from the Swedish Natural Science Research Council.

REFERENCES

Baudry, A., Combes, F., Perault, M., Dickman, R.: 1980, Astron. Astrophys. 85, pp. 244-248
Evans II, N.J., Blair, G.N.: 1981, preprint
Evans II, N.J., Beichman, C., Gatley, I., Harvey, P., Nadeau, D., Sellgren, K.: 1981, preprint
Gottlieb, C.A., Lada, C.J., Gottlieb, E.W., Lilley, A.E., Litvak, M.M.: 1975, Astrophys. J. 202, pp. 655-672
Guilloteau, S., Baudry, A.: 1981, Astron. Astrophys. 97, pp. 213-217
Israel, F., Felli, M.: 1978, Astron. Astrophys. 63, pp. 325-334
Kwan, J., Scoville, N.: 1975, Astrophys. J. 195, pp. L85-L88
McCutcheon, W.H., Dickman, R.L., Shuter, W.L.H., Roger, R.S.: 1980, Astrophys. J. 237, pp. 9-18
Lovas, F.J., Snyder, L.E., Johnson, D.R.: 1979, Astrophys. J. Suppl. 41, pp. 451-480
Rydbeck, O.E.H., Hjalmarson, Å., Rydbeck, G., Elldér, J., Olofsson, H.,

Sume, A.: 1981, Astrophys. J. 243, pp. L41-L45
Sandell, G., Höglund, B., Kislyakov, A.G.: 1981, in preparation
St. Clair Dinger, A., Dickinson, D.F., Gottlieb, C.A., Gottlieb, E.W.: 1979, Publ. Astron. Soc. Pacific 91, pp. 830-839
Tokunaga, A.T., Thompson, R.I.: 1979, Astrophys. J. 233, pp. 127-131
Wannier, P.G., Penzias, A.A., Linke, R.A., Wilson, R.W.: 1976, Astrophys. J. 204, pp. 26-42

DISCUSSION FOLLOWING PAPER BY SANDELL, HÖGLUND AND KISLYAKOV

WALMSLEY: In L673 the F=0-1 line is stronger than it would be with LTE. It seems to be a common property of cold dark clouds that the form of the HCN hyperfine anomaly is quite different from that in clouds close to HII regions.

SANDELL: That is interesting.

TURNER: N_2H^+, whose hyperfine energy splittings and rotational constant are similar to HCN, shows quite different anomalies in the intensities. In dark clouds the line whose relative intensity is 3 is typically too strong relative to the other lines, rather than too weak as for HCN.

PHILLIPS: I think it is about time we sorted out what density is appropriate to HCN excitation zones. This parameter is crucial to abundance determinations and recent papers have used very different values. We obtain $n(H_2) \simeq 10^6$ cm^{-3} from a combination of low and high frequency measures, whereas I note that you use 10^5 cm^{-3}.

SANDELL: The value is quite uncertain. We hope to do some model calculations but the problem is in knowing the amount of beam-filling that one has.

A MODEL FOR THE FORMALDEHYDE MASER NEAR NGC 7538 - IRS 1

Wilfried Boland and Teije de Jong, Astronomical Institute, University of Amsterdam, Roetersstraat 15, 1018 WB Amsterdam, The Netherlands

SUMMARY

The population of the 6 cm H_2CO transition can be inverted by the radio continuum radiation of a nearby compact H II region. The H II region must be very compact with emission measures of 10^8 - 10^{10} $cm^{-6}pc$. Our model does explain the observed maser emission near NGC 7538 - IRS 1 if a large formaldehyde abundance $\sim$ 8 x 10^{-7} is assumed.

OBSERVATIONS

Using the Westerbork Radio Synthesis Telescope at 6 cm Forster et al. (1980) observed two maser features of the 1_{10} - 1_{11} transition of H_2CO near the ultracompact H II region and infrared source NGC 7538 - IRS 1. More recently using the Very Large Array, Rots et al. (1981) have resolved the source into two spots separated by about 0.1 arcseconds. The emission spots have angular sizes smaller than 0.15 arcseconds corresponding to linear dimensions of $\sim$ 8 x 10^{15} cm and implying brightness temperatures $\gtrsim$ 5 x 10^5 K.

MODEL

In a more detailed paper (Boland and de Jong, 1981) we show that the population of the 1_{10} - 1_{11} transition of formaldehyde can be inverted by the free-free radio continuum radiation of a nearby compact H II region. The H II region must be very compact with emission measures of 10^8 - 10^{10} $cm^{-6}pc$ so that it is optically thick at cm wavelengths but still optically thin at mm wavelengths. If that is the case, population inversion of the ground state J=1 doublet of H_2CO occurs because the free-free continuum radiation field induces relatively more downward transitions from level 4 to level 2 than from level 3 to level 1 while the populations of levels 3 and 4 are kept approximately equal by rapid radiative exchange.

Because the H II region is optically thick at 6 cm, the masing gas is in front of the H II region so that it amplifies the radio continuum radiation. To allow sufficiently rapid radiative pumping mm line photons must be able to escape from the cloud which requires a large velocity gradient in the gas. To create sufficient amplification, molecules must

R. S. Roger and P. E. Dewdney (eds.), Regions of Recent Star Formation, 407–408.
Copyright © 1982 by D. Reidel Publishing Company.

be lined up in velocity over long pathlengths. We assume that the compact H II region NGC 7538 - IRS 1 moves through or expands into a molecular cloud with a velocity $V \approx 10$ km s^{-1}. Consistent with CO, HCN, OH, HCO^+ and H 110α lines we further assume that it moves towards us and that the molecular gas in front of the H II region is pushed sideways creating velocity gradients of $V/R_{II} \approx 1300$ km $s^{-1}pc^{-1}$ perpendicular to the line of sight and coherent pathlengths of $\ell = 3 \times 10^{16}$ cm along the line of sight.

In order to reproduce the observed brightness temperature in the H_2CO maser lines, a line optical depth $\tau = -5$ is required. This can be achieved if $n(H_2) \approx 10^4$ cm^{-3}, $T_k \approx 20$ K and $x(H_2CO) \approx 8 \times 10^{-7}$. Our model also explains the observed maser spot sizes. From our calculations we further find that the induced emission rate in the maser transition is approximately equal to the rate of population exchange in the millimeter lines so that the maser is close to saturation. This explains why both maser features have about the same intensity. However our model does not account for the fact that there are two maser spots.

ACKNOWLEDGEMENTS
The work of W.B. is supported by the Netherlands Foundation for Astronomical Research (ASTRON) with financial aid from the Netherlands Organization for the Advancement of Pure Research.

REFERENCES

Boland, W. and de Jong, T. 1981, Astron. Astrophys. 98, 149
Forster, J.R., Goss, W.M., Wilson, T.L., Downes, D. and Dickel, H.R. 1980, Astron. Astrophys. 84, L1
Rots, A.H., Dickel, H.R., Forster, J.R., Goss, W.M. 1981, Astrophys. J. 245, L15

DISCUSSION FOLLOWING PAPER BY BOLAND AND DE JONG

TURNER: Why is there only one formaldehyde maser? You require nothing magical so why aren't there more?

BOLAND: We have observed a few other sources but it seems that they do not have the required HII spectrum to provide sufficient pumping. There is a proposal to observe 25 other compact HII regions with 1665 MHz emission to search for more H_2CO masers.

FORSTER: W3(OH) would appear to have the correct continuum spectrum but it doesn't mase.

BOLAND: Of course it will depend on the density of the gas in the HII region not being too high.

WATSON: The direction of the velocity gradiant which is also the direction from which the pumping radiation is absorbed must be different from the line of sight to the maser. The enhanced continuum emission will be apparent only in a very narrow cone.

H_2CO NEAR COMPACT HII REGIONS - NEW WSRT RESULTS.

J.R. Forster, Radiosterrenwacht Dwingeloo, The Netherlands
W.M. Goss, Kapteyn Astronomical Institute, Groningen, The Netherlands
H.R. Dickel, University of Illinois, Urbana, Illinois, U.S.A.

SUMMARY

Aperture synthesis observations in the 6 cm λ H_2CO line are presented for five fields containing ultra-compact HII regions associated with OH maser sources. The H_2CO optical depths measured towards these sources are large compared to other sources in the fields. Neutral gas densities greater than 10^6 cm^{-3} are suggested.

INTRODUCTION

A necessary precursor to star formation, and the subsequent development of an HII region, is a gravitationally collapsed clump of molecular material. An ultra-compact HII region forms around the ionizing star or stars at the core, which subsequently expands, dispersing the remaining molecular material. Although HII regions of all sizes have been observed, the detection of molecular fragments has proved difficult, largely because of their small size. Indirect evidence for small, dense condensations of neutral material has been obtained by single dish measurements of high excitation molecules (cf. Plambeck and Williams 1979; Matsakis et al. 1977). Recently, aperture synthesis maps of H_2CO absorption against bright background sources in DR21 (Forster et al. 1981) and W3 Main (Arnal et al. 1981) have revealed the presence of small, dense clumps of molecular material. Typical scale sizes and densities are 0.1 pc and 10^5 cm^{-3}, making the clumps reasonable candidates for prestellar objects.

In order to try to detect this dense neutral component in direct association with young stellar systems, observations in the 6 cm line of formaldehyde were undertaken towards four ultra-compact HII regions. Insofar as the distances are known, the linear sizes of these objects are between 0.03 and 0.07 pc. The sources were selected on the basis of their compactness and their 6 cm continuum flux density. (The continuum intensity limits the accuracy to which H_2CO optical depths can be measured in absorption). In addition to these four sources, the H_2CO

R. S. Roger and P. E. Dewdney (eds.), Regions of Recent Star Formation, 409–418.
Copyright © 1982 by D. Reidel Publishing Company.

emission line source near IRS1 in NGC 7538 (Forster et al. 1980) is discussed. All of these sources exhibit strong OH maser emission.

OBSERVATIONS

The observations were made with the Westerbork Synthesis Radio Telescope (WSRT) using the digital line backend in one bit mode to form 31 frequency points. Two orthogonal senses of linear polarization were observed simultaneously. The instrumental parameters for each observation are summarized in Table 1. Maps were made in the standard way, usually with a prolate weighting in the uv plane to reduce the effects of aliasing. The rms noise quoted in the Table is the value measured in a single channel map before cleaning. The bands were centered at the mean velocity of the associated OH emission, using a rest frequency for the $J_{KK} = 1_{10} \rightarrow 1_{11}$ transition of 4829.6596 MHz.

Table 1
Instrumental Parameters for H_2CO Observations

Source Name	Obs. Date	Number of Baselines	Integration time (hrs)	rms (mJy)	Beam (RA"	Dec")	ΔV (km s^{-1})
ON1	15-10-81	32	7.3	8.6	4.0	9.2	0.73
	21-10-81	40	10.8	4.2	3.7	8.7	1.46
ON2	22-10-81	40	12.0	7.0	3.8	6.5	0.73
W58(ON3)	19-10-81	36	12.0	11.3	3.8	7.2	0.73
W3(OH)	27-10-81	36	12.0	6.5	4.0	4.6	0.73
NGC 7538	8-06-79	25	12.0	15.0	8.0	9.2	0.37

RESULTS

The H_2CO line profiles are presented alongside maps of the continuum emission at 6 cm in Figures 1 to 5. Two profiles for ON1 are shown in Figure 1, obtained with 0.7 and 1.4 km s^{-1} velocity resolution. As can be seen in the two channel maps to the left of the profiles, the HII region is not visible above the noise at 8.0 km s^{-1}, whereas it is clearly detected at 5.0 km s^{-1}. The OH position (Hardebeck 1972) is marked by a cross. The optical depth τ, defined by the relation $T_L = T_C e^{-\tau}$ where T_L is the measured line intensity and T_C is the intensity of the continuum, is indicated along the right hand vertical axis of the absorption profile. ON1 is the only source present in the synthesized field and it is unresolved by the WSRT beam.

The field containing ON2 also contains the extended HII regions

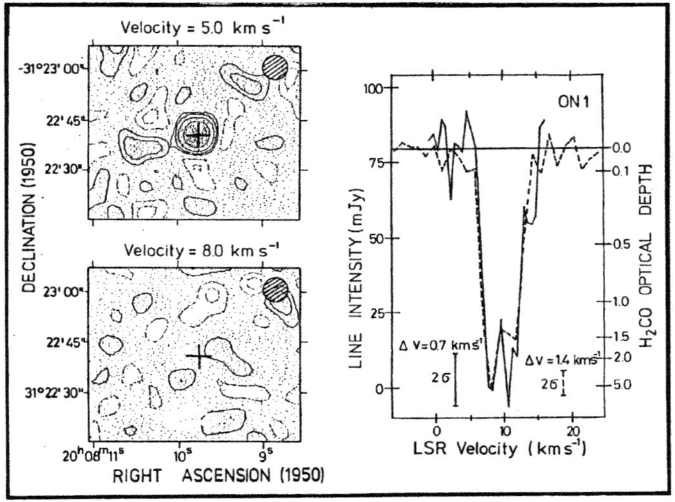

Figure 1. Top left: single channel map of ON1 at a velocity of 5.0 km s^{-1}. The beam is shown in the upper right hand corner. Bottom left: single channel map at 8.0 km s^{-1}. Right: H_2CO absorption profiles with velocity resolution 0.73 km s^{-1} (solid) and 1.46 km s^{-1} (dashed) at the position of peak continuum intensity. The two sigma levels are indicated at the bottom.

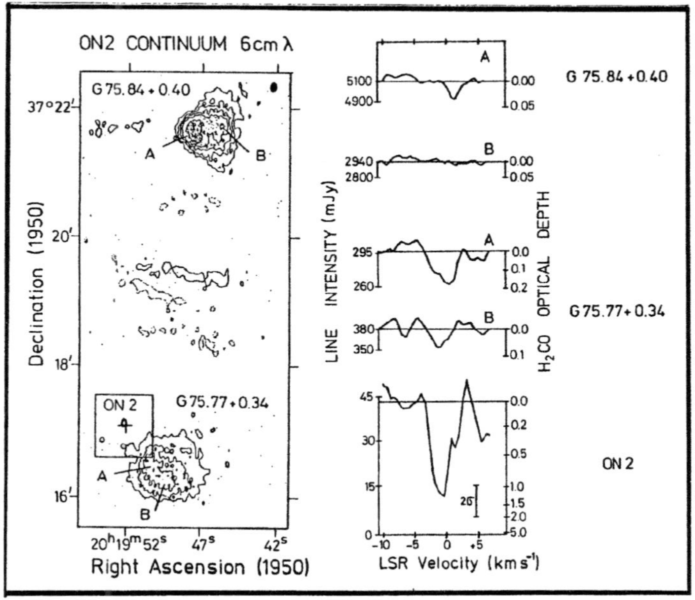

Figure 2. Left: cleaned continuum map of G75.84+0.40, G75.77+0.34 and ON2 at 6 cm. The beam is shown in the upper right hand corner. Right: H_2CO absorption profiles at the positions indicated in the figure at left.

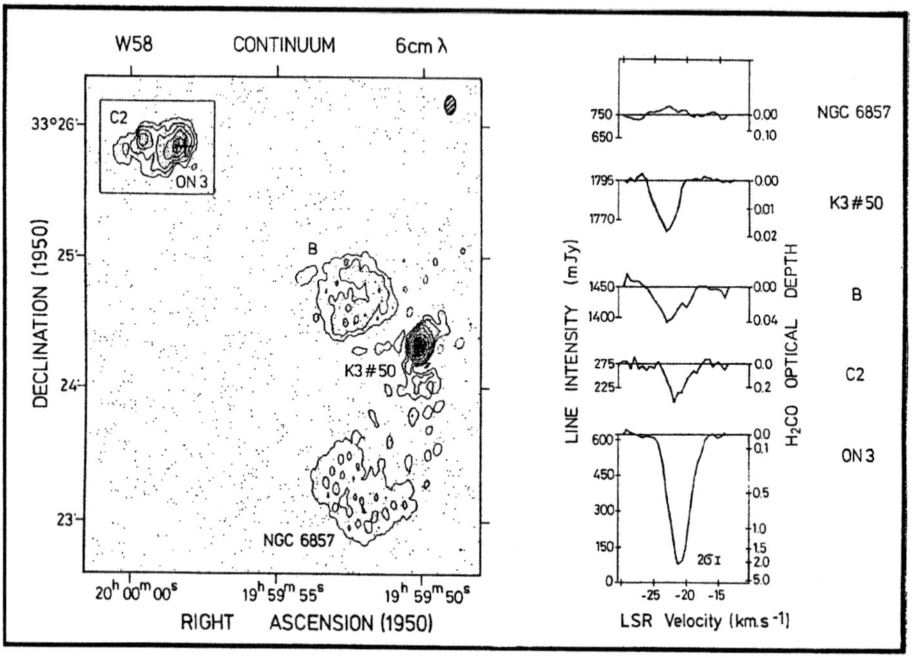

Figure 3. Left: cleaned 6 cm continuum of the W58 region. The OH position is marked by a cross. Right: H_2CO absorption towards continuum components marked in the map.

G 75.84+0.40 and G 75.77+0.34 (Figure 2). Small areas near the positions of peak intensity in these regions have been averaged to form the line profiles shown to the right of the continuum map. Although fairly weak, ON2 is clearly detected in the continuum (box). ON2 lies within the errors of the OH maser position (cross) measured by Hardebeck and Wilson (1971), and is unresolved with the WSRT beam. Similarly to ON1, ON2 does not appear significantly above the noise level of a single channel map at the velocity of maximum absorption. The H_2CO opacity towards ON2 is many times greater than towards G 75.77+0.34 or G 75.84+0.40.

There are many continuum sources in the W58 region (Figure 3) which contains ON3, K3#50 and the optical nebula NGC 6857. By far the strongest H_2CO absorption occurs towards ON3 (cross). The nearby component C2 reaches a peak optical depth of 0.34, while ON3 reaches a value of 2.05. The recombination line velocities of C2 and ON3 are the same within the error of measurement ($\sim$5 km s^{-1}; from unpublished WSRT data) and are about 13 km s^{-1} more positive than K3#50. It therefore seems likely that both C3 and ON3 are located on the far side of the molecular cloud, and that their apparent linear separation of $\sim$1 pc is indicative of their true spatial separation. It should be noted that the apparently strong absorption line measured toward K3#50 is due entirely to the strength of

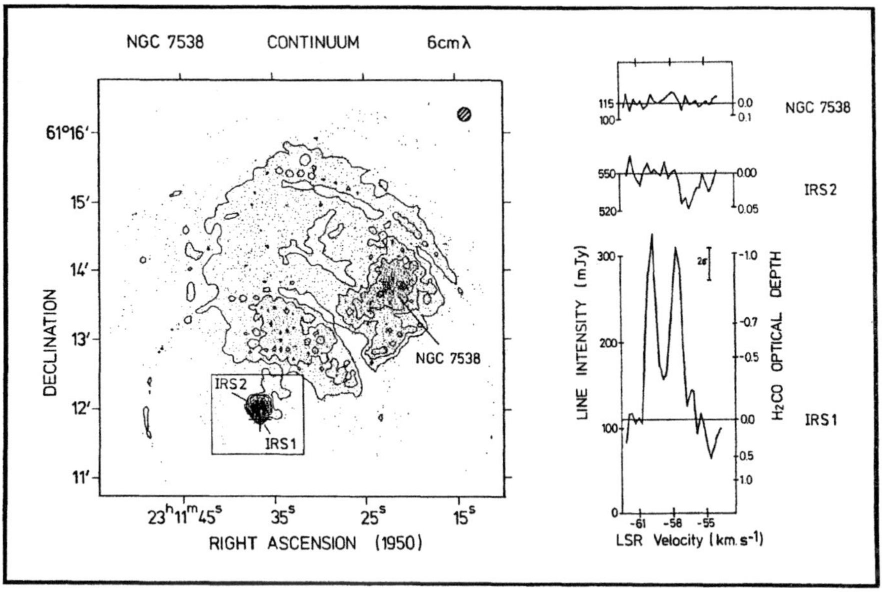

Figure 4: Left: cleaned 6 cm continuum map of NGC 7538 and the IRS 1, 2 and 3 complex (box). Right: H_2CO profiles towards NGC 7538, IRS 2 and IRS 1. The emission line towards IRS 1 (cross) arises in a region unresolved by the WSRT beam.

the continuum source. The absorbed intensity actually amounts to a peak H_2CO optical depth of 0.018. The K3#50 spectrum has been hanning smoothed.

Figure 4 shows the region containing the optical nebula NGC 7538 and the OH/H_2CO maser source IRS1 (Forster et al. 1980). The circular pattern with IRS2 at the center is instrumental. The southern continuum source (box) is composed of a compact HII region (IRS2) of 0.25 pc diameter and two ultra-compact components (IRS1 and IRS3) smaller than about 0.04 pc (Martin, 1973) for an assumed distance of 3.5 kpc. This group of sources is barely resolved with the WSRT beam. The H_2CO profiles observed toward IRS1, IRS2 and NGC 7538 are shown on the right in Figure 4. Weak absorption is seen towards the extended component IRS2, while strong emission arises from IRS1 located at a projected distance of 0.14 pc to the south. The weak component IRS3 is not detected. If we assume that the H_2CO maser amplifies the continuum radiation from IRS1 (Boland and de Jong, 1981) the H_2CO optical depth in the emission lines can be calculated as before. Using 110 mJy for the continuum flux density of IRS1 (Rots et al. 1981) and assuming that the maser region uniformly covers the continuum source, the peak optical depth is -1.03. This is a lower limit to the true optical depth in the line.

The H_2CO absorption profile measured towards W3(OH) is shown in

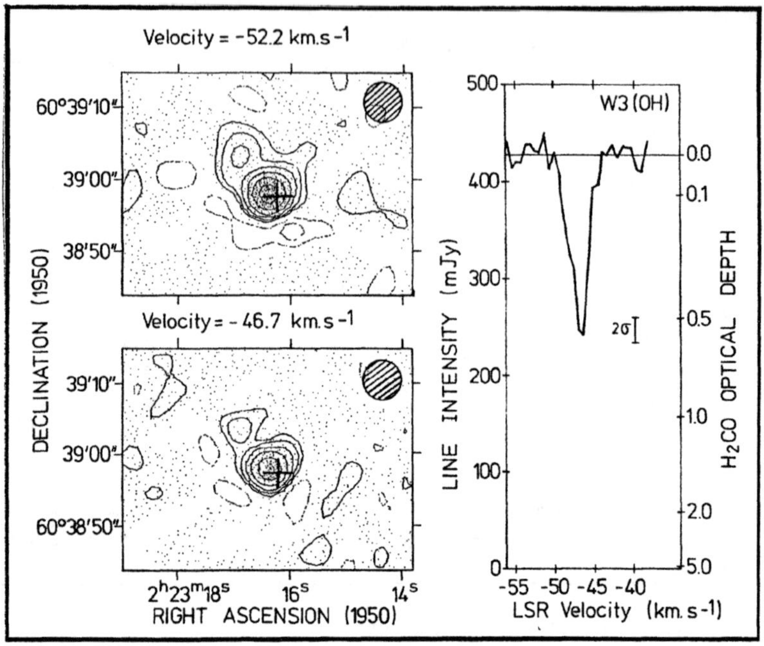

Figure 5: Top left: single channel map of W3(OH) at a velocity of -52.2 km s^{-1}. Bottom left: single channel map at -46.7 km s^{-1}. The contour levels are 12.5, 25, 50, 100, 200 and 400 mJy $beam^{-1}$ for both maps. Right: H_2CO absorption profile at the position of peak continuum intensity.

Figure 5. The two channel maps to the left of the profile show the brightness distributions in the continuum at -52.5 km s^{-1} and at the velocity of deepest absorption, -46.7 km s^{-1}. The H_2CO optical depth is nearly the same for the two continuum components detected. The peak value is 0.56. The cross marks the mean position of the OH maser sources (Norris, Booth and Davis 1980). The main continuum component is shell-like with a diameter of $\sim$0.03 pc (Dreher and Welch 1981; Scott 1981) and is unresolved with the WSRT beam.

DISCUSSION

The H_2CO optical depths observed towards the ultra-compact HII regions range between $\sim$0.5 and $\sim$2.0. The extended HII regions have optical depths which are less than $\sim$0.3. There is a striking difference in H_2CO optical depth between the ultra-compact HII regions associated with OH maser emission, and HII regions whose diameters exceed $\sim$0.1 pc. Figure 6 shows optical depth profiles for all sources observed in this sample. The profiles are directly comparable since continuum levels are

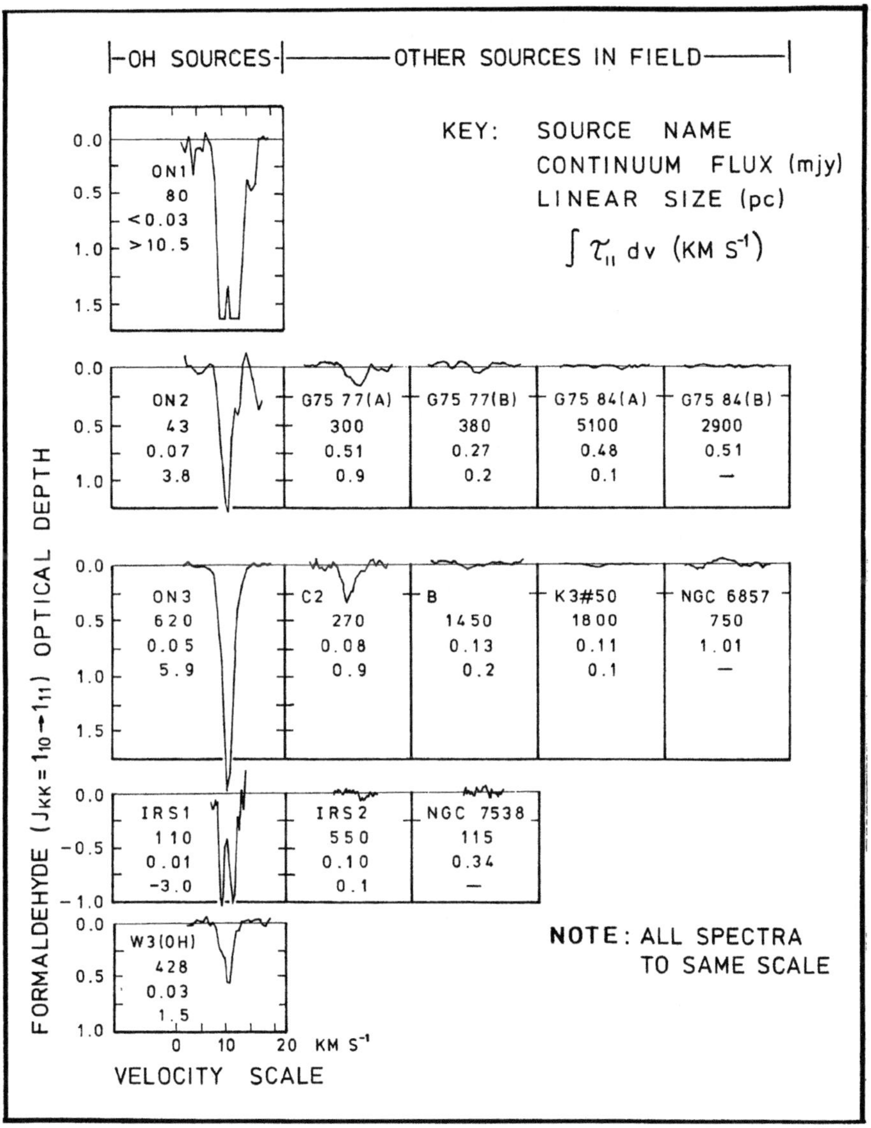

Figure 6: H_2CO optical depth profiles for all sources observed. The sources with OH maser emission are shown at the left. The continuum flux, linear size and H_2CO equivalent width are indicated for each source. The absolute velocities are omitted for clarity. The profile for ON1 has been cut off at the two sigma level. The optical depth scale for IRS1 is inverted.

removed in forming optical depth. The vertical and horizontal scales are the same for all profiles. The continuum flux density, source size and H_2CO equivalent width ($W_{11} = \int \tau dv$) is listed for each source. For sources unresolved with the WSRT beam the size quoted has been obtained from other observations. For resolved sources, the size quoted is the area over which the continuum has been integrated, and the flux is that arising from the area.

H_2CO optical depths less than about 0.5 can be accounted for by distributed absorption in the giant molecular clouds associated with HII regions. For a 25 pc cloud of density $2x10^3$ $cm^{-3}$the hydrogen column density is $N(H_2) \approx 2x10^{23}$ cm^{-2}. For a H_2CO/H_2 abundance ratio fo $5x10^9$ the H_2CO equivalent width is roughly $W_{11} \approx 8x10^{-24}$ $N(H_2)$ for warm clouds. Thus the maximum W_{11} expected is 1.6 km s^{-1} for distributed absorption. For a typical FWHM linewidth of 3.5 km s^{-1} the expected peak optical depth is 0.46. This estimate is strengthened by the results of single dish H_2CO surveys (cf. Scoville, Solomon and Thaddeus 1972, Wilson 1972, Whiteoak and Gardner 1974, Zuckerman et al. 1970). The H_2CO opacities observed towards the extended HII regions in this sample can be easily explained in this way. The high optical depths observed towards the ultra-compact sources are difficult to account for in terms of distributed absorption, even if they are deeply embedded in giant molecular clouds.

In contrast, the optical depth through a clump can become large. Taking the simplest case of uniform, isothermal collapse of M solar masses, the volume density at a given diameter d is $n(H_2) \approx 38$ M/d^3 cm^{-3} for d in pc. The hydrogen column density for a source at the center of the clump is $N(H_2) \approx 2.5x10^{18}n(H_2)d$. In terms of H_2CO equivalent width: $W_{11} \approx 5x10^4$ M/d^2 km s^{-1}. If we take, for example, a 50 $M_\odot$ clump the H_2CO opacities listed in Table 2 result for densities between 10^3 and 10^7 cm^{-3}.

Table 2
H_2CO Opacities Calculated for Uniform
Isothermal Collapse of a 50 $M_\odot$ Fragment

$n(H_2)$ (cm^{-3})	d (pc)	$N(H_2)$ (cm^{-2})	W_{11} (km s^{-1})	τmax
10^3	1.24	2.0 10^{21}	0.02	0.004
10^4	0.58	9.1 10^{21}	0.07	0.02
10^5	0.27	4.2 10^{22}	0.34	0.10
10^6	0.12	2.0 10^{23}	1.56	0.45
10^7	0.06	9.1 10^{23}	7.26	2.07

For this simple model the H_2CO optical depth in a fragment goes as $n^{\frac{2}{3}}$. It exceeds the maximum "background" opacity at about 10^6 cm^{-3}. At such high densities, the upper levels of H_2CO will become populated by collisions; thus it would be hard to achieve these high optical depths of the 6 cm transition without a corresponding increase in the total formaldehyde abundance.

CONCLUSIONS

The 6 cm λ H_2CO lines observed towards ultra-compact HII regions associated with OH maser sources have substantially higher optical depths than the lines observed towards other sources. This is probably due to dense placental material surrounding these young sources. The gas density may be 10^6 cm^{-3} or higher. We feel that neither large amounts of low density gas, nor the chance coincidence of unrelated dense clumps along the line of sight can account for the observations.

ACKNOWLEDGEMENTS

It is a pleasure to thank the WSRT staff, and in particular Tony Willis, for skillful assistance in obtaining these observations. We benefited greatly from discussions with W. Boland. The Westerbork Synthesis Radio Telescope is operated by the Netherlands Foundation for Radio Astronomy with the financial support of the Netherlands Organisation for the Advancement of Pure Research (ZWO).

W.M.G. and H.R.D. gratefully acknowledge support from a joint NATO Research Grant #RG 138.80.

REFERENCES

Arnal, E.M., Goss, W.M., Forster, J.R. and Dickel, H.R.: 1981 (in preparation).
Boland, W. and de Jong, T.: 1981, Astron. and Astrophys. (in press).
Dreher, J.W. and Welch, W.J.: 1981, Astrophys. J. (in press).
Forster, J.R., Goss, W.M., Wilson, T.L., Downes, D. and Dickel, H.R.: 1980, Astron. and Astrophys. 84, L1.
Forster, J.R., Goss, W.M., Dickel, H.R. and Habing, H.J.: 1981, MNRAS (in press).
Hardebeck, E.G. and Wilson, W.J.: 1971, Astrophys. J. 169, L123.
Hardebeck, E.G.: 1972, Astrophys. J. 172, 583.
Martin, A.H.M.: 1973, MNRAS 163, 141.
Masatsakis, D.N., Brandshaft, D., Chui, M.F., Cheung, A.C., Yngvesson, K.S., Cardiasmenos, A.G., Shanley, J.F. and Ho, P.T.P.: 1977, Astrophys. J. Letters 214, L67.
Norris, R.P., Booth, R.S. and Davis, R.J.: 1980, MNRAS 190, 163.
Plambeck, R.L. and Williams, D.R.W.: 1979, Astrophys. J. Letters 227, L43.
Rots, A.H. Dickel, H.R., Forster, J.R. and Goss, W.M.: 1981, Astrophys. J. Letters 245, L15.

Scott, P.F.: 1981, MNRAS 194, 25P.
Scoville, N.Z. Solomon, P.M. and Thaddeus, P.: 1972, Astrophys. J. 172, 335.
Whiteoak, J.B. and Gardner, F.F.: 1974, Astron. and Astrophys. 37, 389.
Wilson, T.L.: 1972, Astron. Astrophys. 19, 354.
Zuckerman, B., Buhl, D., Palmer, P. and Snyder, L.E.: 1970, Astrophys. J. 160, 485.

IS A BN-TYPE OBJECT THE ENERGY INPUT TO THE NH_3 CLOUD IN NGC 2071?

G. Calamai, M. Felli
Osservatorio Astrofisico di Arcetri, Firenze

An extended molecular cloud has been found close to the reflection nebula NGC 2071 in the northern part of the Orion complex. The J,K = (1,1) and (2,2) inversion lines of NH_3 at 23.69 GHz and 23.72 GHz respectively were mapped (see Figure 1) with the 25m radiotelescope of the S.R.C. Appleton Laboratory at Chilbolton Observatory, Hants, U.K. The v_{LSR} values confirm that the ammonia cloud is at the distance of the Orion complex. At this distance the linear resolution in the map is ~0.3 pc, and no clumping has been detected on this scale size. The molecular cloud has an elongated shape with a maximum extension of ~3 pc. The kinetic temperature (T_K~20K), the inferred density ($n(H_2) \lesssim 10^4 cm^{-3}$) and the total derived mass ($M \simeq 10^3 M_\odot$) classify it as a typical molecular cloud in which star formation may occur.

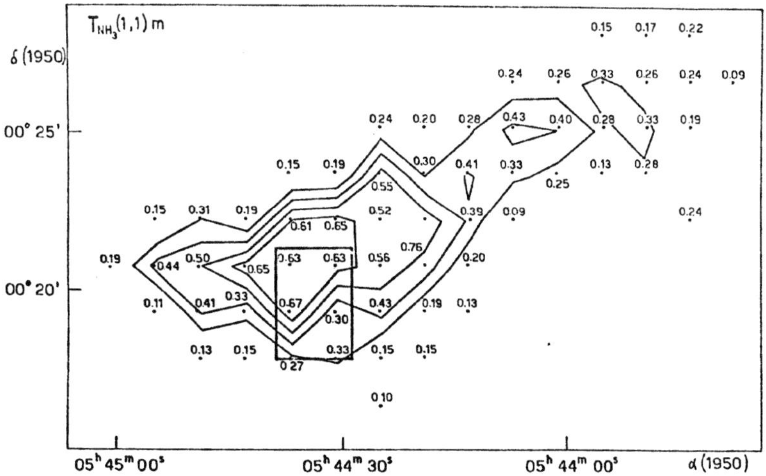

Figure 1. Antenna temperature of the main component of the (1,1) ammonia transition.

R. S. Roger and P. E. Dewdney (eds.), Regions of Recent Star Formation, 419–423.
Copyright © 1982 by D. Reidel Publishing Company.

In addition to the ammonia emission (Macdonald et al., 1980), several molecules have been observed, but the more interesting results are those obtained by CO and by HCO^+ line profile observations (Kutner et al., 1977; White and Phillips, 1980; Loren et al., 1981; Phillips et al., 1981; Loren and Wootten, 1981). At the peak of the CO emission, which is in the same region of the ammonia one, these line profiles are self-reversed, with very broad asymmetrical wings, suggesting an outflow of the gas in the central part of the cloud. Within this region are located an OH maser (Johansson et al., 1974; Pankonin et al., 1977) an H_2O maser (Schwartz and Buhl, 1975; Genzel and Downes, 1979) and a small hot infrared source, NGC 2071 IR (Harvey et al., 1979; Evans et al., 1979). The last one is very close to the 150 μm peak on an extended far infrared emission region (Sargent et al., 1981) and has the typical signposts of a BN-type object. The high-angular-resolution observation of Harvey et al. (1979), which covers a large wavelength interval, from 1.65 μm up to 175 μm, indicates that NGC 2071 IR is formed by a high temperature small core (T_{core} ~ 300K, θ(10 μm) ≃ 3 arcsec), surrounded by a large region of lower temperature ($T_{envelope}$ ≃ 30K, θ(100 μm) ≃ 10 - 20 arcsec).

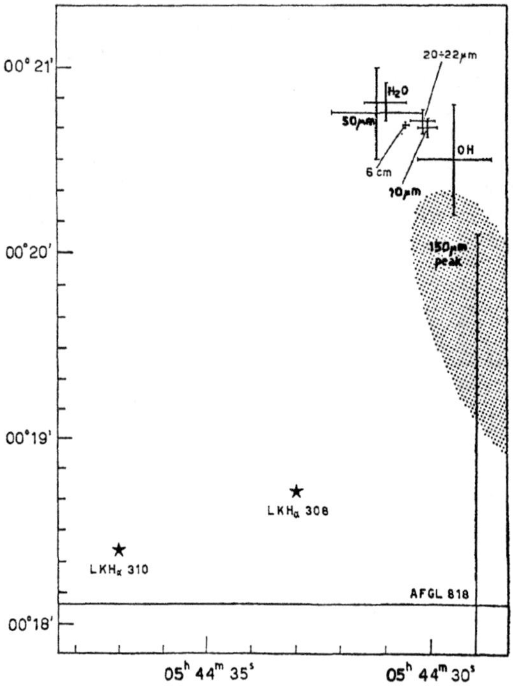

Figure 2. Position and relative error bars of the sources located in the rectangular box of Figure 1.

In this region we have searched for radio continuum emission from possible compact HII components at 6 cm (ν = 4.885 GHz), using the VLA of the National Radio Astronomy Observatory, U.S.A. Over the mapped field of ≃2 x 2 arcmin only one unresolved point source, with peak flux density of 3 mJy, is present and is shown in Figure 3. The coordinates of this source are:

$$\alpha(1950) = 05^h\ 44^m\ 30.6^s \pm 0.1^s$$

$$\delta(1950) = 00°\ 20'\ 40.8'' \pm 1''$$

The diameter of this unresolved source is $\leqslant$ 200 A.U. Beside the component, there is some more extended emission (on a scale size of ≃ 1 arcsec) surrounding the radio peak. The total flux density, inclusive of the point source, is 7 mJy. Although this radio source does not coincide with the point infrared source, it is, however, within its error bars.

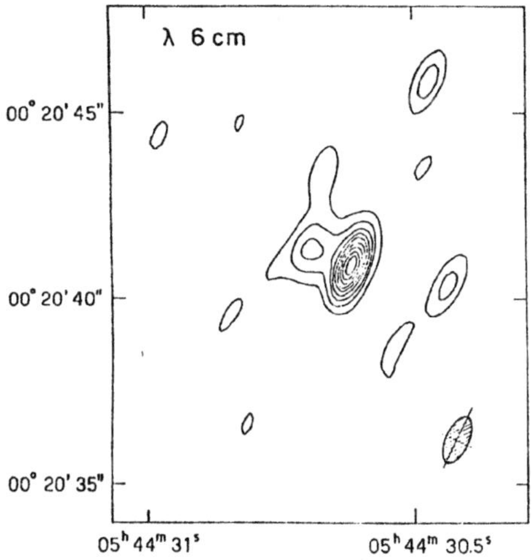

Figure 3. V.L.A. map of the ultra-compact HII regions.

The greatest part of the energy lost by the cloud is emitted in the far infrared from warm dust. Dust can then be considered as the most important agent of energy release in the dust-molecular cloud complex, or conversely, it is the heating of the dust that determines the energy input to the cloud in a steady situation.

To estimate the total infrared emission, we have to set boundaries to the molecular and dust cloud. We shall set as the boundary the outer contour of the NH_3 map of Figure 1. This boundary roughly corresponds to the contour with $T_a(^{12}CO) \geqslant 17K$. By doing this we confine ourselves only to the portion of the molecular cloud containing hot molecular gas, i.e. with temperatures greater than the ones generally found in cold molecular clouds. The higher gas temperature of the chosen cloud, therefore, requires a local discrete source of energy supply.

The total luminosity of the dust cloud may be obtained by the far infrared measurements (where the bulk of the emission occurs) of Sargent et al. (1981). They obtain a temperature, $T_{dust} \sim 40K$, and a luminosity of at least 10^{3} $L_\odot$, which is in agreement with the value of 750 $L_\odot$ obtained by Harvey et al. (1979) for the central region.

Since the detection of an ultra-compact HII region suggests that star formation has already occurred in the core of the molecular cloud, we shall tentatively identify the source of energy input to the dust and molecular cloud with an early type star imbedded within the BN-like IR source. Such a star must then be able to:

i) ionize the ultra-compact HII region

ii) heat the dust to the high temperature derived for the small near-infrared source

iii) supply the necessary energy input to the dust emission and to the high temperature gas in the NH_3 cloud.

The standard procedure generally used to derive the number of Lyman-continuum photons, and then the spectral type of the star necessary to keep the HII region ionized, is to consider the radio emission optically thin. In this case the imbedded object would correspond to a B 3 main sequence star (Panagia, 1973). The total stellar luminosity would then be $\leqslant 1.0\ 10^{3}$ $L_\odot$, barely sufficient to supply the required energy input to the cloud, being close to the lower limit derived from infrared measurements. However, it has been pointed out that, if early type stars imbedded within BN-like objects present mass outflow, the ionized envelope will have a strong density gradient. Due to the high density, the inner part is optically thick to the radio emission, and the observed radio flux will come essentially from outer parts of the envelope. If this is the case, following Felli and Panagia (1981), we have that the number of Lyman-continuum photons is much greater, and corresponds to at least a B 0.5 main sequence star, with a total luminosity of $\sim 10^{4}$ $L_\odot$. From the maximum extent of the CO wings we may assume that a lower limit for the expansion velocity of the ionized flow is 30 km sec^{-1}, and then, according to Felli and Panagia (1981), derive the mass loss rate $\dot{M} \geqslant 1.0\ 10^{7} M_\odot\ yr^{-1}$ and the angular size of the corresponding HII region θ(5 GHz) $\simeq$ 0.06 arcsec. Both values are in agreement with experimental results (Tanzi et al., 1981) and our own observational limit.

Simon et al. (1981) have suggested that different stages in the evolution of BN objects can be established on the basis of the observed properties. The detection of an ultra compact HII region in the BN-type object in NGC 2071 would place this source in the evolutionary phase immediately following that of BN itself, similar to CRL 2591 and S140 IRS 1.

Our conclusions are that the total luminosity of the molecular and dust cloud, the ionization of the ultra-compact HII object, and the large wings of the CO profiles, could all be explained if a B 0.5 star, losing mass at a rate $\dot{M} \simeq 10^{-7} M_{\odot}\ yr^{-1}$ is embedded in the BN-like object. If such is the case, shocked H_2 lines, of the type observed in Orion, should be observable. We are planning observations of Brackett hydrogen recombination lines, which would be of extreme importance in determining the emission measure of the HII region and hence to test the validity of the mass loss hypothesis.

REFERENCES

Evans II, N.J., Beckwith, S., Brown, R.L., Gilmore, W.: 1979, Astrophys. J. 227, 450
Felli, M., Panagia, N.: 1981, Astron. Astrophys., in press
Genzel, R., Downes, D.: 1979, Astron. Astrophys. 72, 234
Harvey, P.M., Campbell, M.F., Hoffmann, W.F., Thronson Jr., H.A., Gatley, I.: 1979, Astrophys. J. 229, 990
Herbig, G.H., Kuhi, L.V.: 1963, Astrophys. J. 137, 398
Johansson, L.E., Höglund, B., Winnberg, A., Nguyen-Q-Rieu, Goss, W.M.: 1974, Astrophys. J. 189, 455
Kutner, M.L., Tucker, K.D., Chin, G., Thaddeus, P.: 1977, Astrophys. J. 215, 521
Loren, R.B., Plambeck, R.L., Davis, J.H., Snell, R.L.: 1981, preprint
Loren, R.B., Wootten, H.A.: 1981, preprint
Macdonald, G.H., Little, L.T., Brown, A.T., Riley, P.W., Matheson, D.N., Felli, M.: 1981, Monthly Notices Roy. Astron. Soc., 195, 387
Panagia, N.: 1973, Astronomical J. 78, 929
Pankonin, V., Winnberg, A., Booth, R.S.: 1977, Astron. Astrophys. 58, L25
Phillips, T.G., Knapp, G.R., Huggins, P.J., Werner, M.W., Wannier, P.G., Neugebauer, G., Ennis, D.: 1981, preprint
Price, S.D., Walker, R.G.: 1976, AFGL Report AFGL-TR-76-0208
Sargent, A.I., van Duinen, R.J., Fridlund, C.V.M., Nordh, H.L., Aalders, J.W.G.: 1981, in press
Schwartz, P.R., Buhl, D.: 1975, Astrophys. J. 201, L27
Simon, M., Righini-Cohen, G., Felli, M., Fisher, J.: 1981, Astrophys. J., in press
Tanzi, E.G., Tarenghi, M., Panagia, N.: 1981, Paper presented at the IAU Colloquium No. 59, "Effects of Mass Loss on Stellar Evolution", ed. C. Chiosi and R. Stalio, D. Reidel, Holland, p. 51.
White, G.J., Phillips, J.P.: 1980, preprint

VLA OBSERVATIONS OF OH MASERS AND ASSOCIATED ULTRACOMPACT CONTINUUM SOURCES

B. E. Turner
National Radio Astronomy Observatory

ABSTRACT

We have observed 22 OH masers and associated continuum sources with the VLA at λ18 cm, and deduce that type I masers are always associated with ultracompact continuum sources, while type II(a) (1720 MHz) masers probably never are. Implications for the type of pumping for each category of maser are discussed.

I. INTRODUCTION

It has long been argued (e.g. Habing et al. 1974; Matthews et al. 1977; Evans et al. 1979) that type I (main line) OH masers tend to be physically associated with ultracompact (thermal) continuum sources. Of 31 such masers adequately studied, 21 are within 10" of a compact ($<$ 4") source. A similar fraction (though not necessarily the same objects) are near IR sources. We have extended the study of OH masers and associated compact objects by using the VLA at λ18 cm both to locate accurately the position of several masers found in the Green Bank OH survey (Turner 1979) and to search, to typically 10 mJy levels, for any nearby continuum objects. In our sample are included 10 type I masers, 2 type I(b) (evolved objects), and 10 type II(a) masers. Neither type I(b) nor type II(a) masers have previously been searched sensitively for compact continuum objects. Types I and II(a) masers are thought to be pumped by quite different mechanisms; the objective is to seek corresponding differences in their association with compact continuum sources.

Since 18 cm is not an optimum wavelength to detect compact continuum objects, we have also used the VLA at λ6 cm to search with higher sensitivity several recently positioned type I masers and several type I's which previously have not shown compact objects. λ6 cm results are still preliminary.

R. S. Roger and P. E. Dewdney (eds.), Regions of Recent Star Formation, 425-432.
Copyright © 1982 by D. Reidel Publishing Company.

II. OBSERVATIONAL RESULTS

The VLA was used in the continuum mode for both line and continuum, by using 1.5 MHz-wide filters in the IF's. Maps were made on the line frequency and offset 1.5 MHz for the continuum, with essentially identical u,v coverage. In this mode, all maser emission for a given source is found to arise from the same position within the uncertainties.

Table 1 gives the results. Columns 1 to 5 list respectively the maser source, VLA-determined maser position, position offset of the continuum source, size of continuum source, and flux of same.

Table 1. OH Masers and Continuum Sources

Source	OH Maser Position 1950.0		Continuum Source		
			Offset arcsec	Size arcsec	Flux mJy
		Type I			
OH343.0+0.0	16 54 43.77	-42 47 32.4	0x0	2.3	168
OH350.1+0.1	17 16 02.05	-37 07 00.2	0x2	3	248
OH351.4+0.6 A	17 17 32.10	-35 44 27.0	1x3	4	162
OH351.1+0.7 B	17 16 35.89	-35 55 05.0	0x1	3	439
OH 10.6-0.4	18 07 30.4	-19 56 28.0	4x1	6	700
OH 12.2-0.1	18 09 48.44	-18 25 13	0x1	1	99
OH 34.3+0.2	18 50 46.07	01 11 11.5	1x1	6	269
OH 43.8-0.1	19 09 30.83	09 30 47.0	.3x.3	1.7	111
S 269	06 11 46.0	13 50 36	< 7	2	< 4
		Type I(b) (Evolved)			
M1-92	19 34 19.62	29 26 04	2x1	5	15.4
OH231.8+4.2	07 39 58.97	-14 35 45.0	---	-	< 3
		Type II(a)			
W28N A	17 58 49.11	-23 18 03	---	-	<10
W28N B	17 58 50.05	-23 19 28.5	---	-	<35
OH11.0+0.0	no maser detected				
OH28.2+0.0	no maser detected				
W44 A	18 53 55.5	01 24 43	---	-	<10
W44 B	18 54 05.0	01 22 20	---	-	<10
OH37.8-0.2	no maser detected				
OH38.4-0.1	no maser detected				
OH76.4-0.6	20 25 34.7	37 12 57.0	1x7	see text	
OH85.5-1.0	20 56 50.38	44 09 19	---	-	< 5
OH189.0+2.9	06 13 25.90	22 30 10	---	-	< 5

Note that the type I(b) category, previously associated with evolved star dust envelopes, here contains the two bipolar nebulae OH231.8+4.2 and M1-92 (cf. Morris and Bowers 1980). Such objects are now also recognized as evolved, and are in no way related to the usual type I's which are associated with very young stellar or protostellar objects.

One of the type II(a) masers (OH76.4-0.6) is seen in the direction of a (non-compact) continuum source. We believe this to be a chance superposition of a foreground OH cloud, and that the relatively strong 1720 emission arises from stimulated amplification of the background source, as is well known to occur with the widespread but usually weak 1720 anomalies (Goss et al. 1973; Turner 1981). Four of the 10 type II(a) OH sources were not detected. These cases all lie in extended regions of enhanced (but not masering) 1720 emission (Turner 1979) and must themselves be spatially extended ($\gtrsim$ 20") to escape detection with the VLA.

The preliminary λ6 cm studies indicate compact continuum sources near at least half of the 10 type I's that previously showed no such sources, as well as continuum sources for several new cases.

We therefore take it as observationally proved that type I OH masers are always associated with ultracompact (thermal) continuum sources, while type II(a) OH masers are not. On this basis, we examine the pumping mechanisms that may apply to each type.

III. PUMPING OF TYPE I OH MASERS

(A) Role of the Compact HII Region

Type I pumping may involve high densities, elevated temperatures, a (strong) IR source, streams of charged particles, a large fractional OH abundance, and a high H/H_2 ratio. The last two of these are provided uniquely by an associated compact HII region while the first four attributes could also occur in an IR source (i.e. protostellar environment). However, efficiency arguments (photon counting) rule out type I pumping by IR (or UV) as a primary agent (cf. Evans et al. 1979) although an external IR pump may certainly affect the resultant maser characteristics. These statements delimit the situation observationally.

Theoretical guidelines are harder to establish, but some seem to emerge from the many models of maser emission of all types. Dominant main-line population inversions may occur purely by collisional pumping. One way involves parity-sensitive propensity rules for rotational collisions (Gwinn et al. 1973; Dixon and Field 1979; Kaplan and Shapiro 1979), while another features streams of charged particles (Johnston 1967; Elitzur 1979). In case these collisional mechanisms are non-operative, "normal" rotational collisional excitation followed by any of several non-LTE effects in the optically thick IR radiation involved in the radiative decay of the OH molecules can produce main-line as well as satellite-line anomalies within several, but especially the ground, rotational state. These radiative effects involve weak asymmetries in the line strengths of pairs of IR transitions operating between upper and lower Λ-doublet levels, overlap of several hyperfine-split IR transitions because of velocity gradients of a few km/s over the cloud, and line overlap of the same sort owing to thermal or microturbulent broadening. All of these radiative effects, acting along with "normal" collisions, have been modelled in detail, though approximately (Bujarrabal et al. 1980a; 1980b; 1980c; Lucas 1980; Guilloteau et al. 1981). While they fail to explain the dominant main-line emission of type I's (possibly because the approximate treatment of the radiation transfer applies least well in this case), all of the models point to two requirements: 1) an enhanced fractional OH abundance, $\geqslant 1 \times 10^{-4}$, seems required; 2) a highly non-spherical geometry is needed. If the opacity in the IR lines is as large as that implied by path lengths along the maser direction, the 1612 MHz line is inevitably the most strongly inverted. A tubular geometry with a length-to-diameter ratio of $\geqslant 10$ solves this problem, which otherwise persists whether or not it is the primary rotational collisions, or the secondary IR decay effects which invert the main lines. Collisional excitation of the ground-state Λ-doublet by streaming particles may avoid this constraint by featuring little or no rotational excitation, although gas temperatures <15K are then implied.

Both an enhanced OH abundance, and a non-spherical geometry occur naturally in the shocked neutral shell surrounding an expanding compact HII region (Elitzur and de Jong 1978; Elitzur 1979). Neither of these conditions would seem to occur in a protostellar (simple IR source) environment. Therefore, using the physical conditions in such a compressed shell as modelled by Elitzur and de Jong (1978), we can assess the prospects for several type I pumping models.

(B) Assessment of Several Type I Pumping Models

(1) Collisional Dissociation of H_2O (Gwinn et al. 1973). OH molecules are created in parity-selected, rotationally excited states, and decay to the ground state, preserving the parity selection. An analysis of the rate equations for this process shows that an equilibrium population inversion can be maintained under conditions in the shocked shell. However the maser photon production rate is some 200 times smaller than the W3(OH) emission rate, and is also insufficient for most other type I masers.

(2) Collisional Excitation by H and H_2 (Gwinn et al. 1973; Dixon and Field 1979; Kaplan and Shapiro 1979). Using the rate constants of Kaplan and Shapiro, and the value of H/H_2 found by Elitzur and de Jong in the shocked shell, the condition for a

population inversion is not satisfied (contrary to the result of Kaplan and Shapiro, who ignore electrons). The stronger parity-selection found for H-collisions by Dixon and Field would satisfy the inversion criterion. H_2 molecules probably do not pump OH as H atoms do (Flower 1980), but in any case their parity-selection efficiency would have to exceed that of H atoms as deduced by Kaplan and Shapiro. Although H atoms are now generally agreed to pump OH, their relevance depends, unfortunately, on the as yet uncertain magnitude of the effect.

(3) Main-Line Pumping by Far-IR Line Overlap Due to Velocity Gradients (Bujarrabal et al. 1980a). This model uses collisions as the primary rotational excitation in the presence of a strong external IR field. It has been applied successfully to OH/IR evolved stellar sources, where the necessary velocity gradients are well established. A somewhat smaller velocity gradient is also expected along the amplifying length of a type I maser confined to a shocked shell surrounding a compact HII region, owing to the overall expansion of the region. The numerical value of $n_{OH}r/v$ expected for the type I case seems to correspond to maximum inversion for the 1612 MHz transition. However the IR trapping in the expected tubular geometry of the type I case is much less than in the spherical geometry appropriate to the evolved star case. This factor may reduce the 1612 inversion relative to that of the main lines. This model needs to be evaluated explicitly for the type I geometry.

(4) Main-Line Pumping by Far IR Line Overlap Due to Thermal Line Broadening (Guilloteau et al. 1981). This model is similar to the one above, except that line overlap arises from thermal or microturbulent broadening, of importance when $T_K > 100K$. For conditions (and geometry) appropriate to shocked HII shells, the 1612 MHz line is found to dominate over the entire acceptable ranges of T_k, n and N_{OH}, owing to the large far-IR opacity. Both IR trapping, and coupling to the maser beam are treated approximately. Inclusion of a modest velocity gradient, and a more exact treatment may rescue this model from its present severe difficulties.

The efficiency of both of these IR-overlap models is adequate to explain the observed maser photon rate of W3(OH), assuming that one maser photon results for every 10 collisional excitations. A 10% efficiency is in fact typical of the conversion efficiency of the IR-trapping models.

(5) Collisional Pumping of Main Lines by Streams of Charged Particles (Johnston 1967; Elitzur 1979). Anisotropic streams of electrons can select particular Zeeman sublevels when exciting the ground state Λ-doublet. Rotational excitation and accompanying trapped IR effects are not involved. It was shown long ago (Johnston 1967; Turner 1967) that the 1665 MHz line is always most strongly pumped. The efficiency of this process can be shown to meet the requirements of W3(OH), using Elitzur and de Jong's parameters for the

shocked shell, and adding the degrading effects of neutral collisions to Elitzur's (1979) estimates. Attaining the necessary streams of charged particles may be a problem. Elitzur (1979) finds that ions can readily acquire streaming by ambipolar diffusion, but it is not clear that ions can invert the OH sublevels. In contrast, inversions via electrons are well established, but the method for producing electron streams is unclear. Finally, for $T_k > 15K$ as is likely, one must worry about the added effects of rotational excitation and trapped IR.

Summary. In order to produce adequate photon fluxes, all of the above models, despite their diversity, require high densities ($\gtrsim 5\times10^5$ cm^{-3}) and high OH abundances ($n_{OH}/n \gtrsim 1\times10^{-4}$) non-symmetric geometry is strongly implied for all models involving trapped IR, and follows naturally (but is not required) by the streaming pump model as well. These characteristics are expected in the shocked shells surrounding compact HII regions, and provide a theoretical basis for what now seems to be a secure observational relationship. Unfortunately this association of type I masers with compact HII's has not yet provided the detailed clues needed to distinguish between the models, but more exact and appropriate calculations may now clarify matters significantly.

IV. TYPE II(a) OH MASERS

The anomalous excitation of the 1720 line is widespread. Many giant clouds exhibit remarkably uniform, though weak, 1720 emission over several degrees, while the other lines are not seen (Turner 1981). In a few cases, such as the 4 non-detected objects in Table 1, the emission is peaked enough to suggest maser activity, which can now be ruled out. There are only 6 bona-fide type II(a) masers at present (Table 1, excluding OH76.4-0.6), of which the pairs in W28N and W44 are separated by 0.75 pc and 3.0 pc respectively. These masers are 10 to 50 times weaker than type I masers (i.e. W3(OH). They are not associated with continuum sources, with compact IR sources, nor with CO peaks, although they do reside in the denser parts of molecular clouds which contain embedded IR sources elsewhere (Wootten 1978; 1981). IR fluxes from these sources are much too weak to account for observed maser photon fluxes, and presumably have nothing to do with the pumping. The W28N and W44 masers are in clouds interacting with supernova remnants, and lie in directions close to the peak (background) continuum. The maser sizes are $<1''$. The gas temperature at the maser positions is ~25K. The maser profiles emulate those of ^{12}CO and ^{13}CO over a ~20 km/s range in W28N, suggesting that the OH pumping is not highly localized.

One immediately thinks of IR pumping to explain any satellite line predominance; in the absence of external IR sources, collisional excitation of the lowest rotational levels and attendant IR trapping would be considered. While excitation of many rotational levels is known to favor the 1612 line, excitation of only the lowest rotational level can be argued (Elitzur 1976; Guibert et al. 1978) to favor the 1720 line. Indeed, Guibert et al. find that, with "ordinary" collisional selection rules, the 1720 line will be strongly excited for $T_k > 22K$, and will have population inversion for $T_k > 37K$, while the 1612 line will be anti-inverted, as observed. These models do require some modifications (e.g. 1720 excitation occurs for unacceptably low particle densities), but the modest column densities and temperatures needed to pump the 1720 line seem to fit the W28N/W44 masers, and also show promise of explaining the much weaker 1720 excitation now seen prevalently in extended clouds throughout the galactic plane. A central consideration in further work on the models will be why the weak excitation should be as remarkably constant as is observed in the extended clouds, and why conditions allow so few type II(a) masers as are observed.

REFERENCES

Bujarrabal, V., Destombes, J. L., Guibert, J., Marlière-Demuynck, C., Rieu, N-Q and Omont, A. 1980a, Astron. Astrophys. 81, 1-7.

Bujarrabal, V., Guibert, J., Rieu, N-Q and Omont, A. 1980b, Astron. Astrophys. 84, 311-316.

Bujarrabal, V. and Rieu, N-Q 1980c, Astron. Astrophys. 91, 283-289.

Dixon, R. N. and Field, D. 1979, Mon. Not. R. Astr. Soc. 189, 583-591.

Elitzur, M. 1976, Astrophys. J. 203, 124-131.

Elitzur, M. 1979, Astron. Astrophys. 73, 322-328.

Elitzur, M. and de Jong, T. 1978, Astron. Astrophys. 67, 323-332.

Evans, N. J., Beckwith, S., Brown, R. L. and Gilmore, W. 1979, Astrophys. J. 227, 450-465.

Flower, D. R. 1980, Astron. Astrophys. 83, 33-37.

Goss, W. M., Johansson, L.E.B., Elldér, J., Höglund, B., Rieu, N-Q and Winnberg, A. 1973, Astron. Astrophys. 28, 89-94.

Guibert, J., Elitzur, M. and Rieu, N-Q 1978, Astron. Astrophys. 66, 395-405.

Guilloteau, S., Lucas, R. and Omont, A. 1981, Astron. Astrophys. 97, 347-358.

Gwinn, W. D., Turner, B. E., Goss, W. M. and Blackman, G. L. 1973, Astrophys. J. 179, 789-813.

Habing, H. J., Goss, W. M., Matthews, H. E. and Winnberg, A. 1974, Astron. Astrophys. 35, 1-5.

Johnston, I. D. 1967, Astrophys. J. 150, 33-45.

Kaplan, H. and Shapiro, M. 1979, Astrophys. J. 229, L91-L96.

Lucas, R., 1980, Astron. Astrophys. 84, 36-39.

Matthews, H. E., Goss, W. M., Winnberg, A. and Habing, H. J. 1977, Astron. Astrophys. 61, 261-274.

Morris, M. R. and Bowers, P. F. 1980, Astron. J. 85, 724-737.

Turner, B. E. 1967, Ph.D. thesis, University of California, Berkeley.

Turner, B. E. 1979, Astron. Astrophys. Suppl. 37, 1-332.

Turner, B. E. 1981, in preparation.

Wootten, A. 1978, Moon and the Planets 19, 163-168.

Wootten, A. 1981 preprint.

DISCUSSION FOLLOWING PAPER BY TURNER

CRUTCHER: Is it possible that the effects of magnetic fields, including in particular ambipolar diffusion, could produce the anisotropic charged-particle flows needed for some OH maser models?

TURNER: I can only echo the sentiments of Elitzur on that. Ambipolar diffusion is, of course, the only way one can clearly see to produce an anisotropic stream of particles. The magnetic field holds up the ions and the neutrals slip through, so that one has a streaming motion towards the center of some collapsing object, or an outburst. The problem is that such ambipolar diffusion can produce streams of ions but does not lend itself well to anisotropic streams of electrons, and it is the electrons which have been demonstrated to pump the OH in this particular way. However, it has not been clearly demonstrated that the ions can't do the pumping.

H. DICKEL: In NGC7538 IRS1 we have a 1720 MHz maser associated with an ultracompact HII region. The H_2CO masers, 1720 MHz OH masers and 1665 MHz OH masers appear as spots on a ring projected against the continuum source. They may exist in a shell which appears very similar to the W3(OH) situation. We don't know whether the shell is expanding or contracting.

CARBON MONOXIDE IN THE MAGELLANIC CLOUDS*

F.P. Israel, Th. de Graauw, S. Lidholm
Astronomy Division, Space Science Department
of ESA, ESTEC, Noordwijk, The Netherlands

H. van de Stadt, C. de Vries
Sterrewacht Sonnenborgh, University of Utrecht
The Netherlands

ABSTRACT

We have detected $^{12}CO(J = 2 - 1)$ emission at four positions in the LMC, in the vicinity of 30 Doradus. Three other positions yielded negative results, as did four positions in the main body of the SMC. The results indicate the presence of a giant molecular cloud complex associated with the HII regions N159 and N160A, comparable to galactic complexes.

INTRODUCTION

A multitude of observations of galactic CO emission have established this molecule as an excellent tracer for dense molecular clouds engaged in the formation of massive stars. Virtually every galactic HII region is associated with a molecular cloud complex. It is of great interest to determine whether a similar situation exists in other nearby extragalactic systems such as the dwarf galaxies LMC and SMC. These galaxies in particular are worth attention because they are very close to the Galaxy (D = 55 kpc, resp. 80 kpc) so that a good linear resolution is obtainable even with modest apertures, and because they are known to have an interstellar medium with characteristics different from our own Galaxy in at least the following ways. a. The LMC has a different extinction law (see e.g. Koornneef and Code, 1981), b. The LMC has a different gas to dust ratio (see e.g. Koornneef, 1981) and c. both LMC and SMC have a relatively low metallicity (see e.g. Lequeux et al., 1979).

Previous attempts to detect $^{12}CO(J = 1 - 0)$ emission only succeeded in finding emission associated with the HII region N159 (Huggins et al., 1975). This same HII region has also been detected as a source of other molecular line emission including OH and H_2O maser emission (a full list is given elsewhere, Israel et al., 1981).

*A more detailed version of this paper has been submitted to the Astrophysical Journal.

R. S. Roger and P. E. Dewdney (eds.), Regions of Recent Star Formation, 433–438.
Copyright © 1982 by D. Reidel Publishing Company.

OBSERVATIONS

Our observations were made in May 1980, with the use of the ESTEC/ Utrecht heterodyne submillimeter receiver, operating at a frequency of 230 GHz (1.3 mm). The receiver had an overall system temperature of about 4000 K (SSB). As a backend we used mainly a 256-channel filter-bank with a resolution of 1 MHz per channel, corresponding to 1.3 km s^{-1} velocity resolution. Calibration was done in the usual manner. Most observations were made at the Cassegrain focus of the ESO/La Silla 3.6 m telescope with a HPBW of 2 arcmin. We also used the Carnegie Foundation 2.5 m DuPont telescope with a HPBW of 3 arcmin. The La Silla beamwidth correspond to a linear resolution of 32 pc for the LMC and 47 pc for the SMC. These resolutions are typically comparable to the dimensions of molecular cloud complexes, rather than to those of individual molecular clouds.

The positions observed were those of obscuring dust near HII regions, with the exception of N159 where we pointed at the HII region itself in order to facilitate comparison with molecular line observations by other authors.

RESULTS

We detected CO emission near the ionized hydrogen velocities $V_{Hel} \cong 255$ km s^{-1}) of the HII regions 30 Doradus core, N159 and N160A. In addition, in all three spectra a second velocity component was seen, at a velocity about 60 km s^{-1} higher (V_{Hel} = 315 km s^{-1}). This component was also seen in a fourth spectrum (again in the 30 Doradus core). Its velocity corresponds to that of some optical absorption lines seen in the direction of 30 Doradus. All results are listed in Tables 1 and 2, and shown in Figure 1.

TABLE 1 $^{12}CO(2-1)$ Detections in LMC

	α(1950)	δ(1950)	T_A^*(K)	V_{Hel} (km s^{-1})	V (km s^{-1})
30 DOR	$05^h38^m50^s$	-69°07'	0.8 ± 0.2	+ 260	7
			0.8 ± 0.4	+ 320	8
N160A	$05^h39^m50^s$	-69°37'	1.1 ± 0.3	+ 256	7
			1.0 ± 0.3	+ 320	6
N159	$05^h40^m30^s$	-69°46'	2.6 ± 0.3	+ 251	6
			1.1 ± 0.3	+ 308	9
30 DOR	$05^h39^m10^s$	-69°05'	0.7 ± 0.3	+ 310	8

The strongest line was seen in the direction of N159 at V_{Hel} = 250 km s^{-1}. On the reasonable assumption that the line is optically thick in both the J = 1-0 and J = 2-1 transitions, comparison of our result with the one obtained by Huggins et al. (1975) with a 4 arcmin beam in-

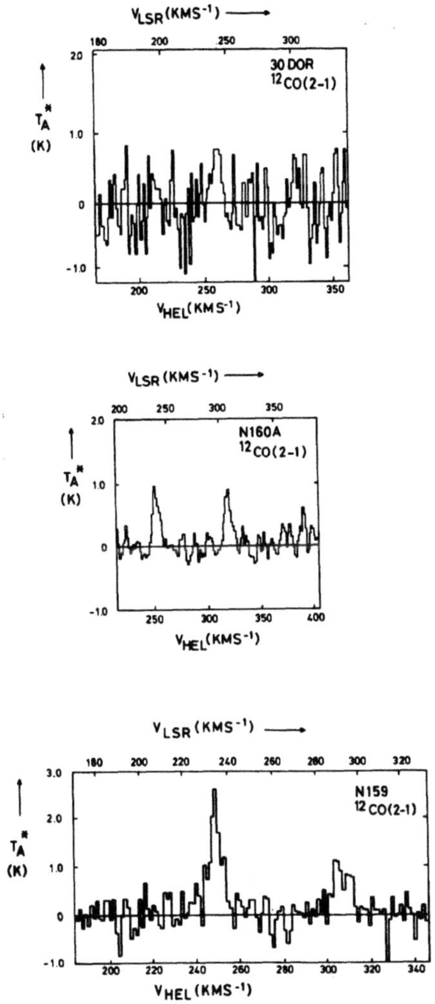

Figure 1. $^{12}CO(J=2-1)$ spectra of 30 Dor, N160A and N159. The spectrum of N160A has been Hanning-smoothed (1:2.4:1). The other two spectra have a velocity resolution of 1.3 km s^{-1}, corresponding to the unsmoothed 1 MHz resolution of the filterbank used.

TABLE 2 $^{12}CO(2-1)$ 3σ Upper Limits in LMC and SMC

LMC -	H51	$05^h38^m45^s$	$-68°59'.5$	≤ 1.6
	30 DOR	05 38 50	-69 17	≤ 0.9
	N158C	05 39 10	-69 27	≤ 0.9
SMC -	N76	01 02 30	-72 15	≤ 0.9
	N66	01 01 00	-72 30	≤ 1.0
	N66	00 57 30	-72 30	≤ 0.7
	N37	00 50 00	-73 00	≤ 1.5

dicates a mean molecular source size of order 3.5 arcmin with a 50 per cent uncertainty. This corresponds to a mean linear size of about 55 pc which is comparable to the mean size found for large galactic molecular cloud complexes. The similarity can also be recognized in the following way. A reasonable estimate for the area-integrated surface brightness of the molecular cloud near N159 is given by $\int T_b dS \geq 7 \times 10^3$ K pc^2; Elmegreen et al. (1980) found $\int T_b\, dS = 1.3 \times 10^4$K pc^2 for seven of the largest galactic molecular cloud complexes. The emission detected in the direction of N160A at $V_{Hel} = 255$ km s^{-1}, at a distance of 10 arcmin (160 pc) to N159 may be due to the same cloud complex in which case the similarity is even more striking.

Although no ^{13}CO observations are available, a column density $N(^{13}CO) = 6 \times 10^{16}$ cm^{-2} may be estimated in analogy with galactic CO cloud complexes. It should be noted that this estimate is strengthened by the value $N(^{13}CO) = 2 \times 10^{16}$ cm^{-2} derived from OH(1667) observations by Whiteoak and Gardner (1976) under the assumption that the [OH]/[CO] abundance ratio is the same in the LMC and in the Galaxy. Taking this result, a molecular mass $M(H_2) = 10\ M_\odot$ may be estimated; it also appears that about ten per cent of the molecular cloud volume is filled with emitting material in good agreement with galactic results (Blitz, private communication).

We conclude that the HII region N159 is associated with a massive and large molecular cloud complex, similar to large molecular cloud complexes in the Galaxy. Previous studies have shown that the same is true for the ionized hydrogen content and exciting star content of N159 (Israel and Koornneef, 1979; Israel, 1980). Both the presence of a molecular cloud complex and OH/H_2O masers indicate that N159 is a 'star factory'; this conclusions has recently been confirmed by near-infrared measurements (Gatley and Becklin, private communication).

In contrast, the CO emission seen in the direction of 30 Doradus at velocities corresponding to those of the ionized gas ($V_{Hel} = 255$ km s^{-1}) are neither very strong, nor very extended. Although several explanations are possible, the most likely one is that most of the molecular material originally present has been destroyed by violent interaction with the giant HII region complex. 30 Doradus exhibits streams of ionized gas with velocities up to 30 km s^{-1} (Elliott et al., 1977) and contains several hundreds of early type stars, creating a very energetic stellar radiation field (see e.g. Israel and Koornneef, 1979). Its age of about 10^7 years implies a sufficiently long interaction time, and is, moreover, very close to the estimated lifetime of giant molecular clouds in the galaxy (Blitz and Shu, 1980). We also note that lack of strong CO emission appears to be a characteristic of giant HII region complexes (Israel and Rowan-Robinson, 1981; see also Elmegreen et al., 1980).

The weak but extended CO emission at high velocity (V_{Hel} = 210 km s^{-1}) can be explained by assuming that it is due to molecular material heated by the ambient radiation field, rather than by HII regions and their ionizing stars directly, so that a low excitation temperature T_k = 15 K is sufficient to explain the observations. This corresponds to cool clouds not associated with star formation processes in our own Galaxy.

Four positions in the main body of the SMC failed to show emission above the (3σ) level of $T_A^* \simeq 1$ K, implying $\int T_b \, dS \leqslant 1.5 \times 10^3$ K pc^2. This is a strong indication that the SMC is genuinely deficient in CO emission, most likely due to the very low metallicity of the SMC (c.f. Pagel et al., 1978). This situation is thus in contrast with the one in the LMC where at least the 30 Doradus sector seems to have CO properties similar to those of the Galaxy. Since the 30 Doradus sector is probably just as representative of the rest of the LMC as California is of the rest of the USA, a general conclusion regarding CO in the LMC must, then, await further observations in the bar and the northern area of the LMC.

ACKNOWLEDGEMENTS

It is a pleasure to thank the director and staff of both the Las Campanas and La Silla observatories for generously granting observing time, and competently supporting the observations sessions. We also thank J. Koornneef for stimulating discussions, and B Fitton and D.E. Page for their continuous encouragement and support.

REFERENCES

Blitz, L., Shu, F.H., 1980, Astrophys. J. 238, 148.

Elliott, K.H., Goudis, C., Meaburn, J., Tebbutt, N.J., 1977, Astron. Astrophys. 55, 187.

Elmegreen, B.G., Elmegreen, D.M., Morris, M., 1980, Astrophys. J. 240, 445.

Huggins, P.J., Gillespie, A.R., Phillips, T.G., Gardner, F.F., Knowles, S.H., 1975, Mon. Not. R.A.S. 173, 69P.

Israel, F.P., 1980, Astron. Astrophys. 90, 246.

Israel, F.P., Koornneef, J., 1979, Astrophys. J. 230, 390.

Israel, F.P., Rowan-Robinson, M., 1981, Astrophys. J., submitted.

Israel, F.P., De Graauw, T., Lidholm, S., Van de Stadt, H., De Vries, C., 1980, Astron. Astrophys., submitted.
Koornneef, J., 1981, Astron. Astrophys., submitted.
Koornneef, J., Code, A.D., 1981, Astrophys. J., submitted.
Lequeux, J., Peimbert, M., Rayo, J.F., Serrano, A., Torres-Peimbert, S., 1979, Astron. Astrophys. 80, 155.
Pagel, B.E.J., Edmunds, M.G., Fosbury, R.A., Webster, B.L., 1978, Mon. Not. R.A.S. 184, 569.
Whiteoak, J.B., Gardner, F.F., 1976, Mon. Not. R.A.S. 174, 51P.

DISCUSSION FOLLOWING PAPER BY ISRAEL ET AL.

KWOK: Both the LMC and SMC are known to be metal deficient from their integrated colors. Therefore, an under-abundance of CO compared to H_2 in both would not be unexpected.

ISRAEL: The deficiency seems to be there for the SMC but not for the 30-Doradus area of the LMC. We do not yet know the situation in the LMC bar.

UNDERHILL: The "deficiencies" of C, N, and O in SMC stars depend on LTE analyses of moderate-dispersion spectra of B1 - O9 supergiants. Several observations point towards the possibility that the spectra are those for several stars superimposed. Furthermore, more appropriate techniques of analysis which include NLTE physics usually lead to a reduction of the abundance "anomalies". Metal deficiencies in the LMC and SMC may well be artefacts of the observational material and methods of analysis that have been used up to now, and they should be treated with scepticism.

ISRAEL: I fully agree with you on the stellar spectra. However, there are values for nitrogen and oxygen based on HII region spectra.

UNDERHILL: Well, I have similar reservations about the techniques used to analyse recombination line spectra from HII regions.

MOLECULAR LINE MAPPING OF OMC-1

F. Peter Schloerb, Paul F. Goldsmith, and N.Z. Scoville
Five College Radio Astronomy Observatory
University of Massachusetts
Amherst, Massachusetts 01003

ABSTRACT

We present new, fully sampled maps of the molecular emission from CO, ^{13}CO, and HCN in the Orion Molecular Cloud. The high resolution of the maps (< 1 arcmin) reveals considerable structure in the molecular gas that is related to the HII region surrounding the Trapezium.

I. INTRODUCTION

Orion A is the best studied example of an HII region/molecular cloud complex. High spatial resolution optical and radio studies of the ionized gas in M42 have shown that considerable structure exists on size scales of 1 arcmin and less. However, much of the molecular line mapping of the region has been done with lower resolution than this. We have therefore obtained fully sampled, high resolution maps of the region in various molecular species in order to better study the relationships between the HII region and the molecular cloud.

II. OBSERVATIONS

In Figure 1, we show fully sampled maps of a $1/2^\circ \times 1/4^\circ$ region of the Orion molecular cloud in the CO and ^{13}CO $J=1\rightarrow0$ transitions. We also show a partial map of the HCN $J=1\rightarrow0$ transition in the central portion of the cloud. These particular molecular species were chosen because each can be considered to be a tracer of a certain property of the molecular cloud. The CO $J=1\rightarrow0$ transition is easily thermalized at densities of only $\sim10^3$ cm^{-3} and optically thick so that the map in Figure 1 can be considered to be a map of the gas kinetic temperature. ^{13}CO is optically thin and the map of ^{13}CO can essentially be considered to be a tracer of the column density of molecular gas.

R. S. Roger and P. E. Dewdney (eds.), Regions of Recent Star Formation, 439–444.
Copyright © 1982 by D. Reidel Publishing Company.

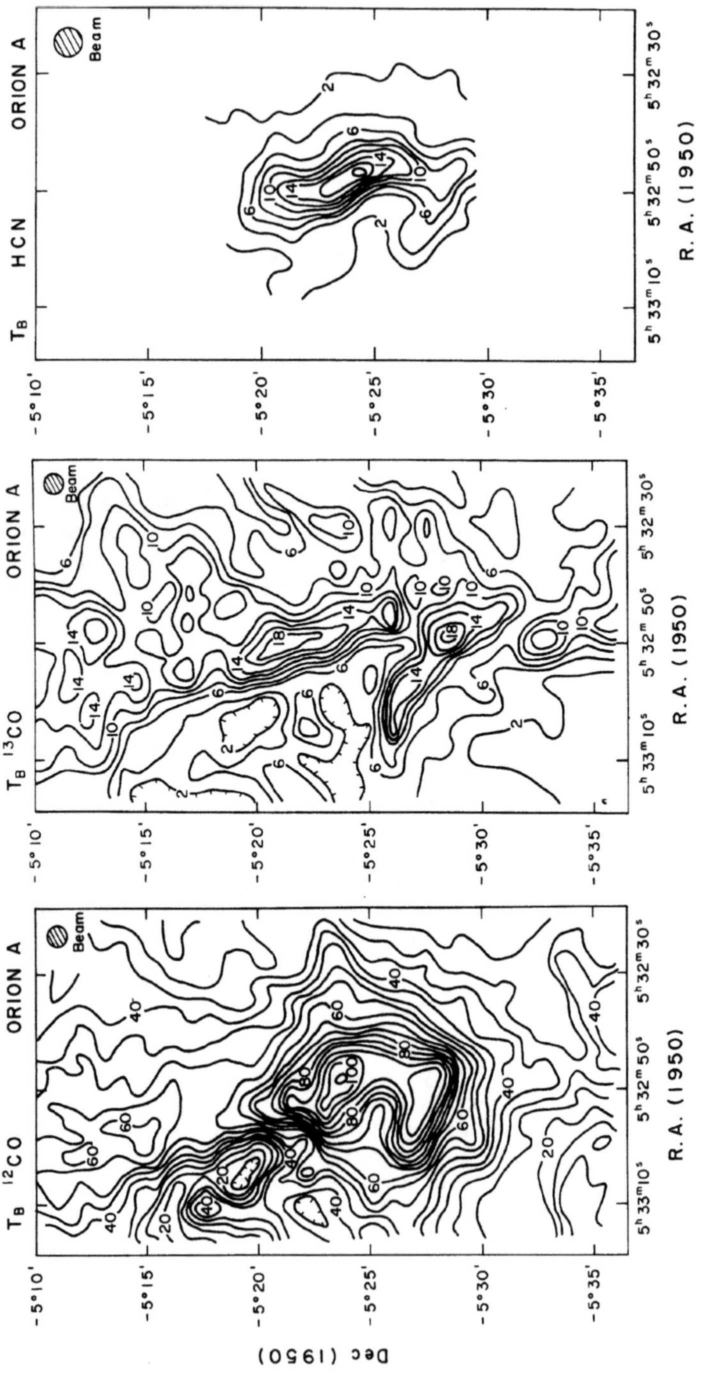

Figure 1 - Maps of CO, ^{13}CO, and HCN brightness temperature in the Orion Molecular Cloud

Finally, HCN has a large dipole moment and requires high densities (10^4–10^5 cm^{-3}) to be significantly excited. Thus the HCN map traces out the high density regions in the cloud.

The data were obtained using the Five College Radio Astronomy Observatory 14m telescope and cooled mixer receiver. The grid spacing in the maps is 45" and the telescope HPBW is 50", 55", and 65" at the CO, ^{13}CO, and HCN frequencies respectively. Spectral resolution is provided by a 250 kHz filter bank, and the spectra were obtained by frequency switching. The receiver noise temperature at all frequencies is 450K and permitted us to obtain good signal-to-noise in relatively short integration times. This in turn permitted the maps to be made quickly with frequent observations of the center position in the map so that the relative calibration of the map points should be very good. The absolute calibration of the observed antenna temperatures to the brightness temperatures reported in Figure 1 is less certain and has been achieved by chopper wheel calibration and observations of the planets and the Moon. In the case of the CO $J=1\rightarrow0$ line, we have adjusted the absolute calibration to give a value of 90K for the brightness temperature of CO at the Kleinmann-Low nebula (R.A.(1950) = 5h32m46.8s; DEC(1950) = $-5^\circ24'28''$) when observed with a 2.3' beam to agree with the observations of Davis and Vanden Bout (1973). For ^{13}CO and HCN, we have adopted forward beam coupling efficiencies for Orion that are between those found for small sources using the planets (0.45) and large sources using the Moon (0.65).

The high resolution and full sampling of the FCRAO maps in Figure 1 makes several interesting structures in OMC-1 apparent. In this paper, we shall concentrate on the portions of the cloud immediately surrounding the HII region, M42.

III. DISCUSSION

There is an interesting correspondence between the molecular emission features and those of the optical and radio emission from the ionized gas in the HII region. For example, we note that the CO map, which traces out the temperatures of the molecular material, follows the shape of the emission from the ionized gas rather well. In general, the peak CO emission occurs in a rim around the ionized gas, as might be expected if it were heated by starlight from the Trapezium stars or by the interaction of the HII region and the neutral molecular cloud. Thus, the distribution of high temperature gas suggests that the energetics of the molecular gas are significantly affected by the Trapezium.

In addition to the morphological similarity between the high temperature molecular emisssion and HII region, there is kinematic evidence to suggest a close relationship between the two. In Figure

2, we show maps of the CO emission in three 1 km s^{-1} resolution bins centered on 6.5, 8.5, and 10.5 km s^{-1}. We note that the peak CO emission in Figure 1 follows a U shape surrounding the ionized gas, and the velocity maps in Figure 2 show that the CO velocity varies systematically with position around the U. The lowest velocities occur to the SW of the Trapezium, near the apex of the U, and the velocity increases along the arms of the U. Such a systematic velocity variation is probably consistent with models of the HII region which place the Trapezium stars in a blister in the neutral cloud (e.g. Zuckerman, 1973; Balick et al., 1974). Expansion of such a region, or shocks resulting from the expansion would push the molecular gas away from the Trapezium. Thus, material between the Trapezium stars and the observer would be blue shifted and material behind the Trapezium would be red shifted, and the geometry of the blister walls surrounding the stars could result in the velocity gradient observed.

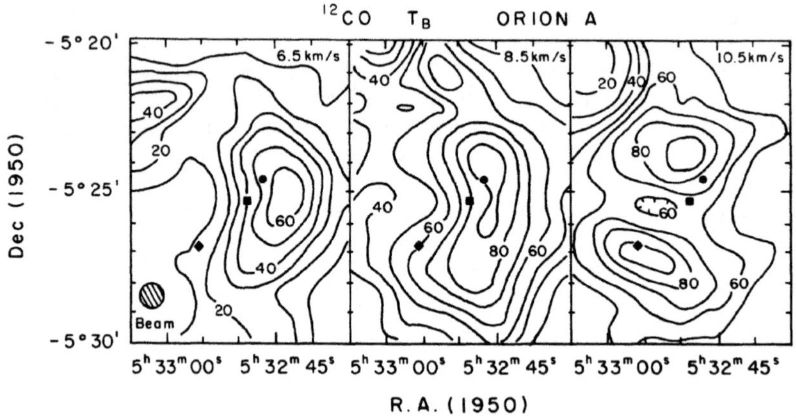

Figure 2. Maps of the CO brightness temperature in three 1 km s^{-1} velocity bins centered on 6.5, 8.5, and 10.5 km s^{-1}. Filled circle locates KL, square locates θ^1C, and diamond locates θ^2A.

An especially interesting example of the relationship between the molecular gas and the HII region is seen in the vicinity of the "bar" feature to the SE of the Trapezium. The optical bar is generally considered to be an ionization front seen edge on (e.g. Zuckerman, 1973; Balick et al., 1974) and has been observed in IR and far-IR maps of the region (Becklin et al., 1976; Werner et al., 1976) as well as in maps of the radio continuum (Martin and Gull, 1976). We note that the bar feature also appears in molecular emission, not only as a region of enhanced temperature, but also as a region of increased column density and molecular density, since it also appears as a prominent feature in the ^{13}CO and HCN maps. In Figure 3 we show cross-sections through the bar, the Trapezium and KL to illustrate the relationship between the optical and radio observations of the ionized gas and the molecular gas. In all cases, the molecular emission peaks outside of the optical bar which occurs at the boundary of the HII region and the neutral cloud.

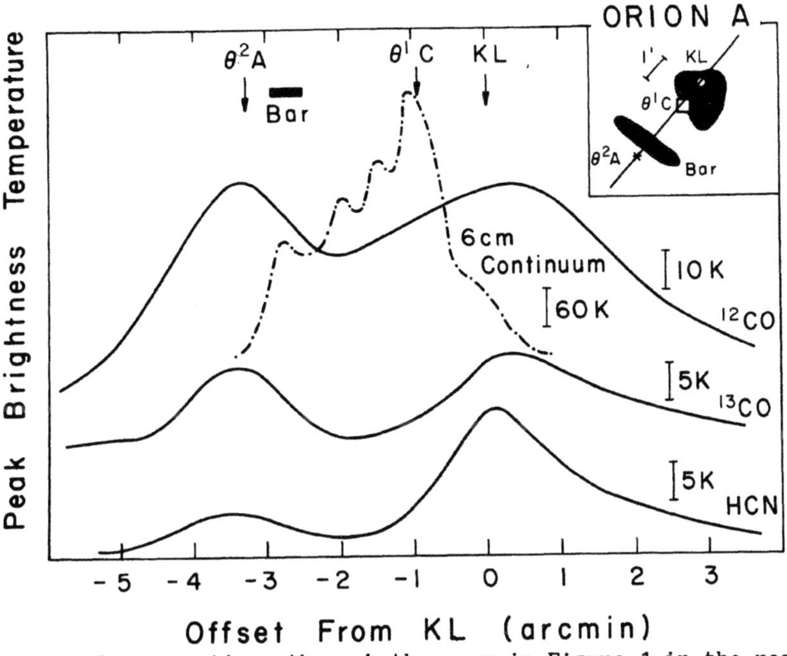

Figure 3. Cross-sections through the maps in Figure 1 in the region of the optical bar, the Trapezium, and KL. The location of the cross-section is shown in the inset to the upper right. Also shown is a cross-section through the 6 cm continuum map of Martin and Gull(1976).

Becklin et al. (1976) argue that the far-IR luminosity of the IR bar feature could be explained by heating from the Trapezium if the linear dimension of the bar is as great along the line of sight as it is perpendicular to the line of sight, and the molecular observations provide some support for this contention. We note that the column density of ^{13}CO at the bar is $\sim 10^{17}$ cm^{-2} so that the H_2 column density should be $\sim 10^{22}$–10^{23} cm^{-2} assuming a reasonable ratio of N(^{13}CO) to N(H_2) (e.g. Dickman, 1978). Since the bar is $\sim 10^{18}$ cm long, the typical density of the region would be $\sim 10^4$–10^5 if the line of sight dimension is also 10^{18} cm. This density is consistent with the excitation requirements of the HCN observed in the bar, suggesting that the bar feature receives enough flux from θ^1C to account for its far-IR luminosity. Densities this high are also sufficient to collisionally couple the gas and dust in the bar (Goldreich and Kwan, 1974) so that the coincidence of the far-IR color temperature (Werner et al., 1976) and the CO brightness temperature indicates that the energy source for the heating of the gas in the bar is probably θ^1C. Finally, it is tempting to explain the density enhancement in the molecular bar by the expansion of the HII region surrounding the Trapezium. Such an explanation might be supported by the observation of a systematic velocity gradient around the U of hot molecular gas in Figure 2. However, we note that a similar feature could have developed if the bar were originally denser than the rest of the cloud since the observed ionization front would follow this pre-existing structure.

The Five College Radio Astronomy Observatory is supported by NSF Grant AST80-26702 and operated with the permission of the Metropolitan District Commission, Commonwealth of Massachusetts. This is contribution number 487 of the Five College Astronomy Department.

REFERENCES

Balick, B., Gammon, R.H. and Hjellming, R.M. (1974) Pub. ASP 86, 616.
Becklin, E.E., Beckwith S., Gatley, I., Matthews, K., Neugebauer G., Sarazin, C. and Werner, M.W. (1976) Ap. J. 207, 770.
Davis, J.H. and Vanden Bout, P. (1973) Ap. Letters 15, 43.
Dickman, R.L. (1978) Ap. J. Suppl. 37, 407.
Goldreich, P. and Kwan, J. (1974) Ap. J. 189, 441.
Martin, A.H.M. and Gull, S.F. (1976) MNRAS 175, 235.
Werner, M.W., Gatley, I., Harper, D.A., Becklin, E.E., Loewenstein, R., Telesco, C. and Thronson, H. (1976) Ap. J. 204, 420.
Zuckerman, B. (1973) Ap. J. 183, 863.

DISCUSSION FOLLOWING PAPER BY SCHLOERB, GOLDSMITH AND SCOVILLE

WOOTTEN: Have you resolved the bar feature in any of the molecular lines you have observed?

SCHLOERB: I think the observed width of the feature is ~1 arcmin and our beam is ~50 arcsec, so it is fairly thin. That is probably the reason it doesn't appear on previous maps.

BALLY: Is there any reason why the overall north-south velocity gradient should not be interpreted as rotation? If the gradient is rotation, it is interesting that the asymmetry reported by White in the J=3-2 transition of CO seen in the Plateau source as well as the position of the $H_2S(0)$ peaks is consistent with ejection of high velocity gas along the rotation axis.

SCHLOERB: I think it has been interpreted as rotation in the past. The whole cloud and the gradient extend over one degree.

AN UPPER LIMIT TO THE ATOMIC CARBON ABUNDANCE IN THE ORION PLATEAU

C.A. Beichman*, T.G. Phillips*, H.A. Wootten*
Department of Physics, California Institute of Technology
M. Frerking*
Jet Propulsion Laboratory, California Insitute of Technology

ABSTRACT

Observations made of the atomic carbon line at 492 GHz toward OMC-1 show no evidence for the high velocity dispersion wings observed for many molecular rotational lines. The 3σ upper limit to the CI column density, N_{CI}, is $6.9 \times 10^{17} cm^{-2}$ for velocities $\geqslant 4$ km s^{-1} from the line center. This upper limit corresponds to a ratio of CI to CO abundances as low as <0.13, depending on the assumed CO column density. Atomic carbon is apparently depleted by a factor as large as five in the hot plateau gas, relative to its abundance in other molecular clouds. The lack of CI in the plateau source may mean that the shocks thought to be present in the region are not dissociative in nature and thus do not produce the UV radiation required to convert CO into CI.

INTRODUCTION

The detection of submillimeter line emission from the $^3P_1-^3P_0$ transition of atomic carbon (Phillips et al. 1980) provided a new probe of the chemistry of the interstellar medium. Both hot (70 K) and cool (20 K) molecular clouds emit strongly in the carbon line with brightness temperatures as high as 25 K in some cases. The abundance of CI must be quite high to produce such strong lines. A recent analysis (Phillips and Huggins 1981) suggests that the CI line is optically thick and that $X(CI)/X(CO) \geqslant 0.5$ in a variety of molecular clouds. The exact location of the emitting carbon and the physical conditions where it is abundant within the clouds are not yet certain,

---

* Visiting Astronomer, Infrared Telescope Facility, operated by the University of Hawaii under contract with the National Aeronautics and Space Administration.

R. S. Roger and P. E. Dewdney (eds.), Regions of Recent Star Formation, 445–452.
Copyright © 1982 by D. Reidel Publishing Company.

however. The lack of very hot CI emission in, say, OMC-1, where the CO brightness temperature is 75 K, but the CI brightness temperature is only 25 K suggests that CI is not found primarily in the hot cores of these clouds, but rather in the cooler envelopes. The abundance of CI in the envelopes may be enhanced by UV radiation capable of dissociating CO diffusing into the clouds from nearby OB stars. Recently Wooten et al. (1981) have found particularly strong CI emission from the bright rim of NGC 1977 near the B2III star 42 ORi.

To learn more about the physical conditions favoring the presence of atomic carbon we searched for CI emission from the so-called "plateau" source in OMC-1, a region characterized by emission from hot (~150 K), dense (~1×10^{5}cm^{-3}) gas over a velocity range of ~100 km s^{-1} (Zuckerman, Kuiper and Kuiper 1976; Phillips et al. 1977). It is generally thought that the high temperatures and velocities are due to shock-wave excitation of the gas (Kwan and Scoville 1976; Kwan 1977; Hollenbach and Shull 1977). Such a process might well provide deviations from quiescent molecular cloud chemistry affecting the abundance of CI. The observations presented in this paper test this concept.

OBSERVATIONS

Observations of the CI line at 492162.3 MHz (Saykallay and Evenson 1980) were made using an InSb hot electron bolometer heterodyne receiver as described in detail by Phillips and Jefferts (1973). The system noise temperature varied between 350 and 500 K. A single 0.6 km s^{-1} wide channel was swept in 1 km s^{-1} intervals across the spectral range of interest by stepping the klystron local oscillator under computer control. The receiver was mounted at the Cassegrain focus of the NASA 3m Infrared Telescope on top of Mauna Kea, Hawaii. Scans across Jupiter showed the beam profile to have a full-width-at-half-maximum (FWHM) of 45" in right ascension (figure 1). The beam efficiency measured using hot and cold loads was approximately 30 percent.

The CI line falls in the submillimeter band of the electromagnetic spectrum where atmospheric water vapor causes considerable absorption. Observations from a high, dry site are essential. During the course of the 5 day observing run zenith optical depths of about 1.0-1.5 were deduced from plots of the sky emission as a function of zenith angle. During the daytime it was possible to measure the amount of water vapor in the line of sight to the sun using a near-infrared absorption technique (Westphal 1974). On three days, values between 0.6-1.0 mm H_2O to the zenith were measured.

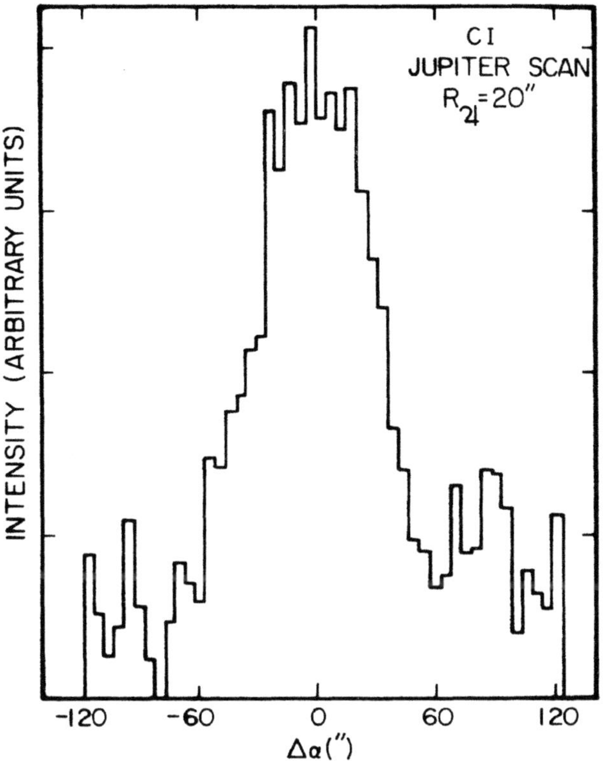

Figure 1. A right ascension scan across Jupiter showing the beam profile of the CI receiver on the 3m telescope at Mauna Kea.

Observations of the plateau source were made by scanning a 100 km s^{-1} wide spectrum centered on V_{LSR}=9 km s^{-1}. The position observed was $\alpha(1950)=5^h32^m46^s.7$, $\delta(1950)=-5°24'21"$, the location of the plateau source as determined from CO 3→2 observations (Phillips et al. 1977) and, within the errors, from CO 1→0 observations (Solomon, Huguenin and Scoville 1981). The "off" positions were one half degree east and west of the "on" position. The data were calibrated assuming that the peak T_A^* of the line from OMC-1 has the value of 11 K as measured from the Kuiper Airborne Observatory (KAO). The fact that a CI strip map across OMC-1 (Phillips and Huggins 1981) made from the KAO with a 3' beam does not show very strong peaking suggests that the peak brightness temperature will be relatively insensitive to beamsize. All of the data were coadded from scans obtained when the zenith optical depth was less than 1.5 and the airmass less than 1.3. A quadratic baseline was fitted to the first and last thirds of the data and subtracted from all of the data. Finally the data were Hanning smoothed. Figure 2 shows the resulting spectrum together with a ^{13}CO

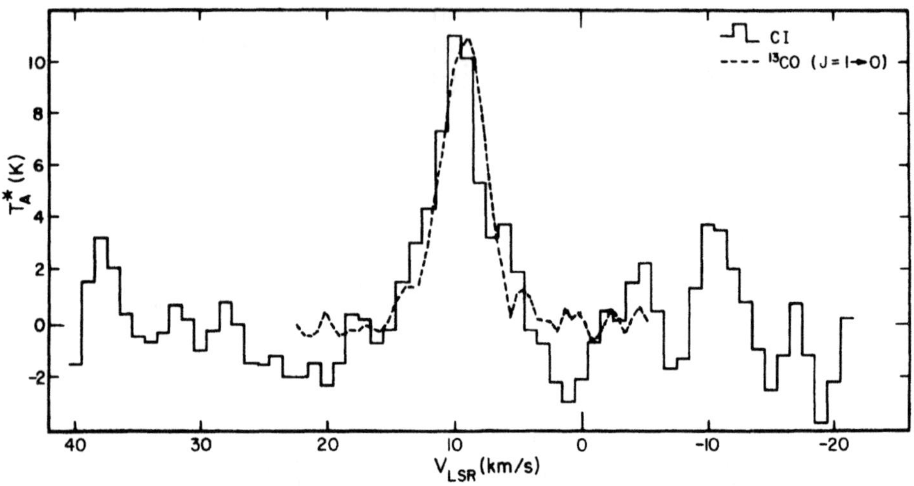

Figure 2. The CI spectrum of OMC-1 taken with a 45" beam. A ^{13}CO spectrum, taken with a 55" beam is shown for comparison (dashed line).

spectrum (from the 10m telescope at the Owens Valley Radio Observatory, Wootten, private communication 1980) for comparison. The ^{13}CO data were obtained with a 55" beam size, quite comparable with that used for the CI measurements.

RESULTS

An examination of figure 2 shows that the CI line profile closely resembles the ^{13}CO profile and shows no evidence for the broad wings that characterize the plateau source. By contrast with the ^{12}CO profile, ^{13}CO has an essentially unobservable broad wing component due to the intrinsically low opacity of the emitting gas and the $^{12}C/^{13}C$ isotope ratio. To set an upper limit to the amount of CI emission from the plateau a 4 km s^{-1} wide (FWHM) gaussian centered on 9.5 km s^{-1} with an amplitude of 11 K was subtracted from the data. As the ^{13}CO profile demonstrates, this line shape is characteristic of the molecular cloud in the absence of plateau emission. The residual intensity (hereafter denoted by $\int T_A^* dv$) between 20 km s^{-1} and -5 km s^{-1}, and excluding velocities within 4 km s^{-1} of the line center was calculated to be -3 ± 6 K km s^{-1} (1σ). If the size of the emitting region is assumed to be the same as that observed in CO ~ 40" (Phillips et al. 1977; Solomon, Huguenin, Scoville 1981; Knapp et al. 1981), then the beam dilution correction factor is 0.44 for a 45" beam. The 3σ upper limit to the plateau emission is 44 K km s^{-1}.

Assuming CI to be optically thin the corresponding limit to the CI column density can be obtained from the LTE relationship given in Phillips and Huggins (1981)

$$N_{CI} = 1.99 \times 10^{15} (3 + \exp\frac{23.6}{T} + 5\exp\frac{-38.8}{T}) \int T_A^* dv (cm^{-2}) \qquad (1)$$

where an Einstein A value of $7.9 \times 10^{-8} s^{-1}$ has been used (Nussbaumer 1971). Adopting an excitation temperature T ~ 150 K yields $N_{CI} < 7.0 \times 10^{17} cm^{-2}$. The result varies by less than 50 percent between 50 and 200 K.

The above upper limit to the CI column density can be compared to that of CO in the plateau. From observations of CO J=2→1 and J=3→2 Phillips et al. (1977) found $N_{CO} \geqslant 6 \times 10^{18} cm^{-2}$, while from observations of CO J=6→5, Goldsmith et al. (1981) deduced $N_{CO} \geqslant 12 \times 10^{18} cm^{-2}$. The latter determination is a lower limit to the CO column density because of the authors' assumption that the line is optically thin. The 1→0, 2→1 and 3→2 data of Knapp et al. (1981) suggest that the higher transitions are quite saturated. Using the above estimates, $N_{CI}/N_{CO} < 0.13$ (3σ) implying that CI is significantly depleted relative to CO in the plateau.

DISCUSSION

The observations show that the overall effect of the physical conditions in a region characterized by high velocity emission, the "plateau" source in OMC-1, is a depletion rather than an enhancement of the CI abundance, relative to the value found in more quiescent clouds. It is necessary to understand this effect both in terms of the chemistry of the interstellar medium and in terms of the special conditions found in the plateau.

Consider first the predictions of standard ion-molecule chemistry for the CI concentration in a dense region. Graedel, Frerking and Langer (1981) found that for there to be a substantial CI abundance it is necessary to have a high electron abundance. The large concentration of electrons destroys the H_3^+ that would otherwise process CI into molecular forms, predominantly CO. Observations show that HCO^+ is abundant in the plateau, implying that the region is both dense and of a low relative electron abundance (Watson 1976; Huggins et al. 1979; Erickson et al. 1980; Kuiper et al. 1980). Under the assumption that the HCO^+ and CO plateaux arise in the same material, the analysis of Wootten, Snell and Glassgold (1979) can be used to set a limit to the relative electron abundance, $X(e) < 5 \times 10^{-8}$, a value similar to determinations of X(e) in other dense clouds. With this value of X(e) it is possible to use Graedel et al.'s equations to show that in equilibrium X(CI)/X(CO)<0.01. Thus after a steady state condition has been achieved all of the carbon in the plateau should be in the form of CO not CI.

While this calculation is consistent with the failure to find CI in the plateau, it is not yet clear that the chemistry is completely understood, particularly in light of the fact that models of this type do not easily explain the high abundance of CI observed toward molecular clouds in general. The only way that ion-molecule chemistry models (Langer 1976; Prasad and Huntress 1980; Graedel et al. 1981) yield substantial CI abundances is for the cloud to be either very young ($<10^6$ years) or very rarified ($<5\times10^3$ cm^{-3}). Phillips and Huggins (1981) have appealed to the former possibility to explain the strength of the CI emission toward molecular clouds. CI is abundant in dense clouds, they suggested, because the gas is continually brought to an early stage of a non-equilibrium chemistry dominated by atomic species. This rejuvenation occurs either because of the shocks and UV radiation produced during star formation (cf. Norman and Silk 1979), or because of churning processes within the cloud which allow the central component to be exposed to the external UV radiation field every $\sim10^6$ yr.

The failure to find CI emission in the plateau poses a problem for the first of these possibilities. If the sole effect of the shock is to process the gas through a region of high, ~2000 K temperatures, then one can expect CI to be depleted, rather than enhanced, due to the onset of neutral-neutral reactions which are unimportant at lower temperatures due to their high activation energies. Preliminary calculations indicate that CI will be transformed rapidly into CH_x family molecules and CO (Hollenbach 1980). If the high temperature associated with a shock is by itself insufficient to replenish the CI reservoir, then perhaps UV radiation arising either from a newly formed star or from the dissociation and subsequent recombination of H_2 in shock could produce CI. Indeed, in the bright rim region near NGC 1977, Wootten et al. (1981) have found that stellar UV does enhance the CI abundance. The fact that CI is depleted in the plateau argues against the presence of UV emission in the region, a result consistent with the failure of Scoville et al. (1981) to find any high velocity wings on the Brackett γ recombination line seen toward OMC-1. Scoville et al. concluded that the amount of ionized material and thus the amount of ionizing radiation in the plateau region must be small (for further discussion see also Chevalier, 1980). The failure to find CI in the plateau is another indicator that the shocks present in this region do not dissociate much H_2 resulting in little or no UV radiation. Processes other than high velocity shocks, mechanisms such as those discussed in Phillips and Huggins (1981), must be responsible for maintaining the surprisingly high CI abundance seen toward molecular clouds generally.

CONCLUSIONS

We have observed the OMC-1 region with high spatial resolution and set an upper limit to the emission from hot, optically thin CI in the plateau source. The upper limit suggests a depletion of CI relative to CO as compared with quiescent molecular clouds. The lack of CI emission implies that there is little UV radiation within the plateau region and that the shocks in this source are not dissociative in nature.

ACKNOWLEDGEMENTS

We would like to thank the entire crew of the IRTF, superintendent Bob Brook and chief scientist Dr. Eric Becklin for their help in making these observations possible. We acknowledge the help of Drs. G. Neugebauer and G.R. Knapp in the initial stages of this project. We thank Drs. D. Hollenbach and M. Allen for a number of useful discussions. This program was supported in part by NASA grant NAGW-107.

REFERENCES

Chevalier, R.A. 1980, Astrophys. Letters 21, 57.

Erickson, N.R., Davis, J.H., Evans II, N.J., Loren, R.B., Mundy, L., Peters III, W.L., Scholtes, M. and Vanden Bout, P.A. 1980, I.A.U. 87, "Interstellar Molecules", ed. B.H. Andrew (Boston, D. Reidel) 25.

Goldsmith, P.F., Erickson, N.R., Fetterman, H.R., Clifton, B.J., Peck, D.D., Tannenwald, P.E., Koepf, G.A., Buhl, D and McAvoy, N. 1981, Ap. J. (Letters) 243, L79.

Graedel, T.E., Frerking, M.A. and Langer, W.D. 1981 preprint.

Hollenbach, D.J., 1980, in I.A.U. 87, "Interstellar Molecules", ed. B.H. Andrew (Boston, D. Reidel) 445.

Hollenbach, D.J. and Shull, J.M. 1977, Ap. J. 216, 419.

Huggins, P.J., Phillips, T.G., Neugebauer, G., Werner, M.W., Wannier, P.G. and Ennis, D.E. 1979, Ap. J. 227, 441.

Knapp, G.R., Phillips, T.G., Huggins, P.J. and Redman, R.O. 1981, Ap. J., in press.

Kuiper, T.B.H., Rodriguez-Kuiper, E.N., Zuckerman, B., 1981, in I.A.U. 87, "Interstellar Molecules", ed. B.H. Andrew, 31.

Kwan, J. 1977, Ap. J. 216, 713.

Kwan, J. and Scoville, N.J. 1976, Ap. J. (Letters) 210, L39.

Langer, W.D. 1976, Ap. J. 210, 328.

Norman, C., Silk, J. 1979, Ap. J. 228, 197.

Nussbaumer, H. 1971, Ap. J. 166, 411.

Phillips, T.G. and Jefferts, K.B. 1973, Rev. Sci. Instr. 44, 1009.

Phillips, T.G., Huggins, P.J., Neugebauer, G., Werner, M.W. 1977, Ap. J. (Letters) 217, L161.

Phillips, T.G., Huggins, P.J., Kuiper, T.B.H., Miller, R.E. 1980, Ap. J. (Letters) 238, L103.
Phillips, T.G. and Huggins, P.J. 1981, Ap. J., in press.
Prasad, S.S. and Huntress, W.T. 1980, Ap. J. 239, 151.
Saykallay, R.J. and Evenson, K.M., 1980, Ap. J. (Letters) 238, L107.
Scoville, N.Z. 1980, in I.A.U. 87, "Interstellar Molecules", ed. B.H. Andrew (Boston, D. Reidel) 33.
Scoville, N.Z., Kleinmann, S.G., Hall, D.N.B., Ridgway, S.T. 1981 preprint.
Solomon, P.M., Huguenin, G.R., Scoville, N.Z. 1981, Ap. J. (Letters) 245, L19.
Watson, W.D. 1976, Rev. Mod. Phys. 48, 513.
Westphal, J.A., "The Infrared Sky Noise Survey", Final Report NASA grant NGR 05-002-185, NASA CR 139693, N74-32782.
Wootten, H.A., Phillips, T.G., Beichman, C.A., Frerking, M.A. this volume, p. 453.
Wootten, H.A., Snell, R.L. and Glassgold, A.E. 1979, Ap. J. 234, 876.
Zuckerman, B., Kuiper, T.B.H., Kuiper, E.N.R. 1976, Ap. J. (Letters) 209, L137.

DISCUSSION FOLLOWING PAPER BY BEICHMAN, PHILLIPS AND WOOTTEN

BLITZ: What do you find when you search for CI emission in the quiescent portions of molecular clouds away from regions of active star formation.

BEICHMAN: The L134 cloud is an example of a dark cloud without active star formation. The ratio of abundance C/CO is about 0.4.

BOLAND: We have calculated the gas temperatures in molecular clouds by solving simultaneously the equations of chemical equilibrium and of thermal balance (Astron. Astrophys. 1980 91, p. 68). For our plane-parallel hydrostatic models we find that the C° (610 μm) line has an optical depth of the order unity. The calculated ratio of the C/CO column densities is about 0.1. This ratio can be enhanced if the photodestruction rate of CO is larger.

BEICHMAN: This is similar to what we are saying.

OBSERVATIONS OF NEUTRAL CARBON IN THE NGC1977 BRIGHT RIM

Alwyn Wootten*, T.G. Phillips*, C.A. Beichman*
California Institute of Technology
M. Frerking*
Jet Propulsion Laboratory, California Insitute of Technology

ABSTRACT

Neutral carbon emission has been observed at 610 μm (492 GHz) from the region of an interface between the HII region NGC1977 and the northern extension of the dense OMC2 cloud in Orion.

INTRODUCTION

Chemical studies of interstellar clouds have reached a modest level of sophistication in modelling both the dense regions of molecular clouds and the more rarified diffuse clouds. In intermediate density regions, the lack of sufficient excitation to produce a rich spectrum and the complexity of photochemical processes occurring within them has hampered investigations. The recent detection of the $^3P_1$-$^3P_0$ line of CI (Phillips et al. 1980) provides a powerful probe of these intermediate regions. The line is well suited to observation. The low Einstein A value for the transition, and the low-lying levels from which it arises ensure robust excitation in regions of moderate density and temperature. Furthermore, neutral carbon should be abundant in moderate density regimes penetrated by ultraviolet radiation capable of dissociating carbon-bearing molecules (Glassgold and Langer 1974). As a consequence of its expected high abundance and easy excitation, it should make an important contribution to the cooling of these regions, especially if CO has been extensively photodissociated.

Steady-state chemical models (e.g. Prasad and Huntress 1980), appropriate for aged clouds whose chemistry has reached equilibrium, predict substantial amounts of CI only in moderate density regions where ultraviolet photons can dissociate molecules. Collisional

* Visiting Astronomer, Infrared Telescope Facility, operated by University of Hawaii under contract to NASA.

R. S. Roger and P. E. Dewdney (eds.), Regions of Recent Star Formation, 453–461.
Copyright © 1982 by D. Reidel Publishing Company.

excitation of CI requires gas kinetic temperatures in excess of 20 K (Phillips et al. 1980) to produce strong emission. Substantial CI abundances require temperatures below 50 K, so that the endothermic charge exchange reaction of O with H^+ is inhibited, as this chemistry transforms CI to CO efficiently. At very high densities, H_3^+ chemistry can also process CI into CO, so a further condition favoring CI emission is moderate density. While CI emission has been observed from a variety of dense clouds, the conditions of temperature, density and ultraviolet flux most conducive to strong CI emission are optimum in bright rims.

Following the philosophy that bright rims might be very strong CI emitters, we have detected and mapped CI emission in the bright-rimmed molecular cloud abutting NGC1977, to the north of OMC2 in Orion. This bright rim is found to be the most intense CI source so far detected. The region has previously been mapped in the lines of CO and ^{13}CO in the J=1-0 transitions by Kutner, Evans and Tucker (1976) and in the J=2-1 transitions by Wootten, Sargent, Huggins and Knapp (1981). Carbon recombination lines found in this region are formed in similar material as are the CO lines (Kutner et al. 1979); furthermore the orientation of the region is favorable as the exciting star and the rim both lie in the plane of the sky, so that a map of the region reveals the abundance as a function of proximity to the rim.

OBSERVATIONS AND RESULTS

Because of the high opacity of the terrestrial atmosphere at submillimeter wavelengths, observations of CI $^3P_1$-$^3P_0$ (492.1623 GHz) must be carried out from high altitudes. The observations described here were obtained at the 3 m NASA Infrared Telescope Facility (IRTF) at Mauna Kea, Hawaii. The receiver was a single channel InSb heterodyne bolometer system of 1 MHz bandwidth. The receiver frequency was swept through the line frequency by stepping the local oscillator to obtain spectra. A typical noise figure for the receiver was 500 K. Measurements of continuum emission from Jupiter allowed a determination of the telescope beamwidth, which was found to be 45". The opacity of the atmosphere is large (τ~1.5) at these wavelengths even with the low column of water vapor (~1.5 mm) encountered. Measurement of the optical depth using sky dips fit to a secant curve was unreliable due to variations in the atmospheric temperature and water vapor content. We therefore refer our calibration to the OMC1 measurement obtained from the Kuiper Airborne Observatory at altitudes where no substantial optical depth is encountered (Phillips and Huggins 1981). The observations reported here were taken in a fixed sequence, integrating on the points sequentially. For each point, therefore, the center time of the observation is the same and variations in sensitivity and opacity should affect each observation in the same way. Observations were also taken of the line at the OMC1 position each night.

The spectra are shown in Figure 1, where they have been arranged as if projected upon the source, alongside ^{13}CO spectra taken at Owens Valley Radio Observatory (OVRO) with a similar (52") beamwidth (Wootten et al. 1981). The spectral resolution of the CI data is 0.6 km s^{-1} and the intensity is given in units of T_A^*, antenna temperature corrected for atmospheric and telescope losses.

The data were taken in a north-south strip, approximately perpendicular to the alignment of the bright rim. A CI line was

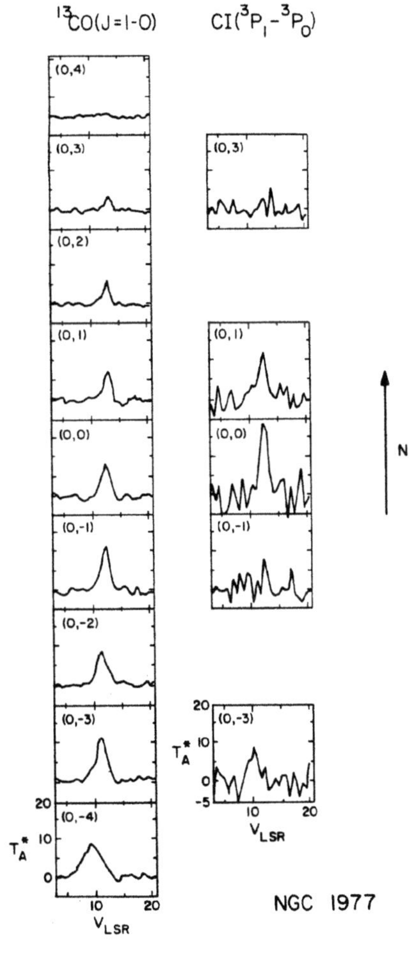

Figure 1. ^{13}CO (J=1-0) and CI($^3P_1-^3P_0$) data near the rim in NGC1977. The position of each data panel is given in arcminutes (α,δ) relative to $05^h32^m47^s$ - $04°58'11''$ (1950). The position of optical rim A coincides with the uppermost panel and that of optical rim B, with the lowermost.

detected at all positions, as expected if the CI emission arises in the molecular cloud material. The linewidth was 1-2 km s^{-1} at a velocity of ~12 km s^{-1}, except that the southernmost point occurs at a slightly lower velocity (~11 km s^{-1}). The line intensity and width peak at the center position of our map (0,0). This behavior is also true of the ^{13}CO line, except that it may be slightly broader. However the CI is much more strongly peaked at the (0,0) position. This is very different from the behavior of the lines in a map of OMC1 (Phillips and Huggins 1981) where ^{13}CO is equally or more strongly peaked than is CI. CI and ^{13}CO temperatures, velocities and linewidths are summarized in Table 1.

Table 1

Observations

	CI			^{13}CO		
Position	T_A^*	VLSR	ΔV	T_A^*	VLSR	ΔV
(0,3)	3.0	13.2	0.9	3.8	13.5	1.5
(0,1)	12.5	12.6	1.5	7.5	12.9	1.5
(0,0)	18.9	12.6	1.7	8.7	12.5	2.0
(0,-1)	8.1	12.6	0.9	8.5	12.6	2.3
(0,-3)	9.0	10.8	1.3	10.6	11.3	2.1

CI EXCITATION

Having observations only of one CI line, it is difficult to determine the optical depth of the line, and consequently the CI column density in the cloud. Phillips and Huggins (1981) have addressed this question in detail. On the basis of comparisons of linewidths of CI, CO and ^{13}CO, they concluded that CI lines are frequently optically thick.

An alternative way to interpret the observations is to compare them with model calculations. For optically thick lines, radiative trapping can affect the excitation of the line. We have used a large velocity gradient (LVG) code (Wootten, Snell and Evans 1980) to model the excitation of CI. Atomic constants were drawn from Nussbaumer (1971) and Saykally and Evenson (1980). Collisional cross-sections calculated by Launay and Roueff (1977) for CI-H collisions were used. No multiplicative correction for molecular hydrogen collisions was used, following the discussion of Flower, Launay and Roueff (1978). The result of this calculation is shown in Figure 2, where a curve indicating the Rayleigh-Jeans approximation to the excitation temperature of the $^3P_1-^3P_0$ line and a curve indicating the ratio R_{21} of the $^3P_2-^3P_1$ to $^3P_1-^3P_0$ lines are plotted against abundance and density, for a temperature of 30 K, and a velocity gradient dV/dR of 1 km s^{-1} pc^{-1}. A dashed line is plotted to show where the optical depth of the 610 μm line reaches unity. It is readily apparent that the intensity of observed lines, typically T_A^*=10-20 K, requires a fairly high abundance in all but the densest regions. The central brightness temperature of 33 K is the largest yet observed for CI. These lines must also have moderate optical depths, probably similar to those found in the ^{13}CO lines, in agreement with the discussion given by Phillips and Huggins (1981). Furthermore, over most of the plane of Figure 2, the CI 370 μm line should be fairly strong.

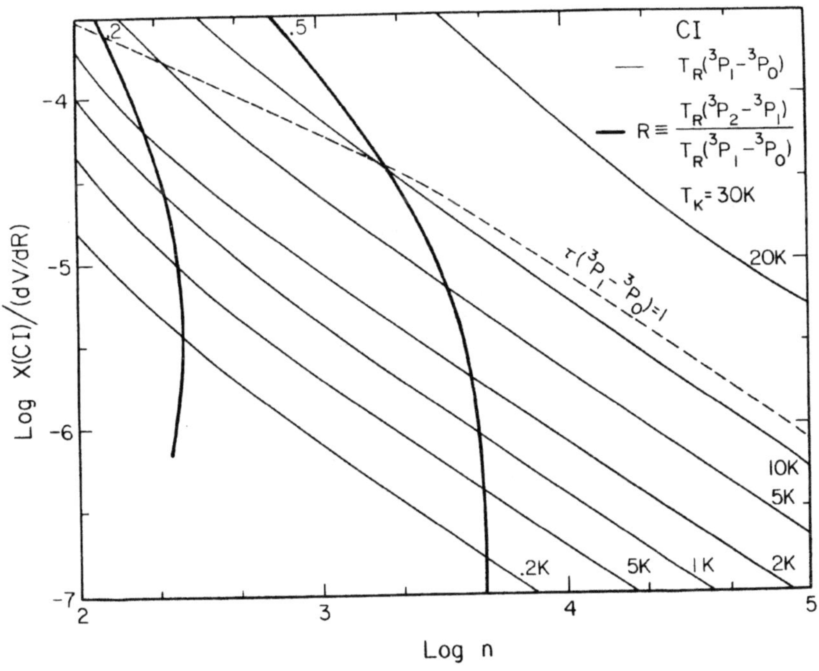

Figure 2. Results of a large velocity gradient model calculation of the excitation of CI. The Rayleigh-Jeans approximation to the excitation temperature of the 610 μm line is given in light curves. The ratio of temperatures of the 370 μm to 610 μm line is given by the heavy curves. The curves are shown in the plane of log total density (n) and log abundance (X(CI)/dV/dR) where the velocity gradient parameter (dV/dR) was set to 1 km s^{-1} pc^{-1} for the calculation. Also shown, by the dashed curve, is the line showing where optical depth in the 610 μm line reaches unity. To the right of this line, optical depth increases above unity.

INTERPRETATION

The observations in Figure 1 clearly show that the CI emission is strongest in the region of the interface between the molecular cloud and the NGC1977 HII region. Unlike the carbon recombination line area mapped by Kutner et al. (1979), it is not restricted to this region. The peak of CI emission corresponds closely to the peak of the ^{13}CO J=2-1 also (Wootten, Sargent, Huggins and Knapp 1981). The velocities and widths of the ^{13}CO and CI lines correspond to within the accuracy of the data at all positions mapped. At the peak of recombination line emission the velocity and width of the CII0α line also match those of the ^{13}CO and CI lines. Because of this correlation in position, width and central velocity of the emission, it is most likely that all arise in a similar portion of the molecular cloud. In this section physical parameters of this bright-rimmed cloud are derived.

The region of the bright rim is best examined on a photograph of the region printed with an unsharp mask (Malin, 1979). Several rimlike structures can be seen. The brightest of these, which we call rim 'A', was first noted by Duncan (1923) and lies at position (0,3) on our map. It is associated with a patch of obscuring matter jutting northward from the large OMC2 molecular cloud. Another fainter rim to the south ('B', at (0,-4)) is associated with the northern extremity of this cloud and can be followed across the dark obscuring matter associated with rim A. The gas in the rim A cloud occurs at velocities of 12-13 km s^{-1}. This is the 'high velocity peak' found in CO observations by Kutner, Evans and Tucker (1976). The exact correspondence between the high velocity CO contours and the obscuring material associated with rim A confirm this identification. At 500 pc distance, the dimension of this cloud is 1 pc, measured between rim A and rim B. Notice that the peak of line emission does not occur at either rim, but at a bright structure near the (0,0) position.

Since the physical conditions in the cloud cannot be determined from CI data alone, we rely upon the discussion of the high resolution CO observations presented by Wootten et al. (1981). The strength of the J=2-1 line of CO indicates that the kinetic temperature of the cloud remains constant at T_k~35 K over the region we have observed. The molecular cloud in the region of the rim shows more extended structure is apparent at visual wavelengths. Both ^{12}CO and ^{13}CO emission falls in strength by a factor of two in a region of extent ~50". The region in which this occurs in ^{12}CO emission is displaced ~30" N of the region where it occurs in ^{13}CO emission. Thus, in the CO spectral lines, two rim structures can be discerned. One, in ^{12}CO, extends through the visible rim. The second, in ^{13}CO, is displaced southward, within the visibly obscured cloud and the rim region. Where ^{13}CO emission is detected, the ^{12}CO line is most likely optically thick. Where the intensity of ^{12}CO emission falls, ^{13}CO

emission is rapidly weakening, suggesting that the decrease in ^{12}CO intensity might be attributed to a decrease in its optical depth, presumably as a result of declining abundance as CO is photodissociated by ultraviolet radiation penetrating the gas. Near the rim, the excitation of ^{13}CO, monitored by the ratio of the J=2-1 and J=1-0 line intensity, increases. This observation, coupled with the intensity decrease of both lines and the constant ^{12}CO intensity, suggests that the density of the gas increases and the ^{13}CO abundance declines as we progress northward through the ^{13}CO rim. On photographs of the rim, the obscuration appears to become more opague close to the rim, which also suggests that a density increase occurs there.

In summary, over a region of extent ~120" centered on the visible rim A, there is a continuous decrease northward in the CO abundance. In the region of the ^{13}CO rim, density increases and kinetic temperature remains fairly constant. In the region of the ^{12}CO rim, the excitation temperature increases, but the separate effects of the temperature and density cannot be disentangled.

It is particularly interesting to contrast the behavior of CO emission just described with that of C^+ and CI. The C110α data of Kutner et al. (1979), observed with a 6'.6 beam, peaks at or just south of the visible rim. The peak of the CI line occurs well to the south of the rim and also to the south of the peak of the C110α line. Since the beamsize used for the OVRO J=1-0 CO observations is comparable to that used for our CI observations, we will assume these parameters can be used to model the excitation of the CI line. Densities and temperatures estimated from the CO observations were used in conjunction with the model mentioned in section 3 to generate CI lines. The intensity of the generated line was matched to that of the observed line and the derived CI abundance listed in Table 2.

Table 2

Physical Parameters in the NGC1977 Rim

Position	n(H2)	T_k	X(^{13}CO)	X(CI)	X(CI)/X(^{13}CO)
(0,3)	~100000	35 K	8.(-9)	3.(-8)	4.
(0,1)	60000	35 K	1.(-7)	2.(-6)	20.
(0,0)	30000	35 K	2.(-7)	8.(-6)	40.
(0,-1)	8000	35 K	6.(-7)	7.(-6)	12.
(0,-3)	10000	35 K	7.(-7)	7.(-6)	10.

Parentheses in columns 4 and 5 enclose exponents of ten by which abundances are to be multipled.

Over the region of the CI observations, both CI and ^{13}CO abundances drop toward the rim. This is the region from which C^+ emission is strongest. In the interpretation of Kutner et al. (1979) the abundance of C^+ in this region is $\sim 10^{-4}$. Since the parameters they used to obtain this estimate are very similar to those we have determined, this number can be compared to the values of $X(^{13}CO)$ and X(CI) we have derived, confirming that C^+ is the more abundant of these three species near the rim.

Isotopic fractionation of CO is expected to be important in regions where C^+ is present (Watson et al. 1976). This process should increase the fractional abundance of ^{13}CO relative to ^{12}CO near the rim. Therefore interpretation of the ^{13}CO abundances in terms of a total CO abundance may require a sharper decline in CO abundance than is apparent from the numbers in Table 2.

The region of high electron abundance exists within .3 pc of the rim (Kutner et al. 1979). Graedel, Frerking and Langer (1981) have shown that in a region of high electron abundance most CI cannot be converted to CO. At a distance of .7 pc from the rim, our observations indicate that the CO abundance has reached a stable value. Since the value we estimate for the ^{13}CO abundance is within a factor or two of that obtained by Dickman (1978), it seems reasonable to suppose that there has been little effect of the ionizing radiation from the exciting star of the rim this far into the cloud. In the region between .3 and .7 pc into the rim cloud, the abundance of CI rises, subsiding only slightly further into the cloud. This confirms a basic prediction of ion-molecule chemical theories of clouds. These theories do not, however, satisfactorily account for the high atomic carbon abundance in cloud interiors found by Phillips and Huggins (1981), also a feature of the cloud we have observed.

REFERENCES

Dickman, R.L., 1978: Astrophys. J. Suppl. 37, 407.
Duncan, J.C., 1923: Ap. J. 57, 137.
Flower, D.R., Launay, J.-M. and Roueff, E., 1980: in "Les Spectres des Molecules Simples au Laboratoire et en Astrophysique" (Liège: Universite de Liege) P. 137.
Glassgold, A.E. and Langer, W.D., 1974: Ap. J. 193, 73.
Graedel, T.E., Frerking, M.A. and Langer, W.D., 1981: preprint.
Kutner, M.L., Evans II, N.J. and Tucker, K.D., 1976: Ap. J. 209, 452.
Kutner, M.L., Guélin, M., Evans II, N.J., Tucker, K.D. and Miller, S.C., 1979: Ap. J. 227, 121.
Launay, J.-M and Roueff, E., 1977: Astr. Ap. 56, 289.
Malin, D., 1979: Sky and Telescope 57, 355.
Nussbaumer, H., 1971: Ap. J. 166, 411.
Phillips, T.G. and Huggins, P.J., 1981: Ap. J., in press.

Phillips, T.G., Huggins, P.J. Kuiper, T.B.H. and Miller, R.E., 1980: Ap. J. (Letters) 238, L103.
Prasad, S.S. and Huntress Jr., W.T., 1980: Ap. J. Suppl. 43, 1.
Saykally, R.J. and Evenson, K., 1980: Ap. J. (Letters) 238, L107.
Watson, W.D., Anicich, V.G. and Huntress Jr., W.T., 1976: Ap. J. (Letters) 205, L165.
Wootten, H.A., Snell, R. and Evans II, N.J., 1980: Ap. J. 240, 532.
Wootten, H.A., Sargent, A., Huggins, P.J. and Knapp, G.R., 1981: in preparation.

DISCUSSION FOLLOWING PAPER BY WOOTTEN ET AL.

J. DICKEL: Is there any evidence for velocity structure across the rim?

WOOTTEN: There is no evidence for either a change in velocity or velocity broadening.

CRUTCHER: Knowing the types of stars producing the UV and the density and other parameters, you should be able to model where the CI, C^+, CO and other transitions should occur.

WOOTTEN: That is the next step. We know there is rough accord. For instance, the C^+ region extends about 0.3 pc into the cloud.

OBSERVATIONS OF CO IN TMC 1

R. Braun, W.H. McCutcheon, and W.L.H. Shuter
Department of Physics
University of B.C.,
Vancouver, B.C.,
V6T 1W5

ABSTRACT

Observations of CO in TMC1, the molecular ridge in the large dust cloud Heiles'2, give evidence for a local contraction or expansion centered in position and velocity on the HC_5N maximum of MacLeod et. al. (1979). The estimated velocity gradient of this motion is $\sim 1.5 kms^{-1}pc^{-1}$ if the cloud is assumed to be at a distance of 115 pc.

Introduction

The region in Heiles' Cloud 2 known as TMC1 has been observed in a wide variety of molecules. Of particular interest was the detection of the $J=4\rightarrow3$ transition of HC_5N by MacLeod et. al. (1979) who found that its distribution defined a rather narrow ridge. Observations of the $J=9\rightarrow8$ transition of HC_5N (Little et. al., 1978; Churchwell et. al., 1978), of HC_9N (Broten et al. 1981), DC_3N (Langer et. al., 1980), C_3N (Friberg et al., 1980), CN (Churchwell, 1980), C_2H (Wootten et al., (1980) and NH_3 (Little et. al., 1979; Avery et al., 1981) all show that these molecules are strongly concentrated towards this ridge.

This work presents results from ^{12}CO observations at 115 GHz in the vicinity of the ridge with approximately the same beamwidth as that used for the HC_5N observations of MacLeod et al., (1979), and Little et al., (1978), undertaken in an attempt to determine if any characteristic of the ridge structure was apparent in the distribution of CO.

Observations

These observations were made in May and July 1979 with the 4.6 m telescope at the University of British Columbia. The system temperature was 1000K SSB, the half power beamwidth was 2'.6 and the velocity resolution was 0.65 km s^{-1}. Data were obtained in a 52 Hz frequency switching mode. Atmospheric opacity was determined from antenna tipping curves made twice a day. Daily observations were made of the ^{12}CO line

R. S. Roger and P. E. Dewdney (eds.), Regions of Recent Star Formation, 463–467.
Copyright © 1982 by D. Reidel Publishing Company.

in OMC-1, for which we assumed a temperature of 60K, to aid in the single sideband calibration. Daily observations were also made of the central position in our map to check the repeatability of our calibration. We believe that the error in our corrected temperature is no greater than 15%.

During these observations we were uncertain of our absolute frequency because of a phase lock problem which shifted the observed line by a constant four channels from what we expected. We overcame this difficulty by observing, in a load switched mode, the Orion line and the Cloud 2 line at the central position, in rapid succession within the same spectrometer bandwidth. The velocities determined for Cloud 2 then are with respect to the ^{12}CO line in Orion for which we took the value 9.5 km s^{-1}. (This value had been determined previously by us using the 4.6 m telescope of the Aerospace Corporation).

Observations were spaced at one beamwidth intervals around the region defined by the HC_5N observations of MacLeod et al., (1979), but away from this region the data were sampled at two to three beamwidth intervals to the point where the intensity had decreased by about 50%. To improve the baselines, reference spectra were taken at an emission free position 1° higher in declination than the central position. The resulting twenty-nine spectra have a r.m.s. noise of 0.7K.

Discussion of Results

The integrated spectrum for the central position of the map, which had 1950 coordinates $\alpha = 04^h 38^m 37^s.8$, $\delta = 25° 35' 00''$, is shown in Figure 1. This profile has velocity width $\Delta\nu$ (FWHM) = 3.3±0.3 km s^{-1} centered on the velocity 5.6±0.3 km s^{-1} (based on the velocity of 9.5 km s^{-1} for the ^{12}CO line in OMC-1) and a peak excitation temperature T_{ex} = 16±1K. Other spectra in the field have similar shoulders.

Maps of individual velocity channel T_A^* and of peak T_A^* were hand drawn from the data and appear in Figure 2. Examination of these reveals a number of regions showing temperature enhancements at particular velocities. The enhancement at about 5.6 km s^{-1} corresponds closely in position and velocity to the HC_5N maximum of MacLeod et al. (1979). A feature roughly 6' south of this position becomes most apparent at about 5.1 and 6.3 km s^{-1} and is near the H_2CO absorption maximum of Sume et al. (1975). To the east of the HC_5N maximum, temperature enhancements appear at about 4.3 and 7.5 km s^{-1}. The peak temperatures of the features to the south of the map center are generally somewhat higher than those to the east.

Phillips and White (1981) have recently published the results of their ^{12}CO and ^{13}CO, J=1→0 observations of this area. An inspection of their ^{12}CO maps made with the same velocity resolution shows overall agreement on the central and eastern features mentioned above. Wilson and Minn (1977) have observed the HI absorption in this region with 8'.7 beamwidth. They find that the dominant absorption feature changes

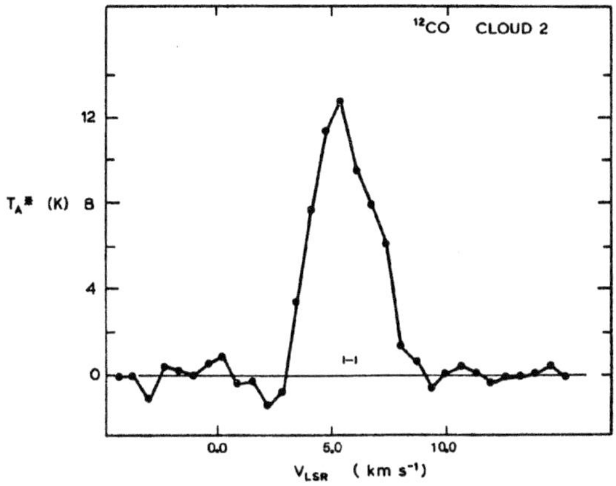

Figure 1. Integrated ^{12}CO profile at the central position of the map: $\alpha(1950) = 04^h38^m37^s.8$, $\delta(1950) = 25°35'00''$. The small bar indicates the range of velocities over which other molecules have been detected (see text).

in velocity from 5.5 to 6.1 km s^{-1} within about 10' in the vicinity of the molecular ridge. They suggest that this may be due to a shock propagating from the south west to the north east.

No evidence for a well defined spatial division of temperature structure of this type is apparent in our data. Instead we find local temperature enhancements on either side of the HC_5N maximum position which become prominent at both positive and negative velocities centered roughly on 5.6 km s^{-1}. What this seems to suggest is an expanding or contracting structure centered on the ridge of complex molecules. Assuming that the cloud lies at a distance of 115 pc (McCuskey 1941), and that the hot spots represent parts of a cylindrical shell of roughly 40' diameter, whose axis is aligned with the molecular ridge, the cloud has linear diameter 1.3 pc, and the supersonic velocity of about 1.0 km s^{-1} corresponds to a velocity gradient of expansion or contraction of $\sim$ 1.5 km $s^{-1}pc^{-1}$.

Conclusions

The CO observations presented here sample the structure of Heiles' Cloud 2 on an angular scale between that of the narrow ridge defined by the concentration of complex molecules and the larger scale structure evident in the HI aborption data referred to above. Within the large scale dynamics of this region, we present evidence for a local contraction or expansion of the volume surrounding the location of maximum emission due to HC_5N and other complex molecules. Using a local

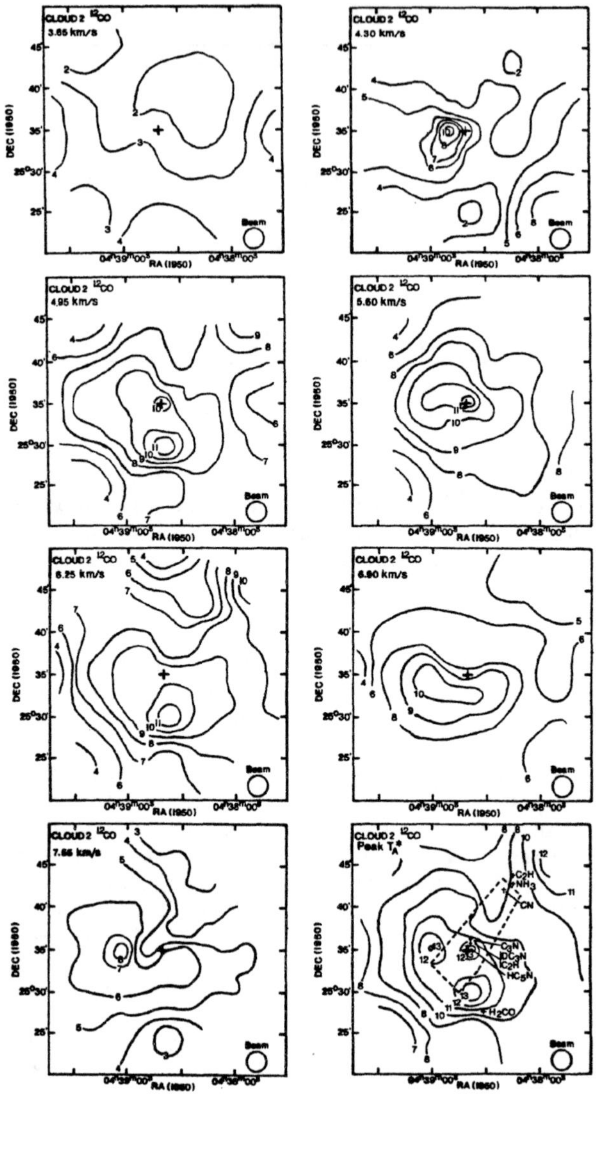

Figure 2. CO contour maps of individual velocity channel T_A^* and peak T_A^*. The dashed outline indicates the region in which HC_5N was detected by MacLeod et al. (1979). The crosses show the peak positions of HC_5N (Little et al. 1978; MacLeod et al. (1979), NH_3 (Little et al. 1979; Avery et al. 1981), C_2H (Wootten et al., 1980), and H_2CO (Sume et al. 1975), and the positions where CN (Churchwell 1980), C_3N (Friberg et al. 1980) and DC_3N (Langer et al. 1980) were detected.

cylindrical model of the source one obtains a velocity of expansion or contraction of about Mach ∿3 and velocity gradient ∿1.5 km $s^{-1}pc^{-1}$.

Acknowledgement

We wish to thank Drs. J.M. MacLeod and L.W. Avery for stimulating our interest in this work and for releasing material prior to publication.

This research was supported by funds from the Natural Sciences and Engineering Research Council of Canada.

References

Avery, L.W., Broten, N.W., MacLeod, J.M. and Oka, T. (1981) in preparation.

Broten, N.W., Oka, T., Avery, L.W., MacLeod, J.M. and Kroto, H.W. (1981) preprint.

Churchwell, E. (1980) Astrophys. J. 240, 811.

Churchwell, E., Winnewisser, G. and Walmsley, C.M. (1978) Astron. Astrophys. 67, 139.

Friberg, P., Hjalmarson, Å., Irvine, W.M. and Guélin, M. (1980) Astrophys. J. Lett. 241, L99.

Langer, W.D., Schloerb, F.P., Snell, R.L. and Young, J.S. (1980) Astrophys. J. Lett. 239, L125.

Little, L.T., Macdonald, G.H., Riley, P.W. and Matheson, D.N. (1979) Mon. Not. R. Astron. Soc. 189, 539.

Little, L.T., Riley, P.W., Macdonald, G.H. and Matheson, D.N. (1978) Mon. Not. R. Astron. Soc. 183, 805.

MacLeod, J.M., Avery, L.W. and Broten, N.W. (1979) Astrophys. J. 233, 584.

McCuskey, S.W. (1941) Astrophys. J. 94, 468.

Phillips, J.P. and White, G.J. (1981) Mon. Not. R. Astr. Soc. 194, 15.

Sume, A., Downes, D. and Wilson, T.L. (1975) Astron. Astrophys. 39, 435.

Wilson, T.L. and Minn, Y.K. (1977) Astron. Astrophys. 54, 933.

Wootten, A., Bozyan, E.P., Garrett, D.B., Loren, R.B. and Shell, R.L. (1980) Astrophys. J. 239, 844.

CO EMISSION ASSOCIATED WITH COLD NEUTRAL HYDROGEN

William L. Peters, III and Frank N. Bash
Astronomy Department, The University of Texas

CO observations were compared with neutral hydrogen observations. It was found that a third of the CO clouds had H I self absorption counterparts. A statistical analysis indicates that the correlation is significant and consistent with the view that molecular clouds contain enough atomic hydrogen to noticeably affect the atomic hydrogen emission line profiles.

This is a preliminary report on a comparison of ^{12}CO (J=1-0) observations and 21 cm H I observations at similar angular and velocity resolution. The purpose of the study was to assess the claim made by Burton, Liszt, and Baker (1978) (hereafter BLB) that CO emission at small angular scales (2' to 20') tends to correspond to places of anomalously low H I emission. That is, they assert that cold atomic gas in the molecular cloud is absorbing emission from warmer H I behind the cloud. This hydrogen is thought to be residual atomic gas which has not yet formed molecular hydrogen or is from molecular hydrogen recently dissociated. The amount of atomic gas required is less than 1% of the molecular hydrogen inferred from the CO emission.

Not all CO clouds have H I counterparts or vice versa. BLB explain that this is possible because special geometric conditions are required for the H I self absorption to be observable: There must be a significant amount of H I emission behind the cloud and not too much in the foreground to fill in the absorption. Also, not every relative minimum in the H I emission need be due to self absorption. The H I line profiles are highly sensitive to the velocity gradient along the line of sight. Some of the minima may simply be due to a steeper than usual gradient

R. S. Roger and P. E. Dewdney (eds.), Regions of Recent Star Formation, 469–477.
Copyright © 1982 by D. Reidel Publishing Company.

at that velocity. This study attempts to evaluate whether the correspondence of CO emission and H I absorption is statistically significant, and, if so, whether there is any difference in observable cloud properties between the group of CO clouds that have H I counterparts and the group that does not, and, finally, whether the data are consistent with the BLB picture.

The H I observations were made by Baker and Burton (1979) with the 305 m Arecibo telescope which has an angular resolution of 3!2 at 21 cm. The CO observations were made with the University of Texas' 4.9 m telescope which has an angular resolution of 2!3 at the CO(1-0) line. The CO data were taken at 2' spacing in the galactic plane over several portions of the longitude range of the H I observations: 40° to 68°. The CO observations cover a total longitude range of $7\overset{\circ}{.}5$, $6\overset{\circ}{.}5$ of which was at $\ell < 54^{\circ}$ where the CO emission is stronger. These observations extend those published by BLB which only cover a range of $2.^{\circ}5$, 1° of which overlaps our observations. (Liszt et al (1981) are publishing additional observations of CO and H I in the range $36^{\circ} < \ell < 40^{\circ}$.)

The data were analysed as follows. The CO data were examined by making v-ℓ contour maps and tabulating the following quantities for each "cloud" with a peak antenna temperature, T_A, of 3 K or more: The peak T_A, the ℓ,v coordinates of the peak, and the points in ℓ and v where the CO intensity fell to half of the peak value. The 3 K cutoff was somewhat arbitrary. Too high a value will cause some clouds to be missed whereas too low a value dilutes the statistics on the prominent clouds by including too many minor "cloudlets". Furthermore, the CO data were not judged to be reliable below 1 K due to low signal/noise and baseline removal uncertainties. Some CO clouds showed multiple peaks and thus an irregular overall shape. These were considered to be two or more clouds for the purposes of the analysis in order to accurately describe their shape and location. The contour maps for the range $40^{\circ} < \ell < 54^{\circ}$ are shown in Figure 1. The contour interval was 1 K with the lowest contour level for $T_A = 1$ K.

The H I data were analysed in a similar fashion by tabulating for each relative minimum, the position and depth of the center, and the ℓ and v extrema of the outermost contour. These quantities were based on the v-ℓ contour map published by Baker and Burton (1979). The contour interval was too coarse to apply a low intensity cutoff as with the CO. (Arrangements have been made to get a contour map of the H I data drawn at higher resolution.)

In the statistical analysis below, the sizes of the H I clouds were adjusted to correspond to half power points like the CO data by assuming that the minima could be represented by gaussians. No H I minimum was included in the analysis if it was larger than 0°.5 in longitude. This was done mainly to exclude those minima caused by perturbations in the Galactic velocity field (e.g. streaming motions due to the spiral density wave) which are

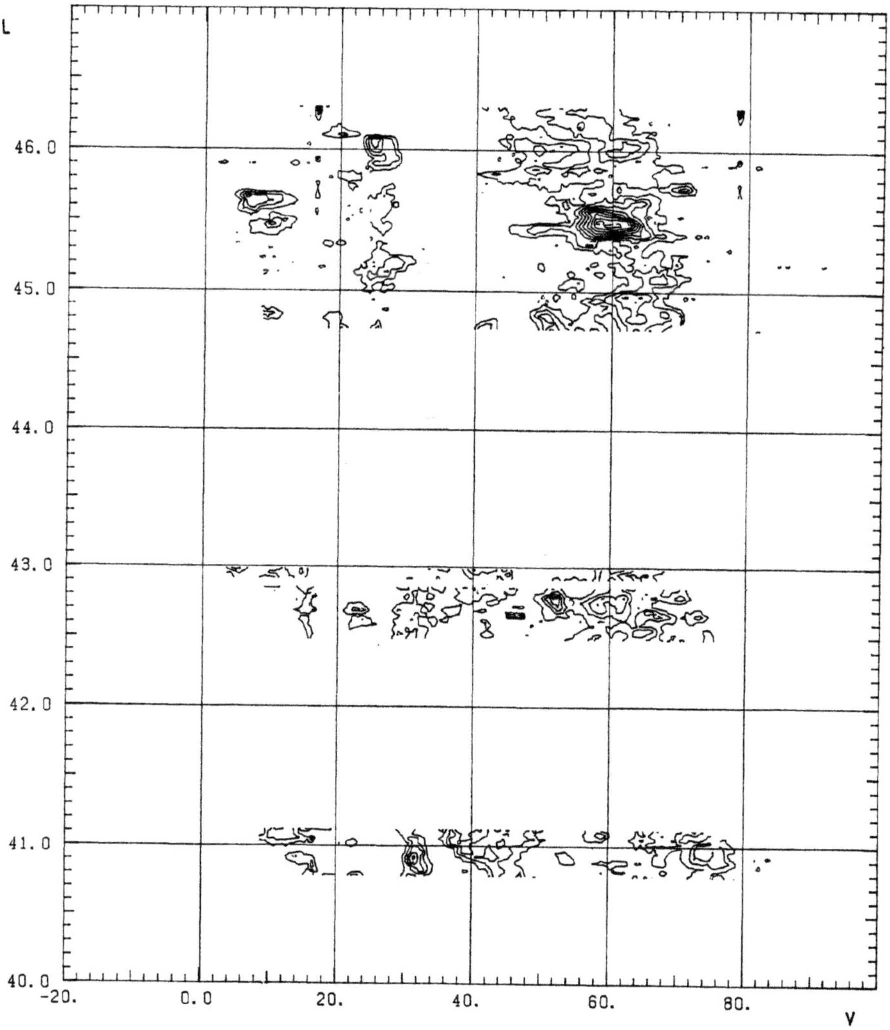

Figure 1a. CO v-ℓ map for $40^\circ < \ell < 47^\circ$

likely to be coherent over a degree or so in ℓ. No CO cloud was as large as this and a histogram showing the distribution of H I sizes showed evidence for two components: a more or less gaussian distribution centered near 0°.2 with a low level tail out to 1°.2. Finally, no H I minima at negative velocities were included. This excludes all clouds located outside the solar circle where there is little CO emission. The boundaries of the H I for the longitude range of Figure 1 are shown in Figure 2.

The mean cloud parameters derived from the above tabulations are given in Table 1. The total v-ℓ area covered by the observations from 0 km s^{-1} to the "terminal

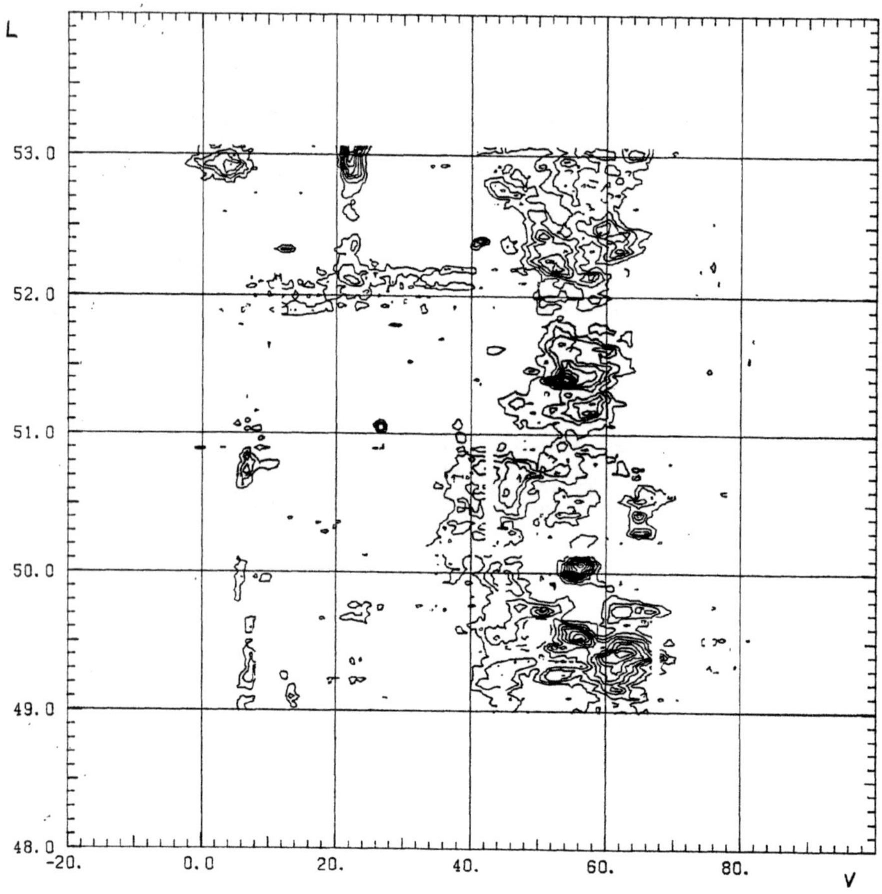

Figure 1b. CO v-ℓ map for 48° < ℓ < 54°

velocity" was 470.5 deg km s^{-1}. (The terminal velocity was taken to be the highest velocity showing significant H I emission on the Baker and Burton map.) In this area, there were 66 CO clouds occupying a total area in the v-ℓ plane of 43.1 deg km s^{-1} or 9% of the total. There were 50 H I minima occupying 42.0 deg km s^{-1} or 8.9%.

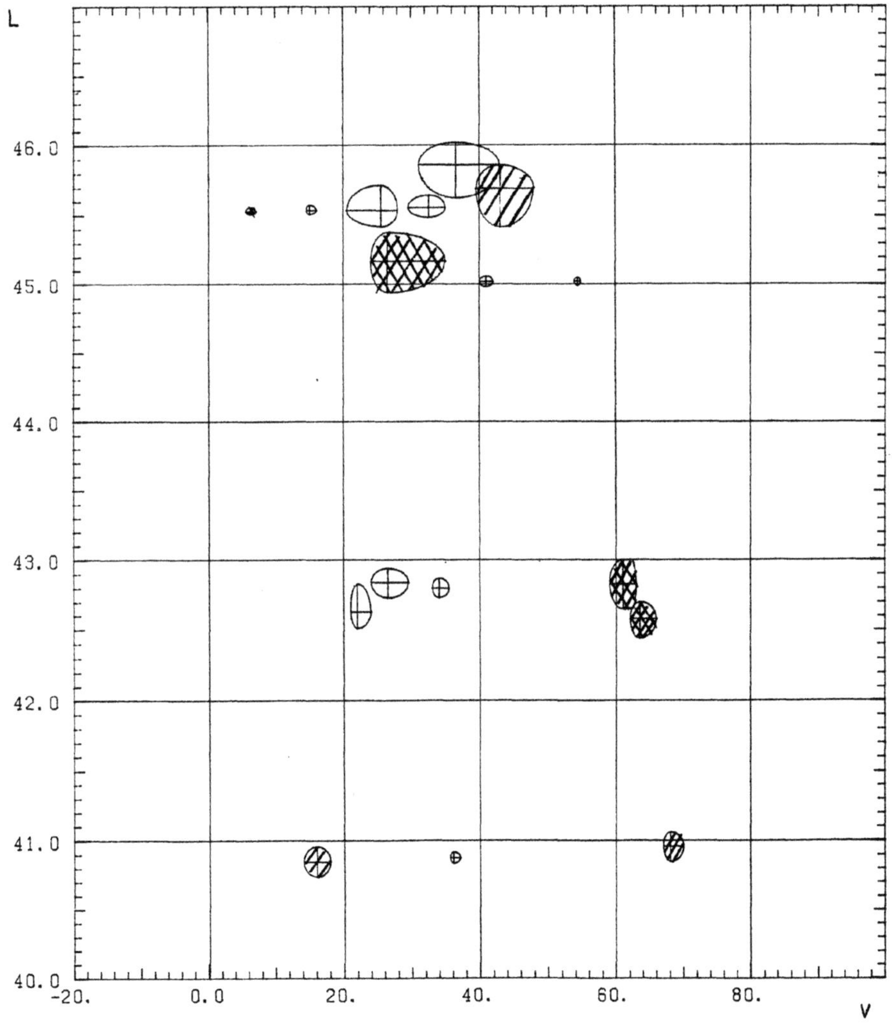

Figure 2a. H I minima boundaries for $40^{\circ} < \ell < 47^{\circ}$

These two lists were then compared for overlapping clouds. If one counts an overlap as any two clouds that intersect, one finds that 20 H I clouds intersect 24 CO clouds. The number of CO clouds one would expect to intersect the H I clouds at random, using the mean sizes listed in the table, is 18. This suggests that most of the intersections are simply due to chance. But upon closer examination, it was discovered that there were fewer "grazing" overlaps than would be expected for random overlaps. If one calls an "overlap" only those intersections that include one or both cloud centers, one finds that 16 H I clouds (32%) intersect 19 CO clouds (29%) producing 19 such intersections. At random, one would

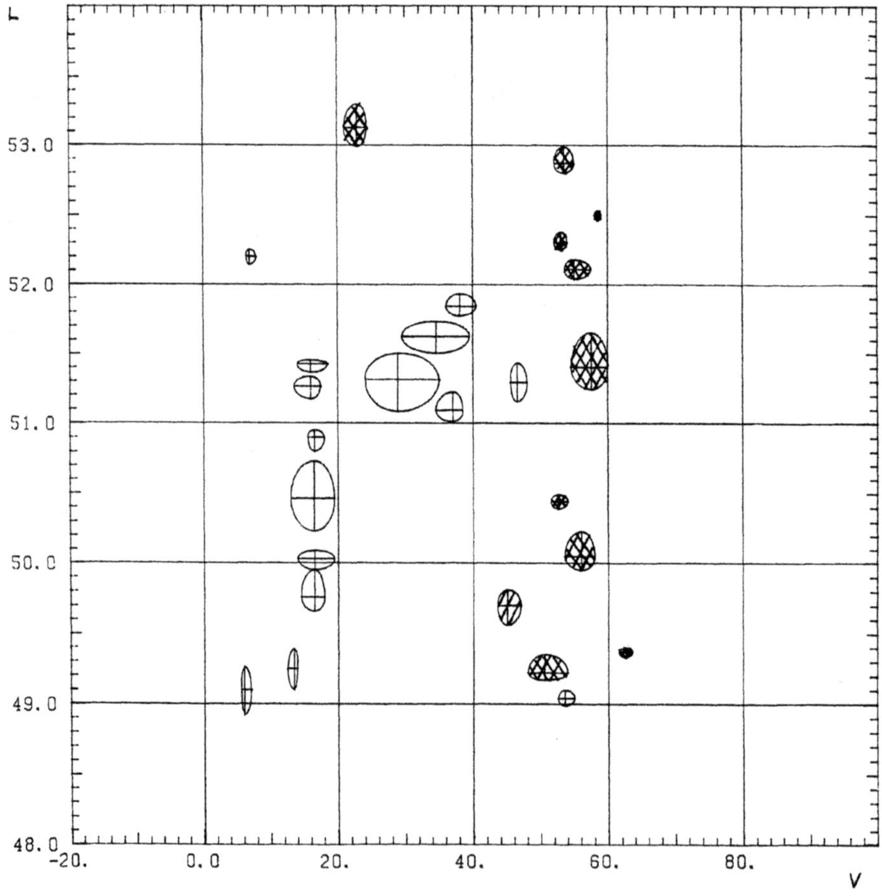

Figure 2b. H I minima boundaries for $48^{\circ} < \ell < 54^{\circ}$

expect only 7 such intersections. These H I clouds are indicated on Figure 2 with double cross hatching. The 4 H I clouds that only overlap the edges of a CO cloud are indicated with single cross hatching. (The latter have been treated as non-overlapping clouds below.)

The mean parameters of these overlapping clouds together with those that didn't overlap are also listed in the table. The errors tabulated are the standard deviations of the averages. The table includes estimates of the length of the line of sight that would be expected to have radial velocities in the range Δv observed in a cloud ("max depth", i.e. the maximum depth expected for the foreground + background gas). Finally, the linear size of a cloud for the two possible distances predicted by its velocity ("chord near" and "far", respectively) are listed. These three quantities were derived under the assumption of pure circular rotation using the rotation curve of Burton and Gordon (1978).

Table 1

	$\Delta \ell$ FWHM ($^{\circ}$)	Δv FWHM (km s^{-1})	T (K)	max depth (pc)	chord (pc) near	far
All CO	.15$\pm$.01	5.2$\pm$.2	4.2$\pm$.2	1850$\pm$150	10$\pm$1	24$\pm$2
Ov. CO	.17$\pm$.02	5.4$\pm$.3	4.7$\pm$.5	2200$\pm$300	12$\pm$2	25$\pm$3
No ov. CO	.14$\pm$.01	5.1$\pm$.4	4.0$\pm$.2	1700$\pm$150	9$\pm$1	23$\pm$2
All H I	.16$\pm$.02	3.0$\pm$.4	4.7$\pm$.6	950$\pm$100	10$\pm$1	41$\pm$4
Ov. H I	.18$\pm$.03	3.0$\pm$.6	6.4$\pm$1.3	1500$\pm$250	15$\pm$2	35$\pm$6
No ov. H I	.15$\pm$.02	3.0$\pm$.5	3.7$\pm$.6	700$\pm$100	7$\pm$1	42$\pm$5

The statistical differences between the various groupings of clouds are marginal but suggestive. The angular sizes of the H I vs the CO are nearly identical. This would argue that the cold H I is probably not located in a large halo surrounding the CO cloud, but is either mixed with the molecular material or confined to a thin skin at the edge of the cloud. The overlapping clouds tend to be slightly larger than those that don't overlap. This can be used to support the argument by BLB that the conditions for making the H I self absorption observable favor clouds located at the near kinematic distance. But it could also be due to chance overlaps which are more likely for the larger clouds. The CO clouds tend to be wider in v than the H I clouds. This is harder to reconcile with the idea that the neutral hydrogen and the CO are well mixed, but might be a result of the fact that the CO lines are usually saturated which would tend to widen them. The overlapping clouds tend to have slightly higher peak T_A's or deeper minima than the non-overlapping clouds. However, no correlation was found between peak CO intensity and H I absorption depth in the overlapping clouds. The "max depth" column shows significantly larger numbers for the overlapping clouds. This is related to the fact that the overlapping clouds tend to be located at the highest v's in the v-ℓ maps (c.f. Figure 2). This is probably due to the "velocity crowding" effect: the velocity gradient along the line of sight is lowest for the highest velocities, thus a given Δv maps to a longer depth along the line of sight. It is an indication that most of the overlaps found are real as opposed to chance overlaps which should be distributed uniformly in v. One could argue, however, that a chance alignment of a CO cloud with an unrelated H I cloud in the background or foreground is more likely when "max depth" is large. But this is inconsistent with the similar H I and CO sizes found as well as the tendency for "central" overlaps.

In summary, we find that the proposal, that many molecular clouds contain enough cold residual H I to cause self absorption in the 21 cm emission profiles, is consistent with the data at hand but cannot be proved conclusively as yet. We intend to improve the above statistics by using higher resolution contour maps of the available H I observations and by including in the sample the CO clouds observed by BLB and by Liszt et al (1981) that lie outside the longitude range of our own CO data. The statistics might be improved further if one could be more selective about which H I minima are likely to be due to self absorption. BLB suggest that a steepness in the velocity sides of the minima might be such an indicator. (We could not use this in the above because we could not estimate this slope reliably due to the coarseness of the H I contours on the published maps.)

We wish to thank Butler Burton, Paul Baker, and Harvey Liszt for providing their data in advance of publication. This work was supported in part by NSF Grant 79-20966 to the University of Texas.

REFERENCES

Baker, P.L. and Burton, W. B.: 1979, Astron. Astrophys. Suppl. 35, 129.
Burton, W.B. and Gordon, M.A.: 1978, Astron. Astrophys. 63, 7.
Burton, W.B., Liszt, H.S. and Baker, P.L.: 1978, Astrophys. J. Letters 219, L67.
Liszt, H.S., Burton, W.B. and Bania, T.M.: 1981 (in press)

DISCUSSION FOLLOWING PAPER BY PETERS AND BASH

BLITZ: Are your data consistent with the notion that all CO clouds have a cold HI component, but that you sometimes don't see it because of observational effects?

PETERS: We will not be able to answer that until we attempt a detailed (e.g. Monte Carlo) analysis.

STARK: If it were less than certain that any absorbing HI cloud had to have background emission to make it visible, then this would explain two points:

a) Near-side clouds would be seen preferentially. They would then be systematically larger than the average. Also, only a fairly small fraction of clouds would be seen in CO-HI absorption coincidence.

b) Tangent velocity clouds would be seen in coincidence more easily than other velocities, since velocity crowding would put HI emission behind the clouds with higher probability.

ON THE CORRELATION OF CH ABUNDANCE AND EXTINCTION IN DARK NEBULAE

G. Sandell
Stockholm Observatory, Sweden and Observatory and
Astrophysics Laboratory, University of Helsinki, Finland
L.E.B. Johansson
Onsala Space Observatory, Sweden

ABSTRACT

The correlation of CH abundance and extinction has been investigated in individual dark nebulae as well as in a sample of clouds. We find a good correlation in individual clouds; however, the relation between CH column density, N_{CH} [cm^{-2}], and blue extinction, A_B [mag], varies from cloud to cloud: $N_{CH} = (2\text{-}7)\ 10^{13} A_B$. The reasons for the cloud to cloud variations are discussed.

1. INTRODUCTION

There are a few species which may be considered as standard tracers of interstellar gas. These appear to have, at least in certain regimes, a significant correlation with interstellar extinction, i.e. HI, ^{13}CO and H_2CO (Knapp and Kerr, 1974; Bohlin et al., 1978; Dickman, 1978; Minn and Greenberg, 1979; Sherwood and Wilson, 1980; Federman and Evans, 1981). Previous studies of CH (Hjalmarson et al., 1977; Lang and Willson, 1978) have indicated that the abundance of this radical also is related to the amount of interstellar extinction. Recent work by Johansson (1979, 1981) of the distribution of CH in the galactic plane shows that CH is useful as a tracer of the large scale characteristics of the Galaxy. Mattila and Sandell (1981) have observed CH in the direction of globular clusters, showing that CH is also present in the more tenuous clouds.

2. OBSERVATIONAL DATA AND STARCOUNTS

Most of the observations were carried out during spring and fall 1977, using the 25.6 m telescope at the Onsala Space Observatory. At 3.3 GHz the beamwidth is 15', the beam efficiency ∿0.6 and the filter bandwidth corresponds to a velocity resolution of 0.9 km s^{-1}. Details of the receivers and observing procedures can be found in Sandell et al. (1981). All spectral line data refer to the CH ground state

R. S. Roger and P. E. Dewdney (eds.), Regions of Recent Star Formation, 479–484.
Copyright © 1982 by D. Reidel Publishing Company.

main line transition at 3335.481 MHz.

Complete maps of L 1172 (45 positions), which includes NGC 7023 (CO map made by Elmegreen and Elmegreen, 1978) and L 1457 (49 positions) were obtained, while only the central region of L 1642 (11 positions; Sandell et al., 1981) was studied. In addition, a few more clouds were observed, essentially on a "single point" basis. However, all of these clouds are not included in the present sample, since they still lack complete extinction data. This implies that our observational material is limited with respect to the sample of clouds. To improve the situation, we have used complementary CH data obtained with the same telescope by Hjalmarson et al. (1977). A further limitation of our data is emphasized by the fact that the sample clouds generally have $A_B < 4^m$.

Star counts have been made on the Palomar Sky Survey blue prints with a square reseau of size 5!6. The counts were converted to photographic magnitudes, A_B, with the aid of the log N(m) tables of van Rhijn (1929); see Sandell et al. (1981) for a more detailed discussion. Due to the coarse spatial resolution of the radio beam, the A_B values used in the analysis are obtained by weighting the extinction with the antenna response. A total of 17 positions in L 134, L 1780 and in the Taurus area utilized star count data by Mattila (1979) and Sherwood and Wilson (1979). The previously mentioned bean weighted extinctions are in these cases basically derived from eye estimates; in all other cases we used a numerical algorithm. The data from Sherwood and Wilson (1979) were converted from A_V to A_B by the relation $A_B = 1.32\ A_V$ (R = 3.1). The star counts and the CH data will be published elsewhere (Johansson and Sandell, 1981).

The CH column densities are derived as in Sandell et al. (1981), assuming a constant excitation temperature $T_{ex} = -15K$. Variations in T_{ex} have only a small effect on the derived N_{CH} as long as $|T_{ex}| <$ the background temperature. No clumping of the CH gas has been assumed, i.e., the beam filling factor is equal to one.

3. DISCUSSION

The column density of CH is plotted against the photographic (blue) extinction in Figures 1 and 2. Fig. 1 gives separate plots for the two dark nebulae L 1457 and L 1172. The dashed lines indicate linear least square fits, obtained by taking into account the facts that both parameters are subject to errors and that the error in A_B is not constant (Sandell and Johansson, 1981). The correlation coefficient, ρ, is indicated in both cases.

Evidently the slopes of the two linear least squares fits differ significantly. For L 1457 we get

$$N_{CH} = 6.8\ 10^{13}\ (A_B - 0.4) \quad [cm^{-2}] \tag{1}$$

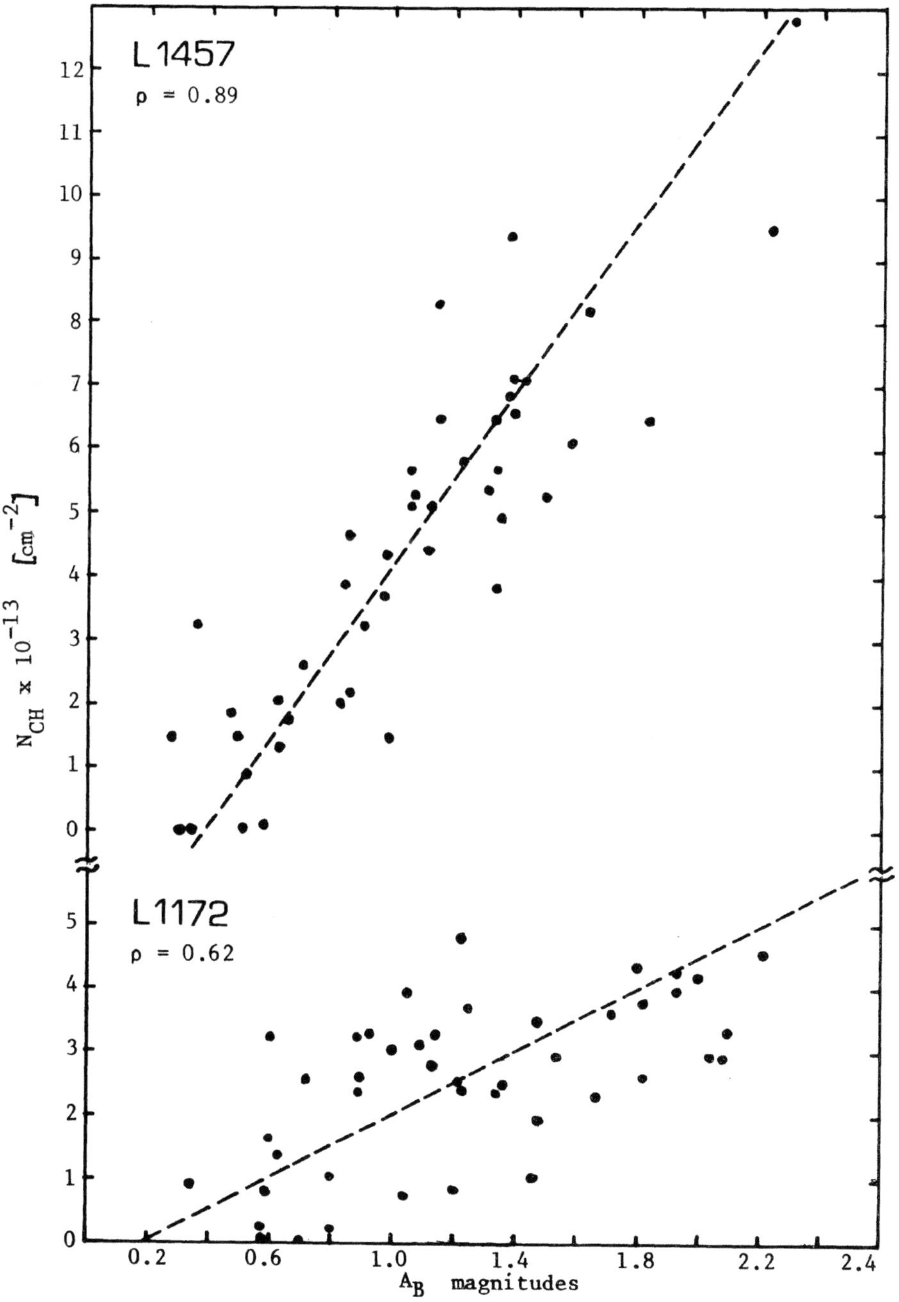

Figure 1. N_{CH} vs. A_B plot for L 1457 and L 1172 (see text).

which is in good agreement with the results of L 1642 (Sandell et al., 1981). L 1172 yields

$$N_{CH} = 2.3\ 10^{13}\ (A_B - 0.14)\ [cm^{-2}] \quad (2)$$

a result similar to that of Lang and Willson (1978), obtained for diffuse clouds toward nearby bright stars.

In Fig. 2 we have collected all CH data for which we have corresponding extinction data. The plot is heavily dominated by L 1172, L 1457 and L 1642 (in total 101 points).

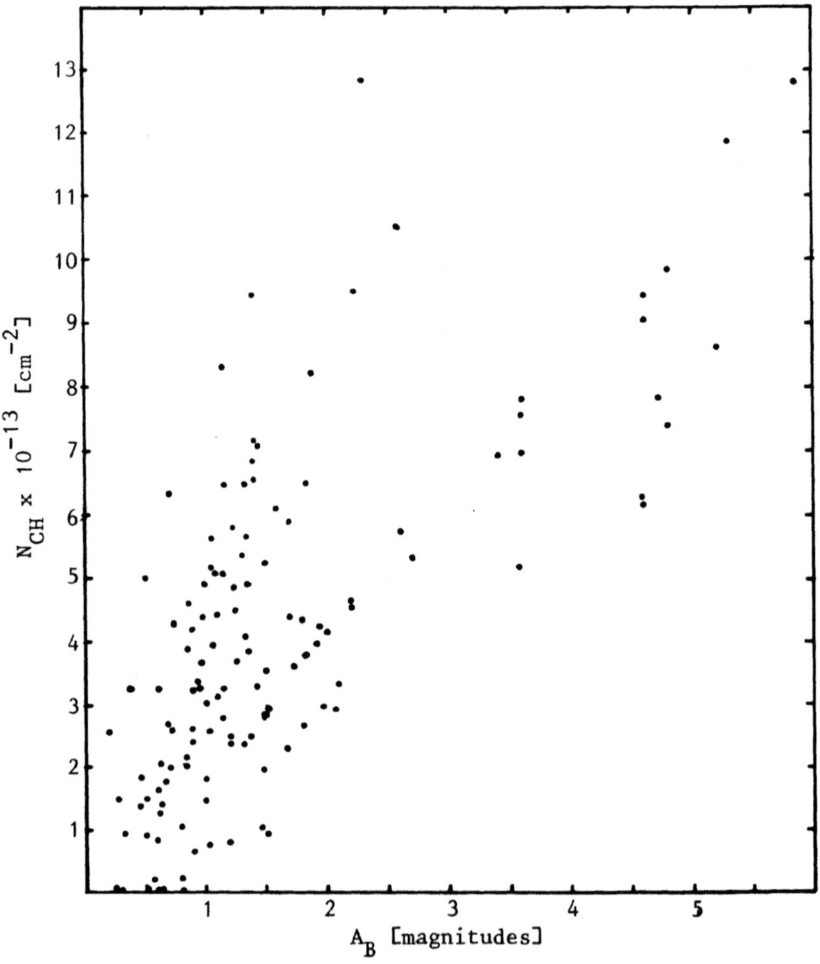

Figure 2. The CH column density, N_{CH}, plotted as a function of photographic extinction, A_B, for our complete sample.

The remaining points are shared by L 134 (4 points), L 1780 (3 points), the Taurus area (10 points) and 9 dark clouds (9 points) from Hjalmarson et al. (1977). Accordingly, the two branches evident in Fig. 2 must not, at this stage, be taken as real but is more likely a reflection of a statistically insignificant sample.

The models of dark cloud chemistry by Clavel et al. (1978) and Viala et al. (1979) indicate that CH is formed primarily by radiative association of C^+ and H_2, possibly with some contribution from grain surface reactions. These models predict that the outer parts (τ_V<1) of a dark nebula should have a CH abundance 10-100 times higher than the interior regions, the actual ratio being dependent on whether surface reactions contribute to the CH formation or not.

If we use the linear N_{CH} vs. A_B relation (Eq. 1 and 2) and the equivalent expression for the total hydrogen column density (Bohlin et al., 1978), we arrive at a fractional CH abundance $[n_{CH}] / [n_{Htot}] \approx (1-5)\ 10^{-8}$. This is a relatively high ratio when compared with the above mentioned theoretical model results.

In our data we see no clear tendency of saturation effects with increasing A_B values. However, data obtained by Mattila and Sandell (1981) toward the opaque core of L 134 (A_B>9$^m_.$7) give a CH column density as expected from a 2^m cloud according to our data. The same effect is also seen in dense molecular clouds, i.e. the CH abundance decreases with increasing gas density (Genzel et al., 1978).

The model by Viala et al. (1979) predicts a "limb brightening": an effect not evident in our data, possibly due to the crude spatial resolution used. Sandell et al. (1980) do find an enhancement of CH abundance near bright rims. In such cases this effect will be stronger due to the higher radiation field causing most of the carbon to be ionized.

Thus the radiation field could to some extent determine the slope of N_{CH} - A_B relation. However, the same effect can be achieved by cloud fragmentation, because several cloud fragments along the line of sight would increase the "surface effects", resulting in a higher CH abundance than in the case of a single cloud with the same extinction. A third possibility which would explain the varying slope of the N_{CH} - A_B curve may be expressed in the terms of clumping. Small dense cloudlets give rise to a lower CH abundance according to the Viala et al. (1979) model and, if the clumps are smaller than the reseau size, the extinction will be underestimated. However, these two effects of clumping would to some degree cancel each other. In this context one should not overlook variations in the excitation temperature, which if large, may alter the observed slope significantly.

Onsala Space Observatory is operated by the Chalmers University of Technology, Gothenburg, Sweden, with financial support from the Swedish Natural Science Research Council.

REFERENCES

Bohlin, R.C., Savage, B.D., Drake, J.F.: 1978, Astrophys. J. 224, pp. 132-142
Clavel, J., Viala, Y.P., Bel, N.: 1978, Astron. Astrophys. 65, pp. 435-448
Dickman, R.L.: 1978, Astrophys. J. Suppl. 37, pp. 407-427
Elmegreen, D.M., Elmegreen, B.G.: 1978, Astrophys. J. 220, pp. 510-515
Federman, S.R., Evans II, N.J.: 1981, preprint
Genzel, R., Downes, D., Pauls, P., Wilson, T.L., Beiging, J.: 1979, Astron. Astrophys. 73, pp. 253-259
Hjalmarson, Å., Sume, A., Elldér, J., Rydbeck, O.E.H., Moore, E.L., Huguenin, G.R., Sandquist, Aa., Lindblad, P.O., Lindroos, P.: 1977, Astrophys. J. Suppl. 35, pp. 263-280
Johansson, L.E.B.: 1979, Research Report No. 136, Res. Lab. of Electronics and Onsala Space Obs.
Johansson, L.E.B.: 1981, in preparation
Johansson, L.E.B., Sandell, G.: 1981, in preparation
Knapp, G.R., Kerr, F.J.: 1974, Astron. Astrophys. 35, pp. 361-369
Lang, K.R., Willson, R.F.: 1978, Astrophys. J. 224, pp. 125-131
Mattila, K.: 1979, Astron. Astrophys. 78, pp. 253-263
Mattila, K., Sandell, G.: 1981, in preparation
Minn, Y.K., Greenberg, J.M.: 1979, Astron. Astrophys. 77, pp. 37-44
Sandell, G., Höglund, B., Friberg, P.: 1980, Astron. Astrophys. 83, pp. 226-233
Sandell, G., Johansson, L.E.B., Nguyen-Q-Rieu, Mattila, K.: 1981, Astron. Astrophys. 97, pp. 317-324
Sandell, G., Johansson, L.E.B.: 1981, in preparation
Sherwood, W.A., Wilson, T.L.: 1979, MPIfR Prepreint Series No. 57
van Rhijn, P.J.: 1929, Groningen Publ. vol 43
Viala, Y.P., Bel, N., Clavel, J.: 1979, Astron. Astrophys. 73, pp. 174-182

AUTHOR INDEX

Ade, P.A.R. 323
Albinson, J.S. 193
Arquilla, R. 295
Arsenault, R. 67

Baas, F. 155
Bally, J. 287, 301, 329, 379
Bash, F.N. 469
Bastien, P. 335
Baudry, A. 81
Beichman, C.A. 445, 453
Bieging, J. 335
Blitz, L. 201, 209
Bodenheimer, P. 15
Boland, W. 407
Braun, R. 463

Calamai, G. 419
Chan, C.P. 343
Christie, R.A. 343
Chu, Y.-H. 53
Crutcher, R.M. 53
Cummins, S.E. 391
Cunningham, C.T. 321

de Graauw, Th. 433
de Jong, T. 407
Deveau, T.J. 107
de Vries, C. 433
Dewdney, P.E. 181
Dickel, H.R. 175, 409
Dickel, J.R. 175
Downes, D. 251

Edwards, S. 133, 141
Emery, R. 73

Falchi, A. 39
Falgarone, E. 315
Felli, M. 25, 419
Fich, M. 201
Fitton, B. 73
Forster, J.R. 409
Frerking, M. 445, 453
Friberg, P. 307
Furniss, I. 73

Geballe, T.R. 147, 155
Genzel, R. 251
Gispert, R. 249
Goldsmith, P.F. 295, 439
Goss, W.M. 409
Grayzeck, E.J. 213
Green, S. 391
Gull, S.F. 193

Harten, R.H. 25, 39
Henkel, C. 335
Heydari-Malayeri, M. 43
Hjalmarson, A. 307
Höglund, B. 399
Hughes, V.A. 349

Israel, F.P. 337, 433

Jackson, P.D. 185, 213, 221
Jennings, R.E. 73
Johansson, L.E.B. 479
Joncas, G. 67

Kahane, C. 43
Kallman, T.R. 385
King, K.J. 73
Kislyakov, A.G. 399
Klein, R.I. 129
Krolik, J.H. 161, 385
Kwok, S. 83

Lane, A.P. 301
Langer, W.D. 379
Lidholm, S. 433
Linke, R.A. 391
Lonsdale, C.J. 155
Loren, R.B. 307
Lucas, R. 43

Martin, R.N. 335
Matthews, H.E. 31
Mazurek, T.J. 61
McCutcheon, W.H. 343, 463
Mitchell, G.F. 107

Nadeau, D. 147
Naylor, D.A. 73

Nittmann, J. 123

Pauls, T. 335
Pérault, M. 315
Persson, S.E. 155
Peters III, W.L. 469
Phillips, J.P. 231, 237, 323
Phillips, T.G. 445, 453
Puget, J.L. 249

Robson, E.I. 323
Roger, R.S. 167
Roy, J.R. 67

Sandell, G. 399, 479
Sandford, M.T. 129
Sandqvist, Aa. 307
Schloerb, F.P. 295, 439
Scoville, N.Z. 295, 439
Serra, G. 249
Sewall, J.R. 213, 221
Shull, J.M. 91
Shuter, W.L.H. 245, 463
Simon, T. 155
Smith, H.A. 161
Snell, R.L. 133, 141
Stacy, J.G. 185
Stark, A.A. 209, 329
Szabo, A. 245

Tenorio-Tagle, G. 1, 15
Testor, G. 43
Thaddeus, P. 391
Tofani, G. 39
Treffers, R.R. 201
Turner, B.E. 425

van de Stadt, H. 433

Walmsley, C.M. 81, 335, 357
Watson, W.D. 357
Watt, G.D. 237, 323
Welter, G.L. 117
Whitaker, R.W. 129
White, G.J. 231, 237, 323
Wilson, T.L. 81, 335
Wilson, W.J. 175
Winnberg, A. 81
Wootten, H.A. 307, 337, 445, 453

Yorke, H.W. 15

Ziurys, L.M. 335

INDEX OF ASTRONOMICAL OBJECTS

Abell 63 209
AFGL490 83, 133, 161, 251, 287, 301, 307
AFGL618 83
AFGL961 237, 251, 287
AS353A 141
AS501/CRL2999 43

B 33 329
Berkeley #59 31
BN 91, 161, 251
BN-KL 357
BN-KL(IR) 251
BN-KL Cluster 251
BN(IRc1) 251

3C10 193
Carina Nebula 231
Cep A 133, 181, 251, 287, 301, 315, 349
Cep B 181
Cepheus Loop 31
Cep IV OB 31
Cep OB3 25, 181
Cep MC-1 307
COM GLOB 231
Cyg OB2 91
Cygnus X-ray superbubble 91
V645 Cyg 83

30 Dor 433
DR 21 61, 175, 237, 409

G109.1-1.0 43
G118.1+5.0 31
G118.6+4.8 31
G269.1-1.1 231
G270.3+0.8 231
G285.3-0.1 231
G309.8+1.8 231
G316.8-0.1 231
G326.6+0.6 231
G329.4-0.6 231
G330.9-0.4 231
G331.5-0.1 231
G333.3-0.4 231
G337.7-0.1 231
GL490 251, 301
GL961 301
GLOB 210 231
Gum Nebula 91

Haffner 18-19 185
HD164794 (9 Sgr) 91
HD271086 181
HD44179 91
Heiles 2 463
HH1-2 91, 133
HH7-11 287
HH25-26 133
HH28-30 91
HH39 91
HH102 91
HH24A 91
HH7-11(IR) 133
HH26(IR) 133
Horsehead Nebula 329

IC434 329
IC443 107, 357
IC1805 349
IC1848 349
IC5146 167
IRC10216 323

K3-50 61
KL 295

L134 479
L183 357
L1172 479
L1457 479
L1551 91, 287, 307
L1642 479
L1780 479
L1551 251
L134N 357
L1551/HH29 237
L1551/(HH29-IR) 251
L1551(IRS5) 133, 251, 343

LMC 433

M17 349
M1-92 425
M42 439
M78 133
Magellanic Clouds 91, 433
M8/E 161
M17(IRS1) 161
R Mon 91, 161
Mon R2 237

N159 433
N160A 433
NGC281 167
NGC1333 237
NGC1555 141
NGC1579 167
NGC1893 39
NGC1977 445, 453
NGC2023 237, 329
NGC2024 1, 237, 329
NGC2068 133, 237
NGC2071 155, 287, 301, 307, 315, 419
NGC2261 91
NGC2264 83, 161
NGC2467 185
NGC7023 479
NGC7535 251
NGC7538 175, 251
NGC7822 31
NGC2071(IR) 419
NGC7538(IRS1) 175, 251, 407
NGC7538(IRS2) 175
NGC7538(IRS5) 175
NGC7538(IRS6) 175
NGC7538-S 251

OH10.6-0.4 425
OH11.0+0.0 425
OH12.2-0.1 425
OH28.2+0.0 425
OH34.3+0.2 425
OH37.8-0.2 425
OH38.4-0.1 425
OH43.8-0.1 425
OH76.4-0.6 425
OH85.5-1.0 425
OH189.0+2.9 425
OH231.8+4.2 425
OH343.0+0.0 425
OH350.1+0.1 425
OH351.4+0.6 A 425
OH351.1+0.7 B 425
OMC 385
OMC-1 147, 237, 357, 445, 463
OMC-2 251, 453
OMC1(IRc2) 335
OMC1(IRc4) 335
OMC/KL 439
OMC1(KL) 155, 323
ON1 251
ON1 OBI 329
ζ Oph 357
ρ Oph 315
Oph 10 315
Oph 12 315
42 Ori 445
Orion Nebula 91, 107, 251, 295
Orion A 53, 61, 231, 439
Orion B 287, 329
Orion Bar 439
Orion H_2O Masers 91
Orion-KL IRc1 251
Orion-KL IRc2 251
Orion-KL IRc3 251
Orion-KL IRc4 251
Orion-KL IRc5 251
Orion Molecular Cloud 357
Orion Plateau 357
Orion Ridge 357

Pelican Nebula 329
Pup OB2 185
Puppis Window 185

RCW38 231
RCW57 231
RCW97 231
RH0 315
RH1 315

S8 185
S14 209
S68 379

S115 25
S128 209
S140 25
S147 43
S148 43
S152 43
S153 43
S155 25, 181
S157 43
S158 175
S170 31
S171 31
S184 25
S201 25
S212 221
S217 221
S222 343
S236 39
S241 213, 221
S242 213
S259 213, 221
S269 425
S271 213, 221
S272 213
S301 221
S305 221
S235A 399
S235B 399
S106(IR) 161
S140(IRS1) 419
HM Sge 83
Sgr B2 391
SMC 433
SNR-G109.1-1.0 43

T Tau 141, 251
TMC1 463
Trapezium Cluster 439
Tycho's SNR 193

W1 25, 31
W3 25, 315
W40 53, 61
W49 251
W51 251
W75 25
W44A 425
W44B 425
W51-(IRS1) 251
W51-(IRS2) 251
W3(IRS4) 307
W3(IRS5) 251, 307
W51-IRS2(H_2O) 251
W3(Main) 409
W51(Main) 251
W49N 251
W28N A 425
W28N B 425
W51north 251
W3(OH) 81, 251
WWH2 185
WWH7 185

SUBJECT INDEX

abundances, general molecular 107, 357, 433
accretion disks 95
Aerospace millimeter telescope 53, 175
Algonquin Radio Observatory 85
ammonia - see NH_3
anticenter, galactic 213
Appleton Laboratory 25-m radiotelescope 419
Arecibo Observatory 39, 213, 469
atomic carbon - see CI
atomic hydrogen - see HI
Auger effect 388

Bell Laboratories 7-m telescope 209, 329, 379, 391
bipolar confinement 137
bipolar mass outflow 133, 272, 287
bipolar nebulae 29, 98, 427
blister model 130, 201
BN-type objects 419
Bordeaux Observatory millimetre telescope 43, 315
bright rims 7, 453
bubbles, interstellar 61, 95, 97

Cambridge Half-Mile telescope 193
carbon, atomic - see CI
CH abundance 479
CH emission 479
CH^+ line 370
CH_3CN emission 391
champagne flow 1, 15, 37, 167
champagne model 1, 26, 43, 61, 178
champagne phase 1, 41
chemistry, gas phase 365
chemistry, ion-molecule 357
chemistry, shock 107, 370
chemistry of sulfur 369
CI abundance 445
CI emission 445, 453
CO clouds 469
CO emission 175, 201, 209, 213, 231, 237, 245, 287, 295, 304, 315, 329, 337, 343, 433, 439, 463, 469
CO isotopes 379
CO near HII regions 175, 213, 379
CO photodestruction 379
CO^+ emission 365
collapsing molecular clouds 15, 129

collisional dissociation 428
collisional excitation 251, 428
collisional ionization 272, 428
color temperatures 253
Columbia University millimeter telescope 39, 337
compact HII regions 83, 251, 315, 337, 409, 421
confinement, bipolar 137
cooling rates, vibrational, rotational 108
cosmic ray ionization 387

dark globules 130
dark nebulae 479
DCO^+ 366
density gradients 62
density waves 123
deuterium enhancement 365
dissociation front 98
dissociation of H_2 150, 167, 179, 371
dissociation of H_2O 428
dissociation of molecules 267
dissociation, shock-induced 371
dust clouds 421
dust extinction 479
dust grains 357
dust temperatures 253

Effelsberg 100-m radiotelescope 31, 335
Einstein Observatory 385
embedded infrared sources 83, 133, 251, 307, 425
evaporation of molecular clouds 21
evolution of HII regions 1, 15, 25, 167
evolution of pre-main sequence stars 137
evolution of shocked gas clouds 123
extinction by dust 347, 479

Fabry-Perot interferometer 53, 67, 147
Five College Radio Astronomy Observatory 133, 439
flow, champagne 15, 37, 167
flow, supersonic 61
formaldehyde - see H_2CO
fractionation of CO 379

galactic plane, IR survey 249
gas phase chemistry 365
giant molecular clouds 245, 433
globules, dark 130

Haute-Provence Observatory 43
HCN emission 399, 439

HCN/HNC ratio 367
HCO^+ emission 307, 315, 363, 364, 366
Herbig-Haro objects 94, 98, 129, 133, 142, 272
HI absorption 181, 193, 464, 469
HI clouds 167, 187, 193, 469
HI emission 167, 181, 185, 193, 345
HI expansion 167, 175
HI infrared 161
HI near HII regions 167, 175, 213
HI zones 167
HI-HII interaction 43, 167
high velocity gas 91, 133, 141, 251, 301
HII bubbles 61
HII region evolution 1, 15, 25
HII region expansion 61, 73
HII regions 15, 31, 39, 43, 53, 73, 81, 129, 175, 181, 201, 209, 213, 231, 315, 349, 399, 407, 439
HII regions, compact and ultracompact 83, 251, 315, 337, 409, 421
HII regions, hydrodynamics 129
H_2 line emission 147, 158, 272, 301
H_2CO emission 176, 335, 409
H_2CO maser 407
H_2-HII interaction 73, 439
H_2O masers 50, 91, 231, 251, 260, 273, 337
H_2S emission 176
Hα emission 54, 201

I-front 1, 15, 129, 167, 329
implosion of cloud clumps 129
infrared sources 83, 133, 251, 307, 425
infrared survey 249
initial mass function 249
instabilities 98, 117, 125, 329
instabilities, Kelvin-Helmholtz 125
instabilities, Rayleigh-Taylor 98, 329
interferometer, Fabry-Perot 53, 67, 147
interferometer, Michelson 73
interstellar bubbles 61, 95, 97
interstellar gas, dynamics of 97
interstellar medium 39, 185, 193
ionization, collisional 272, 428
ionization, cosmic ray 387
ionization, X-ray 385
ionization front 1, 15, 129, 167, 329

ionized gas, velocity 201
ion-molecule chemistry 357
isothermal shocks 1
isotopes of CO 379

Kelvin-Helmholtz instabilities 123
Kitt Peak 2.1-m telescope 301
Kitt Peak 10-m telescope 337

Las Campanas Observatory 231
line wings 287, 307
luminous stars 161
Lyman-Werner band absorption 167, 179

magnetic fields 123, 150, 333
Maryland-Greenbank HI survey 213
maser emission 81, 231, 251, 260, 273, 337, 407, 420, 425
maser pumping 408, 427
masers 98, 129
mass ejection 101
mass loss 92, 137, 161, 251, 276, 419
mass outflow 133, 141, 150, 157, 161, 251, 254, 261, 301, 305, 307, 323, 369, 420
mass outflow, bipolar 133, 272, 287
methyl cyanide 391
Michelson interferometer 73
Millimeter Wave Observatory, Texas 133, 209, 295, 337
molecular abundances 107, 315, 379, 433, 479
molecular cloud cores 256, 287, 301, 307
molecular clouds 73, 83, 91, 129, 181, 201, 231, 237, 337, 343, 349
molecular clouds, ages 16
molecular clouds, collapsing 15, 129
molecular clouds, disruption 21
molecular clouds, kinetic temperature 391
molecular line emission 55

NASA Infrared Telescope Facility, Hawaii 155, 445, 453
nebulae, bipolar 29, 98, 427
nebulae, dark 479
neutral clouds 129
neutral hydrogen - see HI or H_2
NH_3 emission 260, 335, 419
NIII emission 73
non-thermal emission 31
NRAO 11-m radiotelescope 221, 307

NRAO 43-m radiotelescope 337

OB associations 129, 181, 349
OH absorption 81
OH masers 409, 425
OI emission 73
OIII emission 73
Onsala Space Observatory 307, 399, 479
O-star winds 91
O-type stars 15, 92
outflow, bipolar 133, 272, 287
outflow, champagne 1, 15, 37, 167
outflow, mass 98, 133, 141, 150, 157, 161, 251, 254, 261, 301, 305, 307, 323, 357, 420
Owens Valley Radio Observatory 337, 453

Penticton synthesis radiotelescope 167, 181
photo-destruction of CO 379
photo-ionization 386
photostars 251

rarefaction wave 1, 26, 61
Rayleigh-Taylor instabilities 98, 329
recombination lines 40, 255
rims, bright 7, 453
rocket effect 329

shock chemistry 370
shock waves 117, 123, 150, 155, 349, 445
shocked gas 107, 123, 167, 194
shocked region 73
shocked shells 167, 428
shock-induced dissociation 371
shocks 61, 100, 107, 129, 301, 357
slow shocks 107
spiral density waves 123
star formation 15, 117, 129, 155, 175, 218, 287, 349, 357, 433
star formation rate 249
star formation regions 37, 83, 231, 251, 357, 399
stars, new formed 323
stellar winds 91, 130, 136, 256, 272, 287, 349
stellar winds, T Tauri 130, 141, 162
Strömgren sphere 1, 17
successive star formation 29, 37

sulfur chemistry 369
superbubbles 95
supernovae 37, 43, 193
supershells 95
supersonic flows 61

Texas Millimeter Wave Observatory 133, 209, 295, 337, 469
T-Tauri winds 141
turbulence 329

ultracompact HII regions 83, 251, 315, 337, 409, 421
United Kingdom infrared telescope 237, 323
University of British Columbia millimetre wave telescope 245, 343, 463
UV photo-destruction 379

Very Large Array, New Mexico 335, 407

wave, dissociation 175
wave, rarefaction 1, 26, 61
waves, shock 117, 123, 150, 155, 349, 445
Westerbork synthesis radiotelescope 25, 31, 349, 407, 409
winds, O-star 91
winds, radiatively driven 155
winds, stellar 91, 130, 136, 256, 272, 287, 349
winds, T-Tauri 130, 141, 162

X-ray emission 98, 385
X-ray ionization 385

zones, HI 167

GPSR Compliance

The European Union's (EU) General Product Safety Regulation (GPSR) is a set of rules that requires consumer products to be safe and our obligations to ensure this.

If you have any concerns about our products, you can contact us on ProductSafety@springernature.com

In case Publisher is established outside the EU, the EU authorized representative is:

Springer Nature Customer Service Center GmbH
Europaplatz 3
69115 Heidelberg, Germany

Batch number: 10815974

Printed by Printforce, the Netherlands